MACHINE LEARNING AND DEEP LEARNING USING PYTHON AND TENSORFLOW™

MACHINE LEARNING AND DEEP LEARNING USING PYTHON AND TENSORFLOW™

Venkata Reddy Konasani

Shailendra Kadre

New York   Chicago   San Francisco   Athens   London
Madrid   Mexico City   Milan   New Delhi
Singapore   Sydney   Toronto

Library of Congress Control Number: 2020949936

McGraw Hill books are available at special quantity discounts to use as premiums and sales promotions or for use in corporate training programs. To contact a representative, please visit the Contact Us page at www.mhprofessional.com.

Machine Learning and Deep Learning Using Python and TensorFlow™

TensorFlow is a trademark of Google Inc.

1 2 3 4 5 6 7 8 9   LWI   25 24 23 22 21 20

ISBN 978-1-260-46229-6
MHID 1-260-46229-3

Sponsoring Editor Lara Zoble	**Copy Editor** Mohammad Taiyab Khan, MPS Limited
Editing Supervisor Stephen M. Smith	**Proofreader** Lisa McCoy
Production Supervisor Lynn M. Messina	**Indexer** Michael Ferreira
Acquisitions Coordinator Elizabeth M. Houde	**Art Director, Cover** Jeff Weeks
Project Manager Touseen Qadri, MPS Limited	**Composition** MPS Limited

Venkata Reddy Konasani is a data science corporate trainer. He has 10 years of experience as an applied data analyst and data scientist and six years of experience as a corporate trainer. He has experience in credit risk model building, marketing analytics, and social media analytics.

Currently, Venkat conducts regular data science training programs for companies such as Oracle, IBM, CTS, and Bank of America. He has conducted more than 100 corporate batches on machine learning and deep learning topics.

Venkat is the co-founder of the data science e-learning company statinfer.com. He obtained a master's degree in applied statistics and informatics from the Indian Institute of Technology (IIT) Bombay. He lives in Vijayawada, India. You can reach Venkat at 21.venkat@gmail.com or @venkat_io on Twitter.

Shailendra Kadre is a management and information technology consultant based out of Bangalore, India. He has more than 20 years of experience working in the areas of digital transformation, machine learning, presales, and information technology projects and products. He has successfully held many key leadership roles in these areas. He has worked for global blue-chip companies such as HP Inc., Tata Consultancy Services, and Tech Mahindra. Shailendra is the author of the book *Going Corporate: A Geek's Guide* and the co-author of the book *Practical Business Analytics Using SAS.*

He holds a master's degree in design engineering from the Indian Institute of Technology (IIT) Delhi. He currently resides in Bangalore with his wife Meenakshi, daughter Neha, and son Vivek. Shailendra can be reached at shailendrakadre@gmail.com and LinkedIn.

CONTENTS

Chapter 3. Regression and Logistic Regression

61

Chapter 4. Decision Trees

115

Chapter 5. Model Selection and Cross-Validation   **151**

Chapter 6. Cluster Analysis   **227**

Chapter 7. Random Forests and Boosting 251

Chapter 8. Artificial Neural Networks 281

Chapter 9. TensorFlow and Keras 327

Chapter 10. Deep Learning Hyperparameters   **353**

Chapter 11. Convolutional Neural Networks   **401**

Chapter 12. Recurrent Neural Networks and Long Short-Term Memory   **467**

ACKNOWLEDGMENTS

This book, like most other significant efforts, is a result of consistent and tireless teamwork. We, as the authors, would like to express our sincere gratitude toward McGraw Hill editor Lara Zoble for the book. She consistently encouraged us to do quality work and finish this book within the set timelines. Her timely suggestions played a significant role in bringing the book to its current shape. We are also thankful to the copy editor and the entire McGraw Hill editorial team. Without their encouragement and hard work, it would not have been possible to finish this project. The copy editor, particularly, showed an exceptional level of patience and commitment to quality work in giving the finishing touches to this volume. Finally, we express our sincere gratitude toward the entire production, distribution, and marketing team of McGraw Hill worldwide, who have immensely contributed to this project working in the background.

Venkata Reddy Konasani
Shailendra Kadre

I am thankful to the people who helped me shape this book. Vrunda Paranjape, a mathematics student, patiently tested all the code files and made sure that there were no errors. I wish to thank Pradeep Venkataramu, Mohan Silaparasetty, Amiya Ranjan Bhowmick, Bhuvnesh Kumar, and Vijay Krishna (V2K) for encouraging and strengthening me as a corporate trainer.

I sincerely thank my friends Debendra and Sumit and numerous students in my training programs who always cheered me to document my lectures as they appear in this book. The successful completion of the book would not have been possible without the sacrifices made and patience shown by my wife, son, little daughter, and other family members, who saw me working late hours.

These acknowledgments would not be complete without special mentions of Patrick Winston (computer scientist), Andrew Ng (computer scientist), and Nando de Freitas (professor), who inspired me endlessly.

Venkat

I started my consulting career with my 2011 book *Going Corporate: A Geek's Guide*. It's a general management advisory book for geeks. My second book was the 2015 release *Practical Business Analytics Using SAS: A Hands-on Guide*, co-authored with Venkat. The current 2021 book is my third release in the United States. It is my second book project with Venkat. He has been a professional trainer in the field of machine learning for many years. His training experience has come in handy in shaping the content of this book.

For this book, I would like to express my sincere gratitude to my family and friends Srinivas D., Milind Kolhatkar, Laxmi Sahu, and Anup Parkhi, who came up with many useful suggestions during their tireless reviews. All of them were very helpful and encouraging. My brother Shailesh Kadre and his wife Neena Kadre also helped a lot in terms of reviews and suggestions. Sincere thanks to both of them as well. Like in any other technology book, here I have referred to the work of numerous experts and prominent researchers in this field. I must thank each one of these inspiring personalities. Last but not least, I would like to thank all the persons who directly or indirectly helped in this project. There is a long list of such names, which will not be mentioned here due to space constraints.

No book can succeed without the blessings of readers. We have kept in mind the needs of working professionals and the global student community while designing and writing this book. We have kept this book practical and problem-oriented. Nearly the entire book is written with the help of real-life case studies. All useful analytics concepts are explained alongside case studies. The language of the book has been kept as simple as possible for easy comprehension. I hope you will like this book as much as the previous one that was more like the first reading in machine learning. This book covers advanced machine learning topics. It has five complete chapters devoted to neural networks and deep learning.

Any project of this size invariably brings tons of hardships to the author's family. This project was no different. I give the entire credit for completing the book on time to my wife Meenakshi, daughter Neha, and son Vivek. Each one of them firmly stood by me for the entire duration of this project. During the closing months of writing this book, it was the COVID-19-induced lockdown, and no domestic help was available. Meenakshi not only did all her daytime office work, but also beautifully managed all the household tasks, took care of our children, and looked after my aging parents. Without her support and encouragement, it would have been impossible to complete this project on time.

Please feel free to reach out to Venkat or me with any feedback or suggestions. I am available at shailendrakadre@gmail.com and at LinkedIn. You are most welcome to connect with me on LinkedIn.

Good luck!

Shailendra

PREFACE

There are two types of machine learning books in the market now. The first type is the machine learning theory books published mainly for academic purposes. These books have an excellent treatment of the mathematical derivations and equations behind all the algorithms, but real-life implementations with data are minimal. Professionals who do not have a solid background in statistics or mathematics may have a hard time understanding such books. These theory books have nominal content on the real-life challenges faced by practitioners of data science. The books talk very little about the practices used in the field of machine learning. The second category of books is code-oriented cookbooks. These books primarily contain code with its documentation. The reason and logic of why we are performing specific tasks are missing in these books. There is a wide gap in the way machine learning is taught in academic settings and how it is used in industry. So we strongly felt there was a need for a book covering machine learning topics that has a solid theory base, contains the prevalent industry practices, and has a sufficient number of practical business cases with all underlying logic explained. Our book aims to fill this gap.

We set out to write a book that can be easily understood by general readers. Anybody who wants to get started with machine learning should be able to begin with this book. Every topic in the book has an explanation in three phases. In phase 1, we try to develop a topic insight (or intuition) by using some analogies, examples, and visualizations. Phase 2 deepens the understanding using mathematical equations in an academic style accompanied by simplified commentary. In phase 3, we take up a real-life business problem with data and write the code to solve the problem in order to develop an in-depth understanding of the concept.

We wrote *Machine Learning and Deep Learning Using Python and TensorFlow*^{TM} with a clear intention to simplify and explain the concepts of machine learning and deep learning to the common person. As the authors, we can promise you that whatever academic and coding backgrounds you are coming from, you will be able to follow this book from the first chapter until the last. At times the reader may feel the concepts are explained with too many examples. This is because we firmly believe and follow Python's philosophy, which is as follows:

- Explicit is better than implicit.
- Simple is better than complex.
- Complex is better than complicated.
- Sparse is better than dense.

WHAT THIS BOOK COVERS

The book covers the following:

1. Basics of Python and statistics
2. Simple machine learning models
3. Advanced machine learning models
4. Essential deep learning models

Before starting with machine learning, it is essential to know the basics of Python and statistics. This book provides a level playing field to readers with lesser exposure to programming and statistics by covering all the basics.

The book discusses simple machine learning models. Linear regression, logistic regression, decision trees, and cluster analysis have been categorized as the classic machine learning algorithms by many. They are noncomplex methods, easy to build, easy to interpret, and easy to visualize. We have made sure to give you enjoyable theoretical and practical treatments of these topics.

This book provides you with an in-depth treatment of some advanced machine learning methods such as random forests, boosting, and neural networks. You will learn to use these models and develop an understanding of their essential hyperparameters. You will also learn how to build and validate these models while working on real-life examples from industry.

In the book, you will get an introduction to deep learning concepts. The deep learning frameworks are different from the machine learning frameworks. We have covered the programming concepts of TensorFlow™ and Keras. After practicing deep learning concepts, you will be able to work with deep learning models such as CNN, RNN, and LSTM.

This book is intended to be a workbook to gain the skills of machine learning and deep learning using Python. For the best results, our sincere advice to the reader is to practice and execute the code alongside reading the text.

KEY FEATURES OF THIS BOOK

- In-depth coverage of both machine learning and deep learning concepts in a single volume
- Written by industry professionals with many years of experience in the field
- Covers theory, industry best practices, and the issues faced by professionals in the right mix
- Abundant real-life case studies from multiple industry verticals such as banking, insurance, e-commerce, health care, and automobiles
- Easy to follow, even for readers with lesser exposure to statistics, math, and programming
- Complex concepts explained using visualizations and analogies to make them easy to understand
- Self-sufficient volume—you hardly need to refer to or read other sources to work with this book
- Comes with download links for datasets, codes, and sample projects

SKILLS YOU WILL GET AFTER COMPLETING THIS BOOK

- Data handling using Python
- Data exploration using statistical measures and generating reports using Python
- Building and validating linear and logistic regression models and using them for predictions
- Building and validating tree-based models such as decision trees and random forests
- Understanding practical aspects of model building such as feature engineering and model selection
- Expertise in advanced machine learning algorithms such as boosting and ANN
- Writing codes in TensorFlow and Keras
- Handling hyperparameters while building deep learning models
- Understanding computer vision and classifying using CNNs
- Building and validating sequential models such as RNN and LSTM

TARGET AUDIENCE FOR THIS BOOK

- Anybody who wants to get started with machine learning and deep learning
- Data science aspirants and practitioners
- Graduate and undergraduate students with a mathematics or statistics background
- Reporting analysts who want to move into data science
- Predictive modelers who want to learn machine learning and deep learning
- Data visualization experts who want to get started with machine learning and deep learning
- Computer vision enthusiasts
- Deep learning enthusiasts
- Computer science engineering students

PREREQUISITES

- This book is the first course in machine learning. There are no strict prerequisites.
- Anybody with a primary degree can get started with this book.
- Basic high school mathematics is enough to get started.
- Advanced statistical knowledge is *not* required.
- Advanced programming knowledge is *not* required.

SOME OF THE CASE STUDIES IN THIS BOOK

There are numerous examples and case studies in the book. Given below is a representative list.

- Air passengers case study—predicting the number of passengers for an airline company
- Attrition case study—predicting whether a telecom company customer will attrite or not based on usage
- Contact center customer survey case study—predicting customer satisfaction with a contact center
- King County house price prediction—predicting house price in King County based on house features
- Pima Indians diabetes case study—diabetes detection based on diagnostic measures
- Bank loans case study—identifying risky customers before offering a loan
- Retail customer segmentation—performing customer segmentation based on purchases from a retail company
- Car accident prediction—predicting fatal car accidents based on sensor data
- U.S. census income prediction—predicting the people with high income based on census data
- Number detection by taking digit images as input
- Object detection based on taking object images as input
- Next word prediction by taking a sequence of words as input
- Machine translation—from English to a target language

SOFTWARE AND HARDWARE REQUIREMENTS

Software download links:

- Download Anaconda from the website https://www.anaconda.com/distribution/
- Cloud: use Google Colab as an alternative to Anaconda—https://colab.research.google.com/

  Software versions:

- Python version 3.7 or above
- TensorFlow version 2.0.0 or above

DOWNLOADING CODES AND DATASETS

Venkata Reddy Konasani's webpage: https://statinfer.com/venkat/
GitHub link: https://github.com/venkatareddykonasani/ML_DL_py_TF
Google Drive link: https://drive.google.com/open?id=1f3WYLdoQLd4QyccKIGCjnkKAOtJotNyR
Dropbox link: https://www.dropbox.com/sh/9zig86q97vv7xhq/AACfMPFYIRENbfvELxGGobCma?dl=0

UPDATES AND ERRATA

GitHub link: https://github.com/venkatareddykonasani/ML_DL_py_TF_errata
Venkata Reddy Konasani's webpage: https://statinfer.com/venkat/

MACHINE LEARNING AND DEEP LEARNING USING PYTHON AND TENSORFLOW™

CHAPTER 1
INTRODUCTION TO MACHINE LEARNING AND DEEP LEARNING

An artificial intelligence (AI) scientist, popularly called Winston, gave a beautiful definition of AI in 1992. In his words, AI is "The study of the computations that make [it] possible to perceive, reason and act." This definition still holds true. However, almost an upside-down change has taken place on the technology front in the past 25 years. AI has grown at an unprecedented pace. Early humanoids (robots) were standing upright just like we do and performing a few of the day-to-day tasks unique to humans. After 2015, social media is full of humanoids performing all kinds of stunts—be it dancing, jumping, going upside down, or summersaults. Some of the stunts that these robots performed are painful even to human beings. Recently, Sophia, an AI-based humanoid, was awarded Saudi Arabian citizenship. It has become the first-ever robot to have a nationality. Sophia is famous for interviewing world leaders and interviewing celebrities. Sophia even spoke at the United Nations; needless to say, all of it was without any manual intervention. Undoubtedly, it is today's smart AI put to work.

Most robots used in industry and homes are not humanoids, however. Their physical shape depends upon the tasks they are supposed to perform. However, one fact is common—all of them have some degree of intelligence built in. The brain behind all these robots and other intelligent automation machinery is what we call AI [used synonymously with machine learning throughout this book]. We can safely call machine learning, deep learning algorithms as the foundation stones in the study of AI. In the later chapters, we are going to discuss many of these algorithms in detail. However, the scope of this chapter is limited to serve as an introduction to the material presented throughout this book. This chapter may still introduce some terminology that might sound unfamiliar at this stage. The later part of this chapter explains a few of these terms.

AI is a multidisciplinary topic with many foundational faculties. We will take up a few for a brief discussion here. The subject of philosophy helps AI scientists to answer questions like: Can we use formal rules to draw valid conclusions? Can we use today's available knowledge base for the machines to take human-like actions? Math, the central foundational pillar of AI, helps AI professionals to formalize the rules to draw valid conclusions, and how can we reason with ambiguous information? The entire information technology (IT) and computing field (machine learning included) banks heavily on the developments in modern mathematics. AI draws (is dependent) intensely from today's neurosciences, which deals with the study of the nervous system, particularly the brain. All advanced AI systems aim to mimic the human mind as accurately as possible. However, in the current state of AI, we are far away from this goal. It may still take many decades to reach there.

AI systems take much inspiration from the discipline of psychology, cognitive psychology in particular, which sights the human mind as an information or data processing device. Research is going on in the field of how animals and humans think (and act). All this knowledge helps in the advancement of AI systems. The area of control theory and cybernetics also contributes significantly to the field of machine learning and AI. Computer science and engineering help in building efficient computational resources needed to run today's sophisticated and highly resource-intensive AI algorithms (for example, deep learning algorithms with millions of weights).

Advanced studies in AI are becoming very popular among students and professionals nowadays. Many leading educational institutes across the globe are offering graduate degrees in AI under their computer science departments. The Massachusetts Institute of Technology (MIT), one of the leading international tech institutions, has even launched a separate college to advance the field of machine learning and AI.

1.1   A BRIEF HISTORY OF AI AND MACHINE LEARNING

The term "machine learning" was coined in the 1950s. Until the 1980s, various attempts were made to make computer programs work like a small-scale human brain. In the 1990s, machine learning shifted from a knowledge-driven approach to a data-driven approach, as we know it today. The year 1997 was a benchmark when IBM's Deep Blue computer defeated the world champion at chess, a game requiring quite a bit of human-like intelligence to master it. Machine learning was reorganized as a separate field and started to flourish in the 1990s. It began to grow toward the methods loaned out from the field of statistics and probabilistic theory. Machine learning growth was also fueled by the ever-increasing availability of digitized data and the availability of the internet as a distribution channel. Geoffrey Hinton coined the term "deep learning" in the year 2006. One of the early applications of deep learning was in images and videos, in which computers were trained to see and distinguish objects. Since then, AI development has been continuously taken to the next levels by leading universities and technology majors like IBM, Google, Facebook, and Amazon.

In the latter half of the 2010s, we are witnessing a shift from the simple methods of analytics used in corporate decision-making to more complex algorithms of deep learning delivering more complex tasks in a variety of machinery, including more advanced human-like robots. AI is getting termed as one of the most promising fields to get involved with in the coming years.

Given below is a list of popular machine learning algorithms in the industry:

- Regression
- Logistic regression
- Decision trees
- Random forests
- Gradient boosted machines
- ANNs—artificial neural networks
- CNNs—convolution neural networks
- RNNs—recurrent neural networks
- Bayesian techniques
- SVMs—support vector machines
- Evolutionary approaches
- Markov logic networks
- HMM—hidden Markov model
- GANs—generative adversarial networks

Here is a list of widely used programming languages by data scientists around the world:

- Python
- R
- SQL
- Java
- JavaScript
- MATLAB
- Scala
- Julia
- Go
- C/C++
- Ruby
- PHP
- SAS

Based on project demand, a machine learning expert may need to use a variety of tools and programming languages.

1.2   BUILDING BLOCKS OF A MACHINE LEARNING PROJECT

Machine learning, also known as augmented analytics, is considered a subset of AI. It is closely related to computational statistics. Machine learning algorithms are computer programs, which are said to learn from experience concerning some set of tasks. It performs these tasks with an accuracy level that improves with experience. One needs to train machine learning algorithms with training data to get the desired accuracy level. Once trained, these algorithms yield potent insights that can be used to predict future outcomes.

Machine learning is very closely related to statistics in terms of methods, but the underlying goal is different. Descriptive statistical techniques are used to draw inferences on populations (or a set of data), while machine learning algorithms find general predictive patterns. Many of us already know something about primary machine learning concepts, along with its brief history. Machine learning can be used in various industry segments to manage and improve business operations. Even before a single prediction is made using a model, there is much hard work that goes behind it. Just like many other software development projects, for building any useful machine learning model, the data science teams need to source, process, and transform the data to formulate appropriate problem-solving strategies and create a raw model. This model needs to be trained with tons of sample data; results validated before a model can be deployed to an actual production environment. Figure 1.1 is a schematic of the model building process. We will discuss it in detail in the later chapters.

Data science teams generally work with IT departments. There is a fundamental difference between a software development project and a data science project. While software development projects revolve around design, development, and testing of code, data science projects are massive on data and generally need lesser coding. In any machine learning project, as much as 70 percent of time and resources are allocated to data collection, cleansing, and then putting it in usable data structures, for example, in tabular format in many cases. Once the data and model are ready, the next most significant effort may be to train the model, which typically requires a massive amount of training data. Generally, banks and retail websites like Amazon have a lot of reliable user transaction data. Developing and deploying machine learning technologies are an essential part of their day-to-day business operations.

Just like many other projects, every machine learning project needs to solve a business problem. The natural first step is to formulate the business problem and outline strategic business goals; then comes fixing the scope of work. Once it is done and duly reviewed, the project owner, usually from the business side, identifies the genuine stakeholders and finally takes buy-in from top management to secure funds and required resources.

A machine learning project requires many specialists on the project team. Some of them are data experts whose responsibility is to give clean data in a usable format, which can be consumed by the project. While a business analyst represents the business side and helps in formulating the requirements, a solution architect is responsible for the development of the new solution. Data scientists on the project team initially may work on multiple machine learning models and train them to see which model gives the most accurate results. In the model evaluation and testing phase, the goal is to get the simplest model, which provides the target with value fast and with the desired accuracy level.

Once the data scientists finalize on the most reliable model and quantify its performance requirements, it is time to deploy the solution to production, which means it is ready to be taken over by the business. Database administrators typically do the final solution deployment to production. All the team roles discussed in this section depend upon the project size and team structure; it may vary from organization to organization.

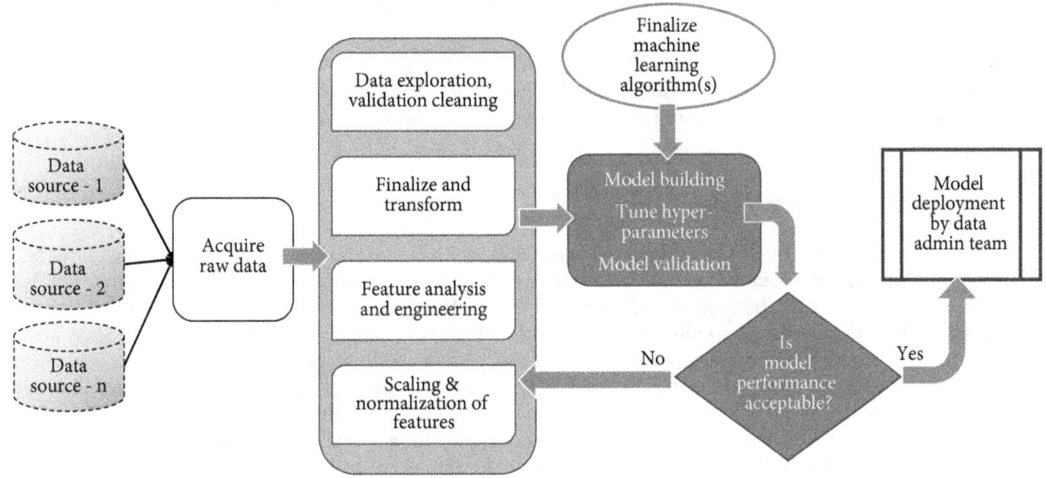

FIGURE 1.1   A typical machine learning project in actual practice.

1.3   *MACHINE LEARNING ALGORITHMS VS. TRADITIONAL COMPUTER PROGRAMS*

Traditional programming, as we all know it, has been around for a while now. We all are familiar with computer programs, and probably each one of us has done some coding (computer program) in our high school or graduation days. Traditional programming, as we know it, is mostly a manual process. A person (programmer) writes the computer code, which is nothing but manually formulated rules in one of many existing programming languages. A computer understands this code; a person then furnishes the input data to the computer program to get the desired output as per the coded rules.

Machine learning algorithms, on the other hand, build a mathematical model based on sample data (known as training data). Once the model is trained, it can be used to make a class of predictions or decisions; too much explicit programming is not required to perform the task. As we input more and more training data, the accuracy of predictions or decisions improves. The accuracy of predictions is measured using historical data, where both inputs and outputs are known. Once we get the desired accuracy level, the trained model can be used to make predictions with new input data. It is much like humans who get better with experience. Machine learning models find applications in almost every industry. Banks and other financial institutions train these models with known historical data, and later use it in the forms like predicting the probability of a new customer defaulting in loan repayments. Here the input data is customer demographics like income, number of credit cards, assets, and previous defaults.

1.4   *HOW DEEP LEARNING WORKS*

While some of the machine learning models are simple statistical exercises, deep learning may be of particular interest to many as it's a relatively new phenomenon—the term "deep learning" itself was coined in 2006. Deep learning, also known as hierarchical learning, is a subset of machine learning algorithms, which given the raw input data, progressively extract higher-level features from multiple layers. A commonly quoted application of deep learning technology is image processing. For example, in a medical image processing application, given an image in the form of a raw matrix of pixels, the first deep learning layer may identify only the edges, and the second, third, and fourth layers may encrypt and compose eyes, nose, and the face. On its own, the deep learning process can decide which layer will extract edges, faces, eyes, nose, and face; in other words, it is intelligent enough to decide on its own the placement of features concerning different layers.

A deep learning process, in general, can learn to solve any problem that otherwise requires human thought to work out. In this technology, artificial neural networks and algorithms mimic the human brain while learning from a large amount of data. We will introduce neural networks and different types of learnings a little later in this book. Deep learning needs tons of data to train the program before it can be put to productive use. That is why it has grown only in recent years with the advanced digital era, where an enormous amount of data and its processing technologies are continually evolving.

1.5   *MACHINE LEARNING AND DEEP LEARNING APPLICATIONS*

Machine learning finds widespread applications across the industry—be it finance and banking, manufacturing, general automation, or robotics, to name a few. By 2021, at least one-fifth of leading manufacturers will rely on embedded intelligence, using AI, and other related technologies to automate processes.

1.5.1   Applications in Daily Life

Self-driving cars are a big hit nowadays. They are fitted with cameras, sensors, lasers, and rangefinders to sense the environment. An autonomous car's AI-based software commands steering, braking, and acceleration. It detects and differentiates objects on the road and adjusts itself much like a human driver. It is anticipated that these autonomous cars are likely to significantly reduce the rate of accidents that are caused by human errors. Speech recognition is another area where machine learning–based technology has a significant potential to affect the way we interact with machines or even with a human being speaking an unfamiliar language. A traveler from the United States can book her air travel on a holiday website with an entire conversation guided by automated speech recognition and dialog management system. It is very much a reality now. Speech recognition is finding tall applications in many conventionally human-dominated areas like customer contact centers for large corporations.

In 1997, IBM's famous Deep Blue computer became the first human-made machine to defeat world chess champion Garry Kasparov in an exhibition match. AI is making much use in gaming and making it more exciting. In your email messaging systems, AI is being used for spam filtering. Billions of messages are being classified as spam every day by learning algorithms. Machine translation is another area where computers are automatically translating from one language to another (with acceptable quality). This translating software uses a sophisticated statistical model built from the samples of translations from native to target languages, for example, from Italian to English. These example samples may run into trillions of words of data.

1.5.2   Machine Learning in Manufacturing

Using machine learning algorithms, factories are reducing unplanned downtime, reducing maintenance costs, increasing production output, and achieving up to a 30 percent increase in product quality. Machine learning is being used to streamline every phase of production—the priority areas are supply chain optimization, improving the yield from raw material, quality control, work schedule, and choosing the most optimal product configuration out of numerous combinations. Using machine learning algorithms, manufacturers can find anomalies in a product and its packaging. It can significantly improve product quality and prevent defective goods from leaving the manufacturing facility. In product quality control, using machine learning can almost double the defect detection efficiency when compared to manual inspection techniques still prevailing in many industries. Machine learning and AI technologies have enormous potential to improve production throughput, maximizing earning potential per hour, and bring energy utilization to the optimal levels by identifying and preventing waste. Early adopters of these technologies in manufacturing are observing even doubling their cash flows as a result of improved efficiencies. In almost every phase of their business cycle, noticeable improvements are observed, which include improving product quality, maximizing plant and human resources utilization, and reducing waste practically everywhere.

Process-based manufacturers like chemical and pharma industries are significantly improving their bottom lines using data and AI technologies like machine learning by improving production efficiencies and improved product quality. Even in office work, machine learning and other AI technologies can take away your repetitive and routine tasks like searching invoice numbers, paper processing, and estimating the order values. In another diverse example, one manufacturing plant in Europe implemented a predictive maintenance solution for all its heavy machinery like pneumatic presses used in the sheet metal production. Before this implementation, engineers were wasting much time attending to breakdowns instead of working for planned maintenance activities. The new maintenance solution allows them to predict equipment failure with more than 90 percent accuracy and do better maintenance planning. It not only improved equipment reliability and uptime but also enhanced overall product quality.

In manufacturing applications, supervised machine learning algorithms are often used in practice as they subsequently lead to a predefined target. Classification and regression types of machine learning algorithms find widespread application in predictive plant maintenance.

1.5.3   Applications in Robotics

The field of robotics is also influenced and directed in some way by machine learning technologies. It finds applications in autonomous vehicles, drones, and other vision-sensitive tasks like detecting and evaluating faults in silicon wafer identification and sorting of objects. To maintain productivity, robots use machine learning technologies for dynamic interaction and hindrance avoidance. Medical world robots, as we know them, extensively use machine learning technologies. Medical robots, today, can perform tasks like surgeries and stitch better than the best available human surgeons. These tasks are generally performed under the medical supervision of trained doctors. Many online retailers like Amazon are very effectively using machine learning–based robotics in their mega warehouses to pick and place hundreds of thousands of objects of various shapes and sizes. There are many such examples of robots all around, which use machine learning to deliver different responsibilities using vision, grasping, motion control, and recognize physical and logical data patterns to act proactively. As far as robotics is concerned, machine learning technologies are still in its beginning stages, but they already have a significant impact.

Systems that are proficient in detecting and measuring faults in silicon wafers use convolutional neural networks (CNNs) that we will discuss in the later chapters. Some robots and control tasks use autonomous learning models, which involve deep learning and unsupervised methods.

1.5.4   Applications in Banking and Finance

The world of banking and finance is probably the most prominent user of machine learning models. They were among the very early adopters of these technologies. Banking and finance companies across the globe generally have a very high volume of accurate historical records of customer transactions and demographics. This industry is major in terms of quantitative computations, so they can effectively use machine learning algorithms for various kinds of financial modeling. Applications are many and ever-increasing with the improvement in accessibility and accuracy of machine learning tools, availability of accurate data, and improved computing systems. Machine learning finds applications in financial risk modeling, assigning credit scores to clients, and loan approvals—to name a few. Algorithms are also used to calibrate a client's portfolio to the set financial goals and her risk appetite. For example, a working woman sets a goal of building up a retirement corpus of $400,000 at the age of 50. With all her current income, assets and the risk appetite, the algorithm advisor can very effectively advise her on how to spread the investments across various asset classes to achieve this set goal. Many leading financial advisors commonly use this kind of machine learning–based software across the globe. The entire process can work more effectively with the assistance of an experienced financial advisor. As most of this process is automated, the advisor can effortlessly suggest many combinations of investment portfolios if the client's business goals, risk appetite, and any governing parameters change over time, which is very natural to expect. Another primary application of machine learning models in banking is fraud detection in credit cards or loan repayments. With input data such as client demographics and the client's historical transactions, these models can alert the security teams of possible defaults or frauds by the client with accuracy of more than 90 percent. To get this level of skill, large financial institutions train their fraud detection models even with millions of samples of consumer data, which includes details like income, number of credit cards, number of loans, job, age, marital status, insurance details, and so on. Now almost all leading banks across the globe have AI functions, which are deploying machine learning–based models and other AI-based technologies to yield great tangible benefits to their organizations.

Simple linear regression finds its widespread use in forecasting (predictions) and financial analysis. Logistic regression is a widely used statistical technique for computing propensity scores. Here, all the selected covariates are simultaneously included in a logistic regression model to forecast the state of the task. The propensity scores are the resultant forecasted probabilities for each unit. Decision trees are often used in shaping a course of action(s) in finance, investing, or business. Deep learning is used in finance for specific strategies such as high-frequency trading (HFT). It also finds its uses in banking fraud detection. We have discussed all these algorithms in the later chapters in detail.

1.5.5   Applications of Deep Learning

Deep learning is a subset of machine learning. Most of the applications discussed above might have already involved deep learning algorithms. Robots and other AI machines use deep learning for vision tasks like handwriting recognition, object detection and sorting, segmentation, and rebuilding of images. News agencies use deep learning algorithms to recognize and filter out negative news. Autonomous vehicles use hundreds of thousands of deep learning–based models (coordinating with each other) to drive on busy roads without accidents effectively. For example, one model may be adept at identifying pedestrians, while additional models in the same vehicle may pinpoint traffic lights and fast-moving objects on the road. All of it requires to work in perfect coordination with each other to avoid any possible mishap; in non-autonomous vehicles, this job is usually done by the human brain with the help of eyes and other sensing organs.

In another case of deep learning application, which you might already be aware of, automated robo-journalists are already writing stories and reports for various news agencies. This phenomenon has been around for more than two to three years now. Automatically translating one language to the other, for example, German to English, has been in existence for many years. Applying deep learning methods to this task is yielding superb accuracy. Now it is possible to translate given words, phrases, or sentences to another language more accurately, at the same time preserving the classicality of translated literature. The more you use the program for translation, the accuracy level improves even further. Deep learning has extensive application in the gaming industry; here, the model learns a very complicated task of playing a computer game based only on the pixels displayed on the gaming screen. Some of the best-performing systems using deep learning technology have originated only in the past ten years in the form of speech recognizers on smartphones like Apple iPhone and Google's latest automatic translator.

1.6   THE ORGANIZATION OF THIS BOOK

In this book, we have tried to cover almost all the essential machine learning and deep learning algorithms. The discussions around various machine learning and deep learning algorithms are divided into 12 convenient chapters. Each chapter is a complete unit in itself, covering the necessary theory and math along with case studies and code to execute the algorithms on real-life data.

Chapter 1: Introduction to Machine Learning and Deep Learning.   Chapter 1 serves as an introduction to machine learning and various stages of developing machine learning models. Examples are discussed on how machine learning models are used in the industry. This chapter lists all popular machine learning and deep learning algorithms and the tools used for developing these models.

Chapter 2: Basics of Python Programming and Statistics.   This chapter introduces Python programming, including basic commands in Python for development. Necessary packages in Python, such as data handling techniques like subsetting, filtering, and creating new variables, are discussed. Some basic concepts of descriptive statistics like mean, median, and percentiles are explained. The chapter also gives some introduction to data exploration in Python.

Chapter 3: Regression and Logistic Regression.   The chapter explains what a regression line is and how to accomplish a regression model building project. The concept of R-squared as a model performance measure is explained. Then we take up on how to build multiple regression models, including the concepts like multicollinearity. This chapter addresses the questions like how to measure the individual impact of the variables. This chapter introduces the logistic regression line and logistic regression model development. Multicollinearity in logistic regression is also discussed along with how to measure the impact of individual variables on the overall model.

Chapter 4: Decision Trees.   This chapter introduces readers to some segmentation techniques, the concept of entropy, and information gain. The decision tree algorithm is discussed in detail along with building a decision tree model using Python. Concepts like validating decision trees, pruning of decision tree models, and pruning parameters are also taken up in sufficient detail.

Chapter 5: Model Selection and Cross-Validation.   This chapter explains model validation techniques, training error, and test error. It introduces readers to receiver operating characteristic (ROC) and area under the curve (AUC) including how to create and measure them. The concepts of model overfitting and underfitting are also taken up. This chapter brings in concepts like bias and variance trade-off and model cross-validation techniques like K-fold cross-validation. The feature engineering tips and tricks explained in this chapter will be very useful in doing real-life projects.

Chapter 6: Cluster Analysis.   This chapter gives an introduction to unsupervised learning. The chapter discusses the K-means cluster analysis algorithm in depth with useful concepts of distance measures and distance matrix.

Chapter 7: Random Forests and Boosting.   This chapter starts with a discussion on the wisdom of crowds and ensemble models. In this chapter, we talk about the random forest algorithm in depth, which is a specific type of bagging algorithm. This chapter provides a detailed discussion on hyperparameters in the random forest and how to fine-tune them. After this a couple of boosting algorithms like Adaboost and Gradient boosting are discussed along with the crucial hyperparameters in boosting.

Chapter 8: Artificial Neural Networks.   This chapter starts with the explanation of the decision boundary, then goes on to topics like hidden layer and backpropagation algorithm. In this chapter, we have discussed all the steps in building a neural network model. This chapter also has an in-depth explanation of gradient descent and how it solves the neural network optimization problem.

Chapter 9: TensorFlow and Keras.   This chapter discusses a couple of deep learning packages with comparisons. We have taken up TensorFlow™ along the building blocks of the TensorFlow programming paradigm. After this we have taken up Keras and explained its features. After finishing this chapter, you will get familiar working with some of the very useful commands in TensorFlow and Keras.

Chapter 10: Deep Learning Hyperparameters.   This chapter discusses some very important hyperparameters that one must know while handling deep learning algorithms. In this chapter, we have discussed some essential deep learning concepts like regularization, learning rate, momentum, and activation functions along with various optimization functions.

Chapter 11: Convolutional Neural Networks.   In this chapter, we will take up the concepts like kernel filter, convolution layers, and pooling layer. We will discuss the detailed working of convolution neural networks. You will learn about various parameters of CNNs and how to fine-tune them. After finishing this chapter, you will be able to construct an optimal convolution neural network with real-time data.

Chapter 12: Recurrent Neural Networks and Long Short-Term Memory.   This chapter starts with a discussion on sequential models. Here you will learn about recurrent neural network models. We will discuss some essential concepts like backpropagation through time along with many others related to RNN. You will learn about the problem of vanishing gradients, and learn the concept of long short-term memory (LSTM). You will also learn other essential concepts like gates and how LSTM works with these gates.

1.7   PREREQUISITES—ESSENTIAL MATHEMATICS

Machine learning is all about data and algorithms. Reasonable math skills are usually desired. To comprehend this book, we expect only a math proficiency of high school level. As you will observe in the later chapters, the most mathematical steps are taken care of by the software; so you should not bother too much about the intricacies or details of it. If you have your basics clear, you will be able to comprehend the entire process better, and at the same time, interpret the results with ease. We are giving a list of essential topics here to gain a reasonably good level of proficiency in this field. For a complete treatment on these topics, you need to refer to standard math textbooks. On a broader scale, you need some ground skills in the following:

1. Functions, equations, and graphs
2. Basic calculus
3. Optimization theory
4. Vectors and matrices
5. Basic statistics and probability

To get started quickly, we will suggest attacking it by specific topic.

- *Linear algebra*: definitions of scalers, vectors, and their operations; eigenvalues and eigenvectors; common types of matrix operations
- *Elementary calculus*: functions, derivatives, partial derivatives, differentiation, gradients, essential integration
- *Basic statistics and probability*: concepts of population, samples, variables and their types, mean, median, mode, parameters, statistics, measures of spread, examples of quantitative and qualitative analysis, the basic theory of probability and related concepts

Apart from all these vital basics, if you can also learn the basics in differential equations, vectors, matrix calculus, and gradient algorithms, it will be advantageous.

1.8   THE TERMINOLOGY YOU SHOULD KNOW

Algorithms: They are probably the second essential constituent in the AI landscape. The first one is reliable data. In the AI context, an algorithm is a set of rules or commands given to machine learning and deep learning models, or neural networks in machines so that it can train itself. Regression, classification, and clustering are termed as the most commonly used types of algorithms.

Cognitive computing: Any computer-based model or algorithm that can run like a human brain by training itself with the help of techniques like data mining, pattern recognition, and natural language processing.

Natural language processing (NLP): These techniques typically use AI to recognize human speech or conversation done in any language like English or German. The language interpretation by NLP algorithms is supposed to be very similar to the way a human understands it.

Supervised learning: It works very much the same way as a teacher training a student in a class—classification- and regression-type algorithms fall in this category. Supervised learning is more commonly used when compared to another machine learning type, unsupervised learning. Just as an example, we train a machine on how an apple looks in color and shape; the next time it encounters a real fruit like it, the device will compare it with the training data and correctly recognize it as an apple.

Unsupervised learning: Unlike supervised learning, here no prior training is provided to the machine. The machine by itself is expected to find the hidden structure(s) in the given dataset. In unsupervised learning, no classified or labeled data is provided to the algorithm for training. The algorithm must act without any guidance. The most commonly used type of unsupervised learning is cluster analysis.

1.9   MACHINE LEARNING—A WIDER OUTLOOK WILL CERTAINLY HELP

Like any other branch of knowledge, having a broader perspective of the AI technology landscape helps you as a smart technocrat. It is especially true for machine learning as it is perpetually disruptive and affects our daily lives, unlike any other technology that has not done before. In this section, we will discuss the current state and the future road map of machine learning (and AI), along with some other issues touching daily life.

1.9.1 AI—The Current State

A few years ago, the companies were rushing analytics—big data to be precise. Now we are observing a similar explosion of businesses running to catch up with the AI and machine learning technologies. Machine learning is still considered as an R&D effort. Until recently, the infrastructure platforms required for product development were not easily accessible.

Machine learning is now undoubtedly recognized as a critical driver of digital transformation. The current widespread availability of cheap parallel computing, large datasets, and better algorithms is fueling much of these developments. Companies like Netflix, Airbnb, Tesla, Uber, Alibaba, Tencent, Xiaomi, and many others have established controlling machine learning capabilities with their internal platforms and models. However, these kinds of in-house cutting-edge machine learning capabilities are not readily available to most companies. Recently, a new generation of available machine learning platforms is likely making machine learning far more accessible to a broader spectrum of businesses—both small and big. As a result, machine learning has now become more accessible. The readily available open-source software allows for developing advanced self-learning systems. TensorFlow is one such popular tool built by Google. We will be using it extensively in the latter half of this book to deal with deep learning problems.

Deep learning is particularly getting much attention. Tech giants like Google, Facebook, Amazon, Baidu, and many other promising start-ups across the globe have achieved some mentionable results by using deep learning–based technologies in the field of vision and language-based chores. Deep learning is here to stay, at least in the short term—until the next three to five years, in our opinion. Interestingly, another area that is catching up very fast in the framework space is reinforcement learning (RL). RL has received lots of attention from researchers over the past decade. The likes of Google and Facebook have announced a variety of RL frameworks. Many of us are aware that Google has recently published Frank on top of TensorFlow. RL has made significant progress with training AI to play games, which are equaling or superior to human performance. Virtual agents, NLP, image recognition, and speech recognition are being considered the thrust areas in the AI-based data-driven companies.

kdnuggets' "Machine Learning Trends of the Future" speaks of the high points in the recent past. It finds the dominance of three machine learning inclinations in businesses existing today: cloud-based analytics, algorithm economy, and data flywheels. Data flywheels are used concerning a massive amount of data leading to better algorithm-based business models and leading to better business efficiency and user experience, ultimately leading to more data. It's a data-algorithm-data cycle. Many experts assert that these significant trends have helped many companies to get converted as the data companies that they are seen today. With the algorithm economy taking over the classical managing style of business processes, eventually, every company will be forced to turn into a data-based company. In these so-called data companies, there is a direct communication of research communities, from both internal and external, like some leading universities, with business leaders to find business solutions jointly.

According to McAfee Labs' 2018 Threats Report, in the future machine learning will be imposed for the detection of cyber-intrusion detection, online frauds, and spam. It can be useful in detecting malware for cybersecurity experts in serverless environments. As the number of cyber-attacks is growing, machine learning is already helping corporations to improve approaches to security. A leading US-based manufacturer of office printers is already using massive data and machine learning technologies for online intrusion prevention (of malware) on its enterprise range of printers and multifunction devices. In the foreseeable future, developers might be able to develop Blockchain-based tools as a viable way to counter network invasion and ensure the safekeeping of data.

1.9.2 AI Future Road Map—How Disruptive Is It?

Digital transformation in enterprises is one big buzzword nowadays. Many business philosophies and technologies drive digital transformation. Of all these technologies driving the change, AI is perhaps considered the most disruptive of all. When it comes to the AI-caused disruptions in enterprises, it is essential to note that only about 5 percent of all the occupations can be fully automated by the currently available AI- or machine learning–based technologies.

IT operations in many organizations have already been automated to a large extent. From a CEO's perspective, IT operations are only a cost center; so if AI can automate it efficiently and cost-effectively, it is a welcome move. The same C-suite perspective holds for finance, marketing, or any other function, which is not strategic to the organization. Given this condition, people will still play an important role, but things will never be the same as they have been for a long time. Professionals will have to work with the assistance of AI (physical robots included) in fast-changing environments. Autonomous cars and trucks are already well within reach. We may soon be able to automate fully all the functions of businesses like a trucking company or an application (app)-based taxi company like Uber. How soon will we see it happening is an open question as of now.

An article by Mike Thomas, "The Future of Artificial Intelligence," beautifully describes how AI is affecting different industry sectors.

Transportation: It may take around a decade more to perfect autonomous cars. By this period, they are likely to replace human drivers fully. There will be much incentive for the researchers and business establishments to bring 100 percent automation (for all business functions) in the companies involving vehicles like self-driving taxis and transportation trucks.

Manufacturing: Human workers work alongside robots powered by AI and perform a limited range of repetitive tasks like assembly and stacking. Machine learning and data-assisted predictive maintenance keep equipment running smoothly. AI takes over many repeated and laborious tasks even in the supply chain to reduce mistakes, improve predictability, and make the entire process more efficient with the assistance of humans.

Health care: AI technologies are comparatively nascent in the field of health care. It helps a quicker and more accurate diagnosis of diseases. Pharma companies are speeding up and streamlining the otherwise time- and resource-intensive drug discovery process. Robo=surgeons are assisting doctors in performing complex surgeries in a better and more efficient manner. Virtual nursing assistants are monitoring patients. More personalized patient experience is created with the help of big data analysis.

Similarly, all other business functions—be it education, media, customer service, to name a few—are all getting affected by AI-powered technologies. Furthermore, we can safely conclude all of it is just the tip of an iceberg. The actual effect is much broader and as significant as one can imagine.

Samsung AI Forum, held in September 2018, unfolded some interesting theories about the future road map for AI. LeCun (AI scientist) argued that unsupervised learning (or self-supervised learning) holds the future of AI. LeCun went on to explain how it is possible. Contrasting to RL models that mainly depend on trial and error, unsupervised learning models could be proficient in signifying mental proficiencies like what we call common sense. Breazeal's speech focused on social robots. He explained in the next 10 to 20 years, we can have socially and emotionally intelligent robots (so-called relational AI technologies)—it presents a wide range of exciting benefits. These robots will significantly improve the quality of our daily lives by becoming an integral part.

AI disruptions and its future road map is an exciting and endless discussion. Specific literature is abundantly available all around that covers this topic in detail from various angles.

1.9.3   Ethical, Social, and Legal Issues Related to AI

In the previous section, we have seen a few glimpses of how disruptive AI can be. It affects us as an individual and society at large. Any discussion on AI and machine learning would be considered incomplete without covering the resulting ethical and social issues.

In an article titled "10 Ethical Issues of Artificial Intelligence and Robotics," Miguel González-Fierro discussed the ten significant ethical issues related to AI and robotics—misinformation and fake news, job displacement, military robots, algorithmic bias, regulation, superintelligence, privacy, cybersecurity, mistakes of AI, and robot rights. Much of it is self-explanatory, but we will still take up a few for a brief discussion in the following section—just to give you a sense of it.

There are all-round discussions on how the modern social media can immensely be misused to spread distorted and fake news. The recent advances in the field of computer vision make it possible to completely fake a video. In essence, you can artificially create almost any video, which can be very tough for a common man to authenticate. Moreover, for the general public, there is hardly any way to find out the truth. For many youths, the sole source of news, facts, and information is social media. We have already taken up the issue of job displacements using AI in the previous sections. In an open letter, 25,000 researcher scientists and AI professionals have called to ban self-governing weapons (military robots included) that do not require human supervision to avoid a worldwide military arms race of AI-based weapons. It is a no-brainer to insist on the need to avoid prejudice and discrimination in the development of AI algorithms. One such case was reported where algorithms were biased—a person with a specific skin color had a lower detection rate when compared to others. We are not getting into the details of that algorithm here, however.

Currently, AI-based products and services are mostly unregulated, as lawmakers never considered this kind of AI explosion while designing laws. New regulations are required to make AI systems more responsible for harm caused by irrational practices. Many internet companies are collecting more information about their users than required for their business. How much information should a company be allowed to gather about their users? How much of it is ethical? We think the society and its lawmakers have not paid much attention to it yet. Another big concern is privacy, the widespread news of the Cambridge Analytica scandal, where 87 million Facebook profiles were compromised and used in influencing the US presidential elections and Brexit campaign. Are our laws clear enough on privacy issues in this internet democracy?

While AI can help us in a big way to protect against cybersecurity attacks, at the same time, hackers can also use it to find new sophisticated ways of attacking establishments. Another ethical issue with AI—robots can make mistakes, and it is tough to fix responsibilities. It gives rise to many more related legal issues. When robots become emotionally intelligent, should they have any rights? How will they relate to humans? Who will define the level of intelligence a robot can have? Do we need any such limits on human-made machines? These included more open questions that need an open debate resulting in clear policy guidelines standard across the globe. All these AI-related ethical, social, and legal issues are not specific to any one country, but it is typical for the entire planet. One country developing harmful AI systems will affect everyone alike. It is a severe global concern similar to today's rapid proliferation of nuclear weapons, and greenhouse gases and pollution causing irreversible environmental changes.

Now, several countries are aggressively forming legal settings for the advancements in technologies using AI. One such example is from South Korea, where the "Intelligent Robot Development and Dissemination Promotion Law" has been in place since the year 2008. This law aims at improving the quality of life and progress of the national economy. It drafts the required policy for creating and promoting an approach for the sustainable growth of the smart robot industry. In another example, the European Union passed a landmark resolution on the civil law rules on robotics that is widely recognized as the first step in the direction of regulating the AI-based technologies. In 2015, the European Union also established a working group on legal queries related to the growth of robotics and AI.

1.10   *PYTHON AND ITS POTENTIAL AS THE LANGUAGE OF MACHINE LEARNING*

Guido van Rossum developed the Python programming language in the 1990s. Today, Python is one of the most widely adopted general-purpose programming languages. Python is a modern scripting language. The programs written in Python are directly fed to its interpreter. This interpreter runs the programs directly without any compiling. Getting feedback (like finding errors) on your Python code is easy and fast.

Python is a cross-platform programming language. You can run Python programs on Mac, Windows, Linux—both on personal computers and large servers. You can use Python even on tablet PCs that use iOS and Android. A lot of third-party libraries are available on the Python platform. It further increases the versatility and usefulness of this language as much of code for a wide variety of programming tasks is already available as ready code written in the form of libraries.

Mind you, Python is free software. There is no need to make any payment for downloading and using Python. Any source code that you write with Python is solely yours, and you can share it at your will. Python is a very versatile language. Software development is faster with Python. You can use Python in whatever you are interested in. It makes programming with Python great fun.

Even the International Space Station's Robonaut 2 robot has deployed Python for its central command system. The 2020 European mission to Mars also plans to use the programs related to the task of collecting soil samples. We, the data scientists, love Python for its inbuilt libraries that make model-building projects a lot easier for us. Python's ease of use results in a very fast-paced development in data science software projects.

1.11   *ABOUT TENSORFLOW*

The TensorFlow library is developed by the Google Brain Team. It is a second-generation system of Google Brain. TensorFlow is built scale to any extent, and it can run on multiple CPUs or GPUs. It runs even on mobile operating systems. It has several wrappings that support languages like Python, C++, or Java. The TensorFlow library integrates different application programming interfaces (APIs) to support deep learning architecture like CNN or RNN.

TensorFlow is an open-source platform that is used mainly for machine learning. It consists of an all-inclusive, flexible ecology of tools, libraries, and community resources. It enables data scientists to use the state of the art in machine learning. Using TensorFlow, developers can quickly build and deploy machine learning–powered systems.

TensorFlow has become a prevalent deep learning library these recent years. It can be used to build any deep learning structure, ranging from a simple artificial neural network to CNN or RNN. TensorFlow is used by Google in practically all Google products including Gmail, Photo, and the Google search engine.

1.12   *CONCLUSION*

In recent years, the top three most searched job titles are machine learning engineers, deep learning engineers, and senior data scientists. A couple of years back, data scientists were most in demand. We observe a clear shift to machine learning and deep learning from elementary analytics that started in the early 2010s. The technology landscape and job titles are changing more every year. However, here is a difference with what was happening about a decade back. Notably, in the past eight to nine years, one technology is not getting replaced by another, but the same field is continually evolving. Accordingly, in 2015, we wrote an analytics book for beginners; in 2020, as authors, keeping pace with the technology evolution, we are responding with an advanced text on machine learning. We can term the 2010s decade as a decade of AI. This technology is here to stay for many more decades to come. It will only evolve with every passing day, week, and year—and you, as a machine learning expert, will always be challenged to keep pace with it.

In the words of Google CEO Sundar Pichai, "AI is one of the most important things humanity is working on. It is more profound than I don't know electricity or fire." We as authors do not have any better words to emphasize the importance of this technology to humanity, and to reiterate further, it will be there forever—continually evolving.

1.13   REFERENCES

1. Stuart J. Russell and Peter Norvig. (2015). *Artificial Intelligence: A Modern Approach*, Englewood Cliffs, NJ.

2. Brendan Scott. What is Python and what can you do with it, Retrieved from www.dummies.com. Accessed on August 7, 2020.

3. What is TensorFlow? Introduction, architecture & example. Retrieved from https://www.guru99.com/what-is-tensorflow.html#:~:text=TensorFlow%20is%20a%20library%20developed,like%20Python%2C%20C%2B%2B%20or%20Java. Accessed on August 7, 2020.

4. Asir Disbudak. *Machine Learning Is Happening Now: A Survey of Organizational Adoption, Implementation, and Investment.* Retrieved from https://www.kdnuggets.com/2019/08/machine-learning-happening-now-survey-organizational-adoption-implementation-investment.html. Accessed on August 7, 2020.

5. Paramita (Guha) Ghosh. (2019, January 29). *Machine Learning and Artificial Intelligence Trends in 2019.* Retrieved from https://www.dataversity.net/machine-learning-and-artificial-intelligence-trends-in-2019/. Accessed on August 7, 2020.

6. Hussain Fakhruddin. (2018, October 9). *Machine Learning in 2019: Tracing the Artificial Intelligence Growth Path.* Retrieved from http://teksmobile.se/2018/10/09/machine-learning-in-2019-tracing-the-artificial-intelligence-growth-path/. Accessed on August 7, 2020.

7. Top 5 latest advancements in artificial intelligence to know. September 2018. Retrieved from https://www.techgenyz.com/2018/09/20/latest-artificial-intelligence-advancements/. Accessed on August 7, 2020.

8. Maria Thomas. (2018, July). The state of machine learning in 2018. Retrieved from https://jaxenter.com/state-of-machine-learning-146002.html#:~:text=Big%20data%2C%20artificial%20intelligence%2C%20and,artificial%20intelligence%2C%20and%20machine%20learning. Accessed on August 7, 2020.

9. Shrikant Srivastava. (2019, August). 37 disruptive AI technology trends for 2019–2020. Retrieved from https://appinventiv.com/blog/ai-technology-trends/. Accessed on August 7, 2020.

10. Jason Bloomberg. (2018, July). Think you know how disruptive artificial intelligence is? Think again. Retrieved from https://www.forbes.com/sites/jasonbloomberg/2018/07/07/think-you-know-how-disruptive-artificial-intelligence-is-think-again/#53291bb3c902. Accessed on August 7, 2020.

11. Mike Thomas. (2019, June). The future of artificial intelligence. Retrieved from https://builtin.com/artificial-intelligence/artificial-intelligence-future. Accessed on August 7, 2020.

12. Samsung AI forum offers a roadmap for the future of AI. (2018, September). Retrieved from https://news.samsung.com/global/samsung-ai-forum-offers-a-roadmap-for-the-future-of-ai. Accessed on August 7, 2020.

13. Miguel González-Fierro. 10 ethical issues of artificial intelligence and robotics. Paper submitted April 2018. Retrieved from https://miguelgfierro.com/blog/2018/10-ethical-issues-of-artificial-intelligence-and-robotics/. Accessed on August 7, 2020.

14. Maksim Karliuk. (2018, April). *The Ethical and Legal Issues of Artificial Intelligence.* Retrieved from https://moderndiplomacy.eu/2018/04/24/the-ethical-and-legal-issues-of-artificial-intelligence/. Accessed on August 7, 2020.

15. Matt Kiser. *Machine Learning Trends and the Future of Artificial Intelligence.* Retrieved from https://www.kdnuggets.com/2016/06/machine-learning-trends-future-ai.html#:~:text=The%20confluence%20of%20data%20flywheels,now%20be%20an%20intelligent%20app. Accessed on August 7, 2020.

16. Sivaramakirshnan Somasegar and Daniel Li. (2016, May). *The Intelligent App Ecosystem (is more than just bots!).* Retrieved from https://techcrunch.com/2016/05/24/the-intelligent-app-ecosystem-is-more-than-just-bots/?guccounter=1&guce_referrer=aHR0cHM6Ly9pbi5zZWFyY2gueWFob28uY29tLw&guce_referrer_sig=AQAAAAGtzHLIHeCDNGiJ9MIRE2Tol_I7RmQvnv0_ZTVgM7FZ3zJbhB8kR0mYWUykhIblXxdtk4TQWnT1l2cKtmclitmkR5dviK0UbF9el67BBTDGUdU-tiuUKGCt38Qe4IDbJh-3hyMzF70zlOEg-oT3QJ3QQZD5F_Vgcx8ShEXWiDHr4. Accessed on August 7, 2020.

CHAPTER 2
BASICS OF PYTHON PROGRAMMING AND STATISTICS

This chapter is organized around three points, the fundamentals of Python, basic statistics, and getting started with data handling. The chapter starts with an introduction to Python followed by basic Python operators. We then proceed to some data handling tips and tricks using Python. We will also learn about some basic descriptive statistics. Finally, we will discuss data exploration and data sanitization techniques.

Knowledge of basic Python would be needed for executing all the hands-on exercises throughout the book. We need to know basics like writing simple code snippets, submitting it for execution, debugging for errors, and some data operations like importing data and preparing it for analysis. In this chapter, we are going to learn the following Python operations:

- Basic operations in Python
- Writing and submitting Python code
- Important packages in Python
- Working with datasets
- Useful tips and tricks while working with data

In addition to basic Python, some fundamentals of statistics are also required for getting started with machine learning algorithms. Certain basics, like central tendencies and dispersion measures, are an absolute essential for any data scientist. We will cover the following topics, which are best covered under the umbrella of basic descriptive statistics in the textbooks on this topic:

- Measures of central tendencies like mean and median
- Measures of dispersions
- Percentiles and quartiles
- Variable distributions

For any data scientist, it is critical first thoroughly to understand the data and be comfortable with it. Raw data is usually extracted from multiple sources and requires a cleansing operation before attempting any statistical analysis. We also need to put it in the format that is convenient for analysis. To make you familiar with these operations, we are going to discuss the following data exploration and cleansing techniques later in this chapter:

- Data exploration using continuous variables
- Data exploration using categorical and descriptive variables
- Data cleaning tips and tricks

2.1   *INTRODUCTION TO PYTHON*

By this time, we are already familiar with Python, and we will use it as a language for analysis throughout this book. Given below are some basics of what we need to know before we start writing the code.

2.1.1   Why Python?

Python is one of the most popular languages in the data science and machine learning ecosystem. Many functions and libraries available in Python can be used for data manipulation, data exploration, data sanitation, machine-learning model building, model validation, and final implementation. Guido Van Rossum is the primary designer of Python. Compared to other languages, Python is simple and easy to learn. The fundamental beauty of Python is its philosophy that is mainly centered around maintaining simplicity and ease of reading. The following are some of the plus points of Python, which will make us fall in love with it automatically:

- Python is available as an open-source tool. It is free for all applications—be it individual learning or commercial use.
- Designers of Python have made a conscious effort to make its syntax simple and easily readable. That is the reason this language is so popular in the entire software development community. Everyone admires Python's power and simplicity.
- Python has a whole lot of functions and libraries devoted specifically to data manipulation and statistical algorithms specifically used in business analytics, making it a preferred language for any data scientist.
- Python is currently used by countless data scientists around the world. Many companies have already started using it with their data science platforms and applications.
- One major advantage with Python, which in no case can be discounted, is that it is well documented and widely discussed. A whole lot of users around the world actively participate in Python-related forums and debates on the Internet. As a result, it is very easy to get answers to our queries related to Python code and libraries. Almost every person related to data science and information technology (IT) would agree on this advantage that Python makes their life a lot easier.

We are mistaken if we assume Python is used just for data science or machine learning. It is a multipurpose programming language. It is also used for developing web applications, regular software development, and even in business applications like enterprise resource planning (ERP).

We are in love with Python because of its simplicity and data handling capabilities. Furthermore, we are sure you will also find other reasons to keep it in your heart.

2.1.2   Python Versions

The first implementation of Python was way back in December 1989. However, Python got famous from 2000 onward. Python 2.0 was released in 2000. Table 2.1 shows some of the important version release dates.

TABLE 2.1   Python Version Release Dates

Release Date	Version
Dec 1989	First implementation
Feb 1991	Python 0.9.0
Jan 1994	Python 1.0
Oct 2000	Python 2.0
Dec 2008	Python 3.0
Jun 2018	Python 3.7

Python Version Release Date

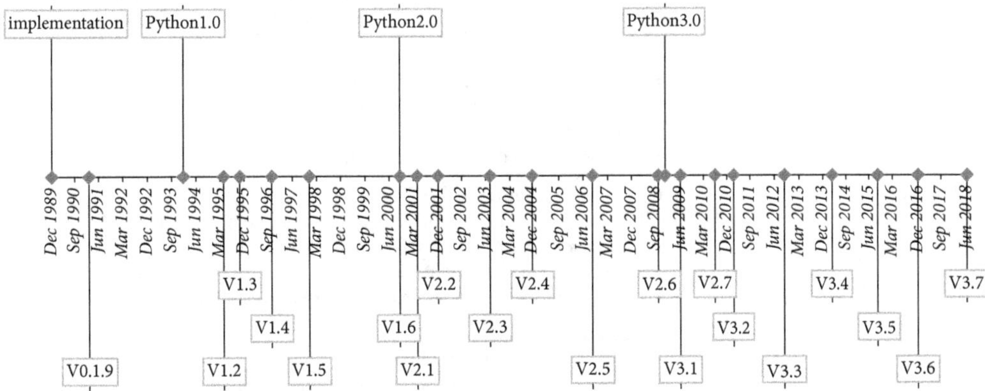

FIGURE 2.1 Major milestones in the journey of Python.

As we can observe in Fig. 2.1, Python 2.0 was released in 2000, and Python 3.0 was released in 2008. Generally, when a new version of a language is released, it would be backward compatible. Any code that we write in Python 2.0 should work in Python 3.0, but it is not the case with Python. Python 3.0 was released to fix some major bugs in Python 2.0. There is a difference in the way some of the calculations and objects are handled in Python 2.0 and Python 3.0. It makes Python 3.0 incompatible with Python 2.0. For example, just have a look at the following code. In the first line, we are storing numeric 10 in the variable x. Now the second line of code works only in Python 2 and throws an error in Python 3, while the third line works in Python 3.0. The takeaway—the second and third lines in the following code mean the same, but one works only with Python 2.0 and the other only with Python 3.0. There are many such contrasts in the second and third releases of Python, because of which the language loses its backward compatibility.

```
x=10
print x
print(x)
```

We need to be aware of these differences and be careful. When we are searching for the code on the Internet, we need to know the Python version that we are using. Currently, there are two versions of Python being developed. We can see the release of 2.7 when Python 3.0 is in existence. Python 2.0 development will eventually be frozen by 2020.

Which version to learn then? If we are learning Python for the first time, we can start with Python 3.0, and the differences with other versions can be picked up later.

In this work, we will be using the latest stable version—Python 3.7—for obvious reasons. By the time you will get to read this book, some syntax might (or might not) have changed, and some of the libraries might be updated. We will try to post all the updates on our book webpage. We suggest you look for the updated version of code if you encounter any syntax-related errors or warnings indicating deprecation of a few functions.

2.1.3 Python IDEs

An integrated development environment (IDE) is a tool used for writing and executing the code. IDEs do exist not only for Python, but for a host of other programming languages also. IDEs usually have an editor to write the code with a console having execution and debugging options. Python has multiple options for writing and executing of code. For a popular multipurpose language like Python, it is natural to expect multiple IDEs, each from a different source. One such popular IDE is Spyder. Each IDE has its special features and strengths. Table 2.2 shows some of the popular IDEs used by data scientists around the world.

TABLE 2.2 List of IDEs and Their Features

IDE	Highlights
Jupyter notebook	<ul><li>Open-source</li><li>Good documentation and presentation features</li><li>Inbuilt markdown and HTML features</li><li>Works within a web browser</li><li>Very popular in data science community</li></ul>
Spyder	<ul><li>Open-source for individual learners</li><li>IDE feels very similar to other popular tools like RStudio and SAS</li><li>Explicitly displays variables, datasets, and objects</li><li>Widely used by data scientists</li><li>Heavy software, you may need a good RAM (recommended 8GB) to launch it</li></ul>
Idle	<ul><li>Open-source</li><li>Developed exclusively for Python</li><li>Auto-completion of the code and syntax highlighting</li><li>Lightweight compared to some other tools</li></ul>
PyCharm	<ul><li>A code editor is smart and fast</li><li>Easy tools for debugging, testing, and deployment</li><li>Not available as an open-source</li></ul>
Sublime Text	<ul><li>Sizable user community; used for multiple other languages</li><li>Faster compared to many other IDEs</li><li>Not available as an open-source</li></ul>

In this book, we will work with only two IDEs—Spyder and Jupyter notebook. Spyder is suitable for beginners and Jupyter notebook is trendy amongst data science professionals.

2.1.4 Installing Python

For a beginner, it is advisable to consider a tool that takes care of admin and software installation–related tasks. In this book, we would prefer using Anaconda as an installation tool for Python. Anaconda is available as an open-source tool for individual learners. An enterprise version is also available. Installing Python using Anaconda is probably the easiest way to get started with the coding—this way, installing Python becomes nothing but installing Anaconda (Fig. 2.2).

- Downloading is easy. Just get it from its website and be careful about what version you choose. You need to first pick up your OS (Mac or Windows) and then choose Python 3.0 to download. Anaconda is freely available from https://www.anaconda.com/distribution/#download-section.
- As you install Anaconda, it takes care of many other tasks. It not only just installs Python, but it also installs IDEs like Spyder, Jupyter notebook, Ipython Console, and RStudio.
- While installing, Anaconda automatically downloads the essential packages required for a data scientist and stores them locally. This feature can come in very handy at times.
- It gives an excellent user interface (UI) navigator to work with installed tools, packages, and environments.

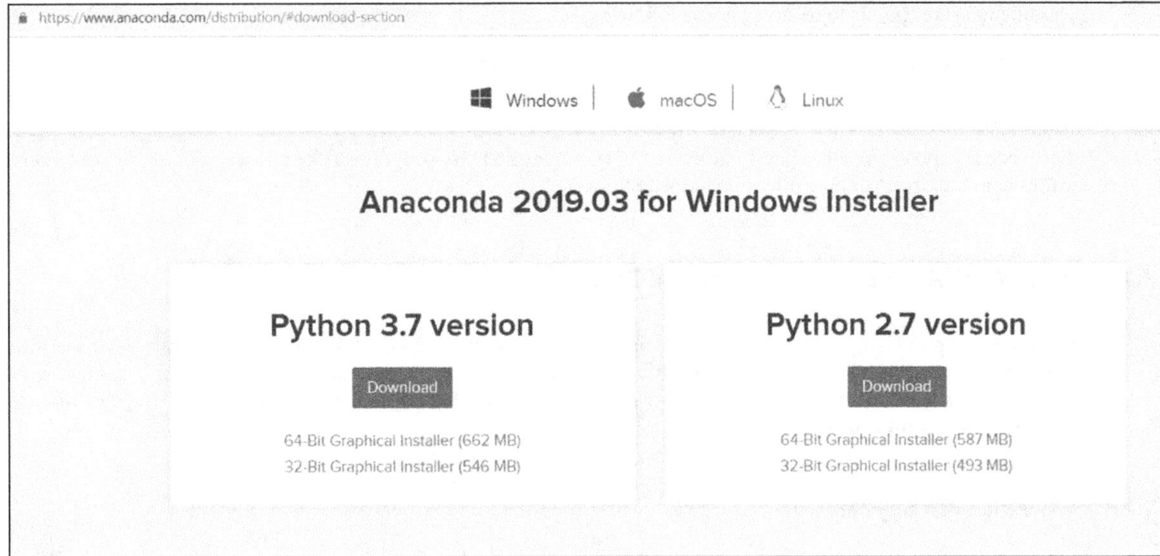

FIGURE 2.2 Anaconda download website.

- We are not detailing the installation process here. It is simple and straightforward. In a rare case, if you face any challenge, you can search the Internet for a solution and choose from many readily available options.
- You can launch Spyder from the Anaconda navigator or its start window. If everything goes fine, you should see the Spyder IDE window, which may take a couple of minutes to load. Once Spyder loads successfully, you would see a layout with three windows (Fig. 2.3).

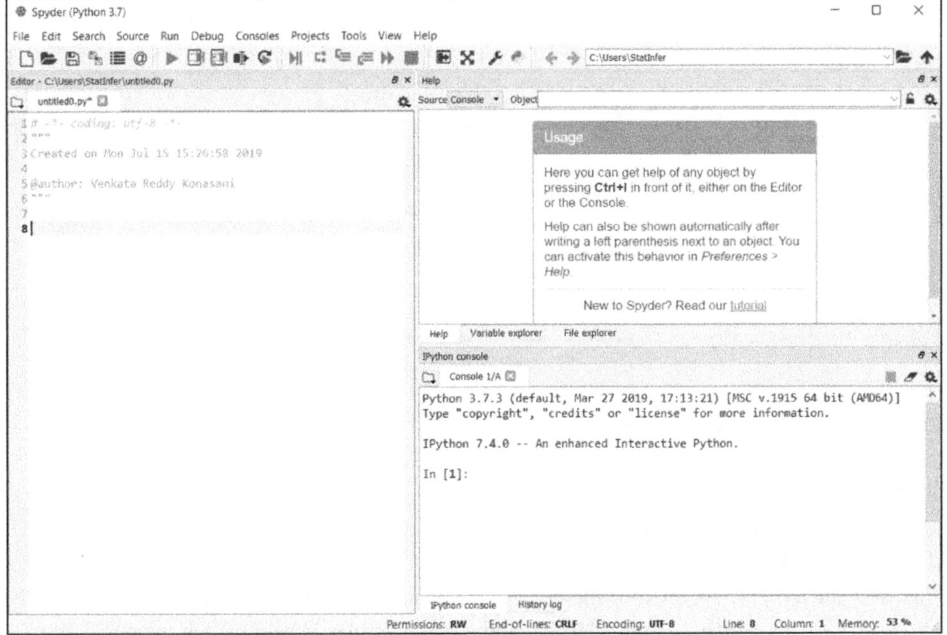

FIGURE 2.3 Spyder IDE windows.

Some important points to be noted are as follows:

- In a few machines, we observed that a restart is required after installing Anaconda.
- We need a minimum of 8GB RAM to run Spyder effortlessly. We can work with less RAM, but Spyder and some other programs may take more time for execution.
- For a better experience, we suggest you close the resource-intensive programs (like virtual machine or containers) running in the background while working with Spyder.

2.2   GETTING STARTED WITH PYTHON CODING

If you know scripting languages like Perl or VB or any other syntax in SAS or R, it will come in handy while learning Python. No worries, even if you do not have any coding background. Python is not hard at all to learn or even to work with on complex real-life problems. You just need to follow the right learning path of gradually increasing the difficulty level. Just be confident and get started with us.

2.2.1   Working with Spyder IDE

As discussed earlier, for the first few chapters, we will be using the Spyder IDE to write and execute the code. When you launch Spyder, you see three windows—editor window on the left-hand side and two more windows on the right-hand side. The right-hand side bottom window is the console. More details of these windows are given in Fig. 2.4.

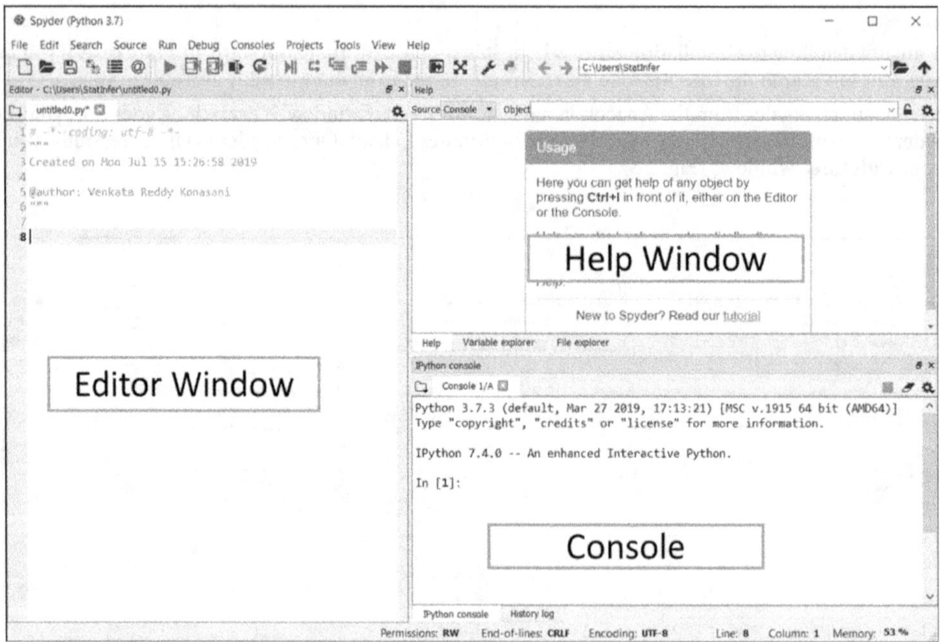

FIGURE 2.4   Spyder IDE windows and tabs.

- Editor window:
  - This is the place where you write the code.
  - Once you write the code, you can submit it by merely using F9 key or Ctrl+Enter.
  - In case if you already have the code file, you can load it using the open file option.
  - Auto-fill for code options are available, but some packages may have some lag time while they give auto-fill suggestions.
  - You may spend a significant portion of your time only in the editor window while working with this IDE.

- Console window
  - This window is on the right-hand side bottom portion of your screen.
  - It is where the code output is shown when you hit Ctrl+Enter in the editor.
  - The console shows you input commands and output results. Errors, if any, will also be shown in the console.
  - It is a usual practice to write a chunk of code in the editor window and then execute to validate if it is working fine. It is very different when compared to any regular software development style, where we write the full code and execute in one go.
- Variable explorer
  - a. The top right side has three windows (Fig. 2.5). There you will notice the help window, variable explorer, and file explorer. Click on variable explorer. It is on the bottom band of the top right-side window.

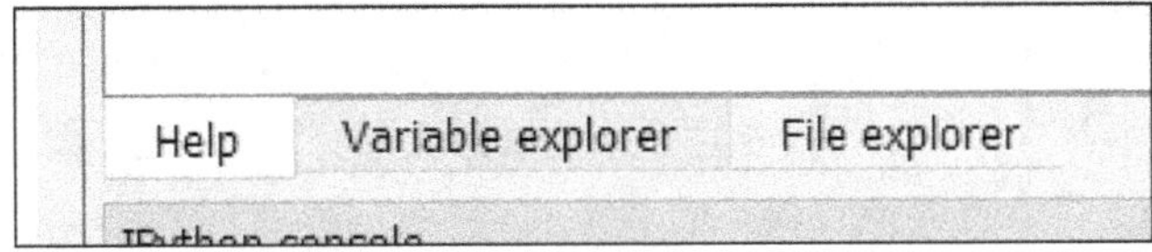

FIGURE 2.5   Variable explorer.

  - b. While working on a project, you will be importing many datasets. You may be creating several variables, objects, etc., and you can see all of it in variable explorer.
  - c. The objects and variables are available only until the current session is on.
  - d. The variable explorer is a handy window in Spyder. It not only shows variables; it also shows variable properties like size, length, and sample values.

2.2.2   First Few Lines of Code

Before you start, please note Python is a case-sensitive language. Everything is case sensitive, be it function names, object names, or strings. For example, `All_data` is not same as `all_data`.

Now it is time to get started with some basic coding with Python.

Go to the Spyder editor window and write the command shown in Fig. 2.6.

```
print(601+49)
```

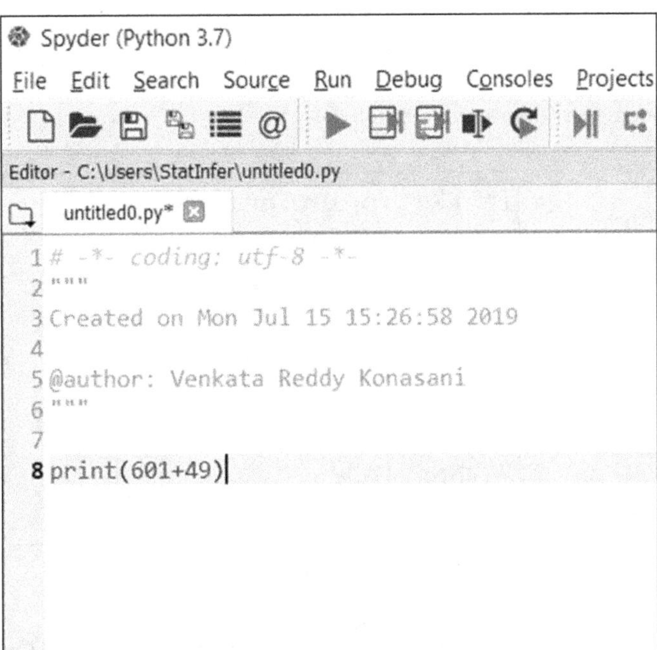

FIGURE 2.6   Editor window.

Select the print line, you just wrote, and hit F9 or Ctrl+Enter. Check the console window. Could you notice the output? What is In[1]? "In" is indicating the input, and the numeric 1 is the input command number (Fig. 2.7). It is our first input command, hence In[1].

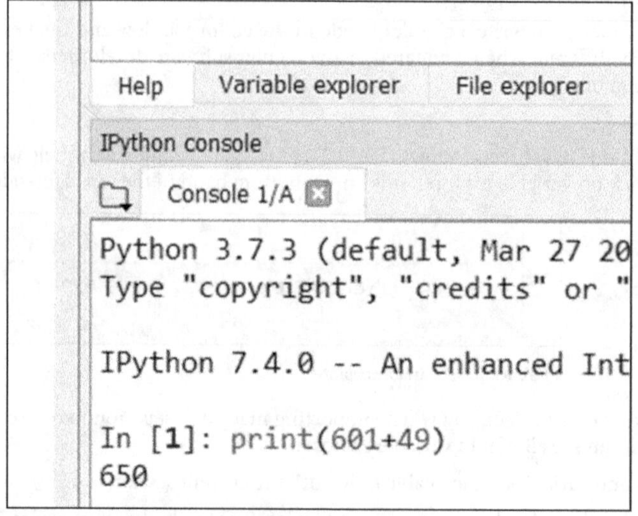

FIGURE 2.7   Printing example.

There are a few menu options to submit the code for execution, but the easier way is to select the code snippet and hit F9 or Ctrl+Enter. The green play button on the top menu bar executes the full code file, but quite often, we are interested in executing a piece of code and not the entire file.

Have you got the output in the console? Now you are ready to write a few more lines of code. Write a line of code and submit it before you go to the next, or you may even execute all of it at once.

```python
print(601+49)
print(19*17)
print("Python code")
x=7
print(x)
```

The above code gives us the output in Fig. 2.8.

FIGURE 2.8   Code output window screenshot.

Variables are stored in variable explorer: Check your variable explorer (Fig. 2.9). Can you see your variables there?

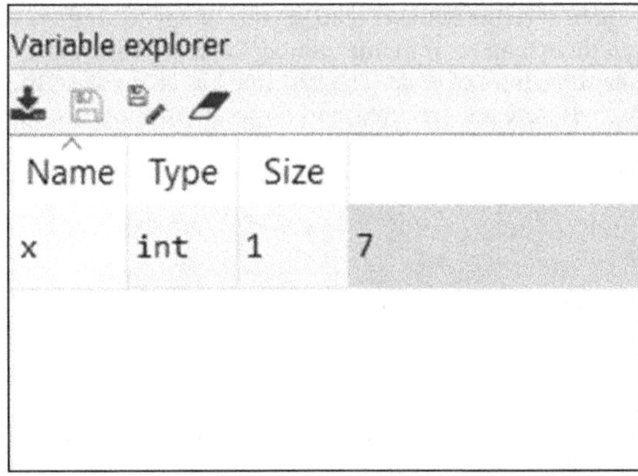

FIGURE 2.9   New variable in the variable explorer.

How are comments written? Use # for single-line comments (Fig. 2.10). You can use three quotes, either single or double for multiline comments.

```
14 #Single line comments
15 """
16 Multi line
17 Commnents
18 """
```

FIGURE 2.10   Comments in IDE screenshot.

2.2.3   Errors and Error Messages

Where are the errors shown? How do you check if there is an issue with the code? You see all the input and output in the console, so the errors are also shown in the console itself. Write the below code and see what the error is.

```
Print(600+900)
```

The above command gives us the below result.

```
Print(600+900)
Traceback (most recent call last):

   File "<ipython-input-2-e6e08d521061>", ine 1, in <module>
     Print(600+900)

NameError: name 'Print' is not defined
```

```
print(576-'96')
```

The above command gives us the below result.

```
print(576-'96')
Traceback (most recent call last):

   File "<ipython-input-3-cfd2e5e6ab20>", line 1, in <module>
     print(576-'96')

TypeError: unsupported operand type(s) for -: 'int' and 'str'
```

It is important to take note of errors while executing code in Spyder. If there is an error, Python code execution will be aborted. For example, if there is an error in line number 26, no code line after it would be executed. In the following example, we try to execute three lines of code. The first line has an error, so the code execution was aborted at line 1 itself. The remaining two lines were not executed, and no result for the second line containing the variable y is shown.

```
print(576-'96')
y=10
print(y)
```

The following is the output of the above code.

```
Traceback (most recent call last):

  File "<ipython-input-4-3501a1318948>", line 1, in <module>
    print(576-'96')

TypeError: unsupported operand type(s) for -: 'int' and 'str'
```

2.2.4   Naming Conventions

- Every object name should start with a letter only. Numbers are not allowed as the starting letters of an object name.
- An object name can contain numbers followed by a letter(s).
- We cannot have all the special characters in the variable names. For example, an underscore is allowed, but the dot and dollar sign are not allowed (Tables 2.3 and 2.4)

TABLE 2.3   Python Naming Conventions

1x=20	Does not work
x1=20	Works
x.1=20	Does not work
x_1=20	Works

TABLE 2.4   Not Even Get a Warning

The variable assignment in Python is dynamic. Look at the code snippet in the right-hand-side cell. We need not declare income as an integer before assigning 12000. Python will dynamically identify it as an integer. The code execution takes place line by line. If we directly try to execute the code line z=x*y without beforehand assigning or creating x and y, an error will be thrown.	`income=12000`
	`print(income)`
	`12000`
	`z=x*y`
	`Traceback (most recent call last):`
	`  File "<ipython-input-8-3d9d541540c2>",`
	`  line 1, in <module>`
	`    z=x*y`

By the time you execute z=x*y you should have x and y stored in your variable explorer.	`NameError: name 'y' is not defined`
	`x=20`
	`print(x)`
	
	`y=30`
	`z=x*y`
	`print(z)`
	`600`

If we store some value in a variable, we can replace it with any other value, even if it is of a different data type. You will not even get a warning before the replacement takes place. One needs to be familiar and be careful with this feature of Python.	`income="March"`
	`print(income)`
	`March`

2.2.5   Printing with a Message

What if you want to print the output along with a message? For example, age=35, instead of using **print**(age), which will just return the value of age. You want to print something like "Age value is 35". You need to use the print function with a comma; **print**("Age value is", age) works. Given below are a few more examples.

```
age=35
print("Age value is", age)

income=5000
print("income is", income , "$")

gdp_percap=59531
Country="United States"
print("GDP per capita of", Country, "is" , gdp_percap)
```

The above code gives us the below output.

```
Age value is 35
income is 5000 $
GDP per capita of United States is 59531
```

2.3   TYPES OF OBJECTS IN PYTHON

Numbers, strings, lists, booleans, tuples, dictionaries, and sets are a few examples of the types of objects in Python. We will cover some of the useful types of objects. It is essential to know the object type as we get the output in a particular format. Ways to access or handle objects in Python are based on their types. For example, handling lists is different from handling dictionaries.

2.3.1   Numbers

Numerical values have a "number" as an object type. Integers and floats are two different types of objects indicating numbers. We do not need to predefine the object type. Based on the value we store in an object, it will automatically be picked up as an integer or a float.

```
sales=30000
print(sales)
print("type of object is ", type(sales))

30000
type of object is  <class 'int'>
-----------------------------------------
Avg_expenses =5000.25
print(Avg_expenses)
print("type of object is ", type(Avg_expenses))

5000.25
type of object is  <class 'float'>
```

2.3.2   Strings

Python uniquely handles strings. There are abundant built-in methods for handling strings in Python; on the other hand, in many other languages, we need to use a separate function to do the same job. As soon as you create a string, it is indexed, and each character will be given an index value internally (Table 2.5). Using this index number, we can access any character value in the string. Please note that Python indexing starts with zero.

TABLE 2.5   Python String Code Snippets

String Operation	Code
Defining strings.	`name="Sheldon"`
	`msg="sent a mail to Jack"`
Accessing values from string.	`print(name[0])`
	`print(name[1])`
For the printing part of a string—called substring— we need to mention the start index and end index in the print statement. Here the last index value will be ignored. For example, while writing print(name[0:4]), the last index number 4 would be ignored, and only the indexes 0, 1, 2, 3 will be used.	`print(name[0:4])`
	`print(name[4:6])`
	`print(msg[0:9])`
	`print(msg[9:14])`
Length of the string can be accessed by using len() function.	`print(len(msg))`
	`print(msg[9:len(msg)])`
For string concatenation we can simply use the plus sign as shown here.	`new_msg= name +" " + msg`
	`print(new_msg)`

A few important points to be noted are here. First, we can retrieve the strings, or better say a substring, simply by using square brackets. For example, msg [0:7] would take out the first seven letters numbered from 0 to 6. In the process, the last index number 7 would be ignored. In Python, as you have also noticed by now, indexes are always numbered starting from zero.

Let us now try practicing the code lines in Fig. 2.11.

```
In [70]: name="Sheldon"
    ...: msg="sent a mail to Jack"

In [71]: print(name[0])
S

In [72]: print(name[1])
h

In [73]: print(name[0:4])
Shel

In [74]: print(name[4:6])
do

In [75]: print(msg[0:9])
sent a ma

In [76]: print(msg[9:14])
il to

In [77]: print(len(msg))
19

In [78]: print(msg[9:len(msg)])
il to Jack

In [79]: new_msg= name +" " + msg
    ...: print(new_msg)
Sheldon sent a mail to Jack
```

FIGURE 2.11   Screenshot; example code snippets to try out.

2.3.3 Working with Lists

A list is a collection of elements. Elements of the list are indexed, which start with zero. Lists are particularly useful for storing and iterating upon sequences. List creation can be done by using square brackets [ ]. We need to be careful, as the same type of brackets are used for string accessing operations as well. A list looks like an array, but it is NOT exactly an array. An array is a collection of similar elements, while a list can contain mixed data types (Table 2.6). Any operation on an array applies automatically on all its individual elements, which is not the case in lists. Would you like to try out the code lines in Table 2.6, which will introduce the List code syntax and operations?

TABLE 2.6   List Operations

List Operation	Code
Defining lists.	```mylist1=["Sheldon","Tommy", "Benny"]``` ```print(type(mylist1))```
Accessing values from a list.	```print(mylist1[0])``` ```print(mylist1[1])```
Number of elements in the list can be accessed by using len() function.	```len(mylist1)```

List appending or combining two lists into one.	```mylist2=["Ken","Bill"]```
	```new_list=mylist1 + mylist2```
	```print(new_list)```

Updating an element in the list.	```print("actual list",mylist1)```
	```mylist1[0]="John"```
	```print("list after updating" ,mylist1)```

Deleting an element from the list.	```print("actual list",mylist2)```
	```del mylist2[0]```
	```print("list after deleting" ,mylist2)```

There is a difference between an array and a list. In this code, if val1 and val2 are two arrays, then val3 would be [7,9,8]; however, here val1 is a list. It is a collection of three elements, and val2 is also a list. When we combine the two to form val3, it becomes a collection of six elements as expected.	```val1=[1,7,6]```
	```val2=[6,2,2]```
	```val3=val1+val2```
	```print(val3)```

A list can contain different data types.	```details=["John", 1500, "LA"]``` ```print(details)```
There can be a list inside a list.	```details_all=["John", 1500, "LA",mylist1]``` ```print(details_all)```

Try out the code lines given in Table 2.6 and compare your output with the results given in the screenshot in Fig. 2.12. Please ignore the input command numbers while you are comparing them through the output. You should be comfortable with lists before attempting any machine learning lessons. We will be using lists later in this book while we do some actual analysis.

```
In [95]: mylist1=["Sheldon","Tommy", "Benny"]
    ...: print(type(mylist1))
<class 'list'>

In [96]: print(mylist1[0])
    ...: print(mylist1[1])
Sheldon
Tommy

In [97]: len(mylist1)
Out[97]: 3

In [98]: mylist2=["Ken","Bill"]
    ...: new_list=mylist1 + mylist2
    ...: print(new_list)
['Sheldon', 'Tommy', 'Benny', 'Ken', 'Bill']

In [99]: print("actual list",mylist1)
    ...: mylist1[0]="John"
    ...: print("list after updaring" ,mylist1)
actual list ['Sheldon', 'Tommy', 'Benny']
list after updaring ['John', 'Tommy', 'Benny']

In [100]: print("actual list",mylist2)
    ...: del mylist2[0]
    ...: print("list after deleting" ,mylist2)
actual list ['Ken', 'Bill']
list after deleting ['Bill']

In [101]: val1=[1,7,6]
    ...: val2=[6,2,2]
    ...: val3=val1+val2
    ...: print(val3)
[1, 7, 6, 6, 2, 2]

In [105]: details=["John", 1500, "LA"]
    ...: print(details)
['John', 1500, 'LA']

In [106]: details_all=["John", 1500, "LA",mylist1 ]
    ...: print(details_all)
['John', 1500, 'LA', ['John', 'Tommy', 'Benny']]
```

FIGURE 2.12   List operations—code output screenshot.

2.3.4 Dictionaries

Dictionaries are very different compared to other data types that we have learned until now. By now, we know that Python default indexing starts from zero. What if you wish to alter this default indexing and need custom-defined indexes? In certain situations, you may wish to have, say, for example, customer_id as an index and one more field as its accompanying value. In some other situations, you may need the account number as index and account balance its is associated value. In this kind of situation, you need to define your key-value pairs by using Python dictionaries. A dictionary is nothing but a collection of key-value pairs. Dictionary keys are usually like the primary keys, as defined in RDBMS textbooks, unique to the data. You need to provide the key to access its associated value. Python dictionaries are defined using curly braces { }. Key and the values are separated using a colon (Table 2.7).

TABLE 2.7   Dictionary Operations

Dictionary Operation	Code

Defining a dictionary.	```city={2:"Los Angeles", 9:"Dallas" , 21:"Boston"}```
	```print(city)```
	```print(type(city))```

Accessing values from a dictionary.	```print(city[9])``` ```print(city[2])```
Printing all keys.	```print(city.keys())```
Printing all values.	```print(city.values())```
Updating an element value in a dictionary.	```city[2]="New York"``` ```print(city)```
Updating a key in the dictionary.	**Not possible**
Deleting a value from a dictionary.	```del(city[2])``` ```print(city)```

Can there be a repetition of the key? No, keys can never repeat. In the case of a repeat key, Python does not throw an error, but it automatically ignores the value associated with the first key.	```country={1:"USA", 6:"Brazil" , 7:"India", 6:```
	```"France" }```
	```print(country)```

Can we have non-numeric keys? The answer is yes, but while accessing it you need to pass the key values like a string—with quotes.	```GDP= {"USA": 20494, "China" : 13407}```
	```print(GDP)```
	```print(GDP["USA"])```
	```print(GDP[USA])#This code does not work```

Can the "values" be a list? Even that is possible.	```cust={"cust1":[19, 9500], "cust2":[21, 10000]}```
	```print(cust)```
	```print(cust["cust1"])```

To proceed effectively with machine learning lessons later in this book, we need to develop a reasonably good understanding of dictionaries. We need to know the differences between a dictionary and a list to use them to our advantage. While working with some of the libraries later, we will get the output as a list in some cases and as a dictionary in others. The way we access lists is very different from the way we handle dictionaries.

Now try executing the code lines in Table 2.7 and compare your output with the output screenshot given below.

```
city={2:"Los Angeles", 9:"Dallas" , 21:"Boston"}
print(city)
print(type(city))
{2: 'Los Angeles', 9: 'Dallas', 21: 'Boston'}
<class 'dict'>

print(city[9])
print(city[2])
Dallas
Los Angeles

print(city.keys())
dict_keys([2, 9, 21])

print(city.values())
dict_values(['Los Angeles', 'Dallas', 'Boston'])

city[2]="New York"
print(city)
{2: 'New York', 9: 'Dallas', 21: 'Boston'}

del(city[2])
print(city)
{9: 'Dallas', 21: 'Boston'}
```

```
country={1:"USA", 6:"Brazil" , 7:"India", 6: "France" }
print(country)
{1: 'USA', 6: 'France', 7: 'India'}

GDP= {"USA": 20494, "China" : 13407}
print(GDP)
print(GDP["USA"])
print(GDP[USA])#This code doesn't work
{'USA': 20494, 'China': 13407}
20494
Traceback (most recent call last):
  File "<ipython-input-39-2de5394f85be>", line 4, in <module>
    print(GDP[USA])#This code doesn't work

NameError: name 'USA' is not defined


cust={"cust1":[19, 9500], "cust2":[21, 10000]}
print(cust)
print(cust["cust1"])
{'cust1': [19, 9500], 'cust2': [21, 10000]}
[19, 9500]
```

Until now, we have discussed some important data types. There are many more in Python, which you may explore based on your need, but it is important that you thoroughly understand all the data types we have discussed so far in this chapter.

2.4   PYTHON PACKAGES

Let us now proceed with packages, which are very much like the soul of Python. How would you compute log(15) or the square root of 256? Just to get started, you can use the syntax **log(15)** or **sqrt(256)**, and you may get the results similar to what is shown in the code below.

```
print(log(10))
Traceback (most recent call last):
  File "<ipython-input-41-a4265d6da271>", line 1, in <module>
    print(log(10))
NameError: name 'log' is not defined


print(sqrt(256))
Traceback (most recent call last):
  File "<ipython-input-42-112b573c917c>", line 1, in <module>
    print(sqrt(256))
NameError: name 'sqrt' is not defined
```

Oops! Python is throwing errors. It is not due to the wrong syntax or erroneous function names. Python is a multipurpose language; it does not have a log() as one of its core functions. Python does not have a square root function, either. Python contains a package called math. Just import the math package, and you can now safely use log and sqrt. Here is how it works.

```
import math

print(math.log(10))
2.302585092994046

print(math.sqrt(256))
16.0
```

Here are some of the frequently asked questions related to packages (Table 2.8). You need to know it all.

TABLE 2.8 Frequently Asked Questions on Packages

Frequently Asked Questions on Packages	Answer
What is a Python package?	A package is a well-complied code bundle containing many reusable functions. Most standard mathematical formulas and many scientific applications are casted as functions. These applications include math, machine learning, and other statistical functions.
What does the package contain?	A package contains sub packages and functions. Packages are made up of a lot of Python code files with prewritten modules, which can be used to solve a specific problem.
How can we include a package in our code?	Use the command `import` is followed by the package name. For example, `import math`.
Once we include a package, can it be used forever?	Once we include a package in any code file, we can use that package and its subpages until we close the session. In a new session, we need to execute the `import` command again if that package is required.
Does the `import` command get the package from the Internet?	Not really, the `import` command just attaches the package to this current session.
How do we install a new package?	We need to open the Anaconda prompt and use the command pip install <package name>. For example, `pip install math`.
How can we print all the installed packages?	Use the command `pip list` for printing all packages. Write this command at the Anaconda prompt.
Are there any preinstalled packages in the local machine?	Yes! Luckily Anaconda takes care of several small tasks for us. When we install Anaconda, it already downloads and installs the most widely used packages in our local system.
There are many packages. How do we know when to use which package? Which package contains what function?	We will get to know it only with practice. We do not need to memorize the package and function names. We can simply google for Python documentation and find whatever we need.

Do we need to write the name of the package every time?	Yes! We need to write the package name, which is followed by the function name.
	
	`import math`
	`print(math.log(10))`
	`print(math.sqrt(256))`
	
	Otherwise, we can use an alias. It is a standard coding practice.
	
	`import math as mt`
	`print(mt.math.log(10))`
	`print(mt.math.sqrt(256))`

Remember, Python is a multipurpose language. It contains many packages for web application development, user interface creation, server management, and many others required by the software development community. Python is also very rich in terms of packages and functions essential for data scientists, which is the main focus of this book. Here are some of the important packages that we frequently need to work with as data scientist.

1. NumPy
2. Pandas
3. Matplotlib
4. Scikit learn
5. nltk
6. TensorFlow
7. SciPy

As said, we do not need to memorize these package names. Each package is created to solve a specific need. As we keep practicing with Python code, these packages will get automatically stored in our memory. The following are some details we need to know. For now, we are just introducing these packages. We will get more details with functions as and when we need it with our analysis.

2.4.1 NumPy

The NumPy package is a must-have package if we are dealing with mathematical calculations. NumPy will give us an option to create arrays. Whenever complex mathematical operations are needed, we need to store data in the form of arrays and matrices. The NumPy package contains many functions, which allow us to do some quick operations on arrays and matrices. It also comes in handy while working with mathematical calculations, sorting, selecting, and reshaping objects. NumPy is the foundation package for several other advanced packages like SciPy, Scikit Learn, and TensorFlow. You might be using NumPy indirectly while working with several other packages. Here is an example application.

In the following code, we are creating an array using `np.array()` function. This function takes a list as an input. We are creating a new array by applying the multiplication operation on the income array. Furthermore, as we can expect, this multiplication operation gets applied to every element.

```
import numpy as np

income = np.array([6725, 9365, 8030, 8750])
print(type(income))
<class 'numpy.ndarray'>

print(income)
[6725 9365 8030 8750]

print(income[0])
6725

expenses=income*0.65
print(expenses)
[4371.25 6087.25 5219.5  5687.5 ]

savings=income-expenses
print(savings)
[2353.75 3277.75 2810.5  3062.5 ]
```

By now, we are aware that an array is very different from a list. An array takes a list of values as input. Just have a look at the following output sample. As we multiply the list by 2, it just doubles the number of elements from four to eight. While we do the same operation on an array, each element gets doubled in value. It is a significant difference to be noted.

```
income_list=[6725, 9365, 8030, 8750]
income_array = np.array(income_list)

print(income_list*2)
[6725, 9365, 8030, 8750, 6725, 9365, 8030, 8750]

print(income_array*2)
[13450 18730 16060 17500]
```

2.4.2 Pandas

In most analytics projects, we often need a convenient tool, usually a package which can read data files into Python and create data frames, also known as datasets, or tables. We need a package that can create subsets. We also need a package that can give us some metadata details and basic summaries on datasets. Analysis demands sorting the data and merging datasets. Is there any package in Python that can take care of all these data manipulation tasks that a data scientist like us needs daily? Fortunately, there is one by the name Pandas. Following is an example of its usage.

Here we will be using it as `pd.read_csv()` function.

```python
import pandas as pd
bank= pd.read_csv('D:/Chapter2 Python Programming/Datasets/Bank Tele Marketing/
bank_market.csv')
print(bank)
```

	Cust_num	age	job	marital	...	pdays	previous	poutcome	y
0	1	58	management	married	...	-1	0	unknown	no
1	2	44	technician	single	...	-1	0	unknown	no
2	3	33	entrepreneur	married	...	-1	0	unknown	no
3	4	47	blue-collar	married	...	-1	0	unknown	no
4	5	33	unknown	single	...	-1	0	unknown	no
...	...	...	...	...	...	...	...	...	...
45206	45207	51	technician	married	...	-1	0	unknown	yes
45207	45208	71	retired	divorced	...	-1	0	unknown	yes
45208	45209	72	retired	married	...	184	3	success	yes
45209	45210	57	blue-collar	married	...	-1	0	unknown	no
45210	45211	37	entrepreneur	married	...	188	11	other	no

```
[45211 rows x 18 columns]
```

A few points to be noted while importing any data file into the Python environment:

- We need to mention the full file path along with the file name and its extension.
- The path and file name are strictly case sensitive.
- One of the most encountered errors may be file-not-found error, which does not always mean the file is not present. It may well mean the given file path is not correct. The very first thing we need to do is to look for any possible typographic errors
- We can conveniently use the Linux style of providing the file path, which involves the use of a forward slash (/) to traverse the path string. We may also use the Windows style by using the following two options.
  - Use a double backward slash (\\).
  - Include "r" in the path.
- When data contains practically countless rows and columns, which often is the case, printing whole datasets on the console may not be possible; hence the console shows the truncated output. We will see better options to explore the data later in the discussion.

Now let us try out the following options used to read datasets.

```python
bank1= pd.read_csv('D:\\Chapter2\\Datasets\\Bank Tele Marketing\\bank_market.
csv')
print(bank1)

bank2= pd.read_csv(r'D:\Chapter2\Datasets\Bank Tele Marketing\bank_market.
csv')
print(bank2)
```

We will do a deep dive into the main commands of the Pandas package a bit later. For now, it is just introduced.

2.4.3 Matplotlib

This package is used for data visualization and plotting. While working on an analysis, we may wish to represent the data in the form of, say, a scatter plot or a bar chart or some other form of visualization. Matplotlib will come in handy here. It contains numerous sub packages and a long list of functions for creating visualization plots. Following is an example of how to use this package for a scatter plot.

```python
#Code for importing the data
bank= pd.read_csv('D:/Chapter2/Datasets/Bank Tele Marketing/bank_market.
csv')
```

```
#Code for printing the column names from the data
print(bank.columns)

#Code for importing the package
import matplotlib as mp

#Code for a scatter plot between two columns "age" and "account balance", where the bank is the dataset name here.

mp.pyplot.scatter(bank['age'],bank['balance'])
```

Here is the output of this code. A scatter plot (Fig. 2.13) can tell us if the relationship between the two variables is strong or weak.

```
print(bank.columns)
Index(['Cust_num',   'age',   'job',   'marital',   'education',   'default',
'balance','housing',   'loan',   'contact',   'day',   'month',   'duration',
'campaign','pdays', 'previous', 'poutcome', 'y'], dtype='object')

import matplotlib as mp

mp.pyplot.scatter(bank['age'],bank['balance'])
Out[64]: <matplotlib.collections.PathCollection at 0x1ad187f27c8>
```

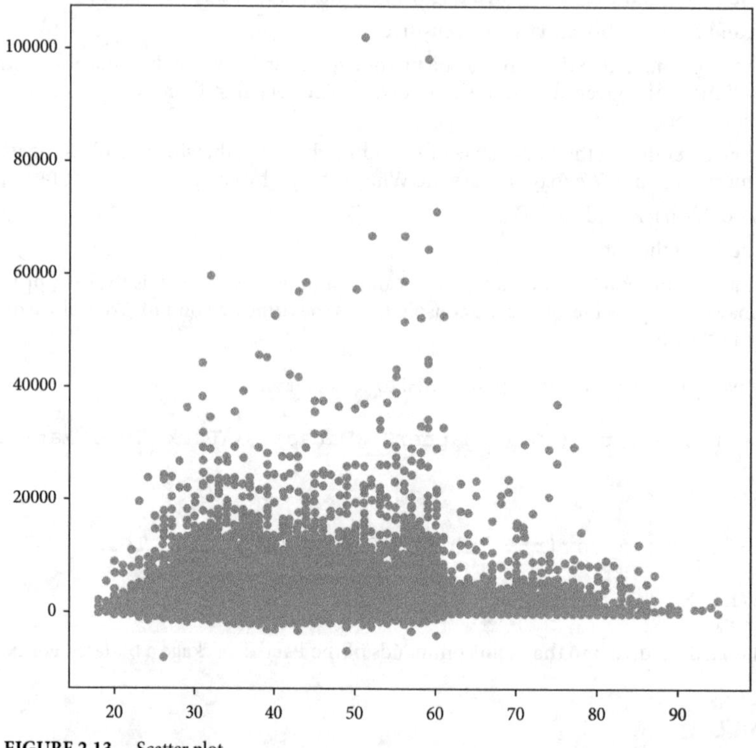

FIGURE 2.13   Scatter plot.

2.4.4   ScikitLearn

Just to introduce, this package is used for building and validating machine learning models. It is also used for fine-tuning machine learning models and calculating critical statistical measures. This package, like most packages, contains many sub packages and several functions that will help us to work with machine learning models and perform specific statistical analysis. We will be using this package quite often.

2.4.5   nltk

In Python, the package name ntlk is the short form of the natural language tool kit. Handling text data is very different from handling numerical data. There is a need for different functions, which can help in preparing data for analysis and model building. nltk is one such package, which contains all the text mining and natural language processing–related features. The package is comprehensive and big. If we are working with text data in Python, we are mostly working with nltk.

2.4.6   TensorFlow

When it comes to deep learning, there is a special focus on coding efficiency and execution time. We need a package that can handle deep learning algorithms efficiently. TensorFlow is one very popular package in deep learning space. Like nltk, it is also gigantic as far as its size is a concerned. It is a complete framework. It follows a different programming paradigm altogether. If we are working with deep learning algorithms, TensorFlow will prove to be a best friend in need. We will deep dive into TensorFlow a bit later.

There are many more packages in Python, which you can explore as you advance in your journey as a data scientist.

2.5   CONDITIONS AND LOOPS IN PYTHON

We will wrap up our basic Python session discussing some more commonly used features like conditions and loops (Table 2.9). There is nothing too complex to comprehend here. You might have already encountered much of it if you have taken any programming language course during your high school or graduation days. You just need to learn Python-specific syntax.

TABLE 2.9   Conditions and Loops in Python

Comments	Code

This is how you write an if-condition. What will be the result of this code?	`level=60`
	`if level<50:`
	`    print("Stage1")`
	`print("Done with If")`

What will be the result of this code? Will Python throw an error or just skip some of it?	`level=60`
	`if level<50:`
	`    print("Stage1")`
	`    print("Done with If")`

Is there a meaning of indentation? Indentation works much like parentheses, as explained below.	`level=40`
	`if level<50:`
	`    print("Stage1")`
	`print("Done with If")`

Using If-else statements	`level=60`
	`if level<50:`
	`    print("Stage1")`
	`else:`
	`    print("Stage2")`
	`print("Done with If")`

For loop	`names=["Tommy", "Benny", "Ken"]`
	`for i in names:`
	`    print("The name is", i)`

For loop again with a different example	`nums=range(1,10)`
	`cumsum=0`
	`for i in nums:`
	`    cumsum=cumsum+i`
	`    print("Cumulative sum till", i ,"is", cumsum)`

We need to be careful with the indentation in Python code. Sometimes we may need to use ready code from other sources like GitHub or StackOverflow. Please check if the code is starting from the beginning of the line or with a tab space. If it is starting with a tab space, it may well be part of some of the condition loops. Sometimes it may even throw an error for the wrong indentation.

Now it is time to validate the results of the code that we have just written for conditions and loops. Given below is the output screenshot as usual.

```
level=60
if level<50:
    print("Stage1")
print("Done with If")


Done with If
----------------------------------
level=60
if level<50:
    print("Stage1")
    print("Done with If")
----------------------------------
level=40
if level<50:
    print("Stage1")
print("Done with If")


Stage1
Done with If
----------------------------------
level=60
if level<50:
    print("Stage1")
else:
    print("Stage2")
print("Done with If")


Stage2
Done with If
----------------------------------
names=["Tommy", "Benny", "Ken"]
for i in names:
    print("The name is", i)


The name is Tommy
The name is Benny
The name is Ken
----------------------------------
nums=range(1,10)
cumsum=0
for i in nums:
    cumsum=cumsum+i
    print("Cumulative sum till", i ,"is", cumsum)


Cumulative sum till 1 is 1
Cumulative sum till 2 is 3
Cumulative sum till 3 is 6
Cumulative sum till 4 is 10
Cumulative sum till 5 is 15
Cumulative sum till 6 is 21
Cumulative sum till 7 is 28
Cumulative sum till 8 is 36
Cumulative sum till 9 is 45
```

Make yourself comfortable with all the Python basics discussed so far in this chapter before moving to the next topic. If you need, just go back and have a relook.

2.6 DATA HANDLING AND PANDAS DEEP DIVE

As a data scientist, you will be working with datasets of different types, which may well be from various disparate sources. Sometimes only subset data is needed, or you may need to create new variables and columns, or merge two datasets, or do sorting operations. There are times when you may need a combination of these. The Pandas package comes in very handy here. We will now learn some important features of Pandas.

2.6.1 Data Importing and Basic Details

To start with, we will discuss some basic file handling, which is invariably needed in almost all projects.

2.6.1.1 Importing from Comma-Separated CSV Files   As discussed earlier, you need to mention the full file path along with the file name and extension to import any file in the Python environment. The path needs to be in a specific format. Just have a close look at the following code and the output screenshot for a quick refresher:

```
import pandas as pd
sales= pd.read_csv('D:\\Chapter2\\Datasets\\Sales\\Sales.csv')
print(sales)
```

The above code gives us the below output:

```
print(sales)
        Cust_id          CustName  ... Net_sales Invoice_Amount
0       SIE39906     Garrett Bauer  ...         1            400
1       EST39196       Rama Norris  ...         2            400
2       AFG39258       Serena Carr  ...         2            400
3       CAN39302     Brendan Daniel ...         2            400
4       POR39323     Judith Beasley ...         2            400
..          ...               ...  ...       ...            ...
972     GUA39889   Hermione Grimes  ...       129           3870
973     TRI39891     Daryl Gilbert  ...       155           4650
974     SOU39893     Upton Lambert  ...       171           5130
975     GEO39888    Jaime Alvarado  ...       188           5640
976     ICE39892       Bert Fowler  ...       240           7200

[977 rows x 7 columns]
```

2.6.1.2 Importing Data from Microsoft Excel Files   While importing Excel files, you need to mention the file name as well as the worksheet name where data is present. If you do not mention the worksheet name, Python will pick up the first worksheet by default, which may not always be the data of interest. Have a look at the following code and output for details.

```
wb_data = pd.read_excel("D:\\Chapter2\\Datasets\\World Bank Data\\World Bank
Indicators.xlsx" , "Data by country")
print(wb_data)
```

The above code gives us the below output:

```
                      Country Name  ...  Finance: GDP per capita (current US$)
0                    United States  ...                                46702.0
1                            China  ...                                 4433.0
2                            Japan  ...                                43063.0
3                          Germany  ...                                39852.0
4                           France  ...                                39170.0
                            ...  ...                                    ...
2349                       Somalia  ...                                    NaN
2350                  South Sudan  ...                                    NaN
2351        St. Martin (French part)  ...                                 NaN
2352       Turks and Caicos Islands  ...                                 NaN
2353         Virgin Islands (U.S.)  ...                                    NaN

[2354 rows x 20 columns]
```

2.6.1.3 *Basic Details of a Dataset*

You may wish to check if the import has happened correctly with all the data columns and rows safely imported. How do you ensure it? Printing the whole data in the console is not the solution. The console may not print all the values and truncate the output. The below commands will help us to look at some important details of the dataset. Let us quickly run through the following code lines and the corresponding output screenshots.

print (sales.shape) gives the number of rows and number of columns present in the data. Sales is the name of the dataset containing 977 rows and 7 columns.

```
print(sales.shape)
(977, 7)
```

print(sales.columns) prints all the column names.

```
Index(['Cust_id',   'CustName',   'Product_code',   'Country_code',   'Sales_
Type','Net_sales', 'Invoice_Amount'],  dtype='object')
```

print(sales.head(10)) prints the first 10 rows.

```
       Cust_id     custName     Product_code ...  Sales_Type Net_sales Invoice_Amount
0    SIE39906   Garrett Bauer          HA1 ...        Direct         1            400
1    EST39196     Rama Norris          GI2 ...       Website         2            400
2    AFG39258     Serena Carr          ER3 ...       Website         2            400
3    CAN39302  Brendan Daniel          BR1 ...       Website         2            400
4    POR39323  Judith Beasley          LU4 ...       Website         2            400
5    NOR39103  Olympia Hewitt          DA1 ...         Other         3            600
6    ITA39581     Fallon Soto          DI3 ...       Website         2            800
7    TUR39194  Aquila Russell          DE1 ...       Website         4            800
8    GRE39678   Teagan Hebert          SO2 ...       Website         4            800
9    MIC39753    Emily Brooks          AN3 ...        Direct         4            800

[10 rows x 7 columns]
```

print(sales.tail(10)) prints the last 10 rows.

```
        Cust_id          CustName  ...  Net_sales Invoice_Amount
967   PHI39885   Cedric Ferguson  ...        420          168000
968   SAM39126     Kitra Hendrix  ...        422          168800
969   UKR39521       Hu Jacobson  ...        422          168800
970   SOM39594    Autumn Mcbride  ...        424          169600
971   NET39890     Ruth Fletcher  ...         15             450
972   GUA39889   Hermione Grimes  ...        129            3870
973   TRI39891     Daryl Gilbert  ...        155            4650
974   SOU39893    Upton Lambert   ...        171            5130
975   GEO39888    Jaime Alvarado  ...        188            5640
976   ICE39892       Bert Fowler  ...        240            7200

[10 rows x 7 columns]
```

print(sales.sample(n=10)) prints 10 rows randomly picked from sales data.

```
        Cust_id          CustName  ... Net_sales Invoice_Amount
777   TAI39396      Hedwig Quinn   ...       374          74800
845   CON39704    Quinlan Hopper   ...       420          84000
911   NOR39213         Bruce Ruiz  ...       318         127200
772   BAH39687       Howard Reed   ...       184          73600
527   NEW39717    Lester Buckner   ...       204          40800
309   NIC39647      Chava Briggs   ...       124          24800
957   SAI39179        Graham Orr   ...       408         163200
794   FRE39498       Mira Clarke   ...       388          77600
350   BUR39174    Blair Mckenzie   ...       143          28600
569   GUI39074  Baker Strickland   ...       232          46400

[10 rows x 7 columns]
```

print(sales.dtypes) displays the data types of columns.

```
Cust_id              object
CustName             object
Product_code         object
Country_code         object
Sales_Type           object
Net_sales             int64
Invoice_Amount        int64
dtype: object
```

print(sales.describe()) gives the summary of numerical variables. This summary contains the minimum value, maximum value, average value, and very useful percentile values. We will discuss percentiles later, however. As of now, just focus on minimum, maximum, and average values.

```
          Net_sales   Invoice_Amount
count    977.000000       977.000000
mean     219.760491     47906.591607
std      750.624569     38362.358333
min        1.000000       400.000000
25%       75.000000     20000.000000
50%      151.000000     37600.000000
75%      224.000000     68000.000000
max    15300.000000    169600.000000
```

print(sales["Invoice_Amount"].describe()) If you are interested in a single variable, use this command. It gives the summary of a single variable "Invoice_Amount" from sales data.

```
count       977.000000
mean      47906.591607
std       38362.358333
min         400.000000
25%       20000.000000
50%       37600.000000
75%       68000.000000
max      169600.000000
Name: Invoice_Amount, dtype: float64
```

print(sales["Sales_Type"].value_counts()) The describe() function works only on numerical variables. The command(value_counts) gives us the frequency count table for the non-numeric variable Sales_Type. The function "value_counts" works on non-numeric variables like customer country, customer type, and region, where there are no minimum and maximum values. We can summarize them by giving unique values taken by the variable and their counts. Following is the frequency count table output for the variable Sales_Type.

```
print(sales["Sales_Type"].value_counts())
Website     494
Direct      402
Other        81
Name: Sales_Type, dtype: int64
```

The following code helps in counting the missing values in a variable:

```
print(sum(sales["Country_code"].isnull()))
print(sum(sales["CustName"].isnull()))
print(sum(sales["Invoice_Amount"].isnull()))
```

Below is the output for the above code.

```
print(sum(sales["Country_code"].isnull()))
21
```

```
print(sum(sales["CustName"].isnull()))
0
```

```
print(sum(sales["Invoice_Amount"].isnull()))
0
```

2.6.2 Subsets and Data Filters

Let us now understand some subset operations using bank telemarketing data.

```
bank= pd.read_csv('D:/Chapter2/Datasets/Bank Tele Marketing/bank_market.csv')
print(bank.shape)
print(bank.columns)
```

The following is the output.

```
print(bank.shape)
(45211, 18)
```

```
print(bank.columns)
Index(['Cust_num', 'age', 'job', 'marital', 'education', 'default', 'balance',
'housing', 'loan', 'contact', 'day', 'month', 'duration', 'campaign', 'pdays',
'previous', 'poutcome', 'y'],
      dtype='object')
```

Let us now learn how we can create new datasets by selecting or excluding a few columns or rows. Given below is the code for creating a new dataset by keeping selected rows. Here we need to mention the count to keep the first few rows.

```
bank1 = bank.head(5)
print(bank1)

   Cust_num  age           job  marital  ... pdays  previous  poutcome   y
0         1   58    management  married  ...    -1         0   unknown  no
1         2   44     technician   single  ...    -1         0   unknown  no
2         3   33  entrepreneur  married  ...    -1         0   unknown  no
3         4   47   blue-collar  married  ...    -1         0   unknown  no
4         5   33       unknown   single  ...    -1         0   unknown  no

[5 rows x 18 columns]
```

In Python, the row index starts with zero. You can mention the index number in iloc(index location) to keep some specific rows. If you are keeping only a specific row, the result will be formatted as a series and not as a data frame.

```
bank2=bank.iloc[2]
print(bank2)
print(type(bank2))
```

Given below is the output.

```
Cust_num                    3
age                        33
job               entrepreneur
marital                married
education             secondary
default                    no
balance                     2
housing                   yes
loan                      yes
contact               unknown
day                         5
month                     may
duration                   76
campaign                    1
pdays                      -1
previous                    0
poutcome              unknown
y                          no
Name: 2, dtype: object
```

```
print(type(bank2))
<class 'pandas.core.series.Series'>
```

In case you wish to mention a greater number of indices, mention them as a list. Either you can define the list beforehand or pass it on directly. You just need to appreciate the syntax. Have a look at [[2,9,15,25]]. Here the outside bracket [ ] is for accessing, while the inside bracket is to define the list.

```
index_vals=[2,9,15,25]
bank3=bank.iloc[index_vals]
print(bank3)

bank3_1=bank.iloc[[2,9,15,25]]
print(bank3_1)
```

Given below is the output screenshot.

```
print(bank3)
    Cust_num  age           job  marital  ... pdays previous  poutcome   y
2          3   33  entrepreneur  married  ...    -1        0   unknown  no
9         10   43     technician   single  ...    -1        0   unknown  no
15        16   51        retired  married  ...    -1        0   unknown  no
25        26   44         admin.  married  ...    -1        0   unknown  no

[4 rows x 18 columns]
```

```
bank3_1=bank.iloc[[2,9,15,25]]
print(bank3_1)
    Cust_num  age           job  marital  ... pdays previous  poutcome   y
2          3   33  entrepreneur  married  ...    -1        0   unknown  no
9         10   43     technician   single  ...    -1        0   unknown  no
15        16   51        retired  married  ...    -1        0   unknown  no
25        26   44         admin.  married  ...    -1        0   unknown  no

[4 rows x 18 columns]
```

What if you need a new dataset by keeping selected columns? Not tough at all, just mention the column names. The code lines given below keep two specific columns. Please note we are not updating any existing data here; we are just creating a new dataset for our regular work. There will not be any changes in the bank dataset and the source CSV file.

```
bank4 = bank[["job", "age"]]
print(bank4.head(5))
```

The following are the results.

```
bank4 = bank[["job", "age"]]
print(bank4.head(5))
            job  age
0    management   58
1    technician   44
2  entrepreneur   33
3   blue-collar   47
4       unknown   33
```

If you need a new dataset by keeping selected columns and rows, just mention the relevant column names and row indices in iloc[ ]. The code given below keeps only two columns of job and age along with the first five rows of the bank dataset. Remember, mentioning [0:5] will include zero and exclude five. A total of five rows will be selected—zero-till four.

```
bank5 = bank[["job", "age"]].iloc[0:5]
print(bank5)
```

The output screenshot will look like as given below:

```
bank5 = bank[["job", "age"]].iloc[0:5]
print(bank5)
            job  age
0    management   58
1    technician   44
2  entrepreneur   33
3   blue-collar   47
4       unknown   33
```

Following is the code for creating a new dataset by excluding selected rows. We need to use the drop function with a mention of the row indices. The code below excludes four rows and creates a new dataset bank6. Here we are excluding row1, row3, row5, and row7. You can verify that by looking at the customer numbers.

```
bank6=bank.drop([0,2,4,6])
print(bank6.head(5))
```

Following is the result:

```
bank6=bank.drop([0,2,4,6])
print(bank6.head(5))
   Cust_num  age           job   marital  ... pdays  previous  poutcome   y
1         2   44    technician    single  ...    -1         0   unknown  no
3         4   47   blue-collar   married  ...    -1         0   unknown  no
5         6   35    management   married  ...    -1         0   unknown  no
7         8   42  entrepreneur  divorced  ...    -1         0   unknown  no
8         9   58       retired   married  ...    -1         0   unknown  no

[5 rows x 18 columns]
```

Now let us see how we can create a new dataset by excluding selected columns. Again, we need to use the drop function and mention the column names, but there is an additional parameter we need to provide as axis=1. The same drop function is used both for dropping rows and columns. While dropping rows, we have to use axis=0, and while doing the same with columns, we have to use axis=1. If you do not mention any axis value, it will be take 0 as the default.

While dropping rows, it is perfectly fine if you do not mention the axis. However, while dropping columns, it is mandatory to mention axis=1; otherwise, you will get an axis-related error.

```python
bank7=bank.drop(["Cust_num"], axis=1)
print(bank7.head(5))
```

Given below is the output:

```
bank7=bank.drop(["Cust_num"], axis=1)
print(bank7.head(5))
   age              job  marital   education  ... pdays  previous poutcome    y
0   58       management  married    tertiary  ...    -1         0  unknown   no
1   44       technician   single   secondary  ...    -1         0  unknown   no
2   33     entrepreneur  married   secondary  ...    -1         0  unknown   no
3   47      blue-collar  married     unknown  ...    -1         0  unknown   no
4   33          unknown   single     unknown  ...    -1         0  unknown   no

[5 rows x 17 columns]
```

If you do not mention axis=1, like what is given below while dropping columns, Python will throw the following error.

```python
bank7_1=bank.drop(["Cust_num"])
```

It is a lengthy error message, the final line shows the actual error.

```
bank7_1=bank.drop(["Cust_num"])
Traceback (most recent call last):

  File "<ipython-input-105-a0b9378992d5>", line 1, in <module>
    bank7_1=bank.drop(["Cust_num"])

  File "C:\ProgramData\Anaconda3\lib\site-packages\pandas\core\frame.py",
line 4102, in drop
    errors=errors,

  File "C:\ProgramData\Anaconda3\lib\site-packages\pandas\core\generic.py",
line 3914, in drop
    obj = obj._drop_axis(labels, axis, level=level, errors=errors)

  File "C:\ProgramData\Anaconda3\lib\site-packages\pandas\core\generic.py",
line 3946, in _drop_axis
    new_axis = axis.drop(labels, errors=errors)

  File "C:\ProgramData\Anaconda3\lib\site-packages\pandas\core\indexes\base.
py", line 5340, in drop
    raise KeyError("{} not found in axis".format(labels[mask]))

KeyError: "['Cust_num'] not found in axis"
```

Following is the code for creating a new dataset using filter conditions on column values. Most of the time, we use these types of filters for subsetting the data rather than using the indices. In the below example, we are trying to get a subset from bank data where age>40. While mentioning the column name in the filter condition, you need to mention the dataset name again.

```python
bank8=bank[bank['age']>40]
print(bank8.shape)
```

In one more example, as given below, we are trying to get a subset from bank data where age>40 and loan status is "no." We just need some Python syntax details before we proceed. Single equal to `"="` is used for assignment, double equal to `"=="` is used for comparison. `"!="` is the symbol for not equal to. In the second example, we are using two filter conditions. While you are using more than one filter, you also need to use the parentheses as shown below.

```
bank9=bank[(bank['age']>40) & (bank['loan']=="no")]
print(bank9.shape) "

bank10=bank[(bank['age']>40) | (bank['loan']=="no")]
print(bank10.shape)
```

Write this code and compare your results with the below output.

```
print(bank8.shape)
(20494, 18)

print(bank9.shape)
(17156, 18)

print(bank10.shape)
(41305, 18)
```

2.6.3   Useful Pandas Commands

We have only two sections here on commands. Once we finish, we will be ready to get started with some basic statistical concepts, which would work as a foundation for our upcoming machine learning lessons.

2.6.3.1   Creating a New Column or Variable Inside a Data Frame   In the code given below, we are trying to create a new variable bal_new by multiplying the balance column from the bank by 0.9. Before creating the new variable, we have a total of 18 columns, and this new variable will be added as the 19th column.

```
print(bank.shape)
print(bank.columns)
bank["bal_new"]=bank["balance"]*0.9
print(bank.shape)

print(bank.shape)
(45211, 18)

print(bank.columns)
Index(['Cust_nu',    'age',    'job',    'marita'',    'education',    'default',
'balance',   'housing',   'loan',   'contact',   'day',   'month',   'duration',
'campaign', 'pdays', 'previous', 'poutcome', 'y'],
     dtype='object')

bank["bal_new"]=bank["balance"]*0.9

print(bank.shape)
(45211, 19)

print(bank.columns)
Index(['Cust_num',    'age',    'job',    'marital',    'education',    'default',
'balance',    'housing',   'loan',   'contact',   'day',   'month',   'duration',
'campaign', 'pdays', 'previous', 'poutcome', 'y', 'bal_new'], dtype='object')
```

2.6.3.2   Joining Datasets   The function used for joining the datasets is pd.merge() We can perform standard joining operations like outer join, inner join, and left and right outer joins. Here is the code with an explanation for performing joins.

In the following code, product1 and product2 are the two datasets. "On" is the primary key that is present in both the datasets. The parameter value "how" will decide the type of join.

```
###Inner Join
inner_data=pd.merge(product1, product2, on='Cust_id', how='inner')
print(inner_data.shape)
```

```
###Outer Join
outer_data=pd.merge(product1, product2, on='Cust_id', how='outer')
print(outer_data.shape)
###Left outer Join
L_outer_data=pd.merge(product1, product2, on='Cust_id', how='left')
print(L_outer_data.shape)
###Right outer Join
R_outer_data=pd.merge(product1, product2, on='Cust_id', how='right')
print(R_outer_data.shape)
```

If the comparing column name is different in two datasets, you can use the `left_on` and `right_on` options.

```
inner_data1=pd.merge(product1, product2,
left_on='Cust_id',right_on='Cust_id', how='inner')
print(inner_data1.shape)
```

Given below is the sample code and the output.

```
product1= pd.read_csv("D:/Chapter2 Python
Programming/Datasets/Orders Products/Product1_orders.csv")
print(product1.shape)
(665, 5)
```

```
print(product1.columns)
Index(['Cust_id', 'Country_code', 'Sales_Type', 'Net_sales', 'Invoice_Amount'],
dtype='object')
```

```
product2= pd.read_csv("D:/Google Drive/Training/Book/0.Chapters/Chapter2 Python
Programming/Datasets/Orders Products/Product2_orders.csv")
```

```
print(product2.shape)
(654, 5)
```

```
print(product2.columns)
Index(['Cust_id', 'CustName', 'Sales_Type', 'Net_sales', 'Invoice_Amount'],
dtype='object')
```

```
inner_data=pd.merge(product1, product2, on='Cust_id', how='inner')
print(inner_data.shape)
(342, 9)
```

```
outer_data=pd.merge(product1, product2, on='Cust_id', how='outer')
print(outer_data.shape)
(977, 9)
```

```
L_outer_data=pd.merge(product1, product2, on='Cust_id', how='left')
print(L_outer_data.shape)
(665, 9)
```

```
R_outer_data=pd.merge(product1, product2, on='Cust_id', how='right')
print(R_outer_data.shape)
(654, 9)
```

```
inner_data1=pd.merge(product1, product2, left_on='Cust_id',right_on='Cust_
id', how='inner')
```

```
print(inner_data1.shape)
(342, 9)
```

By now, we are already familiar with some basic data handling operations using Python. We will be using these commands again and again. Now let us gear up for some basic statistics.

2.7 BASIC DESCRIPTIVE STATISTICS

Suppose you have the data of 10,000 individuals containing their income values: How would you report the income? What would be your first calculation to get a feel for income data? Is it not computing the simple average on the income column as you have learned during your early schooling? It is nothing but a descriptive statistic. If you go one step further, you may also like to have a minimum and maximum of available income values. These are all the descriptive statistical measures. Let us explore it more using Python code.

2.7.1 Mean

Mean is the most widely used descriptive measure. Mean is used for finding the average value of numerical variables. In simple terms, the mean of any N number of numeric values can be calculated by the computing summation of all the values divided by their count N. Following is the code in Python to compute the mean value.

We will work on our example using census income data. We will first import the dataset and find the mean of its numeric column capital-gain.

```
income_data= pd.read_csv(r"D:\Datasets\Census Income Data\Income_data.csv")
print(income_data.shape)
print(income_data.columns)
print(income_data["capital-gain"].mean())
```

What follows is the output. As you can see, the mean value of capital-gain is 1,077.65 after rounding off.

```
print(income_data.shape)
(32561, 15)

print(income_data.columns)
Index(['age',    'workclass',    'fnlwgt',    'education',    'education-num',
'marital-status',    'occupation',    'relationship',    'race',    'sex',
'capital-gain',    'capital-loss',    'hours-per-week',    'native-country',
'Income_band'], dtype='object')

print(income_data["capital-gain"].mean())
1077.6488437087312
```

Mean is often the first measure that we calculate. We use mean to get a feel for its center value, better known as an average. We need to be careful with mean, in any case. Look at the array in Table 2.10. It looks like all values are between 90 and 100.

TABLE 2.10 Example Data Array

Customer ID	ROI	Customer ID	ROI
1	95.86	14	92.42
2	98.01	15	97.65
3	94.71	16	91.61
4	96.02	17	99.96
5	97.46	18	93.94
6	98.45	19	94.84
7	98.79	20	95.91
8	94.84	21	96.26
9	93.63	22	99.22
10	93.94	23	93.45
11	98.49	24	98.31
12	961.3	25	99.75
13	95.21		

When we calculate the mean, we will get it as **130.8**. Most entries in this table appear to be less than 100, but the mean value is still 130.8. If you carefully observe all the entries, you will be able to appreciate why the mean value is above 100. What is causing this mean value to inflate? The culprit is right in the middle.

One entry is very different from the rest. All the entries are less than 100 except 961.3. These entries are termed as outliers. Outliers may be single or multiple, and are significantly different from most records. Outliers may significantly impact our analysis results. One of the most affected measures is the mean. Outliers change the "mean" far away from the actual center. The following are the take-aways:

- Mean is significantly impacted by the outliers.
- In the presence of outliers, mean does not realistically indicate the center value.
- Mean should not be used if outliers are not removed or treated properly.

Is there any alternative measure of central tendency? Just read on.

2.7.2 Median

Median is a positional measure. Arrange the data in the column either in ascending or descending order and take the middle value. Please note, here the focus is on the position of the value in a column. If there are 25 records, the middle entry would be the 13th record, so the median value for this data is 96.02. The median is the actual center of the data. The mean is 130.8, and the median is 96.02 (Table 2.11).

TABLE 2.11   Demonstration of Median

Actual Data	Data after Sorting	Actual Data	Data after Sorting
95.86	91.61	92.42	96.26
98.01	92.42	97.65	97.46
94.71	93.45	91.61	97.65
96.02	93.63	99.96	98.01
97.46	93.94	93.94	98.31
98.45	93.94	94.84	98.45
98.79	94.71	95.91	98.49
94.84	94.84	96.26	98.79
93.63	94.84	99.22	99.22
93.94	95.21	93.45	99.75
98.49	95.86	98.31	99.96
961.3	95.91	99.75	961.3
95.21	**96.02**		

In the previous exercise, we have calculated the mean value for the capital-gain variable. Now let us find out the median value of the same variable. Following is the code and the output.

```
print(income_data["capital-gain"].median())
```

The above code gives us the below output

```
print(income_data["capital-gain"].median())
0.0
```

The median value is 0. The mean value of the same variable was 1,077.6. There is a huge gap between mean and median. It is hot for an outlier search. This difference between mean and median is a hint for the existence of some extreme values or outliers in the data. A deep data dive is required, which we will do a little later.

For now, try to learn from the following short question and answer series:

- How do you calculate the median if the count of records is even? There will now be two middle values.
  - We can arrange the data in ascending or descending order and take a mean of the middle two records, which will be our median. For example, if we have 24 records, the median will be just the mean value of 12th and 13th records.
- Are outliers always on the high side only?
  - No, not necessarily. The data can contain low side outliers as well.
- The highest value here in this data is 961.3. What if it is 9,961? Would the median also get affected?
  - If we repeat the exercise of finding out the median with the highest entry in the data as 9,961, we will get the median again as the old value. The median is not impacted by outliers.
- Can we say the difference between mean and median is the only way to detect outliers?
  - Not necessarily. It just gives us a hint. There are better ways to detect the presence of outliers. We will get into details a bit later.

2.7.3   Variance and Standard Deviation

Imagine there are two companies, company A and company B. In the last 15 quarters, company A has made an average profit of 15 million dollars. Company B also made an average profit of 15 million dollars. The results are shown in Table 2.12.

TABLE 2.12   Example Company Profits

	Mean Profit	Median Profit
Company A	15 million	15 million
Company B	15 million	15 million

If you are a long-term investor and if you are not interested in a high-risk company, which company stock will you buy? We cannot make that decision by looking at mean and median values. Let us look at the actual data (Table 2.13).

TABLE 2.13   Example Company Data

Company A	16	14	13	16	14	16	17	16	14	15	15	14	15
Company B	4	4	20	23	10	15	14	−3	26	26	16	10	30

If we look at the last few quarters' results, we can see that company A is very consistent around 15. For company B the average is 15, but it is very volatile. There are a few quarters where it has shown losses as well. By looking at the central tendency, we cannot decide the overall dispersion or spread in the data, like the way mean and median give us an idea of the central tendency of the data. There is a metric to measure the dispersion in the data. It is known as "variance." Below is the formula for variance calculation.

The variance calculation has two steps. First, calculate the mean of the data and take the deviation of each point from the mean.

TABLE 2.14   Company A Data

Company A	16	14	13	16	14	16	17	16	14	15	15	14	15
Mean	15	15	15	15	15	15	15	15	15	15	15	15	15
(Value-Mean)	1	−1	−2	1	−1	1	2	1	−1	0	0	−1	0

We can see in Table 2.14 that the deviations from mean are very minimal for company A. Company B shows very high deviations from mean (Table 2.15). If these deviations are high, then the variance is high; if these deviations are low, then the variance is less. We cannot merely sum these deviations. A few deviations are positive, and a few deviations are negative. The next step is to square these deviations. Finally, find the average of all these squared deviations. That will give us the variance in the data (Tables 2.16 and 2.17).

TABLE 2.15   Company B Data

Company B	4	4	20	23	10	15	14	−3	26	26	16	10	30
Mean	15	15	15	15	15	15	15	15	15	15	15	15	15
(Value-Mean)	−11	−11	5	8	−5	0	−1	−18	11	11	1	−5	15

TABLE 2.16   Company A Variance

Company A	16	14	13	16	14	16	17	16	14	15	15	14	15
Mean	15	15	15	15	15	15	15	15	15	15	15	15	15
Value-Mean	1	−1	−2	1	−1	1	2	1	−1	0	0	−1	0
(Value-Mean)^2	1	1	4	1	1	1	4	1	1	0	0	1	0
Variance= 1.23													

TABLE 2.17   Company B Variance

Company B	4	4	20	23	10	15	14	−3	26	26	16	10	30
Mean	15	15	15	15	15	15	15	15	15	15	15	15	15
Value-Mean	−11	−11	5	8	−5	0	−1	−18	11	11	1	−5	15
(Value-Mean)^2	121	121	25	64	25	0	1	324	121	121	1	25	225
Variance=90													

Both companies A and B have the same mean value. However, company A has a variance of 1.23, and company B has a variance of 90. Below is the formula for the calculation of a variance.

$$\text{Variance}(x) = \frac{\sum_{i=1}^{n}(x_i - \bar{x})^2}{n}$$

We have considered the square of deviations while calculating the variance. There is another connected measure for capturing the dispersion in the data. We simply take the square root of variance and call it as standard deviation or SD.

$$SD(x) = \sqrt{\text{Variance}(x)}$$

$$SD(x) = \sqrt{\frac{\sum_{i=1}^{n}(x_i - \bar{x})^2}{n}}$$

The below code is used for finding the variance and standard deviation:

```
bank= pd.read_csv('D:/Chapter2 Python Programming/Datasets/Bank Tele
Marketing/bank_market.csv')
house_loan_yes=bank[bank["housing"]=="yes"]
house_loan_no=bank[bank["housing"]=="no"]

print(bank["balance"].std())
print(house_loan_yes["balance"].std())
print(house_loan_no["balance"].std())

print(bank["balance"].var())
print(house_loan_yes["balance"].var())
print(house_loan_no["balance"].var())
```

In the above code, we first downloaded the bank market data. We then created two subsets from it—customers with a house loan and without a house loan. We then calculated the standard deviation of overall data and the two subsets. Below are the results.

```
print(bank["balance"].std())
3044.7658291686002

print(house_loan_yes["balance"].std())
2483.285760899055

print(house_loan_no["balance"].std())
3613.405338934082

print(bank["balance"].var())
9270598.954472754

print(house_loan_yes["balance"].var())
6166708.170283998

print(house_loan_no["balance"].var())
13056698.143437328
```

We can see from the output that the overall data has a standard deviation of 3044. There is a lesser standard deviation in the bank balance for the customers with the house loan. Variance and standard deviation are descriptive measures. They describe the data. We will not be able to reach any conclusion by looking at these measures. We use these measurements just to explore the data and to describe the underlying information.

2.8   DATA EXPLORATION

We can say the usefulness of mean and median is limited to getting some intelligence on the central tendency. Having just the center, minimum, and maximum values is not everything as far as getting a feel for the data. In the next few sections, we are going to discuss how you can explore more and get a better understanding of all available variables. Data variables can be numeric as well as non-numeric. Within the numeric type, we have continuous and discrete variables. Data exploration is different for different types of variables.

Take a note of the following checkpoints before we get more into data exploration.

1. You need to develop an exceptional understanding of the business problem before you even touch the data.
2. Once you have that, try to get complete information on metadata elements like the number of available records, columns, and all column definitions. Then comes finding out the number of missing values or blank spaces. Then you may look for the unique identifier, better known as a primary key like customer ID, machine, account number, and product code. These are just examples. The primary key can take many different forms depending upon the business problem and type of data.

3. Once you have all the variables, divide them into different classes based on the values they take. Exploring different types of variables is not the same. The following are some examples:

   a. Numeric continuous variables, for example, income, sales, debt ratio, loss percentage, quantity, and invoice amount.
   b. Numeric discrete variables, for example, number of credit cards per person, number of loans given to a single client, number of dependents, and feedback ratings (1-5).
   c. Categorical variables with limited classes, for example, gender (M and F), region (E, W, N, S), country code (1, 2, 3, 4, 5), and customer class (A, B, C, D). As you can observe, it can be both numeric and non-numeric.
   d. Non-numeric variables with unlimited classes or string type of variables, for example, customer name, customer feedback, product description.
   e. Date and datetime variables, for example, order date, date of birth, incident time.

4. Make sure that you have the definition of each column. For example, if the data has variable names like x1, x2, and x3 or var1, var2, and var3, we cannot do any meaningful exploration or analysis on such variables without knowing the business context. It pays to spend some time at the start of the analysis and get yourself comfortable with the available data.

Let us now consider how to explore some of the most frequently encountered data types.

2.8.1   Exploring Numeric Continuous Variables

What is a numeric continuous variable? How do we identify it? If you take any numeric variable, take its maximum and minimum values. Then ask yourself a question—can this variable take all the values between these two limits? If the variable can take any value in between the limits, it is a numeric continuous variable.

Nothing can explain better than examples; if you have a variable called loss percentage with its minimum and maximum values as 0.1 and 1, respectively. Can it take any value between these limits? Can we have 0.11 or 0.25 assigned to the loss percentage? The obvious answer is yes, which categorizes the loss percentage as a continuous variable.

Finally, if you have a variable as the number of complaints with minimum values as 0 and the maximum value is 4, can it take any values between 0 and 4? Can there be 2.5 complaints? Of course, this time the answer is no which automatically categorizes it as a discrete variable (not continuous).

To explore continuous variables, we use percentiles and percentile distributions. First, let us try to explore what percentiles are.

2.8.1.1   Percentiles   For example, a student got 68 marks out of 100 marks in an exam. Did she do well? Are we in a position to give any meaningful answer to this question? Does it not depend on the difficulty level of the paper or how other students have fared in the same exam paper? With 68 marks, she may have scored better than 90 percent of the participants. If that is the case, her performance may be appreciated. What if 90 percent of the students have secured more marks than her, and she is in the bottom 10 percent? Both scenarios are represented in the sketches shown in Fig. 2.14.

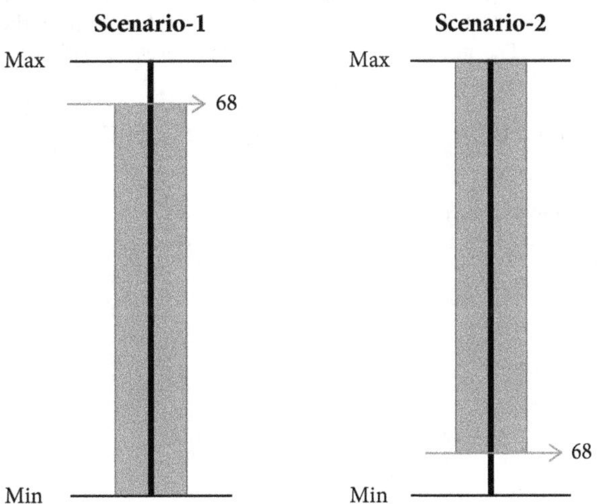

FIGURE 2.14   Student scenarios for percentiles demonstration.

Here we discussed two scenarios. In scenario 1, the value 68 is more than 90 percent of the data population. In scenario 2, exactly the opposite happened. Equipped with this information, called percentiles, you can easily tell if 68 marks is a good performance or not. Obviously, in scenario 1, it will be rated better when compared to scenario 2. In scenario 1, the student stands at 90 percentile and 10 percentile in the other.

To calculate the percentile, we arranged the marks data in descending order and divided it into 100 buckets. We asked a simple question: Where does the number in hand (68) fall? This question is indirectly asking which percentile does it fall into. If 68 hits on the 90 percentile, that means 90 percent of the data is below that number and only 10 percent is above. Sometimes, to get a better idea of the data, it may help to look at the percentile value rather than the exact value. In fact, in many competitive exams across the globe, the focus is on student's percentile marks (relative performance) rather than the actual final marks.

We need to differentiate between the percentile and percentage. They are not the same. Someone getting 95 percentile does not necessarily mean she got 95 out of 100. Try to answer this question. Can someone scoring only 60 out of 100 end up having 95 percentile? It is perfectly fine when only 5 percent of the participants get greater than 60 marks. Can someone scoring 95 out of 100 marks end up getting 70 percentile? It is also possible when 30 percent of the participants score more than 95 marks. There are much wider applications of percentiles. Percentiles can help us in detecting outliers. In the next few sections, we will use it while exploring our numerical variables.

Let us have an example to explore percentiles further. In Python, we have a function quantiles(). This function takes percentiles as input and gives the value as output. That means if a student gives us her percentile value, we can get her absolute score using this function, provided we have score data of all the students.

The variable we are going to use here is capital-gain from income data. In previous discussions, we had already identified something fishy about this variable. The mean value of capital-gain is 1,077.6, and the median is 0. Let us now try to calculate its percentile values. Following is the repeat code and corresponding results to find out its mean and median.

```
print(income_data["capital-gain"].mean())
1077.6488437087312

print(income_data["capital-gain"].median())
0.0
```

The following code is to work with percentiles about capital-gain using quantile the function. We have supplied 0.2 as input, which means we are trying to find out the actual value of the capital-gain variable associated with the 20 percentile.

```
print(income_data["capital-gain"].quantile(0.2))
```

The above code gives us the below output:

```
print(income_data["capital-gain"].quantile(0.2))
0.0
```

The result is reading as 0.0, meaning 20 percent of the data is less than or equal to zero, and 80 percent of the data is more than or equal to zero. Let us do a deep dive and fetch more percentile values. In the following code, we are trying to fetch the 0th percentile, 10th percentile, 20th percentile, and so on until the 100th percentile. By the way, what is the zero percentile value? The minimum value, isn't it? Similarly, the 100 percentile is nothing but the maximum value. Furthermore, the 50 percentile is the median.

```
print(income_data["capital-gain"].quantile([0,0.1,0.2,0.3,0.4,0.5,0.6,0.7,
0.8,0.9,1]))
```

The output from the above code is:

```
0.0        0.0
0.1        0.0
0.2        0.0
0.3        0.0
0.4        0.0
0.5        0.0
0.6        0.0
0.7        0.0
0.8        0.0
0.9        0.0
1.0    99999.0
```

The output is not really what we expected. At first glance, it is confusing and not easy to comprehend. Let us study the output. There are two columns in the output. The first column is the percentile, and the second column is the corresponding value. It is the value of the variable, which is capital-gain in this case (Table 2.18).

TABLE 2.18 Percentile Demonstration Output

Percentile	Value	Comment
0 (0th percentile)	0	Minimum value of capital-gain is zero.
0.1 (10th percentile)	0	Ten percent of capital-gain is less than or equal to zero; maybe the remaining 90 percent is more than or equal to zero, but we are not sure. We need to check.
0.2 (20th percentile)	0	Same analogy as above.
0.3 (30th percentile)	0	Same analogy as above.
0.4 (40th percentile)	0	Same analogy as above.
0.5 (50th percentile)	0	Fifty percent of capital-gain is less than or equal to zero, maybe the rest 50 percent is more than or equal to zero, but we are not sure. We need to check.
0.6 (60th percentile)	0	Same analogy as above.
0.7 (70th percentile)	0	Same analogy as above.
0.8 (80th percentile)	0	Same analogy as above.
0.9 (90th percentile)	0	*Look at this carefully*—90 percent of capital-gain is ≤0 and the remaining 10 percent may be ≥0, but we are still not sure. We need to check further.
1 (100th percentile)	99999	The maximum value is 9999. Can we conclude that 10 percent of this data is more than zero and 90 percent of the data is less than or equal to zero? How are you sure only 10 percent of the data is greater than zero? It can be only 1 percent or 5 percent , which is greater than zero. We have not explored the last 10 percent of the data, and it does not mean all this 10 percent is more than zero. How do we dive deep into the last 10 percent from the 90 percentile to 100 percentile?

Following is the code for exploring the last 10 percentiles. You need to mention values from 0.91 to 0.99.

```
print(income_data["capital-gain"].quantile([0.9,0.91,0.92,0.93,0.94,0.95,
0.96,0.97,0.98,0.99,1]))
0.90         0.0
0.91         0.0
0.92      1458.2
0.93      2885.0
0.94      3818.0
0.95      5013.0
0.96      7298.0
0.97      7688.0
0.98     14084.0
0.99     15024.0
1.00     99999.0
```

As we can see, the 91 percentile is 0, and from the 92 percentile onward we see some positive values. This means 91 percent of the data is simply zeros; only the remaining 9 percent of the data have some positive values.

- When we looked at the mean of capital-gain, nearly 1077, did we ever realize 91 percent of the data points will have a zero value?

- When we looked at the median of capital-gain, standing at 0, did we ever realize 91 percent of the data points will have zero value?

- When we looked at the minimum and maximum of capital-gain, did we ever realize 91 percent of the data points will have zero value?

- Now you appreciate the beauty of percentiles. Right? A more useful companion? They will help you unclutter the data completely and understand its depth.

Now we know that for exploring continuous variables, we need to use percentiles. They will not only show you the complete distribution of a variable, but they will also help you in identifying outliers in the process. You can even guess what percentage of the data outliers are. If you have a continuous variable, it can take almost any value. It helps to use percentiles if you wish to get a feel for its distribution. Simply choose a percentile range, and perform a deep dive to understand data.

Let us now proceed with a sample exercise to test our understanding of percentiles and outlier detection. From the same income data, take the variable "hours-per-week." As you can make out by the name, it is nothing but working hours per week. We will do some data exploration here. We generally know five days of weeks have 40 official working hours. If the data has more than 60 working hours, we can safely conclude the presence of outliers on the higher side. Similarly, for the entries less than 20 hours per week, outliers can be said to be present on the lower side. Let us look at the data and percentiles, and let us find out exactly what is the percentage of outliers on the higher side. Let us also find out what exactly is the percentage of outliers on the lower side. Can you try solving it on your own without referring to the code given below? By this time, you should be able to do this.

Let us first get the standard percentiles as usual and identify any deep-dive areas. For this exercise, more than 60 hours is on the higher side, and less than 20 lower.

```
print(income_data["hours-per-week"].quantile([0,0.1,0.2,0.3,0.4,0.5,
0.6,0.7,0.8,0.9,1]))
0.0      1.0
0.1     24.0
0.2     35.0
0.3     40.0
0.4     40.0
0.5     40.0
0.6     40.0
0.7     40.0
0.8     48.0
0.9     55.0
1.0     99.0
```

From the above output, we can see that the 90th percentile value is 55 and 10th percentile value is 24. However, we are not interested in less than 24 and greater than 55 limits. We are interested in the less than 20 and more than 60 population. Understanding the earlier explanations and output tables, we can make out the need to drill down the first 10 and last 10 percentiles as our limits are 20 and 60. Let us proceed to the detection of outliers on the higher side.

```
print(income_data["hours-per-week"].quantile([0.9,0.91,0.92,0.93,0.94,0.95,
0.96,0.97,0.98,0.99,1]))
0.90     55.0
0.91     55.0
0.92     58.0
0.93     60.0
0.94     60.0
0.95     60.0
0.96     60.0
0.97     65.0
0.98     70.0
0.99     80.0
1.00     99.0
```

The above output clearly shows 92 percent of the data is less than 60, and the remaining 8 percent is more than or equal to 60 hours, so 8 percent can be termed as high-side outliers. If we are looking strictly for more than 60 hours, then it is 4 percent of the data.

Can we now proceed to the outliers detected on the low side? We need to drill down the first 10 percentiles.

```
print(income_data["hours-per-week"].quantile([0,0.01,0.02,0.03,0.04,0.05,
0.06,0.07,0.08,0.09,0.1]))
0.00      1.0
0.01      8.0
0.02     10.0
0.03     14.8
0.04     15.0
0.05     18.0
0.06     20.0
0.07     20.0
0.08     20.0
0.09     21.0
0.10     24.0
```

In the above output, we can observe 8 percent of the data is less than or equal to 20 hours. The lower-side outlier percentage is 8 percent. If you are strictly looking for the data that is less than 20 and ignore values equal to 20 hours, this percentile stands at 5 percent. Finally, we have 4 percent high-side outliers and low-side 5 percent outliers. This is how you detect outliers and explore continuous variables.

Should not these outlier data points be always dropped from the data? For now, it may be enough to know that we separate the outliers and perform a separate analysis. There are different types of treatments that we can do. As of now, we are discussing only exploration and outlier detection.

2.8.1.2   Box Plots   Box plots are a pictorial representation of the important percentiles. The important percentiles are 0p, 25p, 50p, 75p, and 100p. The zero percentile is the minimum value. The 25th percentile is also known as the first quartile. The 50th percentile is called a second quartile or median. The 75th percentile is the third quartile, and the 100th percentile is the maximum value or fourth quartile. These percentiles are sufficient to get a good idea of the variable. The describe () function shows these quartiles in summary.

```
print(income_data["age"].describe())
count    32561.000000
mean        38.581647
std         13.640433
min         17.000000
25%         28.000000
50%         37.000000
75%         48.000000
max         90.000000
Name: age, dtype: float64
```

In the above output, we can see the minimum value of age is 17; the 25 percentile value is 28, the second quartile value is 37, the third quartile value is 48, and the maximum value is 90. The distance between the minimum to the first quartile is 11. Similarly, the difference between the first and second quartile is 9; the difference between the second and third quartile is 11. Until the third quartile, the distribution is equally spread. However, the maximum value jumped to 90. We can say there are a few outliers in this data. A box plot will show this information inside a graph.

A box-plot is drawn by taking these percentiles. By drawing these percentiles on one graph, we get a basic idea of the distribution of a variable between the minimum and maximum value. Box-plot helps us in identifying the outliers quickly. The below code helps us in drawing the box plot:

```
plt.boxplot(income_data["age"])
```

The above code gives us the box plot shown in Fig. 2.15.

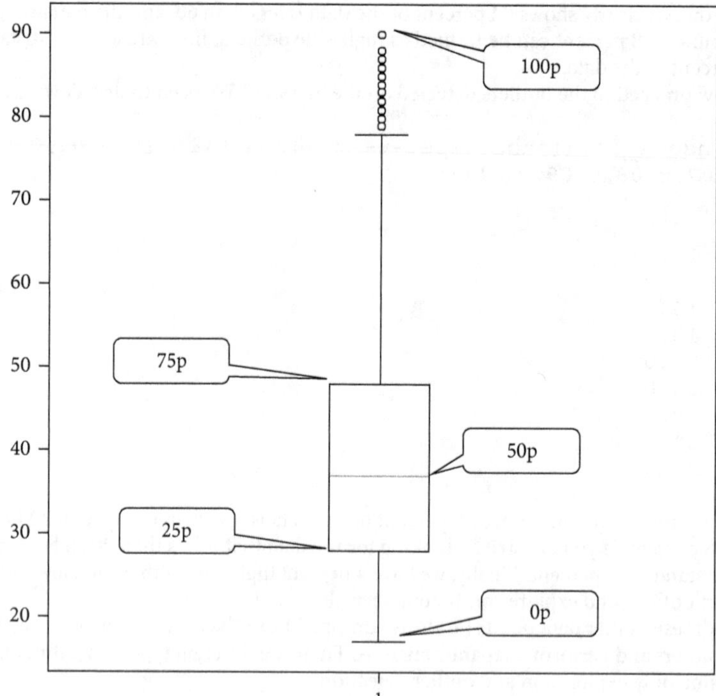

FIGURE 2.15   Box plot for age.

In Fig. 2.15, we can see the equal distribution of quartiles until the third quartile. There are some outliers in the upper quartile. To further analyze and detect the outliers, we can use percentile distribution. Nevertheless, the box plot helps us to see the whole distribution of the variable. If there are extreme outliers, then the box will be compressed on either the lower side or the higher side of the graph. We already did outlier detection on the capital-gain variable. There are extreme outliers on the higher side. The box will be compressed at the lower side. Ninety-two percent of the data is zero in capital-gain. The below code creates the box plot for capital-gain (Fig. 2.16).

```
print(income_data["capital-gain"].describe())
count      32561.000000
mean        1077.648844
std         7385.292085
min            0.000000
25%            0.000000
50%            0.000000
75%            0.000000
max        99999.000000
Name: capital-gain, dtype: float64

plt.boxplot(income_data["capital-gain"])
```

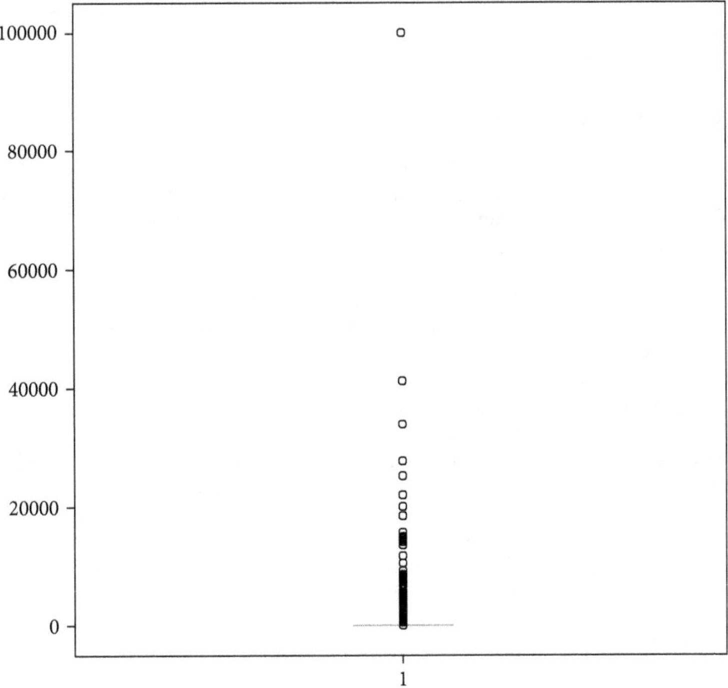

FIGURE 2.16   Capital-gain box plot.

Figure 2.16 does not even show the box and the first three quartiles. There are extreme high-side outliers in this variable. One more example of the box plot for a different variable is Fig. 2.17.

```python
print(income_data["hours-per-week"].describe())
plt.boxplot(income_data["hours-per-week"])
```

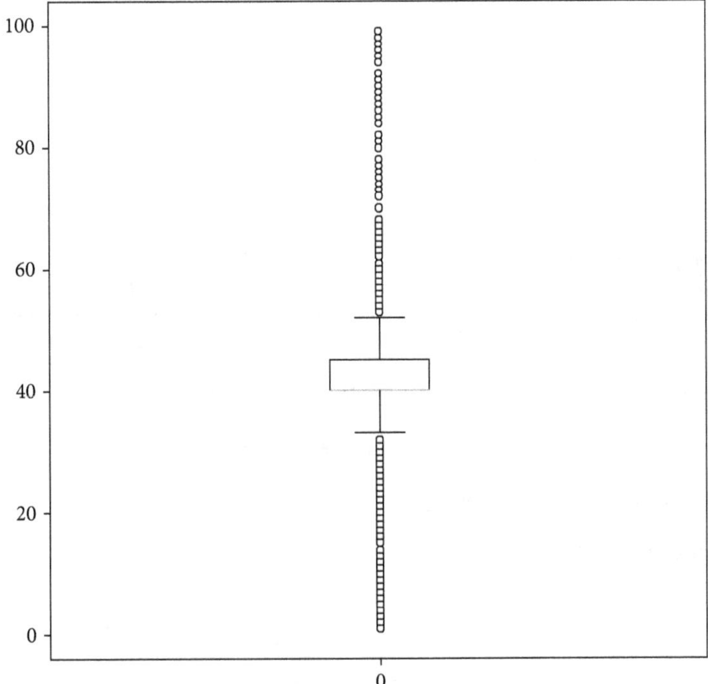

FIGURE 2.17   Hours per week box plot.

The above variable has both high-side and low-side outliers. The box plot just gives us a basic idea of the presence or absence of outliers. It also helps us in visualizing the distribution of a variable.

2.8.2   Exploring Discrete and Categorical Variables

As discussed, discrete numeric variables can take a limited number of values. For example, the number of dependents can only be 0 or 1 or 2. It can never be 1.5. In another example, the number of cars one can have is like 1, 2, 3 or 4—always a whole number. Discrete variables can also take many values, but they are fundamentally different from continuous variables. For easy recall, you may visualize discrete variables as an integer type and continuous variables as floats or a decimal type.

Categorical variables have a limited number of classes. These classes are nothing but the categories. The region as a variable can take four classes: East, West, North, and South. Gender as a variable, in most cases, can be male or female. Payment status is also a categorical variable, which can take values like paid or pending. Sometimes you can have numeric values also as a categorical variable. A categorical variable as country code may take values 1, 2, or 3, but these digits here will still be treated as three different categories. We can never compare and say country code 1 is less than a country 3. It will not make any business sense.

To explore a categorical variable, we need to create a frequency table, which is much like a group by table or a summary table. A frequency table contains all the unique categories (in a variable) and frequency counts or the number of occurrences of each category type. You will get an idea of all unique classes and their weighting in the given data.

Let us work out an example to understand the concept. In income data, there is a variable called "education," which can be used in this exercise. In the following code, we need to write the column name and use the Python function value_counts to create the required frequency table.

```
income_data["education"].value_counts()
Out[144]:
 HS-grad          10501
 Some-college      7291
 Bachelors         5355
 Masters           1723
 Assoc-voc         1382
 11th              1175
 Assoc-acdm        1067
 10th               933
 7th-8th            646
 Prof-school        576
 9th                514
 12th               433
 Doctorate          413
 5th-6th            333
 1st-4th            168
 Preschool           51
Name: education, dtype: int64
```

From this output, you can easily take down your observations like which class has the maximum occurrence, what are the top three classes with more frequencies, and the top classes responsible for 80 percent of the data? Sometimes it helps to print the frequency table with percentage values. To get the percentage values, we need to use the option normalize=True inside the value_counts() function. The following is the sample code.

```
freq=income_data["education"].value_counts()

percent=income_data["education"].value_counts(normalize=True)

freq_table=pd.concat([freq,percent], axis=1, keys=["counts","percent"])

print(freq_table)
```

In this code, we are trying first to create frequency counts, then percentages are calculated using, the normalize=True option. We then concatenate frequency and percentages into a table, axis=1, which indicates we are concatenating columns. The parameter "keys" mention, the actual column names in the resultant table. As usual, the output is given below.

```
freq=income_data["education"].value_counts()
percent=income_data["education"].value_counts(normalize=True)
freq_table=pd.concat([freq,percent], axis=1, keys=["counts","percent"])
print(freq_table)
              counts    percent
HS-grad        10501   0.322502
Some-college    7291   0.223918
Bachelors       5355   0.164461
Masters         1723   0.052916
Assoc-voc       1382   0.042443
11th            1175   0.036086
Assoc-acdm      1067   0.032769
10th             933   0.028654
7th-8th          646   0.019840
Prof-school      576   0.017690
9th              514   0.015786
12th             433   0.013298
Doctorate        413   0.012684
5th-6th          333   0.010227
1st-4th          168   0.005160
Preschool         51   0.001566
```

As we can see, 32 percent of the data population is Hs-grad, 22 percent are some-college grads, 16 percent are bachelors, and 5 percent are masters. These four categories only amount to almost 75 percent of the data.

You can even group this miniature analysis with a different variable and create a cross table to get a better understanding of the data. Following is an example where we attempt to associate the educational qualification with the income band. To keep it simple, we will get the income separately, and then we will combine it with educational qualification. Following is the code snippet:

```
print(income_data["Income_band"].value_counts())
  <=50K     24720
  >50K       7841
Name: Income_band, dtype: int64

print(income_data["Income_band"].value_counts(normalize=True))
  <=50K     0.75919
  >50K      0.24081
Name: Income_band, dtype: float64
```

Only 24 percent are earning more than 50K in the entire available data. Now let us create a cross tab by comparing both the education and income band. After that we will try to get the income band distribution in each class of education.

```
cross_tab=pd.crosstab(income_data["education"],income_data['Income_band'])
cross_tab_p=cross_tab.astype(float).div(cross_tab.sum(axis=1), axis=0)
final_table=pd.concat([cross_tab,cross_tab_p], axis=1)
print(final_table)
```

Income_band	<=50K	>50K	<=50K	>50K
education				
10th	871	62	0.933548	0.066452
11th	1115	60	0.948936	0.051064
12th	400	33	0.923788	0.076212
1st-4th	162	6	0.964286	0.035714
5th-6th	317	16	0.951952	0.048048
7th-8th	606	40	0.938080	0.061920
9th	487	27	0.947471	0.052529
Assoc-acdm	802	265	0.751640	0.248360
Assoc-voc	1021	361	0.738784	0.261216
Bachelors	3134	2221	0.585247	0.414753
Doctorate	107	306	0.259080	0.740920
HS-grad	8826	1675	0.840491	0.159509
Masters	764	959	0.443413	0.556587
Preschool	51	0	1.000000	0.000000
Prof-school	153	423	0.265625	0.734375
Some-college	5904	1387	0.809765	0.190235

In the doctorate and prof-school categories, more than 70 percent of the population is earning more than 50K. This is way higher than 25 percent as the average for the overall population. Similarly, you can see another extreme in the preschool category where no one is earning more than 50K. A similar type of analysis may be possible by grouping other variables.

Take a note of the following checkpoints while exploring categorical variables.

1. Sometimes only a few classes amount to 90 percent of the data. All the others together are 10 percent or less. In such situations, it might considerably simplify the work if all fewer contributing classes are grouped and named, say, as "others." It is just a rule of thumb followed by many seasoned data scientists. You may also use it as the analysis warrants.

2. In some situations, you might have a discrete numerical variable with just too many data values assigned to it (much like a greater number of records than necessary in a column). It is advisable to use percentile in such cases. Use frequency counts if only a limited number of classes are present.

3. Missing values can be treated as a different class. Perhaps you can name it as "Missing" or "NA."

2.8.3   Exploring Other Variables

Remember, there are no shortcuts for exploring the data. One needs to be very comfortable with variables and develop an in-depth understanding before you even think of starting the analysis. Spend enough time and go through every variable present in your target data. The following are some more useful tips.

- *String variables may not be of great value in the analysis.*   You might have variables like customer name, description, or comments in your data. You might find it challenging to apply analysis techniques directly to the pure text data if it is not structured. You need first to evaluate what best can be done based on the available data and the problem statement.

- *Text data needs to be treated differently.*   You might have a variable like customer feedback verbatim in a column. This column alone may require a separate customer sentiment analysis, and better, we convert the free text to something like –1 or 1 based on the sentiment type—positive or negative. Handling text data is very different from the rest of the data. You may need to use special natural language processing (NLP) techniques.

- *Date and time variables cannot be used directly.*   You often have variables like date or date-time in the analysis data. In the original or raw formats, they may not be so useful. They may be very informative and give great insights, but they often need some processing before used in the analysis. You may need to create new variables out of it. If you have a date variable, you may like to create a more relevant (depends upon the type of analysis) weekend indicator variable, quarter variable, or a month variable. If you have a detailed time variable, creating an hour variable might be more useful. These new variables are created based on the need of analysis. For example, the sales revenue of FMCG goods is generally high on weekends. The sales of electronic goods like washing machines may be high in the first week of the month, or festive seasons witness exceptionally high sales revenue on online retail sites.

- *Mapping non-numeric to numeric.*   Do not convert non-numeric variables into numerical variables by just mapping them to numbers, for example, if you have a variable called Region, which takes values East, West, North, and South, simply mapping East as 1, West as 2, North as 3, and South as 4 may backfire. Can you guess why? Which is higher, East or South? It is not the right question to ask, as there is no such ordering. Are we creating any kind of order after mapping them to numbers? There are better ways to do it. You can create four numeric variables from one non-numeric variable like Region, as shown in Table 2.19.

TABLE 2.19   Mapping Non-numeric Variable to Numeric Values

Original Variable Region	Derived Variable1 Region_East	Derived Variable2 Region_West	Derived Variable3 Region_North	Derived Variable4 Region_South
East	1	0	0	0
East	1	0	0	0
West	0	1	0	0
South	0	0	0	1
North	0	0	1	0
North	0	0	1	0
West	0	1	0	0

This method is also known as one-hot encoding or dummy variable creation. Just to repeat, one-hot encoding is a method by which categorical variables are transformed into a form required by machine learning algorithms so that a better job of prediction can be done. In Table 2.19, we have seven rows of data. Just have a close look at all the columns and data; the one-hot encoding operation performed here is very simple and needs no further explanation.

- *Handling geographical data.*   The geographical data may not directly useful as well. To be more useful in analysis, you may have to extract city names, state names, country name, ZIP code, or structured address if only the longitude and latitude information is provided.

There are many other types of data and many more handling tips. In the later chapters, we will try to focus on data cleaning and feature engineering based on the need and the case in hand.

2.9   CONCLUSION

In this chapter, we just introduced you to basic Python programming and statistics. We focused only on those concepts, commands, and packages which are a bare minimum for any data scientist and machine learning enthusiast. We have introduced some useful data manipulation commands in Python. These commands should be your fingertips. We discussed bare basic descriptive statistics and data exploration techniques. It should be treated just as a starting point. This is the minimum you should know before moving on to machine learning algorithms. You may want to explore more around it and related topics. In the later chapters, we are going to get started with machine learning algorithms. Make sure you have set up the Python environment in your system and complete all these exercises before moving any further.

2.10   PRACTICE PROBLEMS

1. Download bank marketing data. The dataset contains the details of the telephone marketing campaigns of a Portuguese bank.
    - Download the dataset and go through the variable descriptions.
    - Import it into Python and perform data exploration tasks on all the variables.
    - Validate the data and identify the missing values and outliers.
    - Clean the data and prepare it for the analysis.
    - Create a detailed report on the data exploration results. Include basic descriptive statistics, necessary data visualizations, and tables.

Dataset Credits—Data can be downloaded from the UCI machine learning repository. https://archive.ics.uci.edu/ml/datasets/Bank+Marketing# . [Moro et al., 2014] S. Moro, P. Cortez, and P. Rita. A Data-Driven Approach to Predict the Success of Bank Telemarketing. *Decision Support Systems*, Elsevier, 62:22-31, June 2014.

2. Download the Pima Indians dataset. The objective is to predict diabetes based on the diagnostic measurements of the patient.

   - Download the dataset and go through the variable descriptions.
   - Import it into Python and perform data exploration tasks on all the variables.
   - Validate the data and identify the missing values and outliers.
   - Clean the data and prepare it for the analysis.
   - Create a detailed report on the data exploration results. Include basic descriptive statistics, necessary data visualizations, and tables.

   Dataset Credits—Login to Kaggle and download Pima Indians dataset https://www.kaggle.com/uciml/pima-indians-diabetes-database. Smith, J. W., Everhart, J. E., Dickson, W. C., Knowler, W. C., & Johannes, R. S. (1988). Using the ADAP learning algorithm to forecast the onset of diabetes mellitus. In *Proceedings of the Symposium on Computer Applications and Medical Care* (pp. 261-265). IEEE Computer Society Press.

2.11   REFERENCES

1. Bank Telemarketing data used in Pandas package explanation—[Moro et al., 2011] S. Moro, R. Laureano, and P. Cortez. Using Data Mining for Bank Direct Marketing: An Application of the CRISP-DM Methodology. In P. Novais et al. (eds), *Proceedings of the European Simulation and Modelling Conference—ESM'2011*, pp. 117–121, Guimarren, Portugal, October 2011. EUROSIS. Available at: [pdf] http://hdl.handle.net/1822/14838, [bib] http://www3.dsi.uminho.pt/pcortez/bib/2011-esm-

2. Census income data used in "mean." This data was extracted from the Census Bureau database found at http://www.census.gov/ftp/pub/DES/www/welcome.html. Donor: Ronny Kohavi and Barry Becker, http://archive.ics.uci.edu/ml/datasets/Census+Income.

CHAPTER 3
REGRESSION AND LOGISTIC REGRESSION

Regression and logistic regression are the two most basic and essential algorithms in machine learning. We will discuss two types of algorithms in this book, which we will call simple algorithms and black-box methods. We can categorize linear and logistic regression as simple algorithms. As you might be aware, machine learning algorithms can solve many complex problems, but at the same time, underlying algorithms can be complex as well. We will discuss such intricate algorithms later in this book, but before that, we need to develop an understanding of simple algorithms. In this chapter, we will put more focus on code explanation and output interpretation. Throughout this chapter, we will also introduce some new words from the machine learning vocabulary.

Linear and logistic regression algorithms are relatively simple to build and easy to interpret even for non-technical business people. They are mostly intuitive and take relatively little time to execute. Businesses across the globe have been using these two machine learning algorithms for >25 years now. Most banks and financial institutions, for example, use logistic regression in their credit risk models.

In this chapter, we will start with the regression model, which will involve building a regression line, selecting or eliminating features from the model, and finally measuring its accuracy. We will then move on to logistic regression and perform similar activities. You are encouraged to run the code snippets encountered in this chapter and verify the output. We are using Python 3.7 in this book. If you are using a different version of Python, there might be a slight change in syntax. We suggest you check the Python documentation for any syntax-related help.

3.1   WHAT IS REGRESSION?

The term "regression" comes from the Latin term "regressus," which means to go back or to return to something. The term "regression" was first introduced by a scientist named Francis Galton to explain the relationship between the height of the father and the height of the son. Regression analysis is about looking at the historical data to find patterns and use them for prediction. In this chapter, we will see how to do it.

We will start with employee profile data, which has three variables in it, namely, monthly income, monthly expenses, and time spent reading books. The code snippet below gives the path of the file containing this data. We will import the data into Python, and draw scatter plots between monthly income and other variables (Figs. 3.1 and 3.2). Before we start with code, try to answer the following questions:

- Is there any association between monthly income and monthly expenses? If there is any, is it positive or negative?
- Is there any association between monthly income and time spent on reading books? If there is any, is it positive or negative?

Let us see if the following code and the subsequent scatter plots help us to answer this.

```python
import pandas as pd

emp_profile=pd.read_csv(r"D:\Chapter3\5. Base Datasets\employee_profile.csv")

#Column names
print(emp_profile.columns)
```

```
#First few rows
emp_profile.head()


#Drawing the Scatter Plot
import matplotlib.pyplot as plt
plt.scatter(emp_profile["Monthly_Income"],emp_profile["Monthly_Expenses"])
plt.title('Income vs Expenses Plot')
plt.xlabel('Monthly Income')
plt.ylabel('Monthly Expenses')
plt.show()

plt.scatter(emp_profile["Monthly_Income"], emp_profile["Time_Spent_Reading_
Books"])
plt.title('Income vs Time_Spent_Reading_Books Plot')
plt.xlabel('Monthly Income')
plt.ylabel('Time_Spent_Reading_Books')
plt.show()
```

The above code gives us the below output:

```
print(emp_profile.columns)
Index(['EmpId', 'Monthly_Income', 'Monthly_Expenses',
'Time_Spent_Reading_Books'],
      dtype='object')

emp_profile.head()
   EmpId   Monthly_Income   Monthly_Expenses   Time_Spent_Reading_Books
0      1             6092            3836.80                          29
1      2             2109            1422.85                          14
2      3             7177            4717.05                          39
3      4             6665            4330.25                           1
4      5             8356            5552.40                          44
```

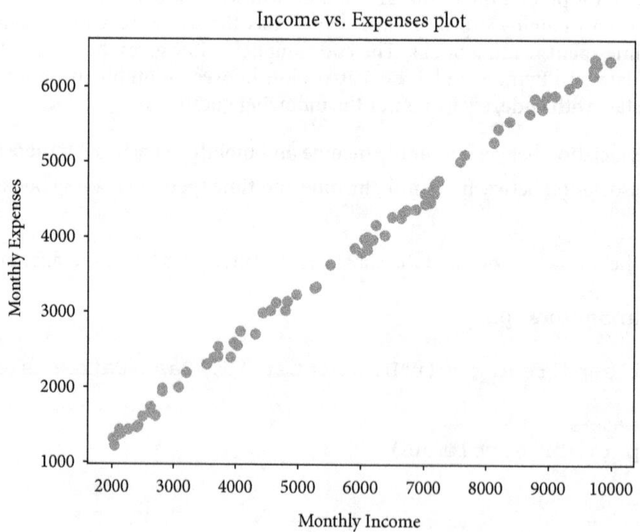

FIGURE 3.1   Monthly income versus monthly expenses plot.

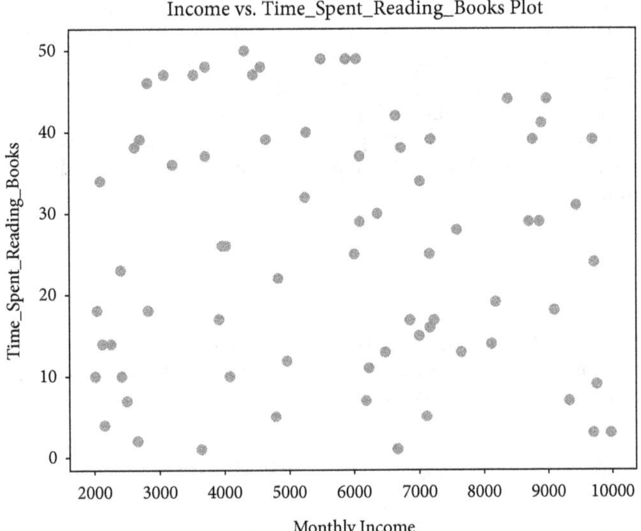

FIGURE 3.2   Monthly income versus time spent reading books.

From the Fig. 3.2 scatter plot, it is evident that there is no relation between income and time spent on reading books. The graph looks scattered. It is also very apparent that there is a strong association between monthly income and monthly expenses (Fig. 3.1). As income is increasing, the expenses are also increasing. It is a clear indication of a strongly positive relationship.

Looking at the scatter plots, can we answer the following questions?

- What is the predicted value of monthly expenses if the monthly income is 6000?
- On the same lines, what is the predicted value of monthly expenses if monthly income is 9000?

As you rightly guessed from the plot, the predicted value of monthly expenses is around 4000 if the monthly income is 6000. Similarly, the predicted value of monthly expenses is about 6000 if the monthly income is 9000 (Fig. 3.3).

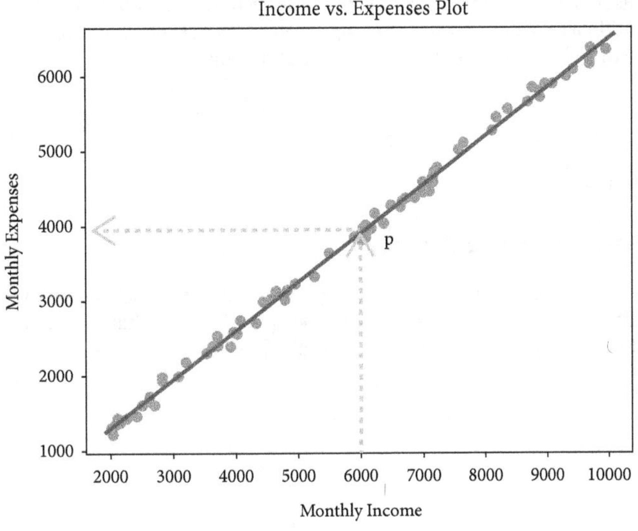

FIGURE 3.3   Income versus expenses.

How did we get these predicted values? For this interpretation, we first imagined a line through the data; we then forgot about the actual data and focused only on the line. Is it correct? In a way, we represented the whole data in the form of a single line. Then we imaged a second line (dotted line) starting at 6000 on the x-axis. This dotted line hits our first line at point p. This way, we find the corresponding y-axis value is around 4000. This imaginary line represents our data, and at the same time, it represents the relation between x and y. As you will also appreciate, this line goes through most of the data points, thus predicting the value of y for a given value of x. It's the best fit for our scatter plot, and it is our target regression line. What is the equation of a line, do you remember it from your school days? Does the following look familiar?

$$y = mx + c$$

In the above equation, c is known as the constant or intercept, and m, the coefficient of x, is known as the slope. We can completely define the straight line if we have the values of m and c. For example, if the value of m is 0.2 and c is 5, the equation of this straight line will be as follows:

$$y = 0.2x + 5$$

We use the same straight-line equation in a regression analysis, but the notations for c and m are slightly different. c, the intercept, is replaced by β_0, and m, by β_1. In statistics, we represent the same straight-line equation as

$$y = \beta_0 + \beta_1 x$$

Here β_0 and β_1 are known as regression coefficients. This statistic straight-line equation involving β_0 and β_1 is also known as the regression model. It is called a model because it takes an input x and gives us the predicted value of y as output. This way, we build a simple model, which is nothing but a relationship between x and y.

3.2   REGRESSION MODEL BUILDING

In the introduction of regression, we discussed that we need to fit a straight line, which will go through our data points. The data in that example was relatively simple; we could easily imagine a line. Can we fit a regression line to any given dataset? We will discuss it later in this chapter. Now the second and straightforward question: Is a scatter plot always necessary to get predicted values? As you can appreciate with a straight-line equation like the following, it is easy to work out the value of y if you know x, even without plotting any graph:

$$y = 0.2x + 5$$

3.2.1   Finding the Regression Coefficients

Look at Fig. 3.4, and imagine the dots are our data points, and you are asked to fit a regression line for this data. As of now, let us try to fit a straight line to this data while keeping in mind that a straight line may not be the best fit in all scenarios. We will discuss the nonstraight line or nonlinear situations later in this book.

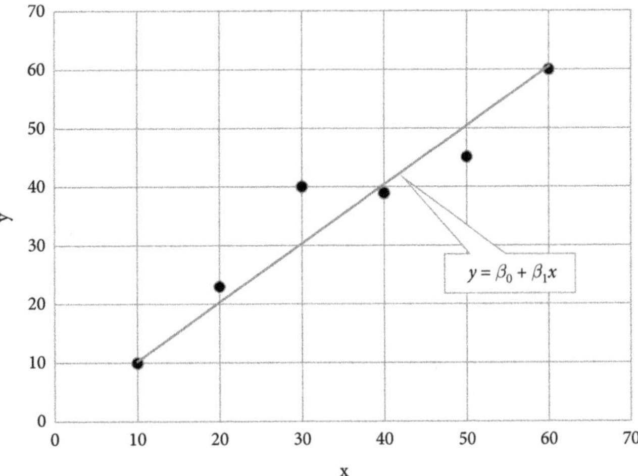

FIGURE 3.4 Regression line fit for the data.

For this data, we need to fit the line $y = \beta_0 + \beta_1 x$. By this time, we know, fitting a line is nothing but finding the values of β_0 and β_1. Once we have these values, we can draw the line and easily get the predicted values. Now the question is: How do we find the values of β_0 and β_1?

Let us have a look at our data points carefully. Look at the third point from the bottom in the graph. At this point, compare the predicted value versus the actual value. We find, at this point, the actual value is at 40 and the predicted value is at 30. In any regression line, the difference between actual versus predicted values at any point is denoted by error e_i. At the third point, as the case here, it will be e_3. Now look at the fifth point; at this point, also, you will note an error between actual and predicted value. Going by the same notation, it will be e_5. As you can notice, there is some amount of error at every data point. Obviously, for the best-fit regression line or the best regression model, this error should be kept to a minimum.

Have a close look at the plots in Figs. 3.5 to 3.7. Do they show all the errors?

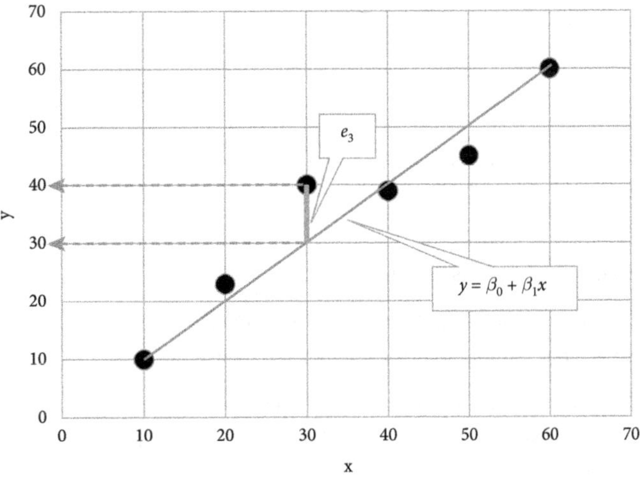

FIGURE 3.5 Regression error example.

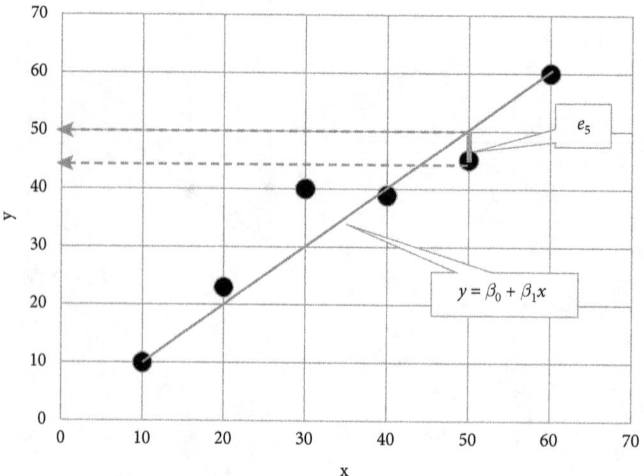

FIGURE 3.6   All errors in regression line.

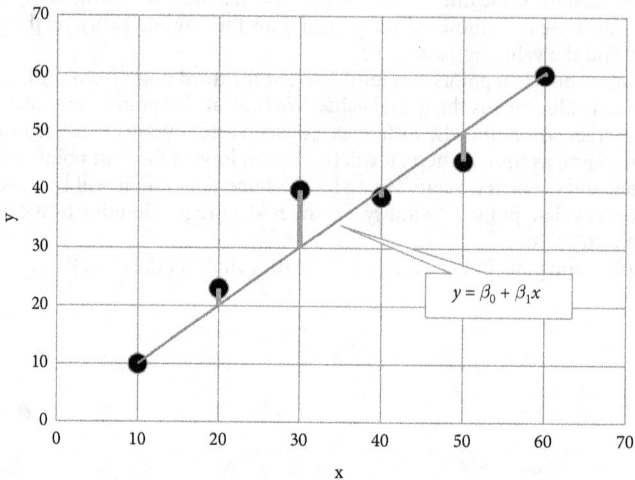

FIGURE 3.7   Regression error example 2.

You will notice the following facts in these figures:

1. Errors are positive at some points and negative at some other points.
2. Error is too high at a few points and error is almost zero for some.

We aim to build a line that has the least amount of errors. Correct? Since a few errors are positive, and a few are negative; we cannot merely sum up errors at all points and try to minimize it. A simple $\min \sum e_i$ will not work here; we need to do $\min \sum e_i^2$.

We need to build a regression line that has the least sum of squares of error. We know finding the best regression fit is finding the values of β_0 and β_1. It all boils down to—at what values of β_0 and β_1 is the value of $\sum e_i^2$ the minimum. Agree? Let us consider the following target function:

$$\min \sum e_i^2$$

$$\min \sum \left(y \text{ actual} - y \text{ predicted}\right)_i^2$$

$$\min \Sigma (y - \hat{y})^2 \text{ where } \hat{y} \text{ is } y\text{-predicted value}$$

$$\min \sum \left(y - (\beta_0 + \beta_1 x)\right)^2$$

We already have the data on x and y. With that, we need to find the values of β_0 and β_1, which will minimize this sum of squares of errors. Now we are working with an optimization problem. We want to find optimal values of β_0 and β_1 so that $\sum \left(y - (\beta_0 + \beta_1 x)\right)^2$ is minimum. Beyond this statement, using optimization techniques in calculus, we can find the values of β_0 and β_1.

For example, if you substitute the values of coefficients as $\beta_0 = 2$ and $\beta_1 = 1.5$, we can find the value of error $\min \sum \left(y - (\beta_0 + \beta_1 x)\right)^2$. Suppose, in this iteration, you get the error as 50. In the second iteration, we change the values of the coefficients as $\beta_0 = 19$ and $\beta_1 = 1.4$ and you may get the new value of error as 45. We keep on playing with the different values of β_0 and β_1 until we find the minimum value for the error. No worries, in actual practice, all of it will be taken care of by Python.

Direct optimization techniques are available to find the parameters that will minimize a function. Whenever we change the values of β_0 and β_1 the regression line changes. By trying different values of coefficients, we are trying different lines. By choosing coefficients with minimum error, we are choosing the best line that optimally goes through most of the points. Figure 3.8 is the sample demonstration of how β_0 and β_1 are optimized in each iteration.

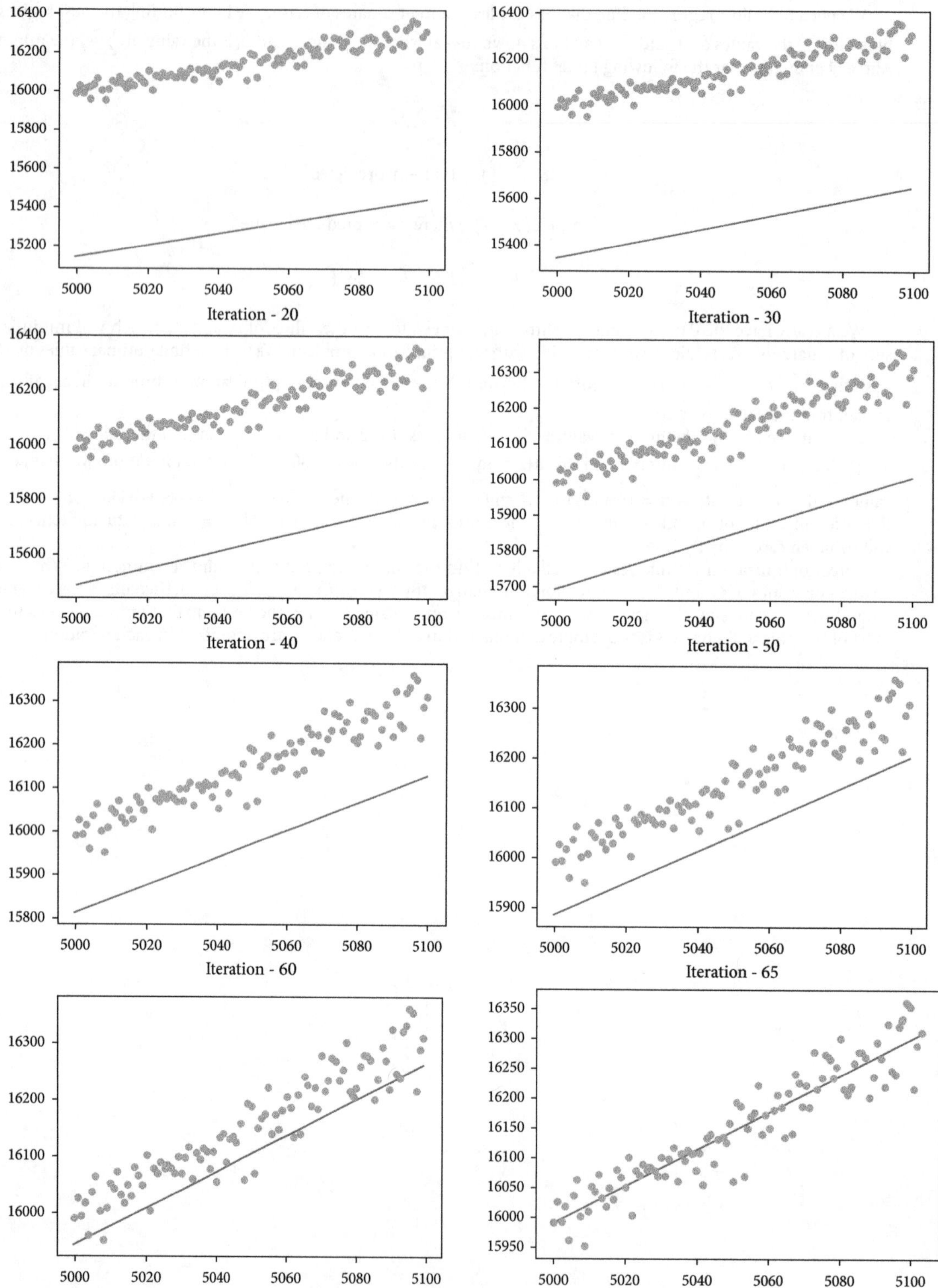

FIGURE 3.8 Demonstration of β_0 and β_1 optimization with each iteration.

In most statistical tools, these methods are inbuilt as library functions. We need to feed in the data on x and y, call the right function, and the rest all is taken care of by the software. The function will automatically calculate the values of coefficients that will minimize the error.

3.2.2 Regression Line Example

Let us take an example and build a regression line to get the predicted values. In this example, we have data collected from several airline companies. Each data point has the number of passengers in a week and some other factors that may directly or indirectly impact the number of passengers. We have gathered random weekly data from different airlines in a year.

The problem statement is to predict the passengers count based on marketing expenses. In particular, we need to find out when the marketing cost is at $4500K, what is the expected number of passengers? We need to build a regression line that will predict the number of passengers using the parameters that impact it. Before we start with problem-solving, it will help us to have a good understanding of the data and the involved variables. It is always a good practice to comprehend the data before making any further steps in problem-solving. Table 3.1 lists some data points on the variables that we are going to work upon.

TABLE 3.1   Data Columns Description

Column Name	Description	Other Details
Passengers_count	Passengers count in a week	Number
marketing_cost	Marketing cost in the last month	In thousands
percent_delayed_flights	Delayed flights percentage in that week	Percentage
number_of_trips	Total number of trips in the week	Number
customer_ratings	Customer feedback	Scale 1-10
poor_weather_index	Poor weather indicator	Percentage
percent_male_customers	Male customers percentage	Percentage
Holiday_week	Holiday week indicator	Boolean
percent_female_customers	Female customers percentage	Percentage

Now it is time to import the data and start fitting the regression line. Following is the code for importing the data and printing the columns.

```
air_pass=pd.read_csv(r"D:\Chapter3\Datasets\Air_Passengers.csv")
print(air_pass.columns)
```

It results in the following output:

```
Index(['Passengers_count',   'marketing_cost',   'percent_delayed_flights',
'number_of_trips', 'customer_ratings', 'poor_weather_index', 'percent_male_
customers', 'Holiday_week', 'percent_female_customers'],
      dtype='object')
```

We are not going to get into the details of data exploration and validation here. The data is already well formatted, and it is clean before we start this analysis, so we will directly go ahead with the regression analysis. In most real-life projects, you may have to first begin with data exploration and data cleaning before you go any further in your analysis.

As discussed earlier, the problem statement involves predicting the passengers count using marketing costs. In this example, our target variable y is passengers count, and our predictor variable x is the marketing cost. As discussed earlier, we will be using the following model equation in our regression model:

$$y = \beta_0 + \beta_1 x$$

Passengers count = $\beta_0 + \beta_1$ (Marketing cost)

In Python, we can use two packages to build our regression model. We will start with "statsmodels" package. In the later part of this book, we will also work with another package by the name "sklearn" to do the same job. Given below is the code snippet used for this job, and the two sections following it will be used for explaining the code and the output interpretation.

```python
import statsmodels.formula.api as sm
model1 = sm.ols(formula='Passengers_count ~ marketing_cost',data=air_pass)

fitted1 = model1.fit()
fitted1.summary()
```

3.2.2.1   Code Explanation

Table 3.2 explains this code in brief.

TABLE 3.2   Explanation of Code

```python
import statsmodels.formula.api as sm
```
- Importing the required subpackage from the statsmodels package.
```python
model1 = sm.ols(formula='Passengers_count ~ marketing_cost',data=air_pass)
```
- We need to mention the variables y and x. It is the configuration of the model, which is always $y \sim x$. In our case, y is the passengers count, x represents the marketing cost.
- Data is another parameter that we have used here to pass on the dataset name, which is termed as `air_pass`.
- In `sm.ols`, sm is our package name, and we use the ols() function to minimize the square of errors. It ultimately gives us the values of both regression beta coefficients. Ols stands for "ordinary least squares." The function name ols() is inspired by the fact that we are minimizing the sum of squares of errors.
```python
fitted1 = model1.fit()
```
- The previous step was for defining the model configuration, and in this step, we are proceeding with the model building.
- In this step, the actual data will be submitted, which will perform the optimization to build the model.
- Every model building process has two steps. The first is to define the model configuration, where we mention the variables (x, y) and datasets. The very next step is to fit or train, where we build the model.
- After running this code, we are ready with the beta coefficients. We store them in the `fitted1` object.
```python
fitted1.summary()
```
- Just now, as we discussed, we have stored the results in fitted1 object.
- `fitted1.summary()`   command gives us the output summary.
- As of now, we are looking only for the beta coefficients. The actual output is lengthy, which we will discuss later.

3.2.2.2   Output Interpretation

Let us have a look at the output of the model we just built (Table 3.3).

TABLE 3.3   OLS Regression Results

```
                          OLS Regression Results
==============================================================================
Dep. Variable:     Passengers_count   R-squared:                       0.761
Model:                          OLS   Adj. R-squared:                  0.760
Method:               Least Squares   F-statistic:                     830.0
Date:              Tue, 07 Apr 2020   Prob (F-statistic):           4.87e-83
Time:                      16:29:13   Log-Likelihood:                 -2453.4
No. Observations:               263   AIC:                             4911.
Df Residuals:                   261   BIC:                             4918.
Df Model:                         1
Covariance Type:          nonrobust
==============================================================================
                  coef    std err          t      P>|t|      [0.025      0.975]
------------------------------------------------------------------------------
Intercept      5186.6868    839.019      6.182      0.000    3534.579    6838.795
marketing_cost    6.3901      0.222     28.810      0.000       5.953       6.827
==============================================================================
Omnibus:                        8.874   Durbin-Watson:                   1.829
Prob(Omnibus):                  0.012   Jarque-Bera (JB):                9.679
Skew:                           0.342   Prob(JB):                      0.00791
Kurtosis:                       3.644   Cond. No.                     1.88e+04
==============================================================================
```

There are several metrics in this output. As of now, we are looking only for the regression coefficients β_0 and β_1. Browse through the output; we can find these regression coefficients in the second table. β_0 is given here as the intercept, while β_1, the coefficient of x, is nothing but the `marketing_cost`. Now we can quickly build our regression model in the form of the following equation involving the variables Passengers count and Marketing cost:

$$y = \beta_0 + \beta_1 x$$

$$\text{Passengers count} = \beta_0 + \beta_1 \text{ (Marketing cost)}$$

From output $\beta_0 = 5186.6868$ and $\beta_1 = 6.3901$

$$\text{Passengers count} = 5186.6868 + 6.3901 \text{ (Marketing cost)}$$

3.2.2.3   *Prediction from the Model*

It is time for making useful predictions using this model. When the marketing cost is 4500, what is the predicted value of passengers?

$$\text{Passengers count} = 5186.6868 + 6.3901 * (4500)$$

$$\text{Passengers count (predicted)} = 33{,}942$$

We are now equipped to predict the number of passengers at any given value of the marketing cost. We need not make these predictions manually. We have a predict function available in Python to do this job for us.

Following is the code for prediction:

```
new_data=pd.DataFrame({"marketing_cost":[4500]})
print(fitted1.predict(new_data))
```

We need first to create a data frame, and then pass it on to the predict function to get the predicted values. Running this code will give you the following output:

```
print(fitted1.predict(new_data))
0    33942.146091
```

We can even make multiple predictions in a single shot by using the following code, where we are trying to predict four values.

```
new_data1=pd.DataFrame({"marketing_cost":[4500,3600, 3000,5000]})
print(fitted1.predict(new_data1))
```

The above code gives us the below output.

```
new_data1=pd.DataFrame({"marketing_cost":[4500,3600, 3000,5000]})
print(fitted1.predict(new_data1))
0    33942.146091
1    28191.054238
2    24356.993003
3    37137.197120
dtype: float64
```

This regression model is our first machine learning model. It involved a simple straight-line relationship between two variables. Similarly, any machine learning model consists in discovering underlying patterns within the given set of data. Later as we work with more complex models, you will see these data patterns becoming much more intricate.

3.3   R-SQUARED

In the previous section, we discussed building a regression model and using it to make some useful predictions. Using our straight-line model, for a given value of the marketing cost as 4500, we predicted the passengers count as 33,942. Now consider the following questions:

- How can we validate the accuracy of our prediction or treat it as final?
- Why should we consider only marketing costs in the model? Why not any other configurations? Can we predict the passengers count using the variables like percent_delayed_flights, number_of_trips, or customer_ratings? Will the model outcome be the same?
- More importantly, have we captured all the possible patterns in the given data?

We need to define a measure of accuracy to answer these questions. We have already seen how to predict values of y using the variable x in the regression equation $y = \beta_0 + \beta_1 x$. Next is to validate the accuracy of the model.

Fortunately, we have some historical data available at our disposal. We already have a ready set of values for both y and x. We can use the values of x from this historical data in our regression model and get the corresponding predicted values of y. To find the accuracy of our regression model, the predicted values of y from our equation can now be compared to the historically available y values. The following code lines can be used:

```
#Predictions for the data
air_pass["passengers_count_pred"]=round(fitted1.predict(air_pass))
```

Given below is the code for printing x, y, and y_predicted values:

```
keep_cols=["marketing_cost", "Passengers_count", "passengers_count_pred"]
air_pass[keep_cols]
```

Have a look at the following output table from this code. It is only a partial snapshot of the actual output, which is much lengthier.

	marketing_cost	Passengers_count	passengers_count_pred
0	3588.1	23291	28115.0
1	3186.3	25523	25547.0
2	3342.0	25620	26542.0
3	2512.5	19625	21242.0
4	3012.1	27231	24434.0
..	...	...	...
258	2929.8	24238	23908.0
259	4024.0	29600	30900.0
260	3003.8	28578	24381.0
261	3327.6	27426	26450.0
262	2052.1	17591	18300.0

```
[263 rows x 3 columns]
```

Looking at the output, can you answer the following simple questions?

- For the model to be reliable, should the predicted values be close to or far away from the actual ones?
- For any model to be ideal, how much of a gap between actual and predicted values is permissible?

If a model is to be trusted, the predicted values should be very close to the actual values. We should consider any gap between actual values and predicted values as an error. For a perfect model, the predicted values should exactly match the actual values. The simple statistical equations in Table 3.4 are used to quantify errors in any regression model. In Table 3.4, y_i is the actual value and $\hat{y}_i$ is the predicted value.

TABLE 3.4   Error Equations in Regression Models

Error at one point	$y_i - \hat{y}_i$
Sum of squares of errors (SSE)	$\sum\limits_{i=1}^{n}(y_i - \hat{y}_i)^2$
Mean absolute deviation (MAD)	$\sum\limits_{i=1}^{n}\dfrac{\left\lvert y_i - \hat{y}_i\right\rvert}{n}$
Mean absolute percent error (MAPE)	$\dfrac{100}{n}\sum\limits_{i=1}^{n}\dfrac{\left\lvert y_i - \hat{y}_i\right\rvert}{y_i}$
Mean square error (MSE)	$\sum\limits_{i=1}^{n}\dfrac{(y_i - \hat{y}_i)^2}{n}$
Root mean square error (RMSE)	$\sqrt{MSE}$

All these error measures tell us the same story. A higher value of the measure indicates an erroneous model. Let us take one measure from this list, SSE—sum of squares of errors. If we are comparing any two given models, the model with the lesser SSE is preferred.

Let us take one data point out of our predictions. In the historical data, we have the actual value of y as 30,000, while the predicted value is 27,000. The error value, in this case, stands at 3000. In actual practice, we usually calculate this error value for all of the data points.

Consider the following simple equation.

$$Y \text{ actual} = Y \text{ predicted} + \text{Error}$$

$$30{,}000 = 27{,}000 + 3000$$

Total sum of squares at Y = Sum of squares of values predicted + Sum of squares of errors
Sum of squares of total = Sum of squares of regression + Sum of squares of error

$$\sum_{i=1}^{n}(y_i - \bar{y}_i)^2 = \sum_{i=1}^{n}(\hat{y}_i - \bar{y}_i)^2 + \sum_{i=1}^{n}(y_i - \hat{y}_i)^2$$

Total variance in y = Explained variance by regression + Unexplained variance

$$SST = SSR + SSE$$

- For a good model, what should be the SSE value? The obvious answer—it should be near to zero.
- For a good model, what should be the SSR value? It should be near to SST.
- For a good model, what should be the SSR/SST value? It should be near to 1.
- The value SSR/SST is known as R-squared (R^2).

$$R - \text{Squared} = \frac{SSR}{SST}$$

$$R - \text{Squared} = \frac{\text{Explained variance}}{\text{Total variance}}$$

We use R-squared for measuring the accuracy or the efficacy or the fit of the model. R-squared is also known as an explained variance by the model.

- For a good model, R-squared should be near to 1.
- When we are comparing two models, the model with the higher R-squared is preferred.
- R-squared is the first measure that we look for in our output.
- The general industry practice is to have a minimum R-squared of 80 percent. This acceptance level may vary for different industry verticals.
- The R-squared value is readily available in the output summary of a regression line.

Let us do one exercise on R-squared. Let us build two models to predict passenger count. In the first model, we will use marketing costs for the prediction, while the second model uses customer ratings to do the same job. Once both the models are ready, we will proceed to find out which model has the higher accuracy. We already have the first model with us. Let us have a look at its output (Table 3.5), which we get after running the code snippet written for this purpose.

TABLE 3.5   OLS Regression Results

```
                            OLS Regression Results
=============================================================================
Dep. Variable:        Passengers_count   R-squared:                      0.761
Model:                             OLS   Adj. R-squared:                 0.760
Method:                  Least Squares   F-statistic:                    830.0
Date:                 Tue, 07 Apr 2020   Prob (F-statistic):          4.87e-83
Time:                         16:29:13   Log-Likelihood:                -2453.4
No. Observations:                  263   AIC:                            4911.
Df Residuals:                      261   BIC:                            4918.
Df Model:                            1
Covariance Type:             nonrobust
=============================================================================
                 coef     std err          t      P>|t|     [0.025      0.975]
-----------------------------------------------------------------------------
Intercept    5186.6868     839.019      6.182      0.000   3534.579    6838.795
marketing_cost  6.3901       0.222     28.810      0.000      5.953       6.827
=============================================================================
Omnibus:                         8.874   Durbin-Watson:                  1.829
Prob(Omnibus):                   0.012   Jarque-Bera (JB):               9.679
Skew:                            0.342   Prob(JB):                     0.00791
Kurtosis:                        3.644   Cond. No.                     1.88e+04
=============================================================================
```

Could you notice the value of R-squared in this output? Following is the code for our predictions with customer ratings. We call it model2.

```
model2 = sm.ols(formula='Passengers_count~customer_ratings',data=air_pass)
fitted2 = model2.fit()
fitted2.summary()
```

Table 3.6 is the output for model2.

TABLE 3.6   OLS Regression Results for Model2

```
                               OLS Regression Results
==============================================================================
Dep. Variable:        Passengers_count   R-squared:                    0.102
Model:                             OLS   Adj. R-squared:               0.099
Method:                  Least Squares   F-statistic:                  29.72
Date:                 Tue, 07 Apr 2020   Prob (F-statistic):        1.16e-07
Time:                         17:50:02   Log-Likelihood:             -2627.4
No. Observations:                  263   AIC:                          5259.
Df Residuals:                      261   BIC:                          5266.
Df Model:                            1
Covariance Type:               nonrobust
==============================================================================
                    coef    std err        t      P>|t|     [0.025    0.975]
------------------------------------------------------------------------------
Intercept        2.261e+04  1192.915   18.955     0.000    2.03e+04   2.5e+04
customer_ratings  894.4643   164.083    5.451     0.000    571.369  1217.560
==============================================================================
Omnibus:                        28.234   Durbin-Watson:                2.024
Prob(Omnibus):                   0.000   Jarque-Bera (JB):            35.541
Skew:                            0.767   Prob(JB):                  1.92e-08
Kurtosis:                        3.944   Cond. No.                      27.0
==============================================================================
```

Do you observe R-squared here?

Looking at both output summaries, we see model1 has an R-squared value of 0.761, while model2 has it as 0.102. These R-squared values tell us that in model1, we have explained 76 percent of the variance in the target variable, whereas using model2, we could explain only 10 percent. Going by what we had discussed while defining R-squared, model1 is better than model2 when it comes to the accuracy of predictions. We had identified a generally acceptable level of R-squared as 80 percent. Model1, at 76 percent, is pretty close to it.

3.4   MULTIPLE REGRESSION

Until now, we used only one predictor variable for the prediction of the target variable. In one model, we used marketing cost for prediction of passengers count; in another, we have used customer ratings to do the same job. Both of these models are simple regression models. In real-life scenarios, we have multiple variables impacting the target variable. In our example also, it is not just the marketing cost that will affect the passengers count; other variables including `marketing_cost`, `percent_delayed_flights`, `number_of_trips`, `customer_ratings`, `poor_weather_index`, `percent_male_customers`, `holiday_week`, and `percent_female_customers` may also have an impact on the overall passengers count. For better accuracy, we should use all these variables in the model. At least our common sense tells us we should try out one model involving all these variables and check its efficacy as well. So now we proceed with building our regression model with multiple predictor variables, which may increase the accuracy of our predictions. Let us start working with our familiar equations:

Simple regression line equation $y = \beta_0 + \beta_1 x$.

The same equation with multiple variables would be

$$y = \beta_0 + \beta_1 x_1 + \beta_2 x_2 + \ldots + \beta_k x_k$$

The result of the multiple regression line will be a plane in multidimensional space. We have numerous input variables; each variable will have its own coefficient. If the coefficient is positive, then the variable has a positive impact on the target; if the coefficient is negative, then the variable is considered to have a negative effect. Most of the time, in real life, we will be building our regression models with multiple variables. Figure 3.9 is an example of a regression line in three-dimensional space. Here, we have two predictor variables along with one target variable. The model looks like a plane, but in principle, still, it is a multidimensional regression line.

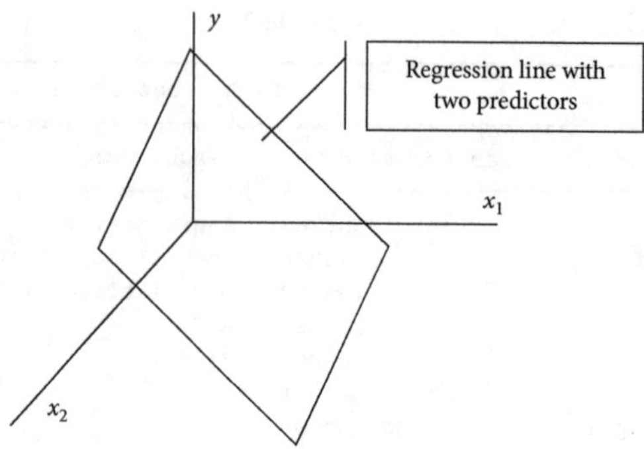

FIGURE 3.9   A regression line in three-dimensional space.

The model-building process and interpretation of the R-squared value remain the same. By now, it is clear that in a multiple regression line, we use numerous variables for predicting any target variable. Using a multivariable regression line, we can expect a higher value of R-squared when compared to a simple regression model that involves a single variable, provided that the variables have some predictive power or influence on our target variable. Next comes how to build a multiple regression line. Following is the code for creating a multiple regression model.

```
# Multiple regression
import statsmodels.formula.api as sm
model3 = sm.ols(formula='Passengers_count ~ marketing_cost+percent_delayed_
flights+number_of_trips+customer_ratings+poor_weather_index+percent_female_
customers+Holiday_week+percent_male_customers', data=air_pass)
fitted3 = model3.fit()
fitted3.summary()
```

As you can observe, we need to mention all the predictor variables on the right-hand side. Please note, here we are not summing the columns; it is just the syntax to supply independent variables. Table 3.7 is the output of this code.

TABLE 3.7   Regression Results for Model3

```
                           OLS Regression Results
=======================================================================
Dep. Variable:        Passengers_count   R-squared:                  0.911
Model:                             OLS   Adj. R-squared:             0.908
Method:                  Least Squares   F-statistic:                325.3
Date:                 Tue, 07 Apr 2020   Prob (F-statistic):     8.93e-129
Time:                         17:50:55   Log-Likelihood:           -2323.3
No. Observations:                  263   AIC:                        4665.
Df Residuals:                      254   BIC:                        4697.
Df Model:                            8
Covariance Type:             nonrobust
=======================================================================
                              coef     std err        t     P>|t|     [0.025     0.975]
-----------------------------------------------------------------------
Intercept                4173.3041    3.71e+04    0.113     0.910   -6.88e+04   7.71e+04
marketing_cost              4.4279       0.168   26.287     0.000       4.096      4.760
percent_delayed_flights  2.187e+04    4827.398    4.530     0.000    1.24e+04     .14e+04
number_of_trips             0.3004       0.270    1.114     0.266      -0.231      0.831
customer_ratings          546.3104      53.897   10.136     0.000     440.168    652.453
poor_weather_index       -919.5035    4520.130   -0.203     0.839   -9821.210    982.203
percent_female_customers  -15.7188     371.808   -0.042     0.966    -747.937    716.499
Holiday_week             6804.5389     598.471   11.370     0.000    5625.942    983.136
percent_male_customers     -7.3113     372.653   -0.020     0.984    -741.195    726.572
=======================================================================
Omnibus:                         0.087   Durbin-Watson:              1.778
Prob(Omnibus):                   0.957   Jarque-Bera (JB):           0.082
Skew:                            0.041   Prob(JB):                   0.960
Kurtosis:                        2.969   Cond. No.               1.43e+06
=======================================================================
```

The output in Table 3.7 is formatted for printing. Figure 3.10 is the actual output screenshot.

```
                          OLS Regression Results
========================================================================
Dep. Variable:        Passengers_count   R-squared:                 0.911
Model:                             OLS   Adj. R-squared:            0.908
Method:                  Least Squares   F-statistic:               325.3
Date:                 Mon, 04 Nov 2019   Prob (F-statistic):     8.93e-129
Time:                         14:31:29   Log-Likelihood:           -2323.3
No. Observations:                  263   AIC:                       4665.
Df Residuals:                      254   BIC:                       4697.
Df Model:                            8
Covariance Type:             nonrobust
========================================================================
                           coef    std err          t      P>|t|      [0.025      0.975]
------------------------------------------------------------------------
Intercept              4173.3041   3.71e+04      0.113      0.910   -6.88e+04    7.71e+04
marketing_cost            4.4279      0.168     26.287      0.000       4.096       4.760
percent_delayed_flights  2.187e+04  4827.398      4.530      0.000    1.24e+04    3.14e+04
number_of_trips           0.3004      0.270      1.114      0.266      -0.231       0.831
customer_ratings        546.3104     53.897     10.136      0.000     440.168     652.453
poor_weather_index     -919.5035   4520.130     -0.203      0.839   -9821.210    7982.203
percent_female_customers -15.7188    371.808     -0.042      0.966    -747.937     716.499
Holiday_week           6804.5389    598.471     11.370      0.000    5625.942    7983.136
percent_male_customers   -7.3113    372.653     -0.020      0.984    -741.195     726.572
========================================================================
Omnibus:                         0.087   Durbin-Watson:             1.778
Prob(Omnibus):                   0.957   Jarque-Bera (JB):          0.082
Skew:                            0.041   Prob(JB):                  0.960
Kurtosis:                        2.969   Cond. No.              1.43e+06
========================================================================

Warnings:
[1] Standard Errors assume that the covariance matrix of the errors is correctly specified.
[2] The condition number is large, 1.43e+06. This might indicate that there are
strong multicollinearity or other numerical problems.
"""
```

FIGURE 3.10   Screenshot—regression results for model3.

Do you observe the R-squared value in this output table? As a matter of fact, in any regression output, the first measure that you should look at is the R-squared value. We can see it is now beyond 90 percent. We have added multiple predictor variables, and we expect the model to get better. We can check the sign of the coefficients and find out whether any variable has a positive or negative impact on the target. It does not necessarily mean all the predictor variables are significant in the model. We will take up judging the importance of individual variables later in this chapter.

3.5   *MULTICOLLINEARITY IN REGRESSION*

A multiple regression line is an excellent tool. It helps us in accurately predicting the target value by using multiple variables. However, we need to observe some caution while building a multiple regression model. Sometimes having too many variables in the model might be a curse instead of a blessing. There is one challenge that may be a cause of worry in multiple regression lines. Let us understand it by using this example. Here we are trying to predict customer expenses based on the income and other factors like the number of credit cards and loans. A sight in expenditures might be of immense use to the sales folks, say in a financial institution, to roll out the right financial product.

The following code will import the data and build the model for us. As usual, you will ensure the file path used here is where you have stored your data file.

```
income_expenses=pd.read_csv(r"D:\Chapter3\5. Base Datasets\customer_income_
expenses.csv")

print(income_expenses.columns)

model4=sm.ols(formula='Monthly_Expenses  ~  Monthly_Income_in_USD+Number_of_
Credit_cards+Number_of_personal_loans+Monthly_Income_in_Euro',
data=income_expenses)
fitted4 = model4.fit()
fitted4.summary()
```

The following is the code (the output is shown in Table 3.8):

TABLE 3.8   Regression Results for Model4

```
                            OLS Regression Results
===============================================================================
Dep. Variable:        Monthly_Expenses   R-squared:                      0.966
Model:                             OLS   Adj. R-squared:                 0.964
Method:                  Least Squares   F-statistic:                    483.1
Date:                 Tue, 07 Apr 2020   Prob (F-statistic):          1.31e-48
Time:                         17:58:59   Log-Likelihood:               -512.77
No. Observations:                   72   AIC:                            1036.
Df Residuals:                       67   BIC:                            1047.
Df Model:                            4
Covariance Type:             nonrobust
===============================================================================
                            coef    std err          t      P>|t|     [0.025      0.975]
-------------------------------------------------------------------------------
Intercept                -72.6691    143.534     -0.506      0.614    -359.164    213.826
Monthly_Income_in_USD      7.5244    121.538      0.062      0.951    -235.066    250.115
Number_of_Credit_cards    30.2664     53.290      0.568      0.572     -76.100    136.633
Number_of_personal_loans 149.2454    104.408      1.429      0.158     -59.155    357.645
Monthly_Income_in_Euro    -7.6337    135.041     -0.057      0.955    -277.178    261.910
===============================================================================
Omnibus:                        29.413   Durbin-Watson:                  2.599
Prob(Omnibus):                   0.000   Jarque-Bera (JB):               5.056
Skew:                            0.104   Prob(JB):                      0.0798
Kurtosis:                        1.719   Cond. No.                    4.68e+04
===============================================================================
```

```
print(income_expenses.columns)
Index(['id', 'Monthly_Income_in_USD', 'Number_of_Credit_cards',
       'Number_of_personal_loans', 'Monthly_Income_in_Euro',
       'Monthly_Expenses'],
      dtype='object')
```

This model is a standard multiple regression model with four predictor variables. The model gives us 96 percent as the value for R-squared. Everything may look fine until here. Let us look at the individual coefficients and their signs in particular. A negative coefficient would mean a negative impact of that variable on the target.

- "`Monthly_Income_in_USD`" has a positive impact on overall expenses. It means a higher value of this variable will ensure a higher value of the target variable (monthly expenses).
- "`Number_of_Credit_cards`" also has a positive impact on overall expenses.
- "`Number_of_personal_loans`" also has a positive impact on overall expenses.
- Finally, we observe "`Monthly_Income_in_Euro`" has a **negative impact** on overall expenses, which would mean the higher this variable, the lower will be our target variable.

The last point appears to go against our intuition. The monthly expenses should be directly proportional to monthly income, irrespective of it being measured in dollars or euros. After all, the monthly income in euros is approximately just 0.9 times the monthly income in dollars. For this example, we have considered one dollar at 0.9 euros. How can the target variable show a positive relation with `Monthly_Income_in_USD` and negative relation with `Monthly_Income_in_Euro`? Is there something wrong here? Let us get into more details. In the next example, we will remove the dollar variable from the model and rebuild the whole model with just three variables. What follows is the code (the output is shown in Table 3.9):

```
model5=sm.ols(formula='Monthly_Expenses   ~Number_of_Credit_cards+Number_of_
personal_loans+Monthly_Income_in_Euro', data=income_expenses)
fitted5 = model5.fit()
fitted5.summary()
```

TABLE 3.9 Regression Results for Model5

```
                            OLS Regression Results
===============================================================================
Dep. Variable:      Monthly_Expenses   R-squared:                      0.966
Model:                           OLS   Adj. R-squared:                 0.965
Method:                Least Squares   F-statistic:                    653.7
Date:               Tue, 07 Apr 2020   Prob (F-statistic):          4.71e-50
Time:                       18:26:47   Log-Likelihood:               -512.77
No. Observations:                 72   AIC:                            1034.
Df Residuals:                     68   BIC:                            1043.
Df Model:                          3
Covariance Type:           nonrobust
===============================================================================
                            coef    std err          t      P>|t|      [0.025      0.975]
-------------------------------------------------------------------------------
Intercept                -67.9274    120.499     -0.564      0.575    -308.379     172.524
Number_of_Credit_cards    30.1840     52.882      0.571      0.570     -75.339     135.707
Number_of_personal_loans 149.1943    103.638      1.440      0.155     -57.611     356.000
Monthly_Income_in_Euro     0.7267      0.017     43.434      0.000       0.693       0.760
===============================================================================
Omnibus:                      29.870   Durbin-Watson:                  2.600
Prob(Omnibus):                 0.000   Jarque-Bera (JB):               5.071
Skew:                          0.099   Prob(JB):                      0.0792
Kurtosis:                      1.715   Cond. No.                    1.87e+04
===============================================================================
```

Observations from this output:

- The model has the same R-squared value. So, dropping `Monthly_Income_in_USD` did not significantly impact the overall accuracy of the model.
- "`Number_of_Credit_cards`" has a positive impact on overall expenses, which means the higher this variable, the higher will be the target variable of monthly expenses.
- "`Number_of_personal_loans`" also has a positive impact on overall expenses.
- Surprisingly, "`Monthly_Income_in_Euro`" now has a **positive impact** on overall expenses.

Our last observation looks utterly contradictory to what we had observed in the previous model. We have not changed the dataset. Neither we have changed any data point anywhere. On the same dataset, how can we see contradictory trends in the same variable (income in euro)? This example is a classic illustration of the adverse effects of "multicollinearity," which is the topic of the following section.

3.5.1   What Is Multicollinearity?

In the previous section, we worked on a couple of models involving multiple variables and also made some interesting observations. Let us get into a bit more detail on what multicollinearity means to an analyst.

- In a layman's language, multicollinearity is an interdependency of predictor variables. It indicates that two or more predictor variables are related to each other.
- Multiple variables are giving the same information. It may also indicate similar information is characterized in several columns of the dataset. For example, if one column variable is x, another column can be $2x$, which does not add any value or gives any new piece of information. They cannot be treated as two different variables. As we encountered in the previous section, income in dollars and the same income in euros are not two different variables at all.
- Predictor variables in our list need to be independent if we want to interpret their impact on the target variable. If they are interdependent, then it is called multicollinearity.
- The coefficients of the regression model cannot be interpreted correctly in the presence of multicollinearity.

Table 3.10 illustrates why multicollinearity is an issue.

TABLE 3.10   Illustration of Why Multicollinearity Is an Issue

For this example, we will use this equation as our regression model.	$y = x_1 + 2x_2 - x_3$
For the time being, let us assume there is some multicollinearity, and this equation expresses the relationship between predictor variables.	$x_1 = 2x_3$
If there is any direct relationship between the two variables, we need not build our model with both these variables included, but we still decided to proceed to see the effect.	$y = x_1 + 2x_2 - x_3$
Just rewrite the regression equation as:	$y = x_1 + 2x_2 + x_3 - 2x_3$
Let us substitute x_1 in the place of $2x_3$. We know the relation between these two variables.	$y = x_1 + 2x_2 + 1x_3 - x_1$
The final equation looks like this. This equation is technically the same as our original regression equation.	$y = 2x_2 + x_3$
In this final equation, there are two challenges: 1. In the original equation, x_3 had a negative impact; now it has a positive impact. 2. x_1 had a positive impact previously, but now it is altogether missing from the analysis.	$y = 2x_2 + 1x_3$
Another way to look at it is:	$y = x_1 + 2x_2 - x_3$ $y = -x_1 + 2x_1 + 2x_2 - x_3$ $y = -x_1 + 4x_3 + 2x_2 - x_3$ $y = -x_1 + 2x_2 + 3x_3$
You will now appreciate that we are facing all these challenges due to this interdependency of the predictor variables x_1 and x_3, also known as multicollinearity.	$x_1 = 2x_3$

From the above examples, it is clear that when multicollinearity is present in the model, we cannot trust the coefficients. The following points are for your further consideration:

- Multicollinearity directly impacts the coefficients of individual predictor variables. We cannot trust these coefficients of the variables in the presence of multicollinearity.
- Multicollinearity can involve multiple variables. It need not be always one variable related to another variable. Sometimes one variable can be impacted by a collection of other variables. For example, $x_1 = 2x_3 + 3x_2$.
- Adding or removing variables based on multicollinearity does not impact R-squared. The prediction accuracy of the model will be the same. It is the individual variable interpretation that goes for a toss.
- Just observing the output and the signs of individual coefficients is not the right way to detect multicollinearity. Sometimes, the signs of the coefficients look correct; multicollinearity can still be present in the model. So we need a scientific method of detecting multicollinearity in any model.

3.5.2 Detection of Multicollinearity

To discover multicollinearity, we follow a straightforward technique. Remember, multicollinearity is related only to predictor variables. The target variable has nothing to do with finding multicollinearity. We derive a method that works for all the combinations of interdependencies. The following is how we can proceed in this direction:

- Ignore the target variable and collect all the predictor variables columns, say $x_1, x_2, x_3 \ldots x_p$.
- Take one predictor variable as the target and build a regression model by using the rest of the predictors in the list. The number of model's will be exactly equal to "p" in this case.
  - Model1: x_1 vs. $x_2, x_3, x_4 \ldots x_p$
  - Model2: x_2 vs. $x_1, x_3, x_4 \ldots x_p$
  - Model3: x_3 vs. $x_1, x_2, x_4 \ldots x_p$
  - Model$p$: x_p vs. $x_1, x_2, x_3 \ldots x_{p-1}$
- Go through each of the above p regression models. Furthermore, note down their R-squared values. If any model has an R-squared value of >80 percent, then it indicates multicollinearity.
- For example, for model1, if the R-squared value is 90 percent, then x_1 can be easily explained by the rest of the variables. That means x_1 is redundant in the presence of other variables. If the R-squared value of model1 is 20 percent, then it indicates that x_1 carries some independent information that cannot be found in other variables. This process is the story of only one variable x_1. Similarly, we need to check the R-squared values of all the p models.
- The real measure for detecting multicollinearity is the variance inflation factor known as VIF. This measure is derived from the individual model R-squared values.

- $\text{VIF} = \dfrac{1}{1 - R^2}$ or in our case $\text{VIF}_i = \dfrac{1}{1 - R_i^2}$
- If the R-squared value of these models is high, then VIF is also high.

  - If the R-squared value is 20%, then $VIF = \dfrac{1}{1 - 0.2} = \dfrac{1}{0.8} = 1.25$

  - If the R-squared value is 50%, then $VIF = \dfrac{1}{1 - 0.5} = \dfrac{1}{0.5} = 2$

  - If the R-squared value is 70%, then $VIF = \dfrac{1}{1 - 0.7} = \dfrac{1}{0.3} = 3.33$

  - If the R-squared value is 80%, then $VIF = \dfrac{1}{1 - 0.8} = \dfrac{1}{0.2} = 5$
  - If the R-squared value is 90%, then $VIF = \dfrac{1}{1 - 0.9} = \dfrac{1}{0.1} = 10$

- There are p-columns, $x_1, x_2, x_3....x_p$. There will be p regression models in total, and there will be p number of VIF values

  - Model1: x_1 vs. $x_2, x_3, x_4 ....x_p$ $VIF_1 = \dfrac{1}{1 - R_1^2}$

  - Model$p$: x_p vs. $x_1, x_2, x_3 ....x_{p-1}$ $VIF_p = \dfrac{1}{1 - R_p^2}$

- Each variable has a VIF value. If the VIF value is >5 points, it means the R-squared value of that individual variable model is >80 percent. It means that other variables can easily explain that variable, which is a clear case of multicollinearity.
- A VIF value of >5 is the general industry standard and currently followed by many companies. In some cases, we also see the VIF threshold set at 4.

3.5.3   Variance Inflation Factor Calculation

Let us again focus on our previous example to detect multicollinearity. We will now try to build a function that takes all the predictor variables and builds individual models. It involves writing a VIF calculation function. The following is the code; its explanation is given in Table 3.11.

```python
def vif_cal(x_vars):
    xvar_names=x_vars.columns
    for i in range(0,xvar_names.shape[0]):
        y=x_vars[xvar_names[i]]
        x=x_vars[xvar_names.drop(xvar_names[i])]
        rsq=sm.ols(formula="y~x", data=x_vars).fit().rsquared
        vif=round(1/(1-rsq),2)
        print (xvar_names[i], " VIF = " , vif)
```

TABLE 3.11   Explanation of Code

`def vif_cal(x_vars):`	Declare the new function name as `vif_cal`. This function accepts the `x_vars` parameter, and we need to pass the data frame containing all the predictor variables.
`xvar_names=x_vars.columns`	Store all column names in `xvar_names`.
`for i in range(0,xvar_names.shape[0]):`	It builds models for all the variables; the loop starts with 0 and takes all the variables iteratively.
`y=x_vars[xvar_names[i]]` `x=x_vars[xvar_names.drop(xvar_names[i])]`	Take ith variable as target y and the remaining variables as x. For example, if $i = 1$, then y will be x_1, and the remaining variables will be stored in x.
`rsq=sm.ols(formula="y~x", data=x_vars).fit().rsquared`	Build a model y vs. x, and store the R-squared value in `rsq`.
`vif=round(1/(1-rsq),2)`	Calculate the VIF value for the variable.
`print (xvar_names[i], " VIF = " , vif)`	Finally, print the VIF value.

Let us use the above VIF function to calculate VIF values for all the predictor variables. Remember, we need to pass only the predictor variables to the VIF function and drop the target variable before passing the data input.

The following code calculates the actual VIF values. `Monthly_Expenses` is our target variable, and we knowingly drop this variable. The variable "id" can also be dropped, but even otherwise, it is not harmful.

```python
vif_cal(x_vars=income_expenses.drop(["Monthly_Expenses"], axis=1))
```

The above code returns the below output:

```
id  VIF =  1.07
Monthly_Income_in_USD  VIF =  65007299.17
Number_of_Credit_cards  VIF =  15.94
Number_of_personal_loans  VIF =  16.15
Monthly_Income_in_Euro  VIF =  65007347.03
```

Given below is the interpretation of this output:

- This output shows high VIF values for all the variables. All the variables stand at >5 VIF. Shall we drop all the four variables from the model? If we do, how can we build a model?
- This output shows high VIF values for `Monthly_Income_in_USD` and `Monthly_Income_in_Euro`. It does **NOT** mean that we drop both of these variables. In the presence of `Monthly_Income_in_USD`, the other variable `Monthly_Income_in_Euro` is redundant. Similarly, in the presence of `Monthly_Income_in_Euro`, the other variable `Monthly_Income_in_USD` is redundant. We can drop any one of them. This observation is essential while dealing with multicollinearity; we should not drop all the variables with VIF >5 in one iteration. We should drop only one variable at a time, and it may auto-correct the VIF value of several other variables.
- We will start with the variable that has the highest VIF. We will drop `Monthly_Income_in_Euro`. It will leave us with three variables in the model. Then we will check multicollinearity among these three variables.

The following code is used for dropping a variable and finding multicollinearity among the rest of the variables:

```
vif_cal(x_vars=income_expenses.drop(["Monthly_Expenses","Monthly_Income_in_
Euro"], axis=1))
```
The above code returns the below output:

```
id  VIF =  1.06
Monthly_Income_in_USD  VIF =  1.01
Number_of_Credit_cards  VIF =  15.94
Number_of_personal_loans  VIF =  16.14
```

There are two significant takeaways from this output. The VIF value of `Monthly_Income_in_USD` has reduced significantly, and it is because we dropped the `Monthly_Income_in_Euro` variable. There is still multicollinearity present in the system. `Number_of_credit_cards` and `Number_of_personal_loans` both have >5 VIF values. We need to drop the highest VIF variable and recalculate the VIF values for the remaining variables.

The below code is used for dropping a variable and finding multicollinearity among the rest of the variables.

```
vif_cal(x_vars=income_expenses.drop(["Monthly_Expenses","Monthly_Income_in_
Euro","Number_of_personal_loans"], axis=1))
```

The above code returns the below output:

```
id  VIF =  1.01
Monthly_Income_in_USD  VIF =  1.0
Number_of_Credit_cards  VIF =  1.01
```

Now both the variables have VIF values <5, which indicates that both of them carry independent information. Now we can conclude we do not need all four variables for building this model; only two are sufficient. Let us develop our final model with these two independent predictor variables.

Following is the code for building our final model after removing multicollinearity.

```
model6=sm.ols(formula='Monthly_Expenses  ~  Monthly_Income_in_USD+Number_of_
Credit_cards', data=income_expenses)
fitted6 = model6.fit()
fitted6.summary()
```

The above code returns the output shown in Table 3.12.

TABLE 3.12 Regression Results for Model6

```
                              OLS Regression Results
==============================================================================
Dep. Variable:       Monthly_Expenses   R-squared:                       0.965
Model:                            OLS   Adj. R-squared:                  0.964
Method:                 Least Squares   F-statistic:                     964.5
Date:                Tue, 07 Apr 2020   Prob (F-statistic):           3.71e-51
Time:                        19:01:24   Log-Likelihood:                -513.85
No. Observations:                  72   AIC:                             1034.
Df Residuals:                      69   BIC:                             1041.
Df Model:                           2
Covariance Type:            nonrobust
==============================================================================
                          coef    std err          t      P>|t|      [0.025      0.975]
------------------------------------------------------------------------------
Intercept              -56.3653    121.148     -0.465      0.643    -298.049     185.319
Monthly_Income_in_USD    0.6522      0.015     43.134      0.000       0.622       0.682
Number_of_Credit_cards 103.7997     13.600      7.632      0.000      76.669     130.931
==============================================================================
Omnibus:                       34.625   Durbin-Watson:                   2.544
Prob(Omnibus):                  0.000   Jarque-Bera (JB):                5.263
Skew:                           0.067   Prob(JB):                       0.0720
Kurtosis:                       1.682   Cond. No.                     2.06e+04
==============================================================================
```

We can observe there is no significant change in overall R-squared. So far, we have dropped only the redundant information so that the model will not get impacted adversely. While building multiple regression models, it is important to check for multicollinearity. In earlier sections, we had built one multiple regression model with air passenger data, model3, to be precise. Let us have another look at its output (Fig. 3.11).

```
                           OLS Regression Results
==============================================================================
Dep. Variable:        Passengers_count   R-squared:                       0.911
Model:                             OLS   Adj. R-squared:                  0.908
Method:                  Least Squares   F-statistic:                     325.3
Date:                 Mon, 04 Nov 2019   Prob (F-statistic):          8.93e-129
Time:                         14:31:29   Log-Likelihood:                 -2323.3
No. Observations:                  263   AIC:                             4665.
Df Residuals:                      254   BIC:                             4697.
Df Model:                            8
Covariance Type:             nonrobust
==============================================================================
                             coef     std err          t      P>|t|      [0.025      0.975]
------------------------------------------------------------------------------
Intercept              4173.3041    3.71e+04      0.113      0.910   -6.88e+04    7.71e+04
marketing_cost            4.4279       0.168     26.287      0.000       4.096       4.760
percent_delayed_flights 2.187e+04    4827.398      4.530      0.000    1.24e+04    3.14e+04
number_of_trips           0.3004       0.270      1.114      0.266      -0.231       0.831
customer_ratings        546.3104      53.897     10.136      0.000     440.168     652.453
poor_weather_index     -919.5035    4520.130     -0.203      0.839   -9821.210    7982.203
percent_female_customers -15.7188     371.808     -0.042      0.966    -747.937     716.499
Holiday_week           6804.5389     598.471     11.370      0.000    5625.942    7983.136
percent_male_customers    -7.3113     372.653     -0.020      0.984    -741.195     726.572
==============================================================================
Omnibus:                         0.087   Durbin-Watson:                   1.778
Prob(Omnibus):                   0.957   Jarque-Bera (JB):                0.082
Skew:                            0.041   Prob(JB):                        0.960
Kurtosis:                        2.969   Cond. No.                     1.43e+06
==============================================================================

Warnings:
[1] Standard Errors assume that the covariance matrix of the errors is correctly specified.
[2] The condition number is large, 1.43e+06. This might indicate that there are
strong multicollinearity or other numerical problems.
"""
```

FIGURE 3.11   Screenshot—regression results for model3.

The following code will find VIF values of model3 predictor variables:

```python
vif_cal(x_vars=air_pass.drop(["Passengers_count","passengers_count_pred"],
axis=1))
```

This code takes all the predictor variables as x_vars data. We are dropping the target column and also one more column that we had created during the prediction exercise—Passengers_count and passengers_count_pred are not part of our list of predictor variables. The above code gives us the below output:

```
marketing_cost  VIF =   1.51
percent_delayed_flights  VIF =   13.4
number_of_trips  VIF =   1.03
customer_ratings  VIF =   1.06
poor_weather_index  VIF =   12.81
percent_male_customers  VIF =   990.52
Holiday_week  VIF =   1.21
percent_female_customers  VIF =   989.91
```

The output shows that multicollinearity is present in this data as well. We need to drop the variables sequentially while removing the multicollinearity. For now, we will drop the variable with the highest VIF. We cannot drop all four variables in one step. The following code will remove the variable `percent_male_customers` and find the VIF values for the rest.

```
vif_cal(x_vars=air_pass.drop(["Passengers_count","passengers_count_pred",
"percent_male_customers"], axis=1))
```

Given below is the output table:

```
marketing_cost  VIF =  1.51
percent_delayed_flights  VIF =  13.34
number_of_trips  VIF =  1.03
customer_ratings  VIF =  1.06
poor_weather_index  VIF =  12.78
Holiday_week  VIF =  1.2
percent_female_customers  VIF =  1.03
```

Observing the VIF values, we need to drop one more variable from this list. `percent_delayed_flights` has the highest VIF on this list. Following is the code.

```
vif_cal(x_vars=air_pass.drop(["Passengers_count","passengers_count_
pred","percent_male_customers", "percent_delayed_flights"], axis=1))
```

The output table is as follows:

```
marketing_cost  VIF =  1.45
number_of_trips  VIF =  1.02
customer_ratings  VIF =  1.06
poor_weather_index  VIF =  1.25
Holiday_week  VIF =  1.2
percent_female_customers  VIF =  1.01
```

A quick look at these VIF values will tell us that this list of predictor variables has no multicollinearity. We can rebuild the model with this final predictor variables list. Our original model with this data had eight variables, while the final model has only six left after removing the variables based on VIF.

```
import statsmodels.formula.api as sm
model7 = sm.ols(formula='Passengers_count ~ marketing_cost+number_of_trips+
customer_ratings+poor_weather_index+percent_female_customers+Holiday_week',
data=air_pass)
fitted7 = model7.fit()
fitted7.summary()
```

Table 3.13 shows the output of our final model.

TABLE 3.13   Regression Results for Model7

```
                        OLS Regression Results
===============================================================================
Dep. Variable:       Passengers_count   R-squared:                      0.904
Model:                          OLS     Adj. R-squared:                 0.902
Method:               Least Squares     F-statistic:                    401.2
Date:             Thu, 16 Apr 2020      Prob (F-statistic):          4.35e-127
Time:                    16:54:57       Log-Likelihood:               -2333.5
No. Observations:             263       AIC:                            4681.
Df Residuals:                 256       BIC:                            4706.
Df Model:                       6
Covariance Type:          nonrobust
===============================================================================
                           coef    std err          t      P>|t|     [0.025      0.975]
-------------------------------------------------------------------------------
Intercept               3659.6674    984.371      3.718      0.000    1721.172    5598.163
marketing_cost             4.5785      0.171     26.797      0.000       4.242       4.915
number_of_trips            0.4177      0.278      1.503      0.134      -0.130       0.965
customer_ratings         547.0027     55.782      9.806      0.000     437.152     656.853
poor_weather_index      1.855e+04   1461.863     12.691      0.000    1.57e+04    2.14e+04
percent_female_customers -15.5571     12.302     -1.265      0.207     -39.783       8.668
Holiday_week            6802.3234    619.101     10.987      0.000    5583.144    8021.503
===============================================================================
Omnibus:                        1.354   Durbin-Watson:                  1.869
Prob(Omnibus):                  0.508   Jarque-Bera (JB):               1.134
Skew:                           0.154   Prob(JB):                       0.567
Kurtosis:                       3.094   Cond. No.                    5.53e+04
===============================================================================
```

The final model is not suffering from multicollinearity, as all the variables in the list are independent. Does it mean all of them are impactful on the target? Only being independent does not tell us anything about a variable's impact on the target. The relationship x-variable versus another x-variable relation is multicollinearity, while x-variable versus y-variable is for measuring impact. Let us explore more about the effects of the individual variables on the target in the next section.

3.6   INDIVIDUAL IMPACT OF THE VARIABLES IN REGRESSION

While building multiple regression model, we try to add as many dimensions (predictor variables) as possible to increase the model prediction accuracy. For example, we can build a model with 30 predictor variables and get a satisfying R-squared value of 92 percent. For now, let us proceed with all 30 variables. The following equation would represent the regression model:

$$y = \beta_0 + \beta_1 x_1 + \beta_2 x_2 + \cdots + \beta_{30} x_{30}$$

Assume we take care of multicollinearity and find all 30 variables are independent. Now the next question is. Are all the 30 variables impactful? There are 30 variables in the model but that does not necessarily mean all the variables have a significant impact on the target variable.

- If we drop a variable x_{10} and still get the same R-squared value of 92 percent, we understand the model has the same accuracy with or without the presence of the variable x_{10}. Do we need x_{10} in the model? We do not. x_{10} is not impactful on the target.
- In the same example, we drop another variable, x_7, and get a significant dip in the R-squared value. We see that the new R-squared value is now <87 percent. We can safely conclude that x_7 has a substantial impact on the target. We must keep this variable in the model to get a higher prediction accuracy level. We are taking up this topic in more detail in the following section.

3.6.1   *P*-value

We can measure the impact of the individual variables on the accuracy of the model by performing the t-test and its *P*-value inference. If the *P*-value for a variable is <0.05, it significantly impacts on the target. If it is equal to or >0.05, we can conclude the variable is insignificant to the model. We can easily find the *P*-value in the code output, which is available in the column P>|t|. We will get into the derivation and theory behind the *P*-value a little later. Given below is the process of applying and dropping less impactful variables using the *P*-value simultaneously.

$$y = \beta_0 + \beta_1 x_1 + \beta_2 x_2 + \cdots + \beta_{30} x_{30}$$

- Look at the *P*-value for all the individual variables:
  - If the *P*-value of a variable is <0.05, the variable is considered impactful. We should keep that variable in the model.
  - If the *P*-value of a variable is ≥0.05, then the variable is NOT impactful. It is safe to drop it.
- We should drop all the nonimpactful variables before finalizing the model.

Let us go back to our air passenger model after removing the multicollinearity (model7). Let us look at its output (Table 3.14).

TABLE 3.14   Air Passenger Model Results after Removing the Multicollinearity (model7)

```
                            OLS Regression Results
=================================================================================
Dep. Variable:        Passengers_count   R-squared:                         0.904
Model:                             OLS   Adj. R-squared:                    0.902
Method:                  Least Squares   F-statistic:                       401.2
Date:                 Thu, 16 Apr 2020   Prob (F-statistic):             4.35e-127
Time:                         16:54:57   Log-Likelihood:                  -2333.5
No. Observations:                  263   AIC:                               4681.
Df Residuals:                      256   BIC:                               4706.
Df Model:                            6
Covariance Type:             nonrobust
=================================================================================
                            coef    std err          t      P>|t|      [0.025      0.975]
---------------------------------------------------------------------------------
Intercept                3659.6674   984.371      3.718      0.000    1721.172    5598.163
marketing_cost              4.5785     0.171     26.797      0.000       4.242       4.915
number_of_trips             0.4177     0.278      1.503      0.134      -0.130       0.965
customer_ratings          547.0027    55.782      9.806      0.000     437.152     656.853
poor_weather_index       1.855e+04  1461.863     12.691      0.000    1.57e+04    2.14e+04
percent_female_customers  -15.5571    12.302     -1.265      0.207     -39.783       8.668
Holiday_week             6802.3234   619.101     10.987      0.000    5583.144    8021.503
=================================================================================
Omnibus:                         1.354   Durbin-Watson:                     1.869
Prob(Omnibus):                   0.508   Jarque-Bera (JB):                  1.134
Skew:                            0.154   Prob(JB):                          0.567
Kurtosis:                        3.094   Cond. No.                       5.53e+04
=================================================================================
```

We can check the individual variable's *P*-value under the heading P >|t|. Two variables have >0.05 *P*-value. For others, the *P*-values are rounded off to 0.000, which may not be exactly zero, but small enough to be rounded off to zero.

To be more precise, `number_of_trips` has a *P*-value of 0.134 and `percent_female_customers` has it as 0.207. We can drop these variables from the model. Following is the code for rebuilding the model after removing the two insignificant variables.

```
import statsmodels.formula.api as sm
model8 = sm.ols(formula='Passengers_count ~ marketing_cost+customer_ratings+
poor_weather_index+Holiday_week', data=air_pass)
fitted8 = model8.fit()
fitted8.summary()
```

Now consider the output in Table 3.15. The output in this table has been formatted; Fig. 3.12 is the actual screenshot.

TABLE 3.15   Unformatted Regression Results after Removing the Two Insignificant Variables (Model8)

```
                            OLS Regression Results
===============================================================================
Dep. Variable:      Passengers_count   R-squared:                        0.903
Model:                           OLS   Adj. R-squared:                   0.901
Method:                Least Squares   F-statistic:                      597.2
Date:               Thu, 16 Apr 2020   Prob (F-statistic):            4.33e-129
Time:                       17:18:48   Log-Likelihood:                 -2335.4
No. Observations:                263   AIC:                              4681.
Df Residuals:                    258   BIC:                              4699.
Df Model:                          4
Covariance Type:           nonrobust
===============================================================================
                     coef    std err          t      P>|t|      [0.025      0.975]
-------------------------------------------------------------------------------
Intercept        3366.1891    664.355      5.067      0.000    2057.941    4674.438
marketing_cost      4.6010      0.170     27.002      0.000       4.265       4.936
customer_ratings  539.0514     55.771      9.665      0.000     429.228     648.875
poor_weather_index 1.852e+04  1465.404     12.638      0.000    1.56e+04    2.14e+04
Holiday_week     6790.5039    618.849     10.973      0.000    5571.866    8009.142
===============================================================================
Omnibus:                       1.156   Durbin-Watson:                    1.894
Prob(Omnibus):                 0.561   Jarque-Bera (JB):                 0.970
Skew:                          0.145   Prob(JB):                         0.616
Kurtosis:                      3.070   Cond. No.                      5.16e+04
===============================================================================
```

```
                        OLS Regression Results
================================================================================
Dep. Variable:       Passengers_count   R-squared:                       0.903
Model:                            OLS   Adj. R-squared:                  0.901
Method:                 Least Squares   F-statistic:                     597.2
Date:                Tue, 05 Nov 2019   Prob (F-statistic):          4.33e-129
Time:                        19:42:38   Log-Likelihood:                 -2335.4
No. Observations:                 263   AIC:                             4681.
Df Residuals:                     258   BIC:                             4699.
Df Model:                           4
Covariance Type:            nonrobust
================================================================================
                        coef     std err          t      P>|t|      [0.025      0.975]
--------------------------------------------------------------------------------
Intercept           3366.1891     664.355      5.067      0.000    2057.941    4674.438
marketing_cost         4.6010       0.170     27.002      0.000       4.265       4.936
customer_ratings     539.0514      55.771      9.665      0.000     429.228     648.875
poor_weather_index  1.852e+04    1465.404     12.638      0.000    1.56e+04    2.14e+04
Holiday_week        6790.5039     618.849     10.973      0.000    5571.866    8009.142
================================================================================
Omnibus:                        1.156   Durbin-Watson:                   1.894
Prob(Omnibus):                  0.561   Jarque-Bera (JB):                0.970
Skew:                           0.145   Prob(JB):                        0.616
Kurtosis:                       3.070   Cond. No.                     5.16e+04
================================================================================

Warnings:
[1] Standard Errors assume that the covariance matrix of the errors is correctly specified.
[2] The condition number is large, 5.16e+04. This might indicate that there are
strong multicollinearity or other numerical problems.
"""
```

FIGURE 3.12   Screenshot—regression results after removing the two insignificant variables.

We can see from the output that R-squared has not changed significantly. It is because we have dropped only the non impactful variables. Let us check what will happen if we drop an impactful variable.

The below code is for dropping `marketing_cost` from the model:

```python
import statsmodels.formula.api as sm
model9 = sm.ols(formula='Passengers_count ~ customer_ratings+poor_weather_index+
Holiday_week', data=air_pass)
fitted9 = model9.fit()
fitted9.summary()
```

Figure 3.13 is the output table.

```
                          OLS Regression Results
===============================================================================
Dep. Variable:        Passengers_count   R-squared:                      0.627
Model:                             OLS   Adj. R-squared:                 0.623
Method:                  Least Squares   F-statistic:                    145.2
Date:                 Tue, 05 Nov 2019   Prob (F-statistic):          3.44e-55
Time:                        19:55:03    Log-Likelihood:               -2511.8
No. Observations:                 263    AIC:                            5032.
Df Residuals:                     259    BIC:                            5046.
Df Model:                           3
Covariance Type:              nonrobust
===============================================================================
                       coef     std err          t      P>|t|      [0.025      0.975]
-------------------------------------------------------------------------------
Intercept            1.45e+04  1016.979      14.257      0.000    1.25e+04    1.65e+04
customer_ratings     834.0677   106.768       7.812      0.000     623.823    1044.312
poor_weather_index   3.379e+04 2639.293      12.801      0.000    2.86e+04     3.9e+04
Holiday_week         1.223e+04 1142.371      10.705      0.000    9979.306    1.45e+04
===============================================================================
Omnibus:                        5.814   Durbin-Watson:                  1.887
Prob(Omnibus):                  0.055   Jarque-Bera (JB):               5.509
Skew:                           0.330   Prob(JB):                      0.0636
Kurtosis:                       3.257   Cond. No.                        95.3
===============================================================================

Warnings:
[1] Standard Errors assume that the covariance matrix of the errors is correctly specified
"""
```

FIGURE 3.13 Regression results to demonstrate what happens if we drop an impactful variable screenshot.

Here we dropped an impactful variable, and as expected, the output shows a significant drop in R-squared value. The value of R-squared has dropped from 90 to 62 percent. So, we must keep this variable in the model. The practice of dropping variables below the P-value of 0.05 is widely followed in the industry.

By this time, it should be clear that the impacts of variable and variable independence are two different things. In this example, we started with six independent variables and found only four variables are impactful.

3.6.2   Theory Behind *P*-value

We derive P-values using the t-test, which is performed on all the variables. To follow this theory, we need some basic understanding of testing an hypothesis. Consider the following equations:

- Null hypothesis H_0: A variable x_p has no significance. $\beta_p = 0$
- Alternative hypothesis H_1: A variable x_p is significant. $\beta_p \neq 0$

- The test statics value $t = \dfrac{\beta_p}{s(\beta_p)}$

- Reject H_0 if $t > t\left(\dfrac{\alpha}{2}; n - k - 1\right)$ or $t < -t\left(\dfrac{\alpha}{2}; n - k - 1\right)$ and α is set at 5 percent

At the end of this test, if the P-value is <0.05, we reject the null hypothesis itself. It means the variable x_p has a significant impact.

3.7   STEPS NEEDED IN BUILDING A REGRESSION MODEL

So far, we have seen several metrics in the process of building a useful multiple regression model. Following is the summary of the steps:

1. Firstly, we need to explore and clean the data. Perform the missing values and outlier treatment.
2. Start with all the predictor variables and build our first model by using all of them.
3. Look at the R-squared value of the model. The general industry benchmark for R-squared is ≥80 percent for a good model.
   - If the value of R-squared is very less, then go back to the data and gather more useful predictor variables.
   - Once we have the satisfactory R-squared value, we can get into the variable selection and elimination steps.
4. Check for multicollinearity using VIF.
   - If VIF of a variable is <5, that variable carries independent information, so keep it in the model.
   - If VIF of a variable is ≥5, we can safely drop it. Drop the variables sequentially. Do not drop all the variables with >5 VIF in one step.
5. Check for the individual impact using P-value.
   - If the P-value of a variable is <0.05, that variable is impactful. We should keep it in the model.
   - On the other hand, if the P-value of a variable is ≥0.05, that variable is not impactful. We can drop it.

We can follow these steps while building a standard regression model for solving any business problem. In some cases, we may be interested in just in the overall model outcome, i.e., only the target prediction and the impact of individual variables is not required. In such cases, we can stop just after R-squared analysis—the second step. The last two stages are for feature selection and elimination.

The regression analysis is our first machine learning model. Machine learning is a machine discovering the underlying patterns in data and capturing them in the form of an equation. A simple linear regression model may not solve all the business problems in the world. We need to learn many more models, which we deal with in the later sections of this book.

3.8   LOGISTIC REGRESSION MODEL

We have discussed linear regression in detail until now. Let us look at the following example where we are trying to predict whether a customer will buy a product or not based on her income. We start with a straightforward example where income is the predictor and buying is the target. The act of buying here is represented by the variable bought, which can take only two values—Yes (1) or No (0).

Given below is the code for importing the data and building the model.

```
product_sales=pd.read_csv(r"D:\Chapter3\5. Base Datasets\Product_sales.csv")
print(product_sales.columns)
```

Below is the output from the above code

```
Index(['Income', 'Bought'], dtype='object')
```

```
import statsmodels.formula.api as sm
model10 = sm.ols(formula='Bought ~  Income', data=product_sales)
fitted10 = model10.fit()
fitted10.summary()
```

Table 3.16 is the usual output.

TABLE 3.16 Regression Results for Model10

```
                           OLS Regression Results
==============================================================================
Dep. Variable:                 Bought   R-squared:                       0.843
Model:                            OLS   Adj. R-squared:                  0.842
Method:                 Least Squares   F-statistic:                     2489.
Date:                Wed, 08 Apr 2020   Prob (F-statistic):           8.50e-189
Time:                        10:12:13   Log-Likelihood:                 96.245
No. Observations:                 467   AIC:                            -188.5
Df Residuals:                     465   BIC:                            -180.2
Df Model:                           1
Covariance Type:            nonrobust
==============================================================================
                 coef    std err          t      P>|t|      [0.025      0.975]
------------------------------------------------------------------------------
Intercept     -0.1805      0.015    -11.712      0.000      -0.211      -0.150
Income      2.095e-05     4.2e-07     49.886      0.000    2.01e-05    2.18e-05
==============================================================================
Omnibus:                       77.189   Durbin-Watson:                   1.886
Prob(Omnibus):                  0.000   Jarque-Bera (JB):             1010.549
Skew:                           0.076   Prob(JB):                    3.65e-220
Kurtosis:                      10.205   Cond. No.                     6.20e+04
==============================================================================
```

Apparently, there is no major issue in the model. For now, we will go ahead with the predictions. This model accepts income as input and predicts whether a customer will make a purchase or not. The following code fetches the predicted values from this model.

```python
new_data=pd.DataFrame({"Income":[4000]})
print(fitted10.predict(new_data))

new_data1=pd.DataFrame({"Income":[85000]})
print(fitted10.predict(new_data1))
```

Consider the following output from this code:

```python
new_data=pd.DataFrame({"Income":[4000]})
print(fitted10.predict(new_data))
0   -0.096753
dtype: float64

new_data1=pd.DataFrame({"Income":[85000]})
print(fitted10.predict(new_data1))
0    1.599893
dtype: float64
```

This output indicated that when income is 4000; the predicted value (of making a buy) is –0.096753; when income is 85,000, then the predicted value is 1.599893.

- Is something wrong with these predictions? Alternatively, is there anything wrong with this model itself?
- The target variable bought takes two values 0 and 1 (called class-0 and class-1). Here class-0 means no buy, and class-1 means buy. It does not take any other value outside these two.
- In our model, the predicted bought value for an income of 4000 is –0.096753. Does a negative value mean anything for a variable like bought? As we just discussed, there is no negative class in the output.
- Similarly, the predicted value for an income of 85,000 is 1.599893. Does it mean anything in this case? We know the predicted value can be either 1 or 0. There is no meaning if bought takes a value of 1.59. We cannot infer if the client made a buy or not.

Let us have a look at some sample data points and draw a scatter plot between predictor and target variables to get a better understanding of the data. Following is the code to display the records and draw the scatter plot.

```
#product_sales data sample
print(product_sales.sample(10))

#Drawing the Scatter Plot
import matplotlib.pyplot as plt
plt.scatter(product_sales["Income"], product_sales["Bought"])
plt.title('Income vs Bought Plot')
plt.xlabel('Income')
plt.ylabel('Bought')
plt.show()
```

The following is the output. (Figure 3.14 is the plot of the output.) Please note the random sample records might be seen differently in your output.

	Income	Bought
24	40637.1	1
86	1679.7	0
386	65441.4	1
54	54463.9	1
219	1792.1	0
371	16418.5	0
308	63374.7	1
213	4527.4	0
378	57542.3	1
45	63725.1	1

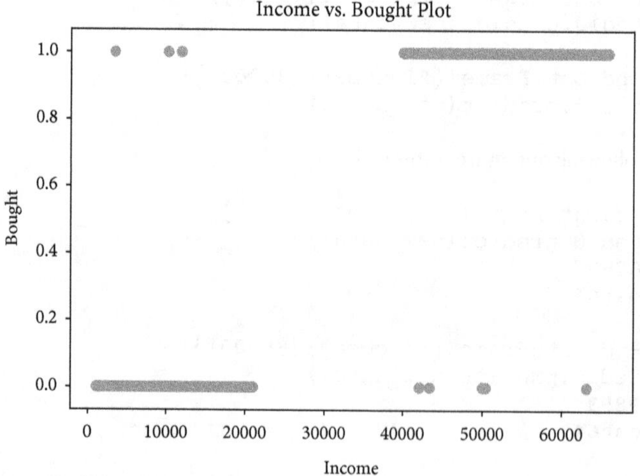

FIGURE 3.14   Income versus bought plot.

From the above output table, it is clear the data is not in the format that we need to draw a linear regression line. The data is concentrated only at two places, class-0 and class-1. For this data, can we fit a linear regression line? A simple straight line will not be a good representation of this data—there is no way a straight line can go through all the data points. Let us try to draw the regression line on top of this data to explore further. The following is the code:

```
pred_values= fitted10.predict(product_sales["Income"])
plt.scatter(product_sales["Income"], product_sales["Bought"])
plt.plot(product_sales["Income"], pred_values, color='green')
plt.title('Income vs Bought Plot')
plt.xlabel('Income')
plt.ylabel('Bought')
plt.show()
```

Figure 3.15 is the output plot for the regression line.

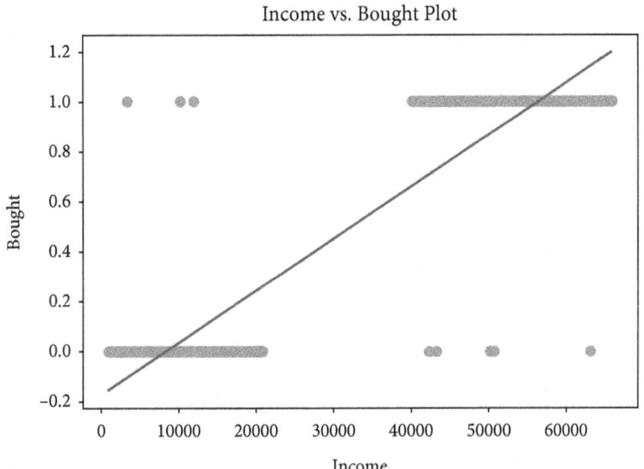

FIGURE 3.15   Income versus bought plot including the regression line.

Can you see the regression line? It is not going through all the points in this data. We can safely conclude it is not possible to build a linear regression model for this data. A linear regression line is not suitable for this kind of discrete classification. The target variable here takes only limited values. In our example, we are just trying to predict a buy or a no-buy. Let us take some cases for the sake of better clarity:

- In credit card and loan models, we try to predict if a customer will turn out to be a loan defaulter or not.
- In some marketing models, we need the customer response to our communication. In such a case, we want to predict whether a customer will respond or not.
- In online advertising models involving web links, we need to know if the viewer will click the link or not.
- In sales models, we need to predict a buy versus no buy.
- In spam email detection, we need to predict spam versus not spam.
- In some corporate HR models involving employee retention, we need to predict attrition versus no attrition.

All these are examples of the binary prediction Yes or No. Computers identify it as 1 or 0. In fact, in many of the business cases, we find this kind of classification task. For example, in some bank fraud detection models, the mission is to predict fraud versus no fraud. The dollar amount of fraud in the transaction may be secondary. A fair majority of problem statements in the real world are related to classification. Simple linear regression works only for continuous output or point predictions. It cannot be applied for solving classification problems. Working with the models that can successfully tackle these classification tasks is the topic of the following sections.

3.8.1   Logistic Function

In the previous section, we saw why we could not use a linear function or linear regression while working with classification problems. We need a different function that can correctly represent our binary classification data. Let us have a look at some nonlinear functions (Fig. 3.16) and check if they are suitable for our data.

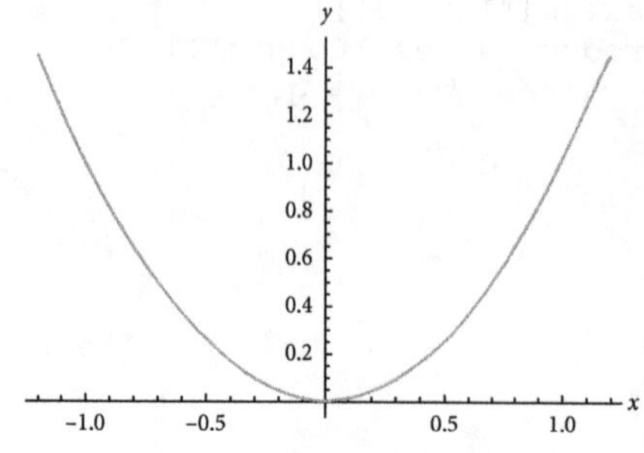

$y = x^2$ *Quadratic Function*

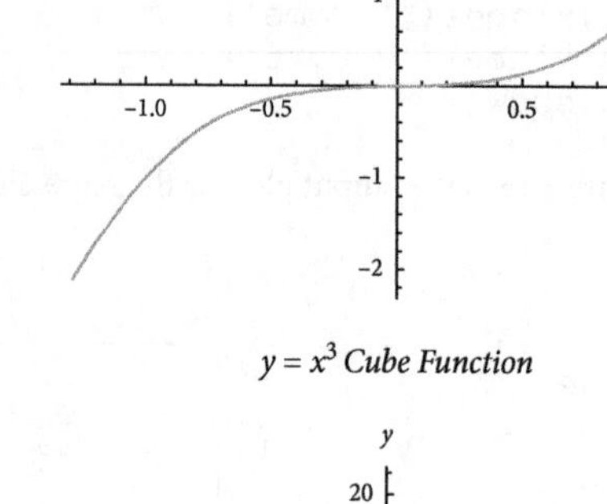

$y = x^3$ *Cube Function*

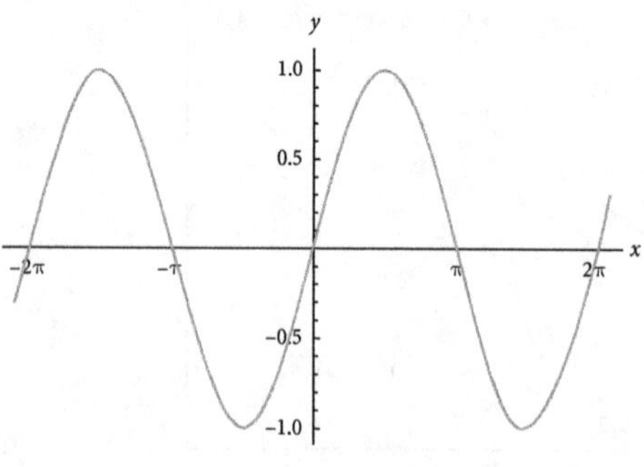

$y = \mathrm{Sin}\,(x)$ *Sine Function*

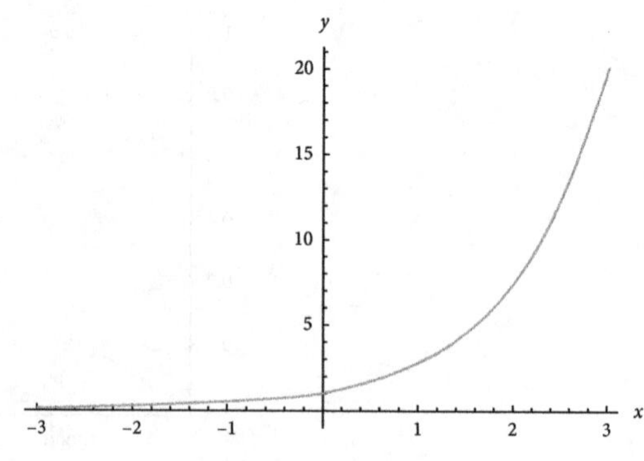

$y = e^x$ *Exponential Function*

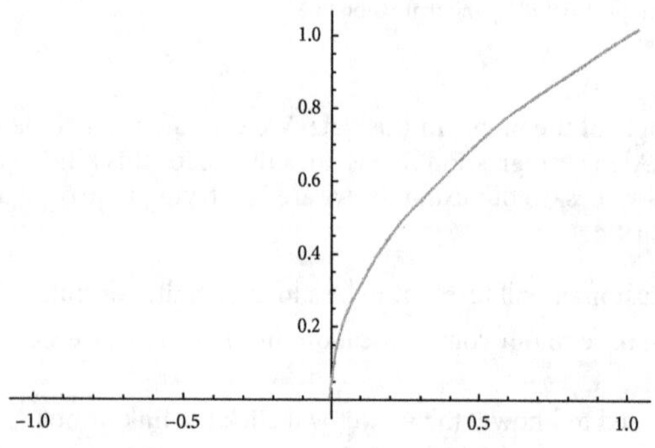

$y = \sqrt{x}$ *Square Root Function*

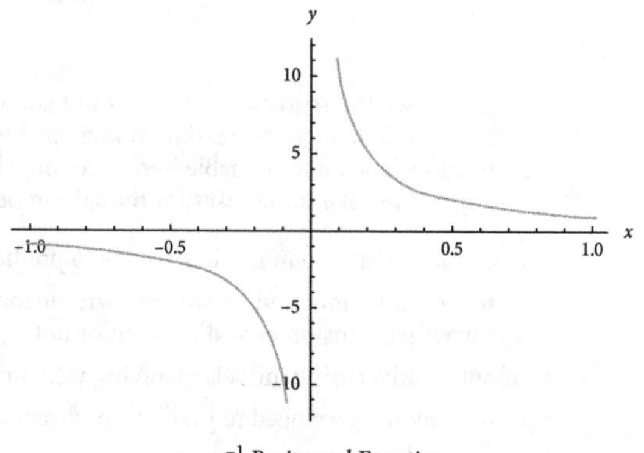

$y = x^{-1}$ *Reciprocal Function*

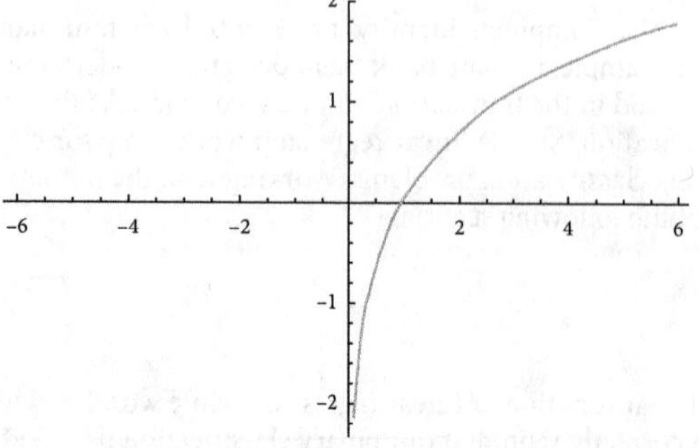

$y = \log(x)$ *Logarithmic Function*

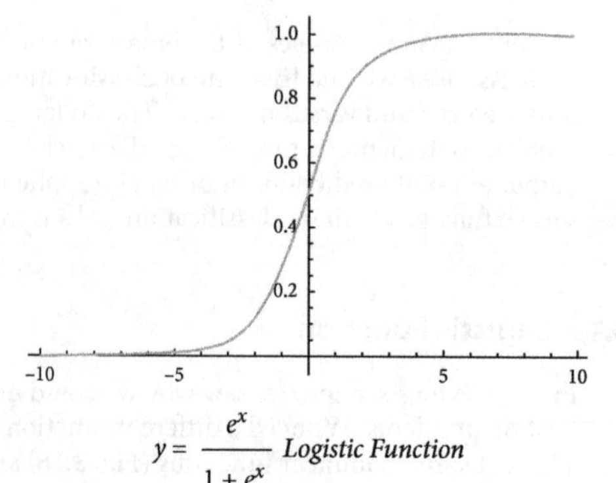

$y = \dfrac{e^x}{1 + e^x}$ *Logistic Function*

FIGURE 3.16   Some example nonlinear function plots.

The available functions are endless. The function that suits best to our dataset is the logistic function. The maximum value of this function is 1, and the minimum is 0. It has longer tails at 0 and 1. So we can use the logistic function to represent our data. By this time, you know a linear function results in a straight line, while the graphical representation of a logistic function gives an "s"-shaped curve.

3.9   LOGISTIC REGRESSION MODEL BUILDING

For our binary target data, we will be building a logistic regression line. Given below is its equation that we will be using. Just like a linear regression line, a logistic regression line also has β_0 and β_1 coefficients that we will need to compute. The equation of logistic regression looks like:

$$y = \frac{e^{\beta_0 + \beta_1 x}}{1 + e^{\beta_0 + \beta_1 x}}$$

```
import statsmodels.api as sm
logit_model=sm.Logit(product_sales["Bought"], product_sales["Income"])
#Model with intercept
logit_model1=sm.Logit(product_sales["Bought"], sm.add_constant(product_sales
["Income"]))
logit_fit1=logit_model1.fit()
logit_fit1.summary()
```

Given below is the code for building a logistic regression line. This code is briefly explained in Table 3.17.

TABLE 3.17   Code for Building a Logistic Regression Line

```
import statsmodels.api as sm
```
• Import the required API.
```
logit_model=sm.Logit(product_sales["Bought"],product_sales["Income"])
```
• It is the model configuration step. We need to mention the target variable in the beginning, followed by predictor variables.
  Logit is the function name for building a logistic regression line.
```
logit_model1=sm.Logit(product_sales["Bought"],sm.add_constant(product_sales["Income"]))
```
• This step is an alternative to the step explained above, which doesn't give us the intercept value.
• The intercept is also known as a constant. We need to use the function `sm.add_constant()` to get the value of the intercept.
```
logit_fit1=logit_model1.fit()
```
• In the previous step, we explained the model configuration, while this step is to build the model.
• In this step, the actual data is submitted and optimization is performed. This step will build the model.
• β_0 and β_1 values are found in this step. The model can be said to be ready in this step.
```
logit_fit1.summary()
```
• It displays the model summary. The model summary contains logistic regression coefficients β_0 and β_1.

After execution, this code will result in the output given in Table 3.18.

TABLE 3.18   Logistic Regression Results

```
                            Logit Regression Results
==============================================================================
Dep. Variable:                 Bought   No. Observations:                  467
Model:                          Logit   Df Residuals:                      465
Method:                           MLE   Df Model:                            1
Date:                Wed, 08 Apr 2020   Pseudo R-squ.:                  0.8525
Time:                        10:21:13   Log-Likelihood:                -47.244
converged:                       True   LL-Null:                       -320.21
Covariance Type:            nonrobust   LLR p-value:                 9.637e-121
==============================================================================
                 coef    std err          z      P>|z|      [0.025      0.975]
------------------------------------------------------------------------------
const         -7.0288      0.739     -9.505      0.000      -8.478      -5.579
Income         0.0002   2.1e-05     10.397      0.000       0.000       0.000
==============================================================================
```

From this output, we can easily find out the values of logistic regression coefficients β_0 and β_1. We can complete the logistic regression line equation. Using the values of these coefficients, we now complete the logistic regression equation that we will use for predictions.

$$y = \frac{e^{\beta_0 + \beta_1 x}}{1 + e^{\beta_0 + \beta_1 x}}$$

$$y = \frac{e^{-7.0288 + 0.0002x}}{1 + e^{-7.0288 + 0.0002x}}$$

In the previous sections, we built a linear regression line for this data, and we tried to predict the target values corresponding to the x-values 4000 and 85,000. The results were not accurate.

Now we will use our freshly done logistic regression line and try to get the predictions for the same input values. Can we expect this line to give us meaningful and accurate predictions?

$$\text{Substitute } x = 4000;\ y = \frac{e^{-7.0288 + 0.0002 * 4000}}{1 + e^{-7.0288 + 0.0002 * 4000}} = 0.01968 \cong 0$$

$$\text{Substitute } x = 85{,}000;\ y = \frac{e^{-7.0288 + 0.0002 * 85000}}{1 + e^{-7.0288 + 0.0002 * 85000}} = 0.99995 \cong 1$$

We can also achieve it by using the `predict()` function in Python. Given below is the code:

```python
new_data=pd.DataFrame({"Constant":[1,1],"Income":[4000, 85000]})
print(logit_fit1.predict(new_data))
```

To accommodate for the intercept term, we need to mention the constant value as 1 in the code. We have added the intercept by using the function `sm.add_constant()`. We do not need to do this when we are using the `statsmodels.formula.api`. However, if we are not using `formula API(statsmodels.api)` we need to mention the intercept explicitly. We have to be aware of both the APIs.

As usual, you can get the output in the following table:

```python
print(logit_fit1.predict(new_data))
0    0.002118
1    0.999990
dtype: float64
```

The predicted value of the target variable for an input income of 4000 is 0, and for 85,000 is 1. The manual calculations by directly solving the logistic regression equation are a little different when compared to the results of Python's `predict()` function. This difference occurs because we rounded the values of β_0 and β_1 in the manual calculations. We always get accurate results when we use the `predict` function.

Following is the code used for the graphical representation of a logistic regression line. In actual practice, you don't need to draw it while solving any business problem. This code is just for the demonstration purposes.

```
#Prepare new data with all x-variables
new_data=product_sales.drop(["Bought"], axis=1)
new_data["Constant"]=1
new_data=new_data[["Constant","Income"]]
#Pass the variables to get the predicted values. Add actual values in a new
column
new_data["pred_values"]= logit_fit1.predict(new_data)
new_data["Actual"]=product_sales["Bought"]
#Sort the data and draw the graph
new_data=new_data.sort_values(["pred_values"])
plt.scatter(new_data["Income"], new_data["Actual"])
plt.plot(new_data["Income"], new_data["pred_values"], color='green')
#Add lables and title
plt.title('Predicted vs Actual Plot')
plt.xlabel('Income')
plt.ylabel('Bought')
plt.show()
```

Figure 3.17 is the resultant graph. You will now appreciate better that a logistic regression line is the right choice for classification-type business problems when compared to a linear regression line.

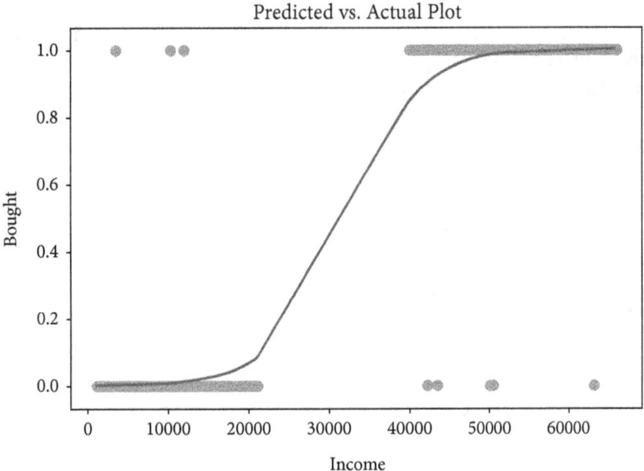

FIGURE 3.17   Resultant logistic regression plot.

3.10   *ACCURACY OF LOGISTIC REGRESSION LINE*

In the above exercise, we built a logistic regression model and also made some predictions. But what about the accuracy? Should we not get an idea of the accuracy of our model before we use the model outcome (predictions) in real-life business situations? In linear regression, the R-squared gives us an idea of the model's accuracy. The R-squared value tells us the explained variation in the target variable. A variance-based measure will not work in the case of logistic regression, as the target variable takes only two values, 0 and 1. We need a different measure of model performance. To explore more on this accuracy measure, let us get back to our dataset and print the first few records.

```
print(product_sales.head(10))
```

The following is the output:

```
      Income   Bought
0     2380.0        0
1     7351.1        0
2    48224.4        1
3     4833.0        0
4    18426.1        0
5    52709.0        1
6    54926.7        1
7    52109.3        1
8     8658.6        0
9    12227.9        0
```

We will now get the predicted values and compare them with the actual values. We will use the following familiar code for getting the predicted values from the model.

```
#Add a new column for intercept. This will be used in prediction
product_sales["Constant"]=1
#Get the predicted values into a new column
product_sales["pred_Bought"]=logit_fit1.predict(product_ sales[["Constant",
"Income"]])
product_sales["pred_Bought"]=round(product_sales["pred_Bought"])
#Data after updating with predicted values
print(product_sales[["Bought","pred_Bought"]])
```

In the above code, we have used the `predict()` function for getting our output. The values predicted using logistic regression are bound between 0 and 1. So we round off the predicted values so that the final output will be either in the class-0 or class-1. Consider the following output from this code.

```
      Bought    pred_Bought
0          0            0.0
1          0            0.0
2          1            1.0
3          0            0.0
4          0            0.0
..       ...            ...
462        0            0.0
463        0            0.0
464        1            1.0
465        1            1.0
466        1            1.0

[467 rows x 2 columns]
```

The table above shows only a part of the output. The output has two columns of actual and predicted values. In the following section, we will use this output to extract the accuracy of our model.

3.10.1 Accuracy Calculations

We will take up a simple example. For now, consider a sample table (Table 3.19) with only 10 records in the dataset.

TABLE 3.19   Sample Dataset with 10 Records

Sr Number	Actual Value	Predicted Value
1	0	0
2	1	0
3	0	0
4	1	1
5	1	1
6	1	0
7	0	0
8	1	1
9	0	0
10	0	1

There are a total of 10 records in the data. Please note that both actual and predicted values contain only 0s and 1s. We go on to create a matrix from this data (Table 3.20).

TABLE 3.20   Data Matrix of Predicted Classes versus Actual Classes

		Predicted Classes	
		0	1
Actual Classes	0	Number of times 0 is predicted as 0	Number of times 0 is predicted as 1
	1	Number of times 1 is predicted as 0	Number of times 1 is predicted as 1

In Table 3.21 let us fill these cells with the numbers based on Table 3.19.

TABLE 3.21   Actual versus Predicted Values

Sr Number	Actual Value	Predicted Value	Cell
1	0	0	00
2	1	0	10
3	0	0	00
4	1	1	11
5	1	1	11
6	1	0	10
7	0	0	00
8	1	1	11
9	0	0	00
10	0	1	01

TABLE 3.22   Predicted Classes versus Actual Class Matrix

		Predicted Classes	
		0	**1**
Actual Classes	**0**	4	1
	1	2	3

Predicting 0 as 0 and 1 as 1 is the correct classification by the model. Out of the 10 records in the data, we have rightly classified only 7 cases, which is an accuracy level of 70 percent.

The matrix discussed in Tables 3.20 and 3.22 is very well known in the classification of algorithms. It is known as a confusion matrix. We create a confusion matrix and calculate the accuracy level of our model as follows:

$$\text{Accuracy} = \frac{\text{Right classifications}}{\text{Overall records}}$$

$$\text{Accuracy} = \frac{cm[0,0] + cm[1,1]}{cm[0,0] + cm[0,1] + cm[1,0] + cm[1,1]}$$

cm is the confusion matrix in the above formula.

$$\text{Error} = 1 - \text{Accuracy} = \frac{cm[0,1] + cm[1,0]}{cm[0,0] + cm[0,1] + cm[1,0] + cm[1,1]}$$

The following code creates the confusion matrix on our data:

```
from sklearn.metrics import confusion_matrix
cm1 = confusion_matrix(product_sales["Bought"], product_sales["pred_Bought"])
print(cm1)

accuracy1=(cm1[0,0]+cm1[1,1])/(cm1[0,0]+cm1[0,1]+cm1[1,0]+cm1[1,1])
print(accuracy1)
```

The following is the output:

```
print(cm1)
[[257    5]
 [  3 202]]

print(accuracy1)
0.9828693790149893
```

The output shows that our logistic regression model has an accuracy level of 98 percent. The general industry standard for an efficient model is to have a prediction accuracy level of >80 percent. Until now, we have discussed only a simple logistic regression line with a single predictor variable. We will discuss multiple logistic regression in the following section.

3.11   MULTIPLE LOGISTIC REGRESSION LINE

By this time, you understand that for solving the classification type of problems, we need to build logistic regression models. In real-life problems, most regression models involve more than one independent variable (x) to help us in predicting the target variable (y). In other words, there can be several factors that impact the target variable. For example, while dealing with a credit risk model, where we predict if a customer is creditworthy, before offering a card or a loan. A client's creditworthiness depends upon several predictor variables like number of credit cards, previous payments, amount due, average spendings, savings bank balance, number of loans, number of transactions, and so on. In another example, whether or not you click on an advertisement is predicted based on the factors like login time, previous pages, duration on each page, number of clicks, and so on. A multiple logistic regression line will be useful in all such cases. It is just an extension of a simple logistic regression line.

$$y = \frac{e^{\beta_0 + \beta_1 x_1 + \beta_2 x_2 + \beta_3 x_3 + \cdots + \beta_k x_k}}{1 + e^{\beta_0 + \beta_1 x_1 + \beta_2 x_2 + \beta_3 x_3 + \cdots + \beta_k x_k}}$$

The target variable here is still either class-0 or class-1. The confusion matrix calculation and accuracy measure remain the same. Figure 3.18 is the visualization of a logistic regression line with two predictor variables.

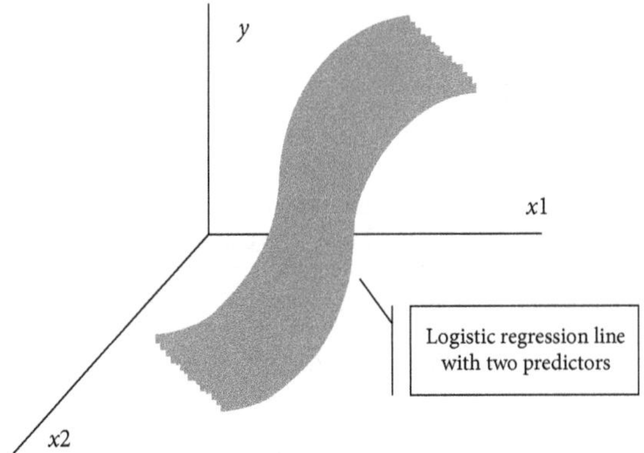

FIGURE 3.18 Logistic regression line with two predictor variables.

Again, we will start with an example. We will first go through the business problem and then try to explore the data and the details of the predictor variables. Finally, we will build multiple logistic regression model, and calculate the accuracy of our model.

Our example takes in a mobile phone network company, which wants to examine its customer attrition. In the last two years, some customers have stopped using their network, while another fraction continues to be loyal. The company has gathered several internal measures to predict attrition. We need to build a logistic regression model, which needs to predict if a customer will continue to use the company's services or leave? In effect, it predicts attrition versus no attrition.

Many companies use this kind of attrition model. Have you ever stopped using a service suddenly—for example, a food delivery app service or a taxi booking service or an e-commerce website? If yes, you might have got some surprise offers, custom-made for you to retain you. It is done using attrition models. Companies run these models to predict whether a customer will switch or stay. If they predict a customer is going to quit, they may try some retention strategies. If the predictions indicate a customer is going to stick, they may attempt some cross-selling or upselling.

Let us get back to the attrition model; we will now explore the variables in Table 3.23. The data, in this case, connects to the past two years. It has the details of 100,000 customers.

TABLE 3.23 Attrition Model Feature Description

Column Name	Description
Id	It is unique. We can use it as `cust_id`.
Active_cust	Active versus inactive customers Inactive—Quit Active
estimated_income	Estimated monthly income of the customer
months_on_network	Customer age on the network. It is the number of months they have been using the network.
complaints_count	Number of complaints in two years by a customer
plan_changes_count	Number of plan changes in two years by a customer
relocated_new_place	Indicator—relocated or not
monthly_bill_avg	Average monthly bill
CSAT_Survey_Score	Internal customer satisfaction score
high_talktime_flag	Indicator—if a customer heavily uses the talk time
internet_time	Monthly internet usage duration

We use the following code for building the multiple logistic regression line. The code is the same as the simple logistic regression line—the only difference is we need to mention all the predictor variables in this case.

```
telco_cust=pd.read_csv(r"D:\Chapter3\5. Base Datasets\telco_data.csv")
print(telco_cust.shape)
print(telco_cust.columns)
```

The above code gives us the below output:

```
print(telco_cust.shape)
(100000, 11)
```

```
print(telco_cust.columns)
Index(['Id', 'Active_cust', 'estimated_income', 'months_on_network',
       'complaints_count', 'plan_changes_count',
'relocated_new_place','monthly_bill_avg', 'CSAT_Survey_Score',
'high_talktime_flag','internet_time'], dtype='object')
```

The code below is used for building the model:

```
import statsmodels.api as sm
logit_model2=sm.Logit(telco_cust['Active_cust'],telco_cust[["estimated_
income"]+['months_on_network']+['complaints_count']+['plan_changes_
count']+['relocated_new_place']+['monthly_bill_avg']+["CSAT_Survey_
Score"]+['high_talktime_flag']+['internet_time']])
logit_fit2=logit_model2.fit()
logit_fit2.summary()
```

The above code gives us the output in Table 3.24. The output in Table 3.24 is formatted for printing. Figure 3.19 is the actual output screenshot.

TABLE 3.24   Logistic Regression Results for Model2

```
                          Logit Regression Results
==============================================================================
Dep. Variable:            Active_cust   No. Observations:            100000
Model:                          Logit   Df Residuals:                 99991
Method:                           MLE   Df Model:                         8
Date:                Wed, 08 Apr 2020   Pseudo R-squ.:               0.5193
Time:                        10:53:05   Log-Likelihood:             -32721.
converged:                       True   LL-Null:                    -68074.
Covariance Type:            nonrobust   LLR p-value:                  0.000
==============================================================================
                        coef    std err          z      P>|z|      [0.025      0.975]
------------------------------------------------------------------------------
estimated_income     5.476e-05  3.15e-05      1.740      0.082   -6.93e-06       0.000
months_on_network      -2.1605     2.473     -0.874      0.382      -7.008       2.687
complaints_count       31.7026    37.097      0.855      0.393     -41.006     104.411
plan_changes_count     -0.5828     0.011    -52.166      0.000      -0.605      -0.561
relocated_new_place    -2.4047     0.047    -51.554      0.000      -2.496      -2.313
monthly_bill_avg       -0.0035     0.000    -17.173      0.000      -0.004      -0.003
CSAT_Survey_Score       3.3119     3.710      0.893      0.372      -3.959      10.583
high_talktime_flag     -0.0354     0.020     -1.763      0.078      -0.075       0.004
internet_time           0.0079  4.68e-05    168.858      0.000       0.008       0.008
==============================================================================
```

```
                        Logit Regression Results
==============================================================================
Dep. Variable:            Active_cust   No. Observations:               100000
Model:                          Logit   Df Residuals:                    99991
Method:                           MLE   Df Model:                            8
Date:                Fri, 08 Nov 2019   Pseudo R-squ.:                  0.5193
Time:                        13:03:26   Log-Likelihood:                -32721.
converged:                       True   LL-Null:                       -68074.
                                        LLR p-value:                     0.000
==============================================================================
                        coef    std err          z      P>|z|      [0.025      0.975]
------------------------------------------------------------------------------
estimated_income     5.476e-05   3.15e-05      1.740      0.082   -6.93e-06       0.000
months_on_network      -2.1605      2.473     -0.874      0.382      -7.008       2.687
complaints_count       31.7026     37.097      0.855      0.393     -41.006     104.411
plan_changes_count     -0.5828      0.011    -52.166      0.000      -0.605      -0.561
relocated_new_place    -2.4047      0.047    -51.554      0.000      -2.496      -2.313
monthly_bill_avg       -0.0035      0.000    -17.173      0.000      -0.004      -0.003
CSAT_Survey_Score       3.3119      3.710      0.893      0.372      -3.959      10.583
high_talktime_flag     -0.0354      0.020     -1.763      0.078      -0.075       0.004
internet_time           0.0079    4.68e-05    168.858      0.000       0.008       0.008
==============================================================================
```

FIGURE 3.19   Screenshot—logistic regression results for model2.

The confusion matrix and accuracy need to be calculated separately. There are other measures like *P*-value in this output as well, which we deal with in the later sections. The following code is used for the calculation of the confusion matrix.

```
#Confuson Matrix and Accuracy
telco_cust["pred_Active_cust"]=logit_fit2.predict(telco_cust.drop
(["Id","Active_cust"],axis=1))
telco_cust["pred_Active_cust"]=round(telco_cust["pred_Active_cust"])

from sklearn.metrics import confusion_matrix
cm2 = confusion_matrix(telco_cust["Active_cust"], telco_cust["pred_Active
cust"])
print(cm2)
accuracy2=(cm2[0,0]+cm2[1,1])/(cm2[0,0]+cm2[0,1]+cm2[1,0]+cm2[1,1])
print(accuracy2)
```

The above code gives us the below output.

```
print(cm2)
[[35985  6156]
 [ 7443 50416]]

print(accuracy2)
0.86401
```

From the output, we can see that the model has a good accuracy of 86.4 percent. The next step is to select all the independent variables and keep all the impactful variables.

3.12   *MULTICOLLINEARITY IN LOGISTIC REGRESSION*

As noticed in multiple linear regression, we cannot trust the beta coefficients in the presence of variables' interdependency. The same is true in the case of multiple logistic regression. To reiterate, in logistic regression, the target variable is a class or categorical, while in the case of linear regression, y is continuous. For predictor variables, in both cases, there is no change—there can be multiple predictor variables.

While dealing with multicollinearity in logistic regression, we ignore the target variable; it is related only to predictor variables. The method for the detection of multicollinearity in logistic regression is the same as linear regression. Here also we can use the same VIF function. We look at the VIF values and drop the variables whose VIF value is >5. While dropping variables, we need to drop them serially—one by one. Not all the variables with VIF values >5 can be dropped in one step. Let us now check the multicollinearity in our attrition model data.

The following is the code for the VIF function.

```python
import statsmodels.formula.api as sm1
def vif_cal(x_vars):
    xvar_names=x_vars.columns
    for i in range(0,xvar_names.shape[0]):
        y=x_vars[xvar_names[i]]
        x=x_vars[xvar_names.drop(xvar_names[i])]
        rsq=sm1.ols(formula="y~x", data=x_vars).fit().rsquared
        vif=round(1/(1-rsq),2)
        print (xvar_names[i], " VIF = " , vif)
```

We will use the VIF function to detect the multicollinearity in our data.

```python
vif_cal(x_vars=telco_cust.drop(["Id","Active_cust","pred_Active_cust"],
axis=1))
```

The following is the code output.

```
estimated_income  VIF =  1.02
months_on_network  VIF =  20991947.38
complaints_count  VIF =  1105768.47
plan_changes_count  VIF =  1.56
relocated_new_place  VIF =  1.63
monthly_bill_avg  VIF =  1.0
CSAT_Survey_Score  VIF =  22885771.92
high_talktime_flag  VIF =  1.0
internet_time  VIF =  1.07
```

From the output, it is very clear that a few variables are interdependent. We need to drop the variable with the highest VIF value. We will drop "CSAT_Survey_Score" and recalculate the VIF values for the rest of the variables.

```python
vif_cal(x_vars=telco_cust.drop(["Id","Active_cust","pred_Active_cust",
"CSAT_Survey_Score"], axis=1))
```

The above code gives us the below output.

```
estimated_income  VIF =  1.02
months_on_network  VIF =  1.03
complaints_count  VIF =  1.02
plan_changes_count  VIF =  1.56
relocated_new_place  VIF =  1.63
monthly_bill_avg  VIF =  1.0
high_talktime_flag  VIF =  1.0
internet_time  VIF =  1.07
```

From the above output, we observe that all the variables have VIF <5. We can now conclude that all these variables are independent; we can safely use all of them in our model. Following is the model after removing the multicollinearity.

```python
import statsmodels.api as sm
logit_model3=sm.Logit(telco_cust['Active_cust'],telco_cust[["estimated_
income"]+['months_on_network']+['complaints_count']+['plan_changes_
count']+['relocated_new_place']+['monthly_bill_avg']+['high_talktime_
flag']+['internet_time']])
logit_fit3=logit_model3.fit()
logit_fit3.summary()
```

TABLE 3.25   Logistic Regression Results

```
                        Logit Regression Results
==============================================================================
Dep. Variable:           Active_cust   No. Observations:              100000
Model:                         Logit   Df Residuals:                   99992
Method:                          MLE   Df Model:                           7
Date:             Wed, 08 Apr 2020   Pseudo R-squ.:                  0.5193
Time:                       11:06:56   Log-Likelihood:                -32721.
converged:                      True   LL-Null:                       -68074.
Covariance Type:           nonrobust   LLR p-value:                    0.000
==============================================================================
                        coef    std err          z      P>|z|      [0.025      0.975]
------------------------------------------------------------------------------
estimated_income     5.457e-05   3.15e-05      1.735      0.083   -7.08e-06       0.000
months_on_network       0.0474      0.001     60.421      0.000       0.046       0.049
complaints_count       -1.4164      0.024    -59.194      0.000      -1.463      -1.370
plan_changes_count     -0.5827      0.011    -52.166      0.000      -0.605      -0.561
relocated_new_place    -2.4051      0.047    -51.561      0.000      -2.496      -2.314
monthly_bill_avg       -0.0035      0.000    -17.172      0.000      -0.004      -0.003
high_talktime_flag     -0.0354      0.020     -1.760      0.078      -0.075       0.004
internet_time           0.0079   4.68e-05    168.861      0.000       0.008       0.008
==============================================================================
```

The output in Table 3.25 is formatted for printing; Fig. 3.20 is the actual output screenshot.

```
                        Logit Regression Results
==============================================================================
Dep. Variable:           Active_cust   No. Observations:              100000
Model:                         Logit   Df Residuals:                   99992
Method:                          MLE   Df Model:                           7
Date:             Fri, 08 Nov 2019   Pseudo R-squ.:                  0.5193
Time:                       14:23:17   Log-Likelihood:                -32721.
converged:                      True   LL-Null:                       -68074.
                                       LLR p-value:                    0.000
==============================================================================
                        coef    std err          z      P>|z|      [0.025      0.975]
------------------------------------------------------------------------------
estimated_income     5.457e-05   3.15e-05      1.735      0.083   -7.08e-06       0.000
months_on_network       0.0474      0.001     60.421      0.000       0.046       0.049
complaints_count       -1.4164      0.024    -59.194      0.000      -1.463      -1.370
plan_changes_count     -0.5827      0.011    -52.166      0.000      -0.605      -0.561
relocated_new_place    -2.4051      0.047    -51.561      0.000      -2.496      -2.314
monthly_bill_avg       -0.0035      0.000    -17.172      0.000      -0.004      -0.003
high_talktime_flag     -0.0354      0.020     -1.760      0.078      -0.075       0.004
internet_time           0.0079   4.68e-05    168.861      0.000       0.008       0.008
==============================================================================
```

FIGURE 3.20   Screenshot—logistic regression results for model3.

This code will compute the confusion matrix and accuracy:

```
telco_cust["pred_Active_cust"]=logit_fit3.predict(telco_cust.drop(["Id",
"Active_cust","pred_Active_cust","CSAT_Survey_Score"],axis=1))
telco_cust["pred_Active_cust"]=round(telco_cust["pred_Active_cust"])

from sklearn.metrics import confusion_matrix
cm3 = confusion_matrix(telco_cust["Active_cust"], telco_cust["pred_Active_
cust"])
print(cm3)

accuracy3=(cm3[0,0]+cm3[1,1])/(cm3[0,0]+cm3[0,1]+cm3[1,0]+cm3[1,1])
print(accuracy3)
```

The following is the output:

```
print(cm3)
[[35983  6158]
 [ 7442 50417]]

print(accuracy3)
0.864
```

This model has an accuracy of 86.4 percent. We need to now look into the impact of the individual variables.

3.13   INDIVIDUAL IMPACT OF THE VARIABLES

Once we finish the multicollinearity analysis using VIF, we have only all the independent variables left with us. In reality, all these independent variables may not be impactful. In other words, if we have a total of 20 variables in the model, it does not necessarily mean that all of them are impactful on the target. If we drop some variables and observe the result, there is no impact on the accuracy of the model; it is safe to conclude that the dropped variables have no significant impact on the target. On the other hand, dropping some variable(s) may adversely impact the accuracy of the model; in such cases, we need to include them in our model, as they have an impact on the target variable.

In linear regression, we had used P-value, which we derived using the t-test. We use P-value in logistic regression as well, but here we calculate P-value differently. In the logistic regression, we use a Z-test in the background. This test examines the null hypothesis—not impactful versus the alternative hypothesis, which is impactful. The Z-test computes the P-value for us. If the P-value of a variable is <0.05, we reject the null hypothesis. It means we reject our initial assumption of the variable being nonimpactful. It amounts to saying that the variable under consideration is impactful.

On the other hand, if the P-value of a variable is >0.05, we accept the null hypothesis. It means we take our assumption of the variable under consideration being not impactful—we can now remove this variable.

In simple terms, in the output, if the P-value of a variable is <0.05, the variable is considered impactful. If the P-value of a variable is ≥0.05, we believe it is not impactful. We need not write any code separately for this purpose. We can look at the summary output and list the nonimpactful variables. Note that to get impactful variables, we need to look at the summary of all the variables that are retained after completing the multicollinearity analysis.

Let us look at Fig. 3.21, the output of our model. We have got it after removing all the multicollinearity.

```
                          Logit Regression Results
==============================================================================
Dep. Variable:           Active_cust   No. Observations:              100000
Model:                         Logit   Df Residuals:                   99992
Method:                          MLE   Df Model:                           7
Date:               Fri, 08 Nov 2019   Pseudo R-squ.:                 0.5193
Time:                       14:30:57   Log-Likelihood:               -32721.
converged:                      True   LL-Null:                      -68074.
                                       LLR p-value:                    0.000
==============================================================================
                         coef     std err          z      P>|z|      [0.025      0.975]
------------------------------------------------------------------------------
estimated_income     5.457e-05    3.15e-05      1.735      0.083   -7.08e-06       0.000
months_on_network       0.0474       0.001     60.421      0.000       0.046       0.049
complaints_count       -1.4164       0.024    -59.194      0.000      -1.463      -1.370
plan_changes_count     -0.5827       0.011    -52.166      0.000      -0.605      -0.561
relocated_new_place    -2.4051       0.047    -51.561      0.000      -2.496      -2.314
monthly_bill_avg       -0.0035       0.000    -17.172      0.000      -0.004      -0.003
high_talktime_flag     -0.0354       0.020     -1.760      0.078      -0.075       0.004
internet_time           0.0079    4.68e-05    168.861      0.000       0.008       0.008
==============================================================================
"""
```

FIGURE 3.21 Screenshot—logistic regression results after removing all the multicollinearity.

In this output, we see two variables, `estimated_income` and `high_talktime_flag`, have *P*-value >0.05. We can safely drop these two variables. This model has an accuracy of 86.4 percent. Can we expect the same accuracy even after dropping these two variables? Given below is the code for the final model.

```
logit_model4=sm.Logit(telco_cust['Active_cust'],telco_cust[['months_
on_network']+['complaints_count']+['plan_changes_count']+['relocated_new_
place']+['monthly_bill_avg']+['internet_time']])
logit_fit4=logit_model4.fit()
logit_fit4.summary()
```

Table 3.26 is the output. This output is formatted for printing; Fig. 3.22 is the actual output screenshot.

TABLE 3.26 Logistic Regression Results

```
                          Logit Regression Results
==============================================================================
Dep. Variable:           Active_cust   No. Observations:              100000
Model:                         Logit   Df Residuals:                   99994
Method:                          MLE   Df Model:                           5
Date:               Thu, 16 Apr 2020   Pseudo R-squ.:                 0.5189
Time:                       18:28:59   Log-Likelihood:               -32752.
converged:                      True   LL-Null:                      -68074.
Covariance Type:            nonrobust   LLR p-value:                   0.000
==============================================================================
                         coef     std err          z      P>|z|      [0.025      0.975]
------------------------------------------------------------------------------
months_on_network       0.0463       0.001     69.979      0.000       0.045       0.048
complaints_count       -1.3804       0.013   -105.503      0.000      -1.406      -1.355
plan_changes_count     -0.5807       0.011    -52.132      0.000      -0.603      -0.559
relocated_new_place    -2.3988       0.046    -51.593      0.000      -2.490      -2.308
monthly_bill_avg       -0.0034       0.000    -17.166      0.000      -0.004      -0.003
internet_time           0.0079    4.67e-05    169.255      0.000       0.008       0.008
==============================================================================
```

```
                        Logit Regression Results
==============================================================================
Dep. Variable:          Active_cust   No. Observations:            100000
Model:                        Logit   Df Residuals:                 99994
Method:                         MLE   Df Model:                         5
Date:             Fri, 08 Nov 2019   Pseudo R-squ.:               0.5189
Time:                      14:34:28   Log-Likelihood:             -32752.
converged:                     True   LL-Null:                    -68074.
                                      LLR p-value:                  0.000
==============================================================================
                       coef    std err          z      P>|z|     [0.025     0.975]
------------------------------------------------------------------------------
months_on_network    0.0463      0.001     69.979      0.000      0.045      0.048
complaints_count    -1.3804      0.013   -105.503      0.000     -1.406     -1.355
plan_changes_count  -0.5807      0.011    -52.132      0.000     -0.603     -0.559
relocated_new_place -2.3988      0.046    -51.593      0.000     -2.490     -2.308
monthly_bill_avg    -0.0034      0.000    -17.166      0.000     -0.004     -0.003
internet_time        0.0079   4.67e-05    169.255      0.000      0.008      0.008
==============================================================================
```

FIGURE 3.22   Screenshot—logistic regression results for the final model.

Given below is the code for creating the confusion matrix, and calculating the accuracy.

```python
telco_cust["pred_Active_cust"]=logit_fit4.predict(telco_cust.
drop(["Id","Active_cust","pred_Active_cust","CSAT_Survey_Score","estimated_
income","high_talktime_flag"],axis=1))
telco_cust["pred_Active_cust"]=round(telco_cust["pred_Active_cust"])

from sklearn.metrics import confusion_matrix
cm4= confusion_matrix(telco_cust["Active_cust"], telco_cust["pred_Active_
cust"])
print(cm4)

accuracy4=(cm4[0,0]+cm4[1,1])/(cm4[0,0]+cm4[0,1]+cm4[1,0]+cm4[1,1])
print(accuracy4)
```

The above code gives us the below output.

```python
print(cm4)
[[36010  6131]
 [ 7474 50385]]

print(accuracy4)
0.864
```

The final model provides us with an accuracy of 86.3 percent, which is the same as before, as we have dropped only the non impactful variables.

3.14   STEPS IN BUILDING A LOGISTIC REGRESSION MODEL

We have seen several metrics while building a logistic regression model. Following is the summary of the steps while building a reliable logistic regression model.

1. Perform data exploration, validation, and data cleaning steps. Prepare the data for analysis.
2. Check the target variable first. If it is a class variable or categorical variable, then we should only choose the logistic regression model.
3. Start with all the predictor variables and build your first model.
4. Create the confusion matrix and calculate the accuracy of your first model. The general industry benchmark is an accuracy level above 80 percent.
   - If the level of accuracy is very less, we need more data; gather more predictor variables that can, you think, affect the target.
   - Rebuild the model with a greater number of predictor variables. Once you have a satisfactory accuracy level, you can turn to the variable selection and elimination steps.
5. Check for multicollinearity using VIF.
   - If VIF of a variable is <5, that variable carries independent information; you need to keep it in the model.
   - If VIF of a variable is ≥5, we can drop it. We need to drop variables sequentially—one after the other. Do not drop all the variables with >5 VIF in one step. Drop only the variable with the highest VIF value.
   - Repeat these steps until you get all the variables with a VIF value of <5.
6. Check for the individual variable's impact using P-value.
   - If the P-value of a variable is <0.05, that variable is impactful. Keep it in the model.
   - If the P-value of a variable is ≥0.05, that variable is not impactful. We can drop it from the model.

Follow these steps while building the standard logistic regression model for solving any business problem involving the categorical target variable. If you are building a model to predict the target and you are not interested in an individual variable's impact, you can stop at step 4. The last two steps are for feature selection and elimination. In some business situations, we just want the model without variable importance, while in others, we need variable impacts as well. The steps until the confusion matrix are mandatory, while the rest of the steps are only based depending on your need.

3.15   LINEAR VS. LOGISTIC REGRESSION COMPARISON

In this chapter, we have discussed linear and logistic regressions in sufficient detail. Given in Table 3.27 are some points of comparison.

TABLE 3.27   Linear versus Logistic Regression Comparison

	Linear Regression	Logistic Regression
Equation	$y = \beta_0 + \beta_1 x$	$y = \dfrac{e^{\beta_0 + \beta_1 x}}{1 + e^{\beta_0 + \beta_1 x}}$
Shape of the line or the curve	Linear regression line.	Logistic regression line.
Data type	Best suited for regression problems where the target is a continuous variable	Best suited for classification problems where the target takes limited classes, like 0 and 1
Examples	Predicting the number of visitors, predicting the number of clicks, predicting the fraud amount, predicting the loss percentage	Predicting buy versus no buy, predicting fraud versus no fraud, predicting click versus no click, predicting attrition versus no attrition
Predictor variables	All are numerical values. They can be continuous or discrete classes.	All are numerical values. They can be continuous or discrete classes.
Predicted range	The predicted value range is the same as the target variable range. There are no hard limits.	Predicted values are bound between 0 and 1
Multivariate model equation	$y = \beta_0 + \beta_1 x_1 + \beta_2 x_2 + \cdots + \beta_k x_k$	$y = \dfrac{e^{\beta_0 + \beta_1 x_1 + \beta_2 x_2 + \beta_3 x_3 + \cdots + \beta_k x_k}}{1 + e^{\beta_0 + \beta_1 x_1 + \beta_2 x_2 + \beta_3 x_3 + \cdots + \beta_k x_k}}$
Accuracy Measure	R-squared Mean absolute percentage error (MAPE) Root mean square error (RMSE)	Confusion matrix accuracy
Multicollinearity	Yes, it exists. VIF is used for detection of multicollinearity.	Yes, it exists. VIF is used for detection of multicollinearity.
Variable impact	t-test and P-value	Z-test and P-value

3.16   CONCLUSION

We started with simple linear regression and discussed its accuracy measure. We also discussed the details of the multiple linear regression and feature selection while building multiple linear regression models. Finally, we discussed the details of the logistic regression model building and feature selection. In this chapter, we discussed only the primary measures. We will introduce some new concepts like overfitting and underfitting (of a model) in the later chapters. Linear and logistic regression methods are straightforward algorithms. A topic like artificial neural networks (ANNs) is a sophisticated algorithm. A good understanding of logistic regression is essential to understand the ANN algorithm. Good knowledge of linear regression is critical to follow logistic regression. In a nutshell, this chapter lays the foundation for the forthcoming topics in this book. Stay tuned.

3.17   PRACTICE PROBLEMS

1. Download the Pima Indians Diabetes dataset. The objective is to predict diabetes based on the diagnostic measurements of the patient.

   - Import the data. Complete the necessary exploration and sanitization of the data.
   - Build a machine learning model to predict diabetes using the predictor variables.

- Perform the model validation and measure the accuracy of the model.
- Check for innovative ways to improve the accuracy of the model.

Dataset credits—https://www.kaggle.com/uciml/pima-indians-diabetes-database. *Source:* Smith, J.W., Everhart, J.E., Dickson, W.C., Knowler, W.C., & Johannes, R.S. (1988). Using the ADAP learning algorithm to forecast the onset of diabetes mellitus. In *Proceedings of the Symposium on Computer Applications and Medical Care* (pp. 261-265). IEEE Computer Society Press.

2. Download the Automobile dataset.

- Import the data. Complete the necessary exploration and sanitization of the data.
- Build a machine learning model to predict the price of the vehicle using the other attributes.
- Perform the model validation and measure the accuracy of the model.
- Check for innovative ways to improve the accuracy of the model.

Dataset credits—from UCI machine learning repository, https://archive.ics.uci.edu/ml/datasets/Automobile. *Creator/Donor:* Jeffrey C. Schlimmer (Jeffrey.Schlimmer@a.gp.cs.cmu.edu). *Sources:* (1) 1985 Model Import Car and Truck Specifications, 1985 Ward's Automotive Yearbook. (2) Personal Auto Manuals, Insurance Services Office, 160 Water Street, New York, NY 10038. (3) Insurance Collision Report, Insurance Institute for Highway Safety, Watergate 600, Washington, DC 20037.

3.18   REFERENCE

1. Galton, Francis (1877). Typical laws of heredity. III. *Nature, 15(389)*, 512–514.

## CHAPTER 4

DECISION TREES

In the previous chapter, while working with logistic regression, we dealt with categorical variables output. Logistic regression is just one of several classification algorithms. In this chapter, we will discuss another classification algorithm called decision trees. Logistic regression works best for a specific type of data, while a decision tree is meant for a different class. Although both are classification algorithms, sometimes you may be confused which model to choose to solve the problem at hand. The best way, in that case, is to work with both logistic regression and decision trees and select the model that gives the most accurate results.

Decision trees are prevalently used in the business community to work on customer segmentation types of problems. Most marketing teams across various industry verticals run their marketing campaigns only after dividing customers into logical subsets, as one size does not fit all. Marketing teams need to run more customized marketing strategies for each of the customer segments to get the best results and ROI from a campaign. For example, if we are running a short messaging service (SMS) campaign for client mobile phones to sell a particular product, sending out the same SMS content for the entire population may not be a great idea. There may be a variety of customers in the data; some of it may be male, some female, some students, some working professionals, and some of them may be even retired persons. Decision trees are most widely used to work with this type of customer segmentation business problems. In this chapter, we will discuss in detail the decision tree algorithm. We will also discuss how to build decision tree models in Python. We will explain how to validate a decision tree–based model, and finally, we will talk about how to select a model which will be deployed by the business in production.

4.1 WHAT ARE DECISION TREES?

Decision trees are one of the most widely used classification algorithms. They follow an entirely different approach for classification. While discussing logistic regression, we worked with a curve fitting–based approach, where we tried fitting the logistic function to a set of data. We observed that the logistic function was an "S"-shaped curve that predicted class0 or class1. While working with decision trees, we will use a tree-based approach for classification. We will try to divide the data into small subsets in such a way that each subset has a dominating class—either class0 or class1. We will also work with an example that appreciates the overall philosophy behind decision trees. The following simple example will help you to grasp the concept. Later in this chapter, we will take a business problem and solve it using decision trees.

We have some historical product sales data from an online website. Here we are discussing about the data in. Table 4.1 lists details on customer gender, income band, and whether the client has ordered the product or not.

TABLE 4.1   Historical Product Sales Data from an Online Website

Historical Data—Use This Data for Analysis			
S. No.	Gender	Income Band	Whether Clients Ordered the Product
1	M	High	No
2	F	Low	Yes
3	M	High	No
4	M	High	No
5	M	High	No
6	M	High	No
7	F	Low	Yes
8	M	Low	Yes
9	F	High	No
10	M	High	No
11	F	High	No
12	M	Low	No
13	F	High	No
14	F	Low	Yes
15	M	High	Yes

The problem statement is prepared to make predictions looking at the dataset. You need to predict if the two new clients, as shown in Table 4.2, will make a purchase. These two new clients are given the credentials to log in the website. We should make our predictions quick enough—as soon as they log in and before they make any other move. Why such a hurry? If we can predict with reasonable confidence they are not going to make a buy or order the product, we can quickly push them an advertisement with a discount offer. If we predict that clients are going to make a buy, then we can even show them advertisements to promote cross-selling (selling a different related product) and upselling (selling a more valuable product in the same category).

TABLE 4.2   New Customers Data

New data – Get predictions for this data		
Gender	**Income Band**	**Will Clients Order?**
M	High	?
F	Low	?

If we look at the entire tabular data, we hardly get any useful inference(s), but if we rearrange the data in the form of a tree, we can very well make these predictions. Let us see how we can make these predictions.

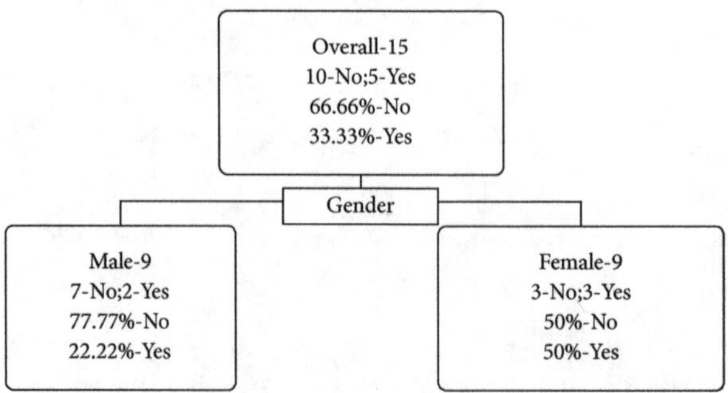

FIGURE 4.1   Tree representation of historical product sales data from a website.

The tree in Fig. 4.1 represents one level of division. There are 15 customers. Out of these 15 customers, 10 did not order the product, while 5 had ordered it. The overall population percentage of class0 (class-No) is 66.66 percent, and the portion of class1 (class-Yes) is 33.33 percent. We look further: out of 15 customers, 9 are male, and 6 belong to the female gender. Where did we get this information? Is it just by visual inspection of the dataset? Looking at the data, we can further discover that out of nine male customers, seven did not order the product. We can do a similar counting for the six female customers. We have the information on their income band as well. With this information, we can further branch the tree into more subsegments, as in Fig. 4.2.

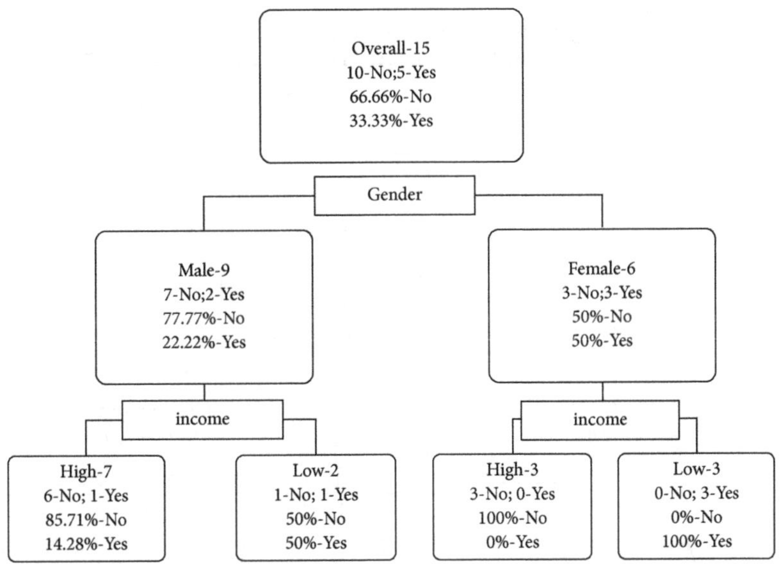

FIGURE 4.2 Further tree classification of historical product sales data from a website.

It looks like we have successfully rearranged all the data points in the form of a tree (Fig. 4.2). Just observe this tree; can you predict the behavior of the new customers? We repeat the new client data table (Table 4.3).

TABLE 4.3 New Customers Data

New data – Get predictions for this data		
Gender	Income band	Will clients order?
M	High	?
F	Low	?

The first row in Table 4.3 represents a male customer with a high income; we can predict this customer will not order the product. Look at the tree in Fig. 4.2 we just created based on the historical data; we can observe that male customers in the high-income group (85.71 percent) did not buy the product. The second row in the table refers to the next customer, who is a female from the low-income band. By applying the same logic, we can predict confidently that the female customer will buy the product—our tree data shows that 100 percent of the female candidates in the low-income category took a buy decision for the product. Table 4.4 is the dataset updated with our findings.

TABLE 4.4   Predictions for the New Customers Data

New Data – Get Predictions for This Data		
Gender	Income Band	Will Clients Order?
M	High	NO
F	Low	YES

The first node in the tree in Fig. 4.3 is the root node. Further segmentation of this root node is known as the parent node, which is connected by child nodes. The last nodes in the tree are leaf nodes.

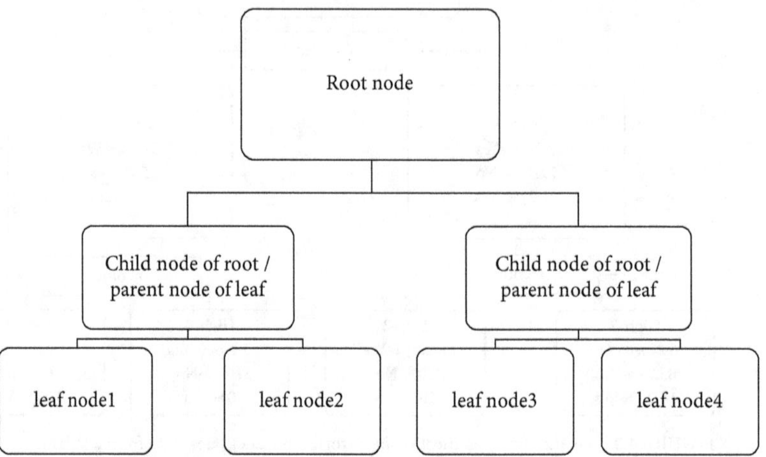

FIGURE 4.3   Root, leaf, and child nodes in a generalized tree.

We can make predictions based on leaf nodes. For this, we need to take a new data point, traverse through the tree, and see which leaf node it fits in. In our example, we have two new data points to make predictions: the first data point (customer1) falls in the leaf node1, where nonbuyers dominate. Similarly, the second data point (customer2) falls in the leaf node4, dominated by buyers. Any prediction is made based on the dominating class in the final leaf node (Table 4.5).

TABLE 4.5   Predictions Made Based on the Dominating Class in the Leaf Nodes

New Data – Get Predictions for This Data			
Gender	Income Band	Will Lead Node?	Will Clients Order?
M	High	Leaf Node1	NO
F	Low	Leaf Node4	YES

The rearrangement of the data into a tree format to get predictions is a new approach known as decision tree technique. We will now take up this technique for discussions. It is known as the decision tree technique. In this example, we manually arranged the data to get predictions. Decision tree algorithms do automation of this process for a dataset with any number of rows and any number of columns. There are many unanswered questions still: How are the variables chosen for splitting? Is it random, or is there any scientific logic behind it? Here we have just two columns, but in reality, we will have multiple columns; how will we choose the columns for splitting in that case? We will answer all these questions in the later sections of this chapter.

4.2 SPLITTING CRITERION METRICS: ENTROPY AND INFORMATION GAIN

A decision tree attempts to divide whole data into subsets in such a way that each subset has a single dominating class. The overall population is impure, with both the classes (class0 and class1) present in it. By dividing it into pure subsets, we make predictions for the new data points for the events like a buy vs. no-buy. As we have already seen in the previous section, we will not be able to predict a customer's decision to buy vs. no-buy only based on the whole population for the reason it has both buyers and non-buyers distributed in it. When we make it into pure subsets, where one class dominates, it is easy for us to predict.

How we define purity and impurity? If a segment has both the classes, it is said to be impure. If we have data with 50 percent buyers and 50 percent non-buyers, then it is an impure segment. We cannot decide anything if a customer belongs to one of these impure segments. On the other hand, if we have 100 percent of buyers and 0 percent of non-buyers, it is a perfect pure segment; we can then easily predict the nature of customers belonging to this group.

In any real-time data, there are many variables; how can we bifurcate impure data into pure subsets? How do we choose the variables for splitting? Can the choice of variable be random? Let us take an example to comprehend it better. We have a population with 100 records, 50 of them are buyers, and another 50 are non-buyers. We have an option to divide the population based on income or gender. Let us explore both these options and find out which variable does the best job of splitting (Table 4.6).

TABLE 4.6 Population Divided Based on Income or Gender

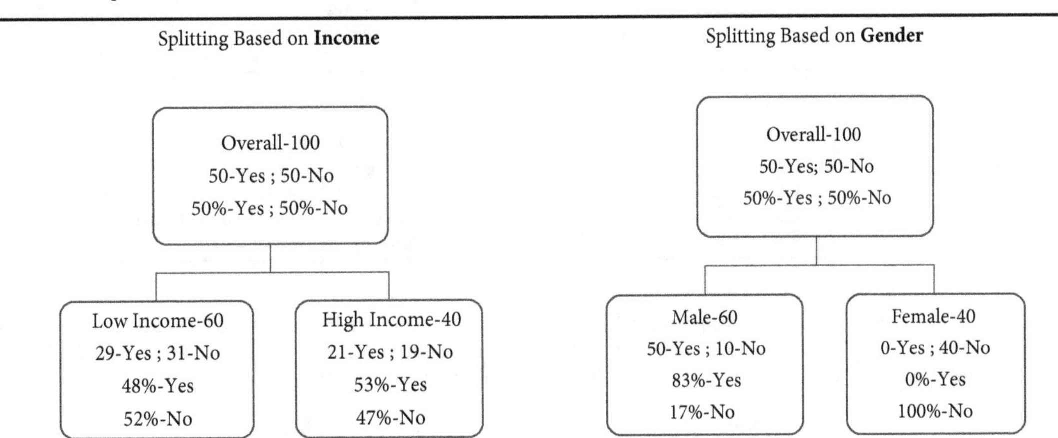

At the root node, the buyers and non-buyers are equally divided (50 percent each). After splitting based on income, we have almost the same situation:

1. The low-income segment has 48 percent buyers and 52 percent non-buyers.
2. The high-income segment has 53 percent buyers and 47 percent non-buyers.

Splitting based on income gives no special information. Whether the customer belongs to a low-income or high-income segment, it does not matter as far as the buy decision is concerned. We conclude that income is not a preferred variable for splitting.

At the root node, the ratio of buyers to non-buyers is 50-50. After splitting based on gender, we can see some interesting patterns:

3. The male segment has 83 percent buyers and 17 percent non-buyers.
4. The female segment has 100 percent non-buyers.

It is much more vibrant and conclusive information that we have gained by splitting based on gender. If the customer belongs to the male category, the chances of making a sale are more when compared with a female, who are always non-buyers. We conclude that gender is the preferred variable for splitting.

In our example it's pretty sure we will go ahead with gender for splitting the population. In this case, there are only two variables; therefore, it is effortless to choose one of them with the best splitting capability. In real-time projects, we have multiple variables: How do we choose the variable with the best splitting capability? Or in other words, how do we choose the variable giving us the best information gain? Is there any mathematical measure or a metric?

Fortunately, yes; the metric name itself is information gain. In our example, if we calculate the information gain value for income and gender, we will get a higher value for gender. Previously we chose the gender variable mostly by visualization, but in the next section, we will do it by quantifying the information gain. First, we will define entropy, the measure of impurity. It will be used later in the information gain formula.

4.2.1   Entropy: The Measure of Impurity

Entropy is the formal measure of impurity in a given segment. Entropy quantifies the impurity present in a segment. Entropy will be highest if a segment has both the classes in nearly equal proportions. Entropy is at its least for pure segments. Following is the formula for entropy, calculated for a segment (S):

$$Entropy(S) = -p_1 log_2(p_1) - p_2 log_2(p_2)$$

Here, S is a segment, p_1 is the probability of class1 in that segment, and p_2 is the probability of class2. Why is log_2 present in this formula? What is the significance of negative signs in the formula? We will further explore entropy by using the example in Table 4.7.

TABLE 4.7   Example Entropy Calculation

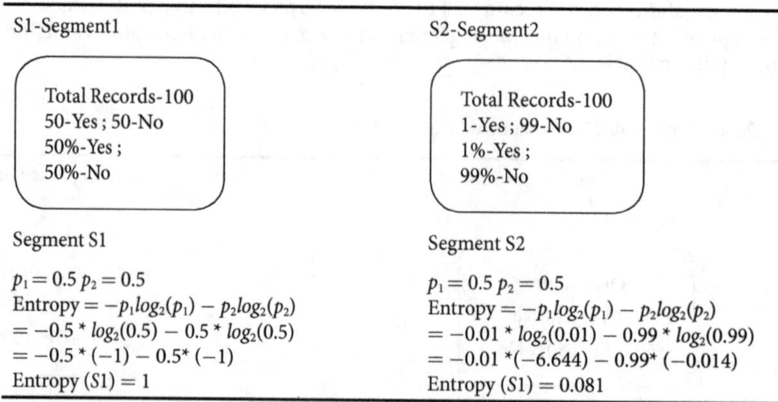

S1-Segment1

Total Records-100
50-Yes ; 50-No
50%-Yes ;
50%-No

Segment S1

$p_1 = 0.5\ p_2 = 0.5$
Entropy $= -p_1 log_2(p_1) - p_2 log_2(p_2)$
$= -0.5 * log_2(0.5) - 0.5 * log_2(0.5)$
$= -0.5 * (-1) - 0.5* (-1)$
Entropy $(S1) = 1$

S2-Segment2

Total Records-100
1-Yes ; 99-No
1%-Yes ;
99%-No

Segment S2

$p_1 = 0.5\ p_2 = 0.5$
Entropy $= -p_1 log_2(p_1) - p_2 log_2(p_2)$
$= -0.01 * log_2(0.01) - 0.99 * log_2(0.99)$
$= -0.01 *(-6.644) - 0.99* (-0.014)$
Entropy $(S1) = 0.081$

The segment with 50-50 proportions has an entropy value of 1, which is the highest value of entropy, as this segment has the highest impurity. If a class dominates in a segment, the entropy will be near zero. In Table 4.8, we show some of the examples of entropy values for both pure and impure segments.

TABLE 4.8   Examples of Entropy Values for Pure and Impure Segments

Segment	P_1	P_2	Entropy
S1	0.0001	0.9999	0.001
S2	0.01	0.99	0.081
S3	0.1	0.9	0.469
S4	0.2	0.8	0.722
S5	0.3	0.7	0.881
S6	0.4	0.6	0.971
S7	0.5	0.5	1.000
S8	0.6	0.4	0.971
S9	0.7	0.3	0.881
S10	0.8	0.2	0.722
S11	0.9	0.1	0.469
S12	0.99	0.01	0.081
S13	0.9999	0.0001	0.001

From the entropy table (Table 4.8) we observe that pure segments S1, S2, S12, and S13 have near-zero entropy. These segments have either P_1 or P_2 near to zero. Impure segments like S6, S7, and S8 have entropy near to 1. Now we go ahead and use the entropy formula for information gain. Please note that entropy has limits of 0 and 1. Hence, entropy is 0 for pure segments and 1 for impure segments.

4.2.2   Information Gain

Information gain gives us an idea of the ability of a variable to split the node. Before splitting a node we can typically expect impurity (high entropy) in that node. Nevertheless, after splitting a node based on a variable, we expect the entropy to reduce at child nodes and child nodes to be purer than the parent node. That is the whole purpose of splitting—the change in entropy measures information gain.

$$\text{Information gain } (x) = \text{Entropy before split} - \text{Entropy after split}$$

Entropy before the split is always a high number, but the right variable will ensure that entropy after a split is low. The overall information gain is higher for the correct splitting variable. Subsequently, there are two segments after splitting; the formula for information gain should be updated as

$$\text{Information gain } (x) = \text{Entropy before split} - \text{Average entropy after sSplit}$$

A variable takes the parent node and makes it into two child nodes; therefore, we have considered the average entropy after splitting. However, in real-life problems, we cannot give equal preference to both the child nodes; the number of records attributed to each node might vary. For example, if we split a node based on gender, we cannot expect the same records in male and female segments. Sometimes as high as 90 percent of records can get into one child node and just 10 percent in the other. We should consider the weighted average after the split, as a simple average would yield the wrong results. We modify the final formula to accommodate the weighted average:

$$\text{Information gain } (x) = \text{Entropy before split} - \text{Weighted average entropy after split}$$

$$\text{Information gain } (x) = \text{Parent node entropy} - \text{Weighted average of child node entropy}$$

Information gain is calculated for each variable. The higher the information gain, the better the variable is for splitting the population into pure subsets. If the information gain for a variable is nonsignificant, the population split will not fulfill the desired aim to get pure segments. Let us come back to the example where we were comparing two variables—income and gender. Using general intuition and our logic, we had picked gender as the desired variable for splitting. Let us calculate the information-gain values for these two variables.

TABLE 4.9   Entropy Calculations Based on Income and Gender

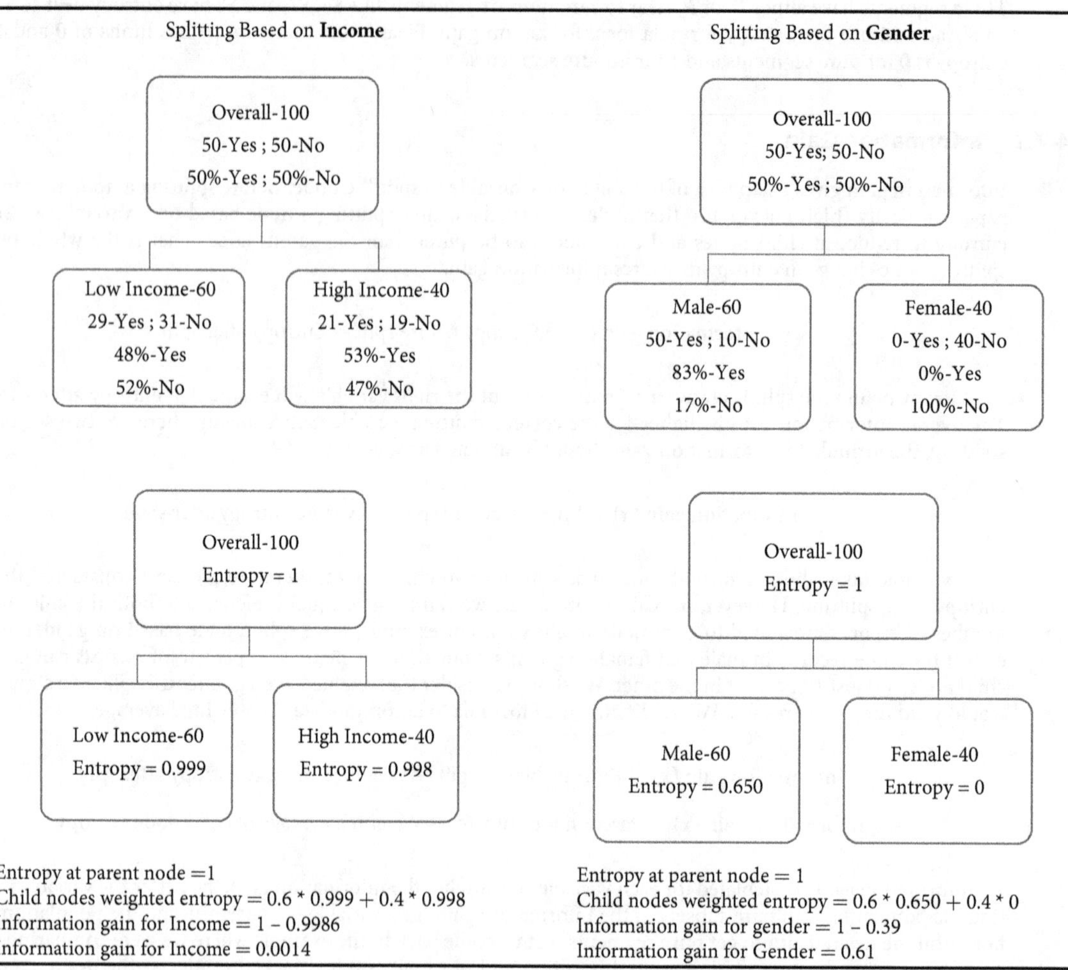

Entropy at parent node $=1$
Child nodes weighted entropy $= 0.6 * 0.999 + 0.4 * 0.998$
Information gain for Income $= 1 - 0.9986$
Information gain for Income $= 0.0014$

Entropy at parent node $= 1$
Child nodes weighted entropy $= 0.6 * 0.650 + 0.4 * 0$
Information gain for gender $= 1 - 0.39$
Information gain for Gender $= 0.61$

The information gain for the variable "income" is 0.0014, while the same value for "gender" is 0.61. Income has minuscule information gain when compared to gender (Table 4.9). We therefore choose the variable "gender" for splitting. If we have many variables, we will select the variable with the highest information gain. Information gain is an important measure to comprehend decision trees.

4.2.3   Gini Index: An Alternative to Entropy

Information gain is sufficient for building decision trees. There are a few more alternatives available to us. The Gini index can be used as an alternative to entropy. Both entropy and the Gini index calculate the impurity in a segment. Both entropy and the Gini index are a measure of impurity. If a segment is pure, both the Gini index and entropy are near to their lower limit and vice versa. The Gini index has a different formula, but the interpretation is the same as that of entropy.

$$\text{Gini index } (S) = 1 - (p_1^2 + p_2^2)$$

We now proceed to calculate the Gini index for most impure and pure segments.
Let S1 be an impure segment, then $P_1 = 0.5$ and $P_2 = 0.5$.

$$\text{Gini index } (S1) = 1 - (0.5^2 + 0.5^2)$$

$$\text{Gini index } (S1) = 1 - (0.25 + 0.25)$$

$$\text{Gini index } (S1) = 1 - (0.5)$$

$$\text{Gini index } (S1) = 0.5$$

Let S2 be a pure segment, then $P_1 = 1$ and $P_2 = 0$.

$$\text{Gini index } (S2) = 1 - (1^2 + 0^2)$$

$$\text{Gini index } (S2) = 1 - (1)$$

$$\text{Gini index } (S2) = 0$$

TABLE 4.10   Comparison of the Gini and Entropy Values

Segment	P_1	P_2	Entropy	Gini Index
S1	0.0001	0.9999	0.001	0.000
S2	0.01	0.99	0.081	0.020
S3	0.1	0.9	0.469	0.180
S4	0.2	0.8	0.722	0.320
S5	0.3	0.7	0.881	0.420
S6	0.4	0.6	0.971	0.480
S7	0.5	0.5	1.000	0.500
S8	0.6	0.4	0.971	0.480
S9	0.7	0.3	0.881	0.420
S10	0.8	0.2	0.722	0.320
S11	0.9	0.1	0.469	0.180
S12	0.99	0.01	0.081	0.020
S13	0.9999	0.0001	0.001	0.000

The limits of the Gini index are 0 to 0.5. Impure segments will have a Gini value near 0.5. Let us compare the Gini index and entropy values (Table 4.10). Let us also represent Gini and entropy values on a graph (Fig. 4.4) to get a better understanding.

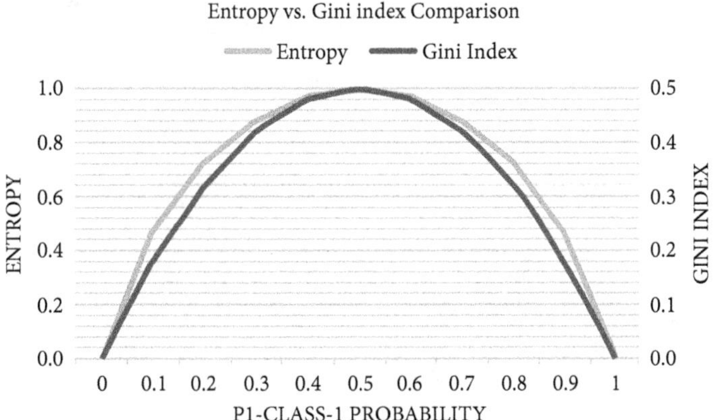

FIGURE 4.4   Entropy and Gini index comparison.

Table 4.10 and Fig. 4.4 show the entropy and Gini index for different subsets. It is clear that both of them work the same way. A rough calculation indicates that entropy is almost twice the Gini value from the segments S3 to S11. For a class of datasets, entropy may work better than the Gini index; for some other datasets the Gini index would be a better choice. Which measure works better for your dataset? You may want to try both entropy and the Gini index and discover which works better. Generally, you can use any one of them, and there will NOT be a significant impact on the model.

There are a few more alternatives to entropy and information gain. All try to do the same job. You can explore the rest, depending upon your needs. Now we are ready to build a decision tree using information gain measure. Let us have a look at the decision tree algorithm that will break the whole population into smaller subsets.

4.3   DECISION TREE ALGORITHM

What are we trying to achieve here using the decision tree algorithm? We want to build a classification model by breaking a whole population into pure subsets. Once we get the pure subsets, we can traverse the new data points through the tree structure and see under which leaf node they fall. We can then make the prediction based on the dominating class in that segment. Given below is the decision tree algorithm.

Step 1: Take a leaf node.

Step 2: Find the best splitting attribute in that node.

Step 3: Split the node using that attribute.

Step 4: Go to each child node and repeat Step 2 and Step 3.

Step 5: Stop growing the tree when we reach a pure leaf node with a single class.

A decision tree algorithm comprises the above five steps. Let us discuss each of these steps in detail.

Step 1: Take a leaf node.	At the start of the algorithm, the root node is the leaf node. We will start with the root node.
Step 2: Find the best splitting attribute in that node.	To find the best splitting attribute, we calculate the information gain of all the variables on this node.
Step 3: Split the node using the attribute.	We will use the variable with the highest information gain to split the node.
Step 4: Go to each child node and repeat Step 2 and Step 3.	The split, as explained above, gives us two child nodes. Go to each child node and calculate the information gain of all the variables again. Repeat it for every child node. Proceed with further splitting independently in each of the child nodes. This method is known as recursion.
Step 5: Stop growing the tree when we reach a pure leaf node with a single class.	If a leaf node has achieved purity, you can stop the growth of that node. The rest of the nodes keep growing until they meet purity. We end when we run out of attributes or when we have achieved the perfect purity in every node.

Let us repeat these five steps with an example in Table 4.11.

TABLE 4.11   Decision Tree Algorithm Steps with Example

Step 1: Take a leaf node.	For example, we have 10,000 records in our data, and we have $x_1, x_{2,} x_3, \dots x_{30}$ variables.
Step 2: Find the best splitting attribute in that node.	Calculate the information gain of $x_1, x_{2,} x_3, \dots x_{30}$. Let us assume x_{14} is the variable with maximum information gain.
Step 3: Split the node using the attribute.	We will use the variable x_{14} to split the node.

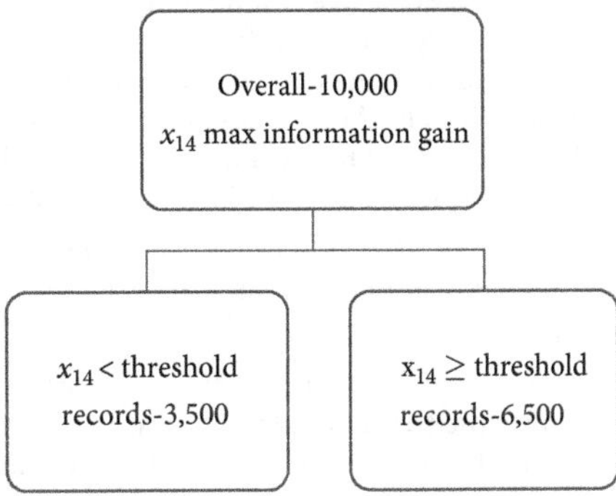

Step 4: Go to each child node and repeat Step 2 and Step 3.	The split explained above gives us two child nodes. We go to each child node and calculate the information gain of all the 30 variables again. We do it separately in the two child nodes. The number of calculations doubles here. This time, let x_{20} be the variable with maximum information gain in leaf node1 and let x_9 be the variable with maximum information gain in leaf node2. We further split the leaves based on these variables.

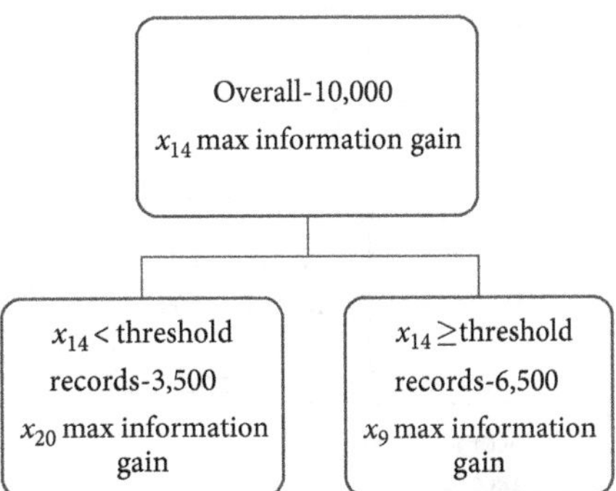

Step 5: Stop growing the tree when we reach a pure leaf node with a single class.	Stop a node if it achieves 100 percent purity. The rest of the nodes will grow further. In this example, the leftmost node stops. The remaining three nodes grow following the algorithm.

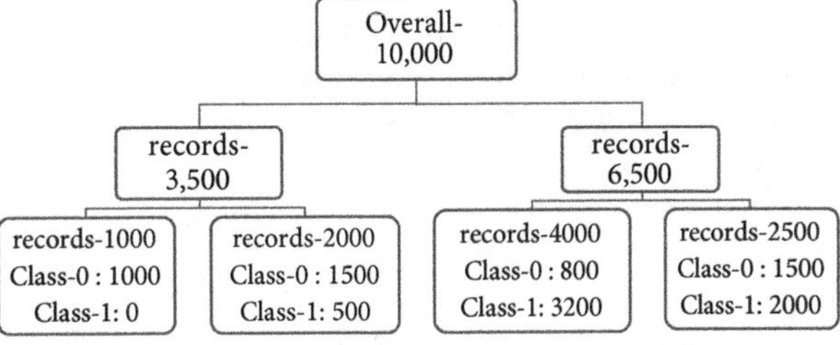

Before getting into a real-life business case, we need to answer a couple of burning questions:

1. How is the split decision made for a continuous numerical variable?
   - We can easily understand the variables such as "gender"; they are straightforward to split. There are only two possibilities: Male and Female. How about a column such as income? If income has the limits of 1000 to 30,000, we can split it at 5000, or we can split it at 6000 or 7000, or any number between 1000 and 30,000 for that matter. How do we get information gain value for such a case?
   - For a continuous variable, we consider all distinct values taken by it and try to calculate individual information gain by considering all the values as split. For example, we will calculate information gain for income at all possible splits. We are going to try splitting income at all the numbers in between, for example, 1200, 1500, 2000… 20,000, 21,000, 21,500, … 29,000. We will do a herculean task of calculating the information gain for all these options. This way, we will have multiple information gain values for one variable. Finally, we will choose the split that has the maximum information gain. For example, we calculate information gain for the split income <5000 and income ≥5000. Let it be 0.61. We will also calculate information gain for a different split income <6000 and income ≥6000. Let it be 0.72. We can consider the split at 6000 as a better option. We will repeat this exercise for all possible splits in income and choose the best splitting threshold.
   - In a decision tree algorithm, we not only choose the best splitting attribute, but also choose the best splitting threshold if any attribute is continuous (Fig. 4.5).
2. Can a variable re-enter the tree?
   - If a variable is categorical (such as gender), it may never re-enter the tree. If a variable is continuous (such as income), it may re-enter the tree at a different split. For example, the first split on income is income <5000 and income ≥5000. Later in the segment income <5000, income may enter again at a different split—for example, income <2000 and income ≥2000. We always have to consider the variable and the split, which has the maximum information gain. It does not matter whether it has been used in the tree already at some level above.

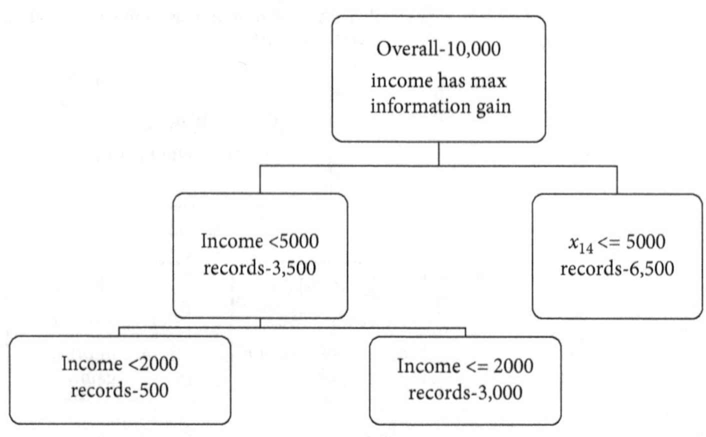

FIGURE 4.5 Decision tree split.

   - Why is there always a binary split? Can there be more than two child nodes? Why can't there be three or four child nodes? In one example, we are very sure the split in Fig. 4.6 with three child nodes will have the maximum purity and information. Can we do it? So far, we have seen only binary trees.

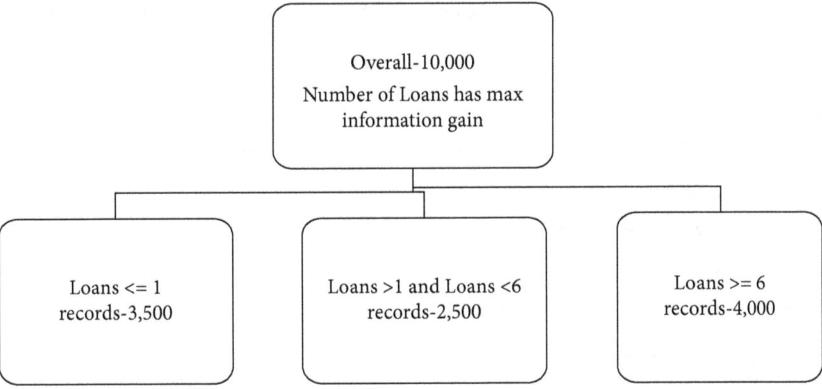

FIGURE 4.6   Decision tree with more than two child node splits.

- This kind of split is possible. We can achieve it in two steps in our binary tree in the manner shown in Fig. 4.7. As can be observed in the tree in Fig. 4.7, we reached the same three segments, but in two steps.

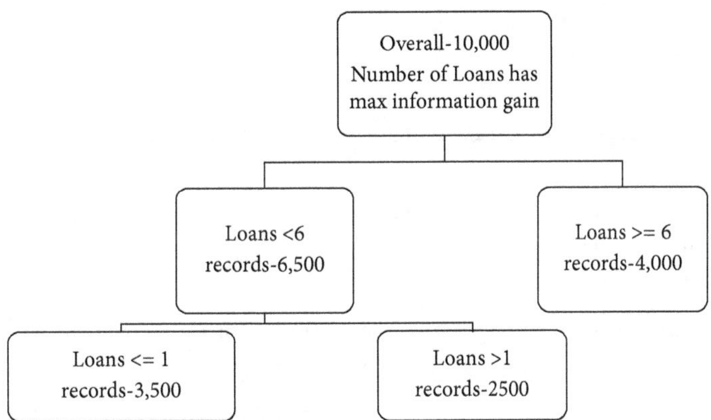

FIGURE 4.7   Three nodes in two steps.

4.4   CASE STUDY: CONTACT CENTER CUSTOMER SEGMENTATION

Let us look at an example of building the decision tree model. This dataset has been collected from a customer care contact center for a bank. Customers call for resolving some queries related to their savings and loan accounts. A survey form is sent to customers after the call. A few customers are satisfied, and a few customers are dissatisfied.

4.4.1   Objective and Data Exploration

We have survey data with us, and we want to build a decision tree model that will predict whether a customer will be satisfied or dissatisfied. We want to predict it as soon as he makes a call before the survey. Based on the customer attributes, if we can predict a customer is going to be dissatisfied, we can route the call to more capable agents having higher skill ratings. If a customer has a higher chance of being satisfied, we can route him to relatively inexperienced agents. This strategy will help us in optimum resource planning and, at the same time, increase the problem resolution rate.

The code below is used for importing the data and getting basic details such as the number of rows and column names.

```python
#Import Data
import pandas as pd
survey_data = pd.read_csv('D:\Chapter4\Datasets\Call_center_survey.csv')

#total number of customers
print(survey_data.shape)

#Column names
print(survey_data.columns)

#Print Sample data
survey_data.head()
```

This code gives us the following output:

```python
print(survey_data.shape)
(12330, 7)


print(survey_data.columns)
Index(['Cust_id', 'Age', 'Account_balance', 'Personal_loan_ind',
       'Home_loan_ind', 'Prime_Customer_ind', 'Overall_Satisfaction'],
      dtype='object')


survey_data.head()
Out[69]:

     Cust_id  Age  Account_balance  Personal_loan_ind  Home_loan_ind  \
0   CX01-001   49            23974                  1              0
1   CX01-002   25            72374                  0              1
2   CX01-003   32            65532                  0              0
3   CX01-004   70            28076                  0              1
4   CX01-005   23            38974                  1              1

   Prime_Customer_ind Overall_Satisfaction
0                   1        Dis Satisfied
1                   1            Satisfied
2                   1            Satisfied
3                   1        Dis Satisfied
4                   1            Satisfied
```

This data is for a banking service contact center. The output tells us the column names are `['Cust_id', 'Age', 'Account_balance', 'Personal_loan_ind', 'Home_loan_ind', 'Prime_Customer_ind', and 'Overall_Satisfaction']`. We have `customer_id` as the primary key. We also have some banking-related information such as `account_balance`, if the customer is having a personal loan, customer home loan indicator, and the type of customer. All these are predictor variables. The final target variable, in this case, is `Overall_Satisfaction`. We will look at some basic summaries before going ahead with model building. We will also look at the summary of each column and perform basic data exploration steps.

Below is the code for printing a summary of all the variables:

```python
summary=survey_data.describe()
round(summary,2)
```

This code yields the following output:

```
             Age  Account_balance  Personal_loan_ind  Home_loan_ind  \
count  12330.00         12330.00            12330.0        12330.0
mean      44.77         41177.14                0.5            0.5
std       13.91         26432.60                0.5            0.5
min       19.00          4904.00                0.0            0.0
25%       35.00         20927.00                0.0            0.0
50%       43.00         34065.00                0.0            0.0
75%       55.00         60264.00                1.0            1.0
max       75.00        109776.00                1.0            1.0

             Prime_Customer_ind
count                  12330.00
mean                       0.58
std                        0.49
min                        0.00
25%                        0.00
50%                        1.00
75%                        1.00
max                        1.00
```

The following are some useful observations based on the above output table.

- The age variable has an average value of 44.
- Account balance has an average of 41,177 with a minimum of 4904 and the maximum of 109,776.
- Personal Loan Indicator, Home Loan indicator, and Prime Customer Indicator are categorical variables. They take only two values, 0 and 1. A better measure to summarize these variables will be a frequency table using the function `value_counts()`.
- `Overall_Satisfaction` is not shown in this output. This variable also takes only two values: "Satisfied" and "Dis Satisfied."
- We will look at the frequency counts table for these indicator and categorical variables.

The following is the code to get the frequency counts tables.

```
survey_data['Overall_Satisfaction'].value_counts()
survey_data["Personal_loan_ind"].value_counts()
survey_data["Home_loan_ind"].value_counts()
survey_data["Prime_Customer_ind"].value_counts()
```

It yields the following output:

```
survey_data['Overall_Satisfaction'].value_counts()
Dis Satisfied     6707
Satisfied         5623
Name: Overall_Satisfaction, dtype: int64

survey_data["Personal_loan_ind"].value_counts()
0     6213
1     6117
Name: Personal_loan_ind, dtype: int64

survey_data["Home_loan_ind"].value_counts()
0     6176
1     6154
Name: Home_loan_ind, dtype: int64
```

```
survey_data["Prime_Customer_ind"].value_counts()
1     7113
0     5217
Name: Prime_Customer_ind, dtype: int64
```

Some observations are as follows:

- Overall, 6707 customers are dissatisfied and the rest are satisfied. More customers are in the dissatisfied category than satisfied customers.
- Almost 50 percent of the customers have personal loans.
- Nearly 50 percent of the customers have an existing home loan.
- Nearly 58 percent of the customers are prime-category customers.

Now we are ready to shape our decision tree model. We will be using the variables `'Age'`, `'Account_balance'`, `'Personal_loan_ind'`, `'Home_loan_ind'`, and `'Prime_Customer_ind'` as predictor variables while keeping `'Overall_Satisfaction'` as the target variable.

4.4.2   Model Building Code in Python

We need to convert the non-numeric variables to numeric variables for building the decision tree model. All non-numeric columns can be easily mapped to 0 and 1. A variable such as "gender" can be very easily converted to the numeric values by mapping Male and Female to 0 and 1, respectively. If any categorical variable has several values populated in it, we need to convert that into multiple dummy variables. For example, a variable such as "Region" may take four values: East, West, North, and South. We cannot map these values to 1, 2, 3, and 4. We need to create all four new columns East_ind, West_ind, North_ind, and South_ind as binary numerics.

In this example, all the predictor variables are already numeric. We also need to convert the target variable from categorical to numeric. The following code is used for mapping the non-numeric values to numeric values.

```
survey_data['Overall_Satisfaction']  =  survey_data['Overall_Satisfaction'].
map( {'Dis Satisfied': 0, 'Satisfied': 1} ).astype(int)
```

Once the values are mapped to numbers, we can check the updated frequency table.

```
survey_data['Overall_Satisfaction'].value_counts()
```

This code gives us the following output:

```
survey_data['Overall_Satisfaction'].value_counts()
0     6707
1     5623
Name: Overall_Satisfaction, dtype: int64
```

We will store the predictor variables in a list called features.

```
features=list(survey_data.columns[1:6])
print(features)
```

The above code gives us the below output:

```
print(features)
['Age',      'Account_balance',      'Personal_loan_ind',      'Home_loan_ind',
'Prime_Customer_ind']
```

We now proceed further to prepare the final features and target matrix using the following code:

```
X = survey_data[features]
y = survey_data['Overall_Satisfaction']
```

We are going to use these matrices in building the model. The following is the code for configuring the model.

```
from sklearn import tree
DT_Model = tree.DecisionTreeClassifier(max_depth=2)
DT_Model.fit(X,y)
```
This code is explained in Table 4.12.

TABLE 4.12   Code Explanation

```
from sklearn import tree
```
Import tree subpackage from sklearn
```
DT_Model =tree.DecisionTreeClassifier(max_depth=2)
```
`DT_model`: This is the model name. It can be any name you like.

`DecisionTreeClassifier()` is the function to build the decision trees. This function will execute the decision tree algorithm.

`max_depth`: This is a pruning parameter. Though essential, we have not explained it yet. There are a few associated concepts to this parameter, which we will take up in the later sections. As of now, you can take it as syntax.

```
DT_Model.fit(X,y)
```

The step explained above is the model configuration. In this step, we supply the actual data of X and y. Once we make a call to the `model.fit()` function, the algorithm will start the information gain calculations and other steps of building the decision tree model.

The code above builds the decision tree and stores it in the model object. In our case, `DT_Model` is the decision tree model. This code does not give us any output summary; all the values are calculated and stored in the model object. The following is the output generated by this code.

```
DT_Model.fit(X,y)
DecisionTreeClassifier(class_weight=None, criterion='gini',
max_depth=2,max_features=None, max_leaf_nodes=None,
min_impurity_decrease=0.0, min_impurity_split=None,
min_samples_leaf=1, min_samples_split=2,
min_weight_fraction_leaf=0.0, presort=False,
random_state=None, splitter='best')
```

This output is not the actual model output that we are expecting; it lists just the function and all its parameters. We still need to draw the tree to comprehend the model stored in `DT_Model`.

4.4.3   Drawing the Decision Tree

Now that all the measures are calculated and stored in `DT_Model`, we can access all the values and draw the decision tree. The following is the code to sketch the decision tree.

```
from IPython.display import Image
from sklearn.externals.six import StringIO

import pydotplus
dot_data = StringIO()
tree.export_graphviz(DT_Model, #Mention the model here
                     out_file = dot_data,
                     filled=True,
                     rounded=True,
                     impurity=False,
                     feature_names = features)

graph = pydotplus.graph_from_dot_data(dot_data.getvalue())
Image(graph.create_png())
```

We need two Python packages to draw this decision tree—namely "Graphviz" and "pydotplus"—and also need to supply the model name. This code piece extracts all the values from the model and returns the tree image with all the details. In some cases, you may get an error—"Graphviz executables not found." If it occurs, you need to update the path variable to fix the error. Alternatively, you can use a scikit-learn–based package for drawing the decision trees. Below is the code using the scikit-learn package.

```python
import matplotlib.pyplot as plt
from sklearn.tree import plot_tree, export_text
plt.figure(figsize=(15,7))
plot_tree(DT_Model, filled=True,
                    rounded=True,
                    impurity=False,
                    feature_names = features)

print(export_text(DT_Model, feature_names = features))
```

What follows is the code output (Fig. 4.8).

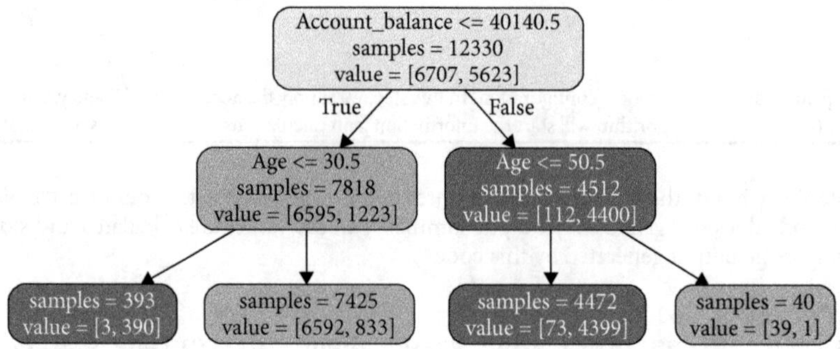

FIGURE 4.8   Decision tree output by using Python code.

4.4.4 Tree Output Interpretation

We will now learn to interpret the tree depicted in Sec. 4.4.3. The following is a detailed explanation. Let us start with the root node.

4.4.4.1 *The Root Node*

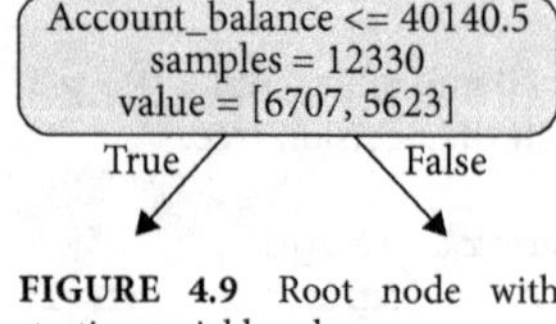

FIGURE 4.9 Root node with starting variable values.

Sample

- The sample indicates the number of samples or records in this node. The root node contains the entire data. In our dataset, there are 12,330 records; the root node correctly displays it.

Value

- The value indicates the target variable's frequency counts. In our data, the target variable takes only two values: 0 and 1. The count of 0s is displayed first, followed by the count of 1s. The original dataset has the target variable as Dissatisfied and Satisfied. We had mapped Dissatisfied to 0 and Satisfied to 1 before building the model. The count of 0s in this node equals 6707, while the count of 1s is 5623.

Account_balance ≤40,140.5 (Fig. 4.9)

- The decision tree algorithm has also calculated the information gain of all the variables on the root node. As we know, the variable with maximum information is picked up for splitting this node. If a variable is continuous, then the algorithm also calculates the best splitting point.

- In this example, we have multiple predictor variables; after examining all, `"Account_balance"` has been found to have the highest information gain, which stands at 40,140.5. This point, 40,140.5, may not exist in the data. Instead of showing the split as Account_balance 40,140 and Account_balance >40,140, Python shows it as Account_balance ≤40,140.5. Let us proceed further and interpret what is True and False.

True and False

- The root node will be split based on Account_balance at 40,140.5; there will be two child nodes created out of it.

- Account_balance ≤40,140.5 is the first child node. In this tree, it is displayed as True, which means Account_balance ≤40,140.5 condition = True.

- Similarly, there will be one more child node where Account_balance >40,140.5, which means the Account_balance ≤40,140.5 condition = False.

4.4.4.2   The Left-Side Child of the Root Node   In this section, we will analyze the left-side node, which is formed by applying the first filter condition (Account_balance ≤40,140.5) on the root node (Fig. 4.10).

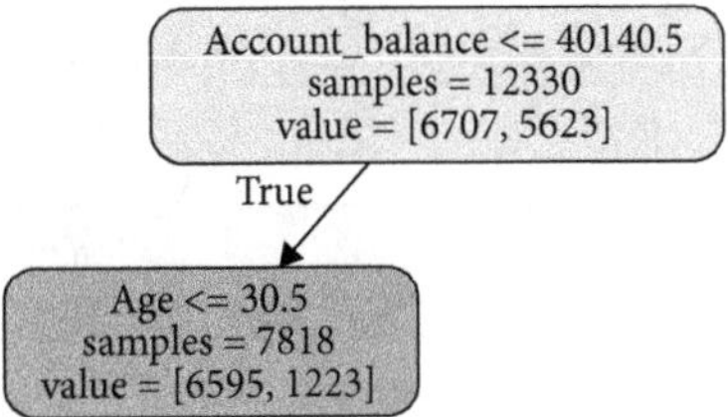

FIGURE 4.10   The left-side child of the root node.

- We can name this node as low account balance customers. There are 7818 customers in this subset. It means that out of all the 12,330 customers, there are 7818 customers with Account_balance ≤40,140.5. Out of 7818 customers in this node, 6595 are dissatisfied and 1223 satisfied.

- We can see better purity in this segment. Initially, in the root node, we had 54 percent as dissatisfied and 46 percent as satisfied. After this split, this segment has 84 percent as dissatisfied and only 16 percent as satisfied.

- It does not mean we stop the tree growth here. The decision tree algorithm looks for better purity. The variable that has the highest information gain in this subset is Age, and it has 30.5 as the maximum information gain point.

4.4.4.3   The Leftmost Leaf Node   In this section, we will examine the node formed by applying filter condition (Age ≤30) on its parent node (Fig. 4.11).

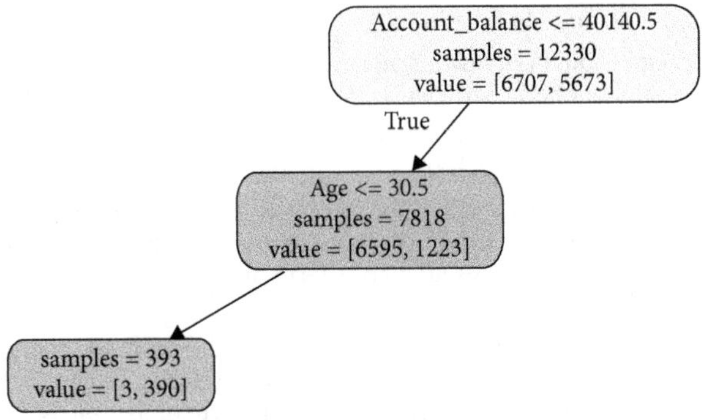

FIGURE 4.11   The leftmost leaf node.

- This node can be formed by applying two filter conditions, Account_balance ≤40,140.5 and Age ≤30.5, on the root node. We can give a name to this node as well. Let us call it low account balance and young customers. There are 393 customers in this subset. That means out of all the 7818 customers with low balance, 393 are with Age ≤30.5.

- Out of 393 customers in this node, there are 3 dissatisfied and 390 satisfied. We can see far better purity in this segment. When compared to the original values of 54 percent dissatisfied and 46 percent satisfied at the root level, in this node, we have 1 percent dissatisfied and as high as 99 percent satisfied customers. This segment is almost pure; hence, we can stop the tree growth here.

4.4.4.4   All the Other Nodes   You should be able to interpret the rest of the nodes using a similar explanation given in the previous section. The complete tree is sketched in Fig. 4.12.

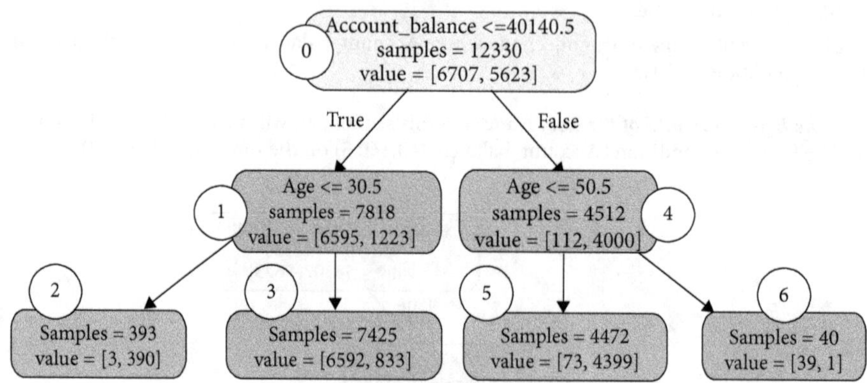

FIGURE 4.12   The complete tree sketch with node numbers.

In the output diagram, we added numbers to the nodes for better understanding and the ease of discussion. The numbering is done from left to right.

1. Node 0 represents the root node, node 1 is the left-side child node of node 0, node 2 on the left side is the child node of node 1, and finally, node 3 is the second child node of node 1. It completes the left side of the tree.

2. Node 4 is the right-side child node of the root node, with node 5 and node 6 as its child nodes. It completes the entire tree.

4.4.4.5   The Leaf Nodes and the Rules   The most critical nodes in the decision tree are the leaf nodes, which frame the final rules to make predictions. The decision tree model is nothing but a set of rules. These rules come from the leaf nodes. The number of rules is equal to the number of leaf nodes. Consider the following points to explore it further:

- As we are aware, in linear regression, the regression equation is the model $y = \beta_0 + \beta_1 x_1 + \beta_2 x_2 + \cdots + \beta_k x_k$. The regression coefficients are found in the process of building the model.

- Similarly, in logistic regression, the following equation is the final model: $y = \dfrac{e^{\beta_0 + \beta_1 x_1 + \beta_2 x_2 + \beta_3 x_3 + \cdots + \beta_k x_k}}{1 + e^{\beta_0 + \beta_1 x_1 + \beta_2 x_2 + \beta_3 x_3 + \cdots + \beta_k x_k}}$. Here also, the key is to find the values of beta coefficients.

- In our decision tree model, there are a total of four leaf nodes, namely, node 2, node 3, node 5, and node 6. You can interpret these leaf nodes as follows (rules):

- Rules from node 2: This node has the conditions Account_balance ≤40,140.5 and Age ≤30. If a customer from our dataset meets these conditions, he or she belongs to node 2. The predicted class for this particular customer is 1, which is the "Satisfied" category. The pseudo-code form of the rule is *IF* (Account balance ≤ 40,140.5) and (Age ≤ 30) *then predicted* = "1."

- Rules from node 3: This node has the conditions Account_balance 40,140.5 and Age >30. Any customer following these rules will be classified in node 3. As per the dominating subclass of node 3, he or she will be predicted as 0. For this example, 0 represents "Dissatisfied." The pseudo-code form of the rule is *IF* (Account balance ≤ 40,140.5) and (Age > 30) *then predicted* = "0."

- Rules from node 5: This node has the conditions Account_balance >40,140.5 and Age ≤50.5. Any customer following these rules will be classified in node 5, and the prediction will be 1, which represents "Satisfied." The pseudo-code form of the rule for this node is *IF* (Account balance > 40,140.5) and (Age ≤ 50.5) *then predicted =* "1."

- Rules from node 6: It has two rules as Account_balance >40,140.5 and Age >50.5. If a customer follows these rules, he or she will be classified in node 6, with the predicted value as 0, representing the "Dissatisfied" category. The pseudo-code form of the rule is *IF* (Account balance > 40,140.5) and (Age > 50.5) *then predicted =* "0."

Combining all these rules, we get the final decision tree model as follows:

$$y = \begin{cases} 0, \textit{IF} \begin{pmatrix} ((\text{Account balance} \leq 40{,}140.5) \text{ and } (\text{Age} > 30)) \\ \text{or} \\ ((\text{Account balance} > 40{,}140.5) \text{ and } (\text{Age} > 50.5)) \end{pmatrix} \\ 1, \textit{IF} \begin{pmatrix} ((\text{Account balance} \leq 40{,}140.5) \text{ and } (\text{Age} \leq 30)) \\ \text{or} \\ ((\text{Account balance} > 40{,}140.5) \text{ and } (\text{Age} \leq 50.5)) \end{pmatrix} \end{cases}$$

On the decision tree, you can see all the rules that we have just discussed in the list. The predicted values 1 and 0 are color-coded. All Satisfied segments are displayed in "Blue," and Dissatisfied sections are in "Orange." We can automatically extract all these rules from the model leaf nodes using the following simple command:

```
print(dot_data.getvalue())
```

Given below is the output table from this code line.

```
print(dot_data.getvalue())
digraph Tree {

node [shape=box, style="filled, rounded", color="black", fontname=helvetica]
;edge [fontname=helvetica] ;

0 [label="Account_balance <= 40140.5\nsamples = 12330\nvalue= [6707, 5623]",
fillcolor="#fbebdf"] ;

1 [label="Age <= 30.5\nsamples = 7818\nvalue = [6595, 1223]", fillcolor=
"#ea985e"];

0 -> 1 [labeldistance=2.5, labelangle=45, headlabel="True"] ;

2 [label="samples = 393\nvalue = [3, 390]", fillcolor="#3b9ee5"] ;
1 -> 2 ;

3 [label="samples = 7425\nvalue = [6592, 833]",fillcolor="#e89152"];
1 -> 3 ;

4 [label="Age <= 50.5\nsamples = 4512\nvalue = [112, 4400]", fillcolor=
"#3e9fe6"] ;
0 -> 4 [labeldistance=2.5, labelangle=-45, headlabel="False"] ;

5 [label="samples = 4472\nvalue = [73, 4399]",fillcolor="#3c9fe5"] ;
4 -> 5 ;

6 [label="samples = 40\nvalue = [39, 1]", fillcolor="#e6843e"] ;
4 -> 6 ;
}
```

After careful observation, you can detect the leaf nodes and all the rules. This output not only has all the rules but also contains all the other formatting details of filling the tree—such as color, edges, and distance.

4.4.4.6 ***Predictions from the Model***   Now we are all set to make predictions by using decision tree rules. Take a new customer with age = 45 and account balance = 65,000, and navigate with these details through the tree or the tree rules; you will end up at node 5. The predicted output, considering the dominating subclass of this node, is "1," representing "Satisfied." In Table 4.13, we have given a few more examples with their nodes and predicted values.

TABLE 4.13   Additional Examples with Their Nodes and Predictions

Age	Account Balance	Predicted Value	Node
25	15,000	1—Satisfied	Node 2
56	80,000	0—Dissatisfied	Node 6
48	2,000	0—Dissatisfied	Node 3
30	76,000	1—Satisfied	Node 5
28	25,000	1—Satisfied	Node 2

We will use the predict function to get all our predictions by supplying the predictor variables (X matrix) data points. We will use this command in the next section.

4.4.5   Tree Validation and Accuracy

Once we are done building the decision tree model, we need to get its rules; like every other statistical modeling exercise, we also need to take note of the accuracy of the model before going ahead with predictions. In our model, the actual values of the target variable are in 0s and 1s. We can get the predicted values and create a confusion matrix to derive accuracy. We have already discussed the details of the confusion matrix and accuracy measures in Chap. 3. We can use the following code to get the predicted values and feel for the model's accuracy.

```
predict1 = DT_Model.predict(X)
print(predict1)

from sklearn.metrics import confusion_matrix
cm = confusion_matrix(y, predict1)
print(cm)

print(sum(cm))
print(sum(sum(cm)))
total = sum(sum(cm))
##### calculate accuracy from confusion matrix
accuracy = (cm[0,0]+cm[1,1])/total
print(accuracy)
```

Given in Table 4.14 is a line-by-line explanation of this code.

TABLE 4.14   Code Explanations

```predict1 = DT_Model.predict(X)``` ```print(predict1)```	Supply X data which we created using the features and get the predicted values.

```from sklearn.metrics import confusion_matrix```
```cm = confusion_matrix(y, predict1)```
```print(cm)```
	Supply actual and predicted values to the confusion matrix function. In this case, "y" represents the actual target value from the data.

```print(sum(cm))```
```print(sum(sum(cm)))```
```total = sum(sum(cm))```
	sum(cm) gives us the column-wise sum, while (sum(sum(cm))) is the actual total of the confusion matrix.

```accuracy = (cm[0,0]+cm[1,1])/total``` ```print(accuracy)```	Calculate the accuracy by taking the sum of diagonal elements/total records.

The code output is given in the following table.

```
predict1 = DT_Model.predict(X)
print(predict1)
[0 1 1 ... 1 1 1]

from sklearn.metrics import confusion_matrix
cm = confusion_matrix(y, predict1)
print(cm)
[[6631   76]
 [ 834 4789]]

total = sum(sum(cm))
#####from confusion matrix calculate accuracy
accuracy = (cm[0,0]+cm[1,1])/total
print(accuracy)
0.9261962692619627
```

From this output, you can appreciate that the accuracy of our decision tree model is 92.6 percent.

4.5   THE PROBLEM OF OVERFITTING

In this section, we will work with a small example to understand the concept of overfitting.

4.5.1   Huge Decision Trees

Imagine a dataset with 1000 records and several continuous variables in the data. We start building the decision tree and growing it. We should stop it at a particular point, but we somehow bypass all the stopping criteria conditions and keep on growing the tree. Can the tree grow infinitely large, or is there a hard stop point where the tree cannot grow any further? Given in Fig. 4.13 is an example of one branch of a fully grown tree. Can you find any hard stop point(s) in it?

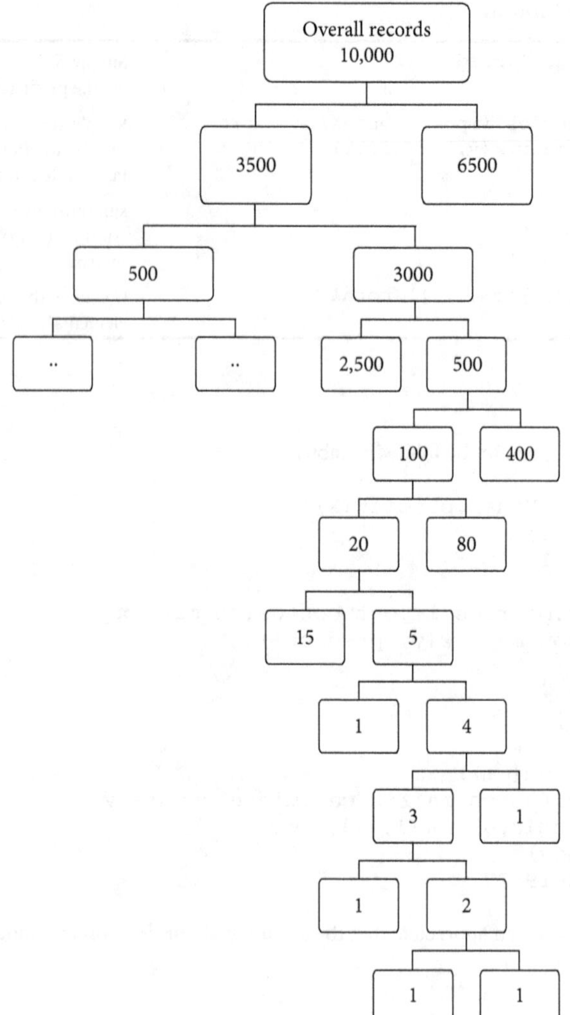

FIGURE 4.13   An example of one branch of a fully grown tree.

Looking at the tree in Fig. 4.13, we can make out there is a hard stopping point. When every leaf node has only one record or one class (pure), there is no further segmentation possible. This example shows only one branch; the same process will be repeated in each branch. If we do not control the growth of a tree, it will grow to its full extent. But the maximum number of leaf nodes will never exceed the number of records (N) in the data. In our example, the tree can have a maximum of 1000 leaf nodes before it reaches the hard stop point. Most of the trees will stop long before that. We should have several continuous variables in the data to grow the tree until the hard stop point. If we have a categorical variable(s) in the data, it cannot enter the tree again at a different level. In that case, the tree growth will stop long before it reaches the highest level of N-leaf nodes.

Imagine a massive decision tree that has grown until its hard stopping point. If we have N records in the data, the tree will nearly have N leaf nodes. Each leaf node will give us rules, and we might end up with almost N-rules. What will be the accuracy of such a decision tree? It will be 100 percent. We can accurately predict almost every record. Can you call it a good tree or a decent model for that matter? If we have 1000 records, then we will have practically 1000 rules. The first record will be the first rule; the very next record will be the second rule. We will formulate the rules as $x_1 =$ value of x_1 in row1, and x_2 equals the value of x_2 in row1, and so on. The predicted value, in this case, will be the value of y in row1. Just as a reminder, $x_1, x_2, \ldots x_n$ are column variables, and y is the target variable. In the name of the rules, we are returning only the dataset. If we have a relook at the fully grown tree, shown in Fig. 4.13, we find that there is no model. We take the vertical format of the data and arrange it in a horizontal tree format. However, ironically, this model still has a 100 percent accuracy level. We are just returning the data as rules, and it is bound to be 100 percent accurate.

In any fully grown tree, we only fetch very specific or definite rules related to a dataset that work well only for that particular dataset. Our real aim is to model generic patterns and get generic rules associated with the input data. An example of a generic rule would be as follows: income ≤5000 are "buyers" and income >5000 are "non-buyers." An example of a specific or definite rule will be income = 5000, then prediction = "buyers," or income = 5100, then prediction = "non-buyers."

We need to build a model that captures generic patterns in the data. These generic patterns will not only give good predictions for the input dataset but also work well on any other data coming from the population under consideration. The focus of a model is quite beyond the current dataset, which was used to build the model; we prepare the models to apply it to new or future datasets.

A fully grown decision tree tries to memorize the data instead of finding the generic patterns. It therefore works well only on the data used to build the model and fails for all other datasets, even if the new datasets are sampled from the same population. These observations make us define two different types of datasets in the practice of model building: "train data" and "test data." The train and test data are discussed next.

4.5.2   The Train and Test Data

The dataset used for building the model is known as train data. The model tries to learn the patterns from this train data. This dataset is wholly exposed to the model. Relying only on the training data alone is not the correct way to assess the accuracy of a model. We need a different dataset that has not been exposed to the model; this dataset is used to validate the robustness of the model. It is therefore called test data.

Test data is sampled from the same population, but it is kept aside while building the model. The actual target values are also known in the test data. A model with a very high training accuracy may not always ensure it will have the same accuracy level on test data as well. We build and calculate initial accuracy levels with train data, but we are required to also maintain the acceptable accuracy level in the models with test data. The model is considered reliable and useful only if it supports similar high accuracy levels both in train and test datasets.

Sometimes in practice, we have only one dataset available. In such a case, we need to make two datasets out of it. Typically, we take 80 percent of the randomly sampled data as train data, which we use to build the model; the remaining 20 percent is kept aside. Once completed, we apply the model on the test data to validate its accuracy. Eighty percent train data and 20 percent test data are a general industry rule of thumb. Depending upon the situation, you can also consider 90 percent train data and the remaining 10 percent test data, or 85 percent train data and 15 percent test data, or even 75 percent train data and 25 percent test data. The following points should be noted:

- An equal split of 50 percent train data and 50 percent test data is not preferred. We do not want to use only 50 percent of the data for building the model. We want the model to learn as many patterns as possible.

- Do not consider the initial 80 percent of the records for training and the last 20 percent for testing. We need the train and test data to be random samples that truly represent the data population. The first 80 percent of the data may not always be a random sample. Sometimes our data is sorted by a column. It is best to take random samples and have 80 percent data for the train and the remaining 20 percent for test.

Test data is a representation of the future data points that our model is going to get in production as input for making the predictions. If a model fails to show good accuracy on test data, we cannot use it for making predictions with the production data. If a model shows high accuracy on the train data and fails miserably on the test data, it is called an overfitted model. That brings us to the next topic on "the problem of overfitting."

4.5.3   Overfitting

We will explore overfitting more in the following format to emphasize the individual points.

1. An overfitted model has high accuracy on the train data and significantly low accuracy on the test data.

2. Instead of generic patterns if a model is learning only the specific patterns, it is memorizing the training data.

3. In overfitted models, for a relatively small change in the training data, there are significant changes in the final rules. In essence, the models and their parameters show a massive difference (models with much variance).

4. Overfitted models are overcomplicated with too many parameters and need simplification. If it is a decision tree—a large tree with too many branches and rules—it must be pruned.

5. Overfitting is a generic concept. It can happen to any model, be it regression or logistic regression, or a decision tree.

How do we detect overfitted models? We do it by finding their accuracy with train data and test data. Any difference of >5 percent is a significant difference. For example, if we have 90 percent accuracy on train data, we expect it to be 85 percent or more with the test data as well; otherwise, it is not advisable to use that model in the production.

Let us learn further with an example of dividing the data into train data and test data, building the model on train data, followed by validating it with the test data. We will check the accuracy level of the model with both test data and train data. The following code will perform all these tasks for us.

```python
import pandas as pd
overall_data = pd.read_csv(r"D:\Book\0.Chapters\Chapter4\5. Base Datasets\
Customer_profile_data.csv")

##print train.info()
print(overall_data.shape)

#First few records
print(overall_data.head())

# the data have string values we need to convert them into numerical values
overall_data['Gender'] = overall_data['Gender'].map( {'Male': 1, 'Female': 0} ).
astype(int)
overall_data['Bought']   =   overall_data['Bought'].map({'Yes':1,   'No':0}).
astype(int)

#Defining features, X and Y
features = list(overall_data.columns[1:3])
print(features)

X = overall_data[features]
y = overall_data['Bought']

print(X.shape)
print(y.shape)
```

This code imports the data, maps non-numerical columns to numbers, creates features, and finally generates X and y for us. Given below is its output.

```python
print(overall_data.shape)
(109, 4)

print(overall_data.head())
    Sr_no  Age   Gender  Bought
0       1   45     Male     Yes
1       2   56     Male     Yes
2       3   49   Female     Yes
3       4   50   Female      No
4       5   75   Female      No

print(overall_data.head())
    Sr_no  Age   Gender   Bought
0       1   45        1        1
1       2   56        1        1
2       3   49        0        1
3       4   50        0        0
4       5   75        0        0
```

```
print(features)
['Age', 'Gender']

print(X.shape)
(109, 2)
print(y.shape)
(109, )
```

Now we derive the train and test datasets from X and y data; the function `train_test_split()` takes X and y data as input and it divides the data X into `X_train`, `X_test`. The corresponding `y_train`, and `y_test` are also generated. The train dataset size is based on the `train_size` parameter. Here we have set it to 0.8, meaning 80 percent.

```
#Dividing X and y to train and test data parts.
from sklearn import model_selection
X_train, X_test, y_train, y_test = model_selection.train_test_split(X,y,
train_size = 0.8 , random_state=5)

print(X_train.shape)
print(y_train.shape)
print(X_test.shape)
print(y_test.shape)
```

Given below is the code output. There are 109 records in total; expect nearly 87 records (~80 percent) in the train data. We can use the optional `random_state` parameter to generate the same random sample— always the same data when you run it the next time—again and again; otherwise, it will be a different train sample whenever we rerun the code. We are mentioning it so that you can also get the same results as we have shown here.

```
X_train.shape (87, 2)
y_train.shape (87,)
X_test.shape (22, 2)
y_test.shape (22,)
```

As expected, 87 records are in the train data, and 22 in the test data. Now we will go ahead and build the model based on the train data followed by accuracy calculations.

```
from sklearn import tree
#training Tree Model
DT_Model1 = tree.DecisionTreeClassifier()
DT_Model1.fit(X_train,y_train)
#Plotting the trees
from IPython.display import Image
from sklearn.externals.six import StringIO
import pydotplus
dot_data = StringIO()
tree.export_graphviz(DT_Model1,
                     out_file = dot_data,
                     feature_names = features,
                     filled=True, rounded=True,
                     impurity=False)

graph = pydotplus.graph_from_dot_data(dot_data.getvalue())
Image(graph.create_png())
```

Figure 4.14 shows the output of this code.

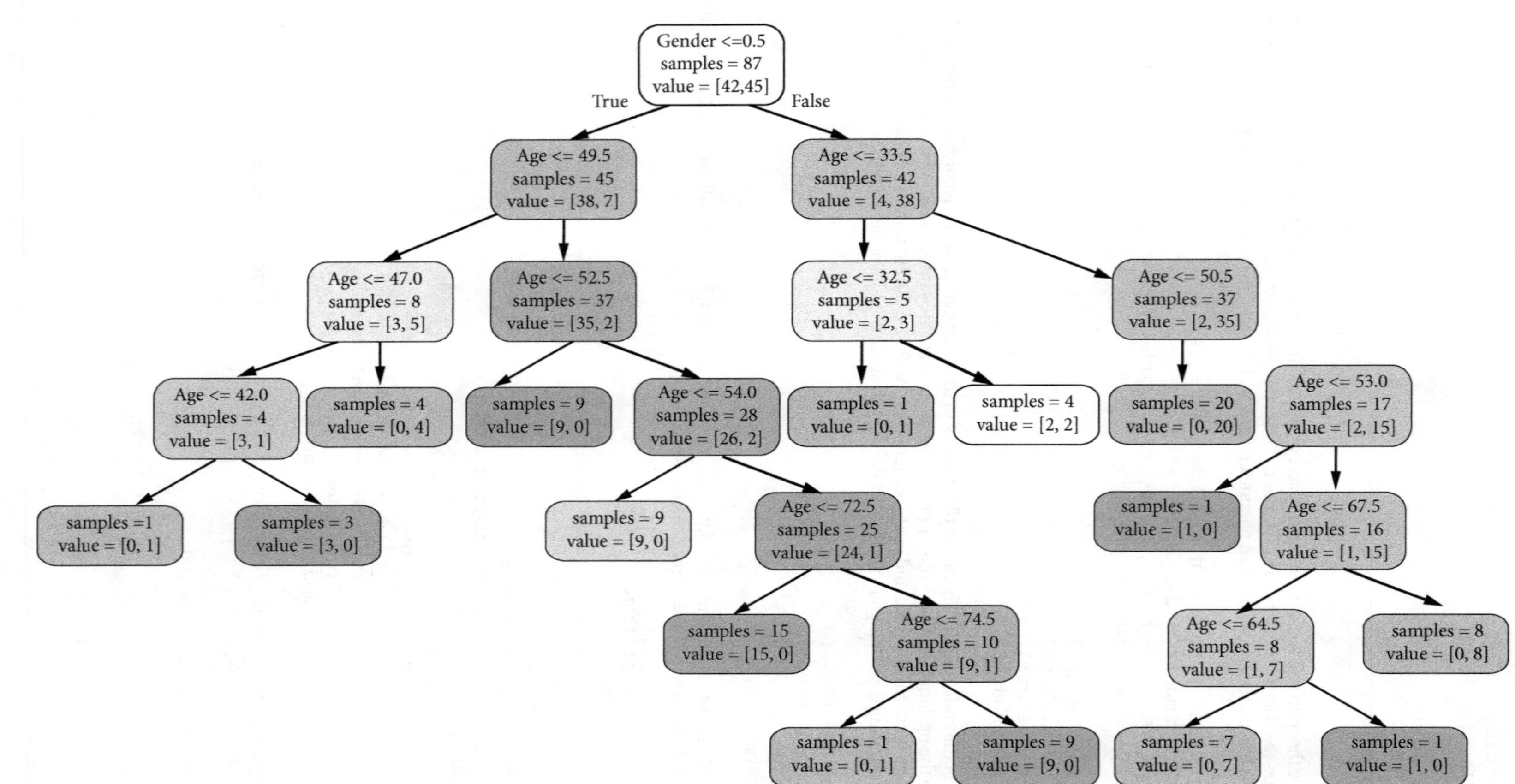

FIGURE 4.14   Output tree result of the code for a model based on train data.

From the tree output, we can see that the tree is grown to its full extent. Almost all the leaf nodes are pure in this tree. Sometimes the tree is so large that it is challenging to draw and interpret the required values from it. We will go ahead and find the accuracy with train data and test data. As usual, the following code does it for us.

```python
#Accuracy on train data
from sklearn.metrics import confusion_matrix

predict1 = DT_Model1.predict(X_train)
cm1 = confusion_matrix(y_train,predict1)
total1 = sum(sum(cm1))
accuracy1 = (cm1[0,0]+cm1[1,1])/total1
print("Train accuracy", accuracy1)

#Accuracy on test data
predict2 = DT_Model1.predict(X_test)
cm2 = confusion_matrix(y_test,predict2)
total2 = sum(sum(cm2))
#####from confusion matrix calculate accuracy
accuracy2 = (cm2[0,0]+cm2[1,1])/total2
print("Test accuracy", accuracy2)
```

This code gives us the below output.

```
Train accuracy 0.9655172413793104
Test accuracy 0.7727272727272727
```

In the above output, we can see that the accuracy with train data is 96.5 percent, so we expect the accuracy with the test data to be 92 percent or more. The accuracy of the test data should be in the range of 92 to 96 percent. Any accuracy of less than 92 percent on test data is a clear indication of overfitting. In our case, the accuracy of the test data is 77.3 percent—the model is overfitted. We should NOT assume that an accuracy of 77.3 percent on the test data is good enough—it may just be chance. We may not get the same level of accuracy on the future datasets. For the model to be called robust, we should have gained nearly 96 percent accuracy on test data, not just 77 percent.

After discussing the problems of overfitting, we next focus on how to avoid overfitting in decision tree models.

4.6   *PRUNING OF DECISION TREES*

Why does overfitting happen in decision trees? If a tree grows enormously to its full extent, as we know, the train data will show 100 percent accuracy by memorizing the input train data. We aim to learn generic patterns from the data and not remember it. We have to learn how to stop the undesired growth of a decision tree.

We need to "prune" the decision trees to avoid overfitting by supplying the pruning parameters. Pruning parameters will not allow the tree to grow beyond the mentioned size. `Max_depth`, `max_leaf nodes`, and `min_samples_in_leaf` are some examples of the pruning parameters. These pruning parameters are discussed in detail next.

4.6.1 Max_Depth

This parameter puts an upper bound on the depth of the tree. If `max_depth` = 3, then the tree cannot grow beyond the size of depth 3. To generalize it, if we put a limit on depth, the tree will not grow beyond that level. At a depth of one, the tree will have two leaf nodes. At a depth of two, the tree will have a maximum of four leaf nodes. Similarly, at a depth of three, the tree will not have more than eight leaf nodes. This way, we can stop the tree from growing beyond the required complexity. Let us take the tree example in Fig. 4.15.

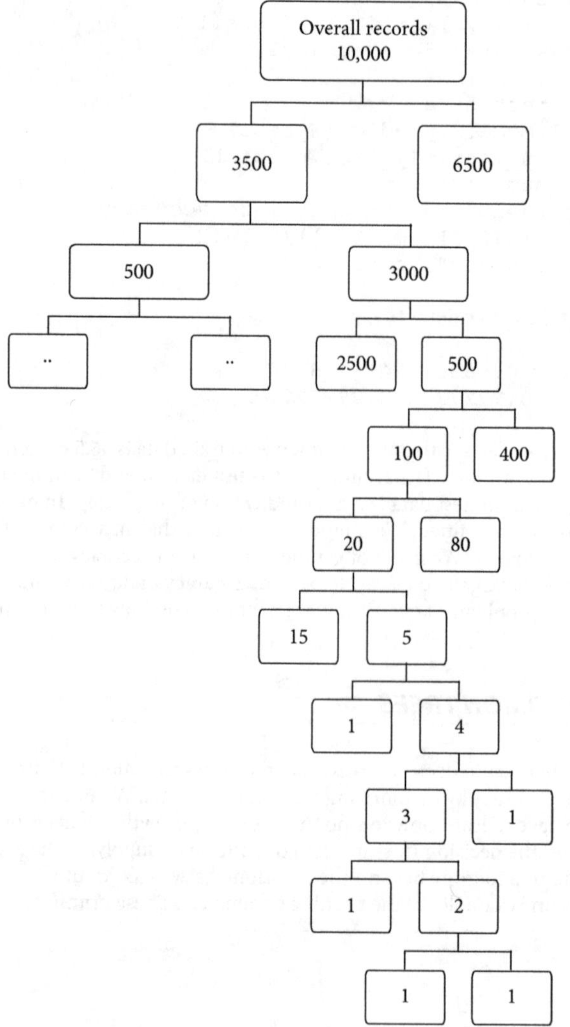

FIGURE 4.15   The tree example to discuss max_depth.

While building the decision tree, if we set `max_depth` = 2 or `max_depth` = 3, the resulting tree would be much smaller (Fig. 4.16). We are showing only one branch for display purposes in this figure; the `max_depth` will work equally on all the branches.

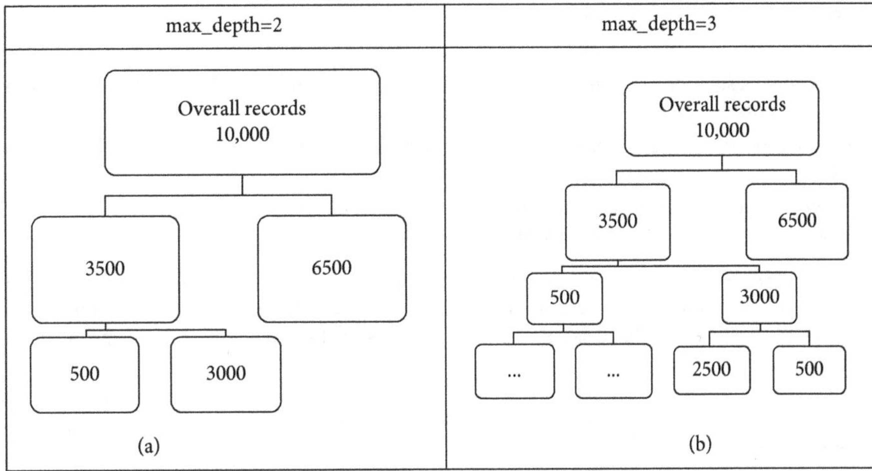

FIGURE 4.16   (*a*) Tree with max_depth = 2; (*b*) tree with max_depth = 3.

We need to mention max_depth while configuring the tree. Given below is the sample code using the `max_depth` pruning parameter. Have a quick look at the output of the decision tree model `DT_Model1`; we can make out that the depth is 6, which means we got an overfitted model at a depth of 6. We will try a lower number and recheck the results.

```
#### max_depth parameter
DT_Model2 = tree.DecisionTreeClassifier(max_depth= 4)
DT_Model2.fit(X_train,y_train)
```

We will now get the train and test data accuracy. Below are the train and test data accuracies.

```
max_depth4 Train Accuracy 0.9425287356321839
max_depth4 Test Accuracy 0.7727272727272727
```

As we can see, the model is still overfitted at a max_depth of 4. We will try `max_depth` =2 in the next attempt.

```
#### max_depth=2
DT_Model2 = tree.DecisionTreeClassifier(max_depth= 2)
DT_Model2.fit(X_train,y_train)
```

Given below is the model with `max_depth=2`.

```
max_depth2 Train Accuracy 0.896551724137931
max_depth2 Test Accuracy 0.8636363636363636
```

Now the train data accuracy is 89.6 percent, and the test data accuracy is 86.3 percent. It is much better than the previous results of 94 percent on the train data and 77 percent on the test data. We can finalize the model with `max_depth` = 2; at this depth, the decision tree is not overfitted. We need to be careful while choosing the max_depth. What will happen if max_depth is less than 2? Let us discuss it in the next section.

4.7   THE CHALLENGE OF UNDERFITTING

In an attempt to avoid overfitting, if we truncate the tree to a small size, then the model might not work well for the train data. It means if the model has real potential for 90 percent accuracy by unduly limiting its size, we may end up getting lesser accuracy, say only 70 percent, on train data. Such models are called underfitted models. If a model exhibits less accuracy on train data, there is no way it will have higher accuracy with the test data and future datasets.

Let us get into more details of underfitting.

- An underfitted model has less accuracy on train data itself.
- If a decision tree is smaller than the optimal depth, it is said to be oversimplified—it always has some scope for further improvements.
- An underfitted model learns fewer patterns in the data. It is a decision tree with less-than-optimal leaf nodes and fewer rules.
- These types of models are also termed as the models with bias. It will have many errors in predictions. Underfitted models have an inherent bias in their parameter estimates.

If a model is underfitted, we need to include more features to improve its accuracy. In the case of a decision tree, we must increase the size of the tree. To achieve it, we can relax the condition of pruning parameters.

In our example, we have finalized `max_depth = 2`. Let us see the model results if we prune the tree further. Let us try with `max_depth = 1`.

```
DT_Model2 = tree.DecisionTreeClassifier(max_depth= 1)
DT_Model2.fit(X_train,y_train)
```

Let us calculate the accuracy with the train and the test data.

```
max_depth1 Train Accuracy 0.8735632183908046
max_depth1 Test Accuracy 0.8181818181818182
```

We can see the accuracy of the train data is 87.3 percent, and for the test data, it is 81 percent. Previously, at a depth of two, we had achieved an accuracy level of 89.6 percent on the train data; the test accuracy was also very near to it. If we stop at `max_depth = 1`, we will end up in an underfitted model. It might perform poorly on future datasets. The example `max_depth = 1` is not an extreme case of underfitting, but undoubtedly an instance of underfitting.

While building a model, we have to make sure it is neither overfitted nor underfitted. If we build a massive tree by configuring the model with a large value of max_depth, the model will be overfitted. If we set it with a minimal value of max_depth, we will end up with an underfitted model. How do we fine-tune the hyperparameter "max_depth"? How do we build the most optimal model for given data? It is achieved by using the binary search method, which is explained in the following section.

4.8   BINARY SEARCH ON PRUNING PARAMETERS

It will help if you appreciate the fundamental fact that you could not build and finalize any decision tree model in the first attempt. You always need to create several models and choose the most optimal one. No one can guess what should be the optimal depth of a decision tree of any given set of data. It has to be discovered by trials. You can try the following:

1. To find the optimal depth, start by building an enormous tree in the first attempt. Depending on the training data, try to get the maximum possible depth. For example, if the max_depth is 20, then we will have nearly a million (2^{20}) leaf nodes, which will create a massive decision tree by any standards. Taking the data size into consideration, you can start with a large max_depth in the first attempt. Call it model_1, which will be overfitted most of the time. You can validate it by observing the training and test accuracy.

2. In the second attempt, build a small tree—a tiny tree with `max_depth = 1`. It will most probably be underfitted. Again, you can look at the training accuracy to confirm. Call it model_2. Both these models have challenges—model_1 is overfitted and model_2 is underfitted. Nevertheless, we have got boundaries. Now we need to search within these two extremes. Just have a quick look at the illustration in Fig. 4.17.

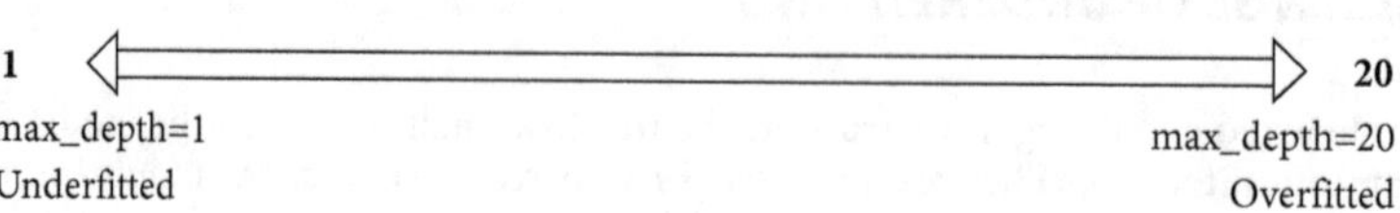

FIGURE 4.17   Illustration of two extremes of max_depth to choose from.

3. In the third attempt, choose a number right in the middle of 1 and 20; try `max_depth = 10`. Build the model on the train data, and call it model_3. If the model is overfitted even at this level, it means we need to reduce the size of the tree further. If you find model_3 is underfitted, we need to grow the tree. Let us assume the tree is overfitted at max_depth = 10 (Fig. 4.18).

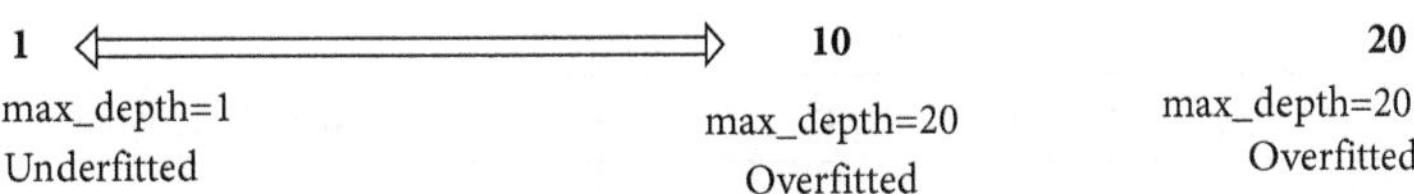

FIGURE 4.18   Illustration of choosing max_depth range.

4. In the fourth attempt, choose a number, right in the middle of 1 and 10; try `max_depth = 5`. Again build the model on the train data and call it model_4. If it is still overfitted, further reduce the depth; otherwise, increase the depth.
5. If you follow this binary search method you will be able to finalize your optimal decision tree model, having the most optimum depth. It may take four or five iterations to arrive at the final results.
6. In our example, we would follow the same approach. Try with `max_depth = 6`; you find that the model is overfitted. With `max_depth = 1`, the model is underfitted as expected. We can now try with `max_depth = 3`; we find the model still somewhat overfitted. Finally, we work with `max_depth = 2` and find it an optimal one. We can use it for future predictions.

Keep on trying with the binary search, and you can finalize the model when you get the highest accuracy on the train data and a matching accuracy on the test data. In the process, you need to avoid a common mistake. If the training data exhibits an accuracy of 75 percent and the test data also shows the same 75 percent accuracy, we tend to finalize the model straight away. Though the train data and test data accuracies are matching, we are still not sure if it is the highest accuracy level as per the true potential of the model. We have to check if there is any chance of getting still higher accuracies, as who knows, even 85 percent accuracy may be possible and the train and test data accuracy levels are yet comparable. The two essential criteria for finalizing a model for production are as follows:

1. The model should have the highest possible accuracy on the train data.
2. The test data accuracy levels should match with the train data.

What would happen if we do not mention any pruning parameter in Python code? The tree will automatically grow to its maximum length. Python functions will not make any effort to stop the growth of the decision tree. Max_depth default value is set to "None," which means the tree will be fully grown. In other words, if we do not mention the pruning parameter, the tree will be overfitted by default. It is the reason why we got an overfitted model `DT_Model1`. The following is the code snippet of the model.

```
DT_Model1 = tree.DecisionTreeClassifier()
DT_Model1.fit(X_train,y_train)
```

In this case, the decision tree will grow to its full extent. While building our first model (`DT_Model`) on Call_center_survey data, we had set `max_depth = 2` to limit the growth of the tree.

```
DT_Model = tree.DecisionTreeClassifier(max_depth=2)
DT_Model.fit(X,y)
```

The set max_depth parameter avoided overfitting.

4.9   MORE PRUNING PARAMETERS

We used `max_depth` as a pruning parameter to control the growth of the tree. It also avoided the challenges of overfitting and underfitting. Is this pruning parameter sufficient, or are there some other useful pruning parameters in existence too?

4.9.1   Maximum Leaf Nodes

Instead of mentioning the depth, we can put a condition on the maximum number of leaf nodes. If we mention `max_leaf_nodes = 2`, the tree will not grow beyond a depth of 1. It is another way to control the growth of a tree. If we say `max_leaf_nodes = 8`, the tree will still be under control, as it has to stop when it reaches eight leaf nodes (Fig. 4.19).

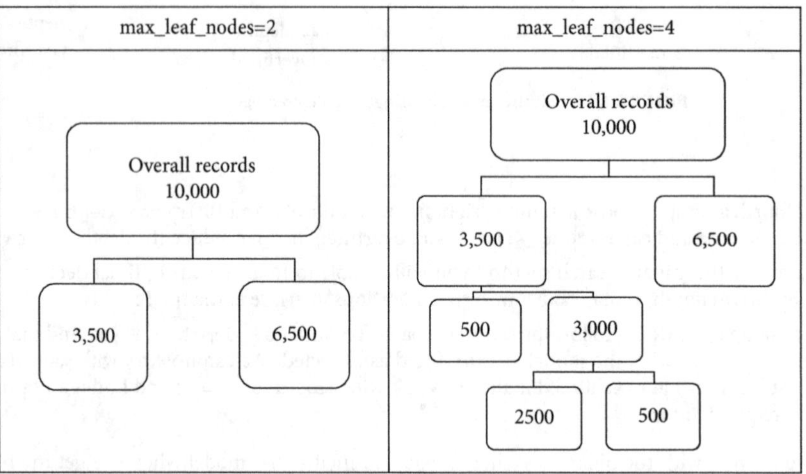

FIGURE 4.19   Illustrations of max_leaf_nodes parameter.

Given below is the code for mentioning the max_leaf nodes.

```
DT_Model3 = tree.DecisionTreeClassifier(max_leaf_nodes= 3)
DT_Model3.fit(X_train,y_train)
```

As far as overfitting and underfitting are concerned, max_leaf_nodes has the same effect as the max_depth parameter. Too large a value of max_leaf_nodes would lead to overfitting and vice versa. We can fine-tune this hyperparameter in the same way as max_depth by using the binary search method. If the value of max_leaf_nodes is not specified, it will be taken as "None," which would lead to a full-blown tree—an overfitted model.

In any model-building exercise we can use any one of these two pruning parameters. What happens if both are mentioned? The tree will stop growing at the condition that occurs first. Consider the following code:

```
tree.DecisionTreeClassifier(max_depth=1, max_leaf_nodes=4)
```

This tree will finally have only two leaf nodes that will satisfy both the conditions of depth and maximum leaf nodes. Now just for investigation purposes, we reverse this order in the code line given below.

```
tree.DecisionTreeClassifier(max_depth=3, max_leaf_nodes=2)
```

This tree will finally have a depth of one and two leaf nodes. It will satisfy both the conditions of maximum depth and maximum leaf nodes. Now consider one more example.

```
tree.DecisionTreeClassifier(max_depth=2, max_leaf_nodes=16)
```

The above tree will have a maximum of four leaf nodes only.

You can use both the pruning parameters, but it is not easy to fine-tune both of them in one go. It is hard to judge which parameter is stopping the tree. We suggest you first fix the maximum depth and then play around with max_leaf_nodes. Apart from these two, there are many more pruning parameters. Minimum samples in the leaf node and minimum sample split are two more such pruning parameters. All of them do the same job of trying to control the growth of a decision tree. Let us consider the pruning parameter "minimum samples in the leaf node" next.

4.9.2 Minimum Samples in the Leaf Node

As the name suggests, it is the least number of samples any leaf node should contain. If it is set to 1, the tree will grow massive. If samples in the leaf node parameter are set to a large number, the tree will shrink. If we set this condition as follows: every leaf should contain at least 3000 records, and we do not want any node with fewer records. If this parameter is set to a tiny number, the tree size will be large. This phenomenon is illustrated in the two charts in Fig. 4.20.

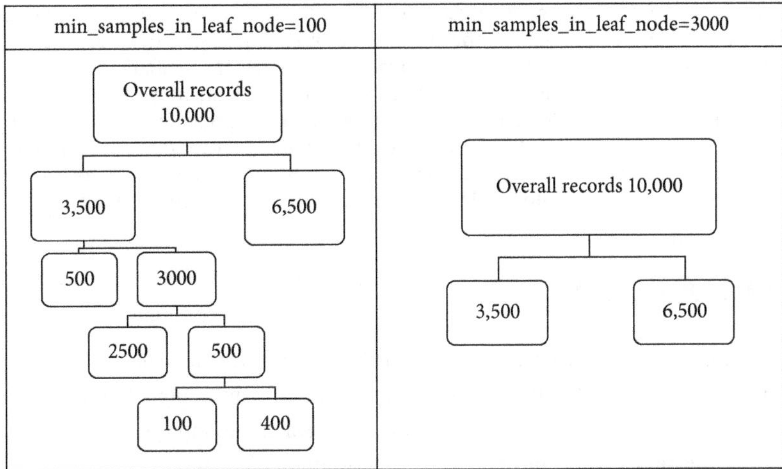

FIGURE 4.20   Illustrations of min_samples_in_leaf_node parameter.

Typically, we do not use this parameter for pruning and fine-tuning. It is used if we need to have a higher number of records in the leaf nodes. Every leaf node gives us the rules. When considering the rules from any tree, we would like to see sufficient data to have confidence in these rules. If we want to avoid rules formed by, say, only 30 records or 10 records, we can mention this parameter. There is another way of putting it: If we want at least 100 records to form a rule, we can set min_samples_in_leaf_node = 100. We do not fine-tune this parameter. We leave it as a constant. As a general rule of thumb, every leaf should have at least 5 percent of the data. It means if we have N records in the data, set min_samples_in_leaf_node = 0.05*N. The default for this parameter in Python is 1, which means the tree will be overfitted if we do not mention this parameter.

4.10   STEPS IN BUILDING A DECISION TREE MODEL

We have discussed several measures and metrics in this chapter. Given below are the steps that we need to follow while building the decision tree models.

1. Import the data. Explore the data, and validate and clean the data.

2. Divide the overall data into two parts. Take 80 percent of the data as the train data and use the remaining 20 percent for testing.

3. Build the model on the training data.

4. Create the confusion matrix to find the accuracy on the train data.

5. Validate the model with test data; create the confusion matrix and find the accuracy on the test data.

6. Compare the train and test data accuracies; fine-tune the model using pruning parameters. Do not forget the following facts in the process.

    a. max_depth(default = "None" – Infinity)
        i. Higher max_depth leads to overfitting.
        ii. Very low max_depth leads to underfitting.

    b. max_leaf_nodes(default = "None" – Infinity)
        i. Higher max_leaf_nodes leads to overfitting.
        ii. Very low max_leaf_nodes leads to underfitting.

7. Build several models with different configurations. Perform a binary search to find the most optimal model. Finalize the model when you get the highest accuracy on the train data and matching accuracy on the test data.

4.11   *CONCLUSION*

The decision tree algorithm is a popular classification technique. A decision tree will not only give us predictions but at the same time also provides us with the customer segments. Decision trees are one of the most widely used algorithms in the customer segmentation type of business problems. Most marketing teams use decision trees for this purpose. The best part of a decision tree is the simplicity and explainability of the model. It is not a black-box model. Decision tree models give us rules that are straightforward to comprehend. These rules are very easy to deploy on any platform. When model deployment and simplicity are of topmost priority, a decision tree is a preferred algorithm. A good understanding of the decision tree is required to understand the algorithms such as random forests and boosting. In the next chapter, we will get into the details of model selection and cross-validation.

4.12   *PRACTICE PROBLEMS*

1. Download the IBM HR Analytics Employee Attrition & Performance dataset.
   - Import the data. Complete the necessary exploration and sanitization of the data.
   - Build a machine learning model to predict the attrition in the employees.
   - Perform the model validation and measure the accuracy of the model.
   - Check for innovative ways to improve the accuracy of the model.

   Dataset credits—https://www.kaggle.com/pavansubhasht/ibm-hr-analytics-attrition-dataset. License—Open Database (DBCL).

2. Download the Breast Cancer Wisconsin (Diagnostic) dataset.
   - Import the data. Complete the necessary exploration and sanitization of the data.
   - Build a machine learning model to classify breast tumors as malignant or benign.
   - Perform the model validation and measure the accuracy of the model.
   - Check for innovative ways to improve the accuracy of the model.

   Dataset credits—http://archive.ics.uci.edu/ml/datasets/breast+cancer+wisconsin+(diagnostic). *Sources:* (1) Dr. William H. Wolberg, General Surgery Department, University of Wisconsin, Clinical Sciences Center Madison, WI 53792 (wolberg@eagle.surgery.wisc.edu). (2) W. Nick Street, Computer Sciences Department, University of Wisconsin, 1210 West Dayton St., Madison, WI 53706 (street@cs.wisc.edu). (3) Olvi L. Mangasarian, Computer Sciences Department, University of Wisconsin, 1210 West Dayton St., Madison, WI 53706 (olvi@cs.wisc.edu).

CHAPTER 5
MODEL SELECTION AND CROSS-VALIDATION

It is not advisable to jump straightway on to model building as soon as you have finished acquiring the raw data. Rarely do we get the data good enough to build the model. Almost in all cases, we need to explore and validate the data. During the data validation process, if we find any problems such as outliers and erroneous values, we should first go for data cleansing before attempting any model building.

Similarly, we never jump onto deploying a model just after creating it; we need to validate it thoroughly. We should make sure the model adheres to all the validation conditions and demonstrates consistent accuracy with in-hand train data, test data, and the production datasets. While building a model, we use some standard model validation techniques such as R-squared in regression and accuracy in classification models. In some specific cases, these standard metrics may fail; in such cases, alternative validation measures are adopted. In this chapter, we will discuss all the critical steps and parameters required to validate models.

Assume that we had completed our model-building and deployment exercise last year and that the model is running well and yielding good results. Can we reuse the same model in the current year as well? If there is not much change in the data patterns, we can continue using the same model. We need to monitor the model performance by validating it at regular intervals. In other words, we need to prove the model accuracy has not deteriorated, having completed one year in production. We periodically revalidate the models on new data, which is known as out-of-time validation data. Remember that the data we had used for building the model was different; we called it the train data. In this chapter, we are going to discuss the steps involved in managing and monitoring models.

In some companies, there are dedicated teams for performing the task of model monitoring and validation. This task is carried out by an independent team which was not involved in model building. This team monitors and validates models to ensure that they are robust and ready for deployment. In the model-building life cycle, model validation is a separate phase. If we find any issues in this phase, we go back to model building and fix problems to make sure only the right model is selected for implementation. The content of this chapter mostly comes from the practical experience of the authors.

5.1 STEPS IN BUILDING A MODEL

Model-building life cycle involves the following five steps:

1. Define the objective(s).
2. Explore, validate, and prepare the data.
3. Build the model.
4. Validate the model.
5. Deploy the model.

These five steps are primarily followed in most of the model-building projects. There might be some industry-specific variations. For example, some of the banking models need to get approvals from regulatory authorities before going ahead with deployment. Some other industry professionals take data gathering as a separate step. Let us get into more details on the above five steps.

1. The first step, i.e., define the objective(s), accomplishes the following:
   - Defining the aim of the entire model building exercise.
   - Answer the question: What are we trying to predict or classify using this model?
   - Identify the input and output features.

For example, the aim of a project can be fraud detection in financial transactions using the associated features.

2. Explore, validate, and prepare the data step involves the following:
   - Do essential data exploration on rows and columns, and identify fundamental issues like data formatting, missing values, and many more.
   - Perform univariate analysis on all the columns, and understand each variable distribution to identify outliers.
   - Clean the data and prepare it for the model building.
   - Convert non-numerical data to numerics using dummy variables.
   - Use imputation techniques to deal with outliers and missing values.
   - Make the necessary transformations for skewed data.

For example, if a variable name is "amount_of_transaction," and it is missing in 2 percent of the data rows, we can replace the missing values with the mean or median of the rest of the data.

3. The third step is model building, which involves the following:
   - Choose the right model-building technique.
   - Check if the target variable is continuous or a class is of variable type to choose a suitable regression or classification technique.
   - Try two or three different algorithms to choose the best fit.
   - Try to include new or derived features to improve the accuracy of the model.

For example, build the logistic regression or decision tree model for predicting fraud versus non-fraud target.

4. The fourth step is model validation which involves the following:
   - Perform necessary validation checks using R-squared or accuracy.
   - Perform validation checks with train and test data.
   - Find out the possible issues or cases where the model may fail.

For example, many project teams use the confusion matrix and accuracy to validate fraud prediction models.

5. The fifth step involves deployment of the model, which is carried out as follows:
   - Decide the deployment platform; it may be a website, or a SAP application, or SQL Server, or Android, or some other cloud platform.
   - Export the final logic of the model to the deployment platform, which in logistic regression is the model equation; in tree models, it is a set of rules. Take care to code the logic separately.
   - Some platforms have readymade applications like flask to deploy the model quickly using the REST API.

For example, we may want to deploy a decision tree model on the SQL Server, where we need to code the decision tree rules logic in SQL language. Data scientists are intensely involved in the three phases of data exploration, model building, and validation. In practice, often we may get the objectives set by clients or business partners. The final deployment may not be on the Python platform; we may have platform experts to perform this step.

5.2   MODEL VALIDATION MEASURES: REGRESSION

In regression models, the output is a continuous variable; the target variable may take any value in its range. We have already discussed the R-squared measure in Chap. 3. R-squared value is a good measure of model accuracy. It talks about the overall model accuracy and the total variance explained by the model.

There are a few more measures just like R-squared. They are also used to get an intuition on how close or far the predicted values are from the actual values. Let us discuss three such measures.

1. Mean absolute deviation (MAD) measure, as defined by the following equation:

$$MAD = \sum_{i=1}^{n} \frac{|y_i - \hat{y}_i|}{n}$$

- As stated, here also we find the deviation of predicted value from the actual value. Both negative and positive deviations are errors, so we go for the absolute deviation. The average of all absolute deviations is known as mean absolute deviation (MAD).
- We calculate MAD on both types of data—train data and test data. A good model will have near to zero MAD on both types of data. MAD gives the average deviation of predictions from the actual values.
- While working with MAD, we are not sure if any calculated value, use 1000, for example, is higher or lower, unless we see the scale of target variable y. If y is in millions then a MAD of 1000 is very less; on the other hand, if y value is in the thousands then a MAD of 1000 is considered high.

2. Mean absolute percentage error (MAPE) is given by the following equation:

$$MAPE = \frac{100}{n}\sum_{i=1}^{n}\frac{|y_i - \hat{y}_i|}{y_i}$$

- Here, we tweak the MAD formula and convert each deviation into the percentage of the actual value.
- If we are interested in knowing the deviation percentage instead of actual deviation, then we can use MAPE.
- In MAPE, we do not need to worry about the scale of the variable. A MAPE value of 2 percent is always lower than MAPE of 10 percent, no matter what is the scale of y.

3. We have an alternate measure as the root mean squared error (RMSE):

$$RMSE = \sqrt{\sum_{i=1}^{n}\frac{\left(y_i - \hat{y}_i\right)^2}{n}}$$

When we compare two modes, a lower value of RMSE is preferred.

All these measures use different formulas to explain errors. Depending on an organization or a professional, different measures are used in various projects. One must be aware of the prevalent practices.

5.3 CASE STUDY: HOUSE SALES IN KING COUNTY, WASHINGTON

In this example, the dataset contains house sale prices for King County, Washington, which includes Seattle. The data contains the details of all houses sold between May 2014 and May 2015. This data is available under License CC0: Public Domain.

5.3.1 Objective and Data

The problem statement is to predict the house price using features such as the number of bedrooms, the number of bathrooms, age of construction, square feet area, and location of the house. In this exercise, we will discuss model building and calculate all the validation metrics.

As a best practice, let us look at the dataset and some basic details before we attempt any further. The following code imports the data and gives us the basic details.

```
kc_house_data = pd.read_csv(r'D:\Chapter5\
Datasets\kc_house_data.csv\kc_house_data.csv')

#Get an idea on the number of rows and columns
print(kc_house_data.shape)

#Print the column names
print(kc_house_data.columns)

#Print the column types
print(kc_house_data.dtypes)

#Additional Details
kc_house_data.info()
```

```python
#Summary
all_cols_summary=kc_house_data.describe()
print(round(all_cols_summary,2))
```

The code output is given below. Let us get into the details.

```python
print(kc_house_data.shape)
(21613, 21)
```

```python
print(kc_house_data.columns)
Index(['id', 'date', 'price', 'bedrooms', 'bathrooms', 'sqft_living', 'sqft_
lot', 'floors', 'waterfront', 'view', 'condition', 'grade', 'sqft_above',
'sqft_basement', 'yr_built', 'yr_renovated', 'zipcode', 'lat', 'long', 'sqft_
living15', 'sqft_lot15'], dtype='object')
```

There are 21,613 rows and 21 columns in the dataset. The column names are self-explanatory, which describe the home and get us the details such as the number of bedrooms, number of bathrooms, number of floors, living room area, and year of construction. With all this data as input, the target variable we are trying to predict here is "price." The following table depicts the data types of all the columns.

```python
print(kc_house_data.dtypes)
id int64
date object
price int64
bedrooms int64
bathrooms float64
sqft_living int64
sqft_lot int64
floors float64
waterfront int64
view int64
condition int64
grade int64
sqft_above int64
sqft_basement int64
yr_built int64
yr_renovated int64
zipcode int64
lat float64
long float64
sqft_living15 int64
sqft_lot15 int64
```

All the variables are numeric except for the date variable. Keeping aside the date variable, we will use the rest of all variables to build our model. Let us now check for the missing values using the info() function.

```python
kc_house_data.info()
<class 'pandas.core.frame.DataFrame'>
RangeIndex: 21613 entries, 0 to 21612
Data columns (total 21 columns):
id            21613 non-null int64
date          21613 non-null object
price         21613 non-null int64
bedrooms      21613 non-null int64
bathrooms     21613 non-null float64
sqft_living   21613 non-null int64
sqft_lot      21613 non-null int64
floors        21613 non-null float64
waterfront    21613 non-null int64
```

```
view              21613 non-null int64
condition         21613 non-null int64
grade             21613 non-null int64
sqft_above        21613 non-null int64
sqft_basement     21613 non-null int64
yr_built          21613 non-null int64
yr_renovated      21613 non-null int64
zipcode           21613 non-null int64
lat               21613 non-null float64
long              21613 non-null float64
sqft_living15     21613 non-null int64
sqft_lot15        21613 non-null int64
```

All the columns have data populated; none of the columns have any missing values. Next, we will go through the summary of each column.

```
all_cols_summary=kc_house_data.describe()
print(round(all_cols_summary,2))
                    id          price   bedrooms   bathrooms   sqft_living    sqft_lot\
count    2.161300e+04    21613.00    21613.00    21613.00      21613.00      21613.00
mean     4.580302e+09   540088.14        3.37        2.11       2079.90      15106.97
std      2.876566e+09   367127.20        0.93        0.77        918.44      41420.51
min      1.000102e+06    75000.00        0.00        0.00        290.00        520.00
25%      2.123049e+09   321950.00        3.00        1.75       1427.00       5040.00
50%      3.904930e+09   450000.00        3.00        2.25       1910.00       7618.00
75%      7.308900e+09   645000.00        4.00        2.50       2550.00      10688.00
max      9.900000e+09  7700000.00       33.00        8.00      13540.00    1651359.00

            floors    waterfront        view    condition       grade    sqft_above \
count     21613.00      21613.00    21613.00     21613.00    21613.00      21613.00
mean          1.49          0.01        0.23         3.41        7.66       1788.39
std           0.54          0.09        0.77         0.65        1.18        828.09
min           1.00          0.00        0.00         1.00        1.00        290.00
25%           1.00          0.00        0.00         3.00        7.00       1190.00
50%           1.50          0.00        0.00         3.00        7.00       1560.00
75%           2.00          0.00        0.00         4.00        8.00       2210.00
max           3.50          1.00        4.00         5.00       13.00       9410.00

          sqft_basement    yr_built   yr_renovated    zipcode        lat       long \
count          21613.00    21613.00       21613.00    21613.00    21613.00   21613.00
mean             291.51     1971.01          84.40    98077.94       47.56    -122.21
std              442.58       29.37         401.68       53.51        0.14       0.14
min                0.00     1900.00           0.00    98001.00       47.16    -122.52
25%                0.00     1951.00           0.00    98033.00       47.47    -122.33
50%                0.00     1975.00           0.00    98065.00       47.57    -122.23
75%              560.00     1997.00           0.00    98118.00       47.68    -122.12
max             4820.00     2015.00        2015.00    98199.00       47.78    -121.32

          sqft_living15    sqft_lot15
count          21613.00      21613.00
mean            1986.55      12768.46
std              685.39      27304.18
min              399.00        651.00
25%             1490.00       5100.00
50%             1840.00       7620.00
75%             2360.00      10083.00
max             6210.00     871200.00
```

Overall the data seems to be in good shape. There are a few columns with outliers. However, we will go ahead with the model building to see the accuracy of version 1. Given below is the code.

```python
import statsmodels.formula.api as sm
model1 = sm.ols(formula='price ~ bedrooms+bathrooms+sqft_living+sqft_lot+flo
ors+waterfront+view+condition+grade+sqft_above+sqft_basement+yr_built+yr_
renovated+zipcode+lat+long+sqft_living15+sqft_lot15', data=kc_house_data)
fitted1 = model1.fit()
fitted1.summary()
```

Until now, we have used the "statsmodels" package to perform regression tasks. In this exercise, we will try to use an alternative package "sklearn." The syntax of sklearn is different, but the results are the same. The interpretation of R-squared and all other related regression measures remains the same. The summary() function is available in statsmodels package; that is the reason we started with it. There is no summary function available in the sklearn package. We manually need to fetch the summary. Many data scientists use the "sklearn" package; better, we also know it.

By now, we know which metrics are essential in the regression output. We can fetch all those metrics from the model object individually in this package. First, we will write the code to create train data and test data as follows:

```python
#Defining X data
X = kc_house_data[['bedrooms', 'bathrooms', 'sqft_living', 'sqft_lot',
'floors', 'waterfront', 'view', 'condition', 'grade', 'sqft_above', 'sqft_
basement', 'yr_built', 'yr_renovated', 'zipcode', 'lat', 'long', 'sqft_liv-
ing15', 'sqft_lot15']]

y = kc_house_data['price']

from sklearn import model_selection
X_train, X_test, y_train, y_test = model_selection.train_test_split(X, y,
test_size=0.2, random_state=55)

print(X_train.shape)
print(y_train.shape)
print(X_test.shape)
print(y_test.shape)
```

The above code gives us the below output.

```python
print(X_train.shape)
(17290, 18)

print(y_train.shape)
(17290,)

print(X_test.shape)
(4323, 18)

print(y_test.shape)
(4323,)
```

5.3.2 Model Building and Validation

We have already discussed that 80 percent of the overall data is considered for training and the remaining 20 percent for testing. Now we are ready to build the model using the sklearn package.

```python
import sklearn
model_1 = sklearn.linear_model.LinearRegression()
model_1.fit(X_train, y_train)
```

When we execute the above code, the model will be fit and stored in model_1. There is no summary() function in this package. We can fetch the coefficients and R-squared values using the below code.

```
#Coefficients and Intercept
print(model_1.intercept_)
print(model_1.coef_)

#Rsquared value on train and test data
from sklearn import metrics
y_pred_train=model_1.predict(X_train)
print(metrics.r2_score(y_train,y_pred_train))

y_pred_test=model_1.predict(X_test)
print(metrics.r2_score(y_test,y_pred_test))
```

The above code gives the below output.

```
print(model_1.intercept_)
8080822.666112712

print(model_1.coef_)
[-3.76187142e+04  4.39929752e+04 1.11927627e+02  1.12260521e-01
  7.86848634e+03  5.82851207e+05 5.24147307e+04  2.56475517e+04
  9.63780999e+04  7.02315647e+01 4.16960620e+01 -2.66443082e+03
  2.22630357e+01 -5.83768676e+02 6.04058556e+05 -2.04643130e+05
  1.84979337e+01 -3.70481687e-01]

print(metrics.r2_score(y_train,y_pred_train))
0.7004310823997761

print(metrics.r2_score(y_test,y_pred_test))
0.6964362880041228
```

We see that the intercept is a large value, mainly because of the scale of the target. The rest of the coefficients are shown in scientific number format. As a refresher, if the number contains e+04, then it means you multiply the number by 10,000; if the number contains e-02, then you divide by 100. The R-squared value of the model on train data is 0.700, i.e., 70 percent, and the R-squared value on the test data is 0.696, i.e., 69.6 percent. The following code calculates the rest of the validation metrics.

```
#MAD
print("MAD on Train data : ", round(np.mean(np.abs(y_train - y_pred_train)),2))

print("MAD on Test data : ", round(np.mean(np.abs(y_test - y_pred_test)),2))

#MAPE
print("MAPE on Train data : ", round(np.mean(np.abs(y_train - y_pred_train)/
y_train),2))

print("MAPE on Test data : ", round(np.mean(np.abs(y_test - y_pred_test)/
y_test),2))

#RMSE
print("RMSE on Train data : ", round(math.sqrt(np.mean(np.abs(y_train -
y_pred_train)**2)),2))

print("RMSE on Test data : ", round(math.sqrt(np.mean(np.abs(y_test
- y_pred_test)**2)),2))
```

The following table gives us the output.

```
MAD on Train data : 125920.32
MAD on Test data : 126818.8
MAPE on Train data : 0.26
MAPE on Test data : 0.26
RMSE on Train data : 202295.52
RMSE on Test data : 196693.42
```

Let us have a look at the average price and overall percentiles.

```
round(kc_house_data.price.describe())
count   21613.0
mean    540088.0
std     367127.0
min      75000.0
25%     321950.0
50%     450000.0
75%     645000.0
max    7700000.0
Name: price, dtype: float64
```

Recall our discussion of MAD, MAPE, and RMSE from the previous sections. From the two tables given above, it is easy to make the following observations:

- The average value of house prices is 540,088.
- The mean absolute deviation of the house price is 125,920.
- The percentage deviation between the actual and predicted value is 26 percent.

This model is just a basic benchmark model. While building this model, we have not cleaned the data for outliers. Also, we have not put any effort into improving the model. Notice that we have used some variables such as zipcode as they are. If we use data cleaning and feature engineering techniques, we can further improve the overall accuracy of the model with the same data and build on the same algorithm. The process of extracting the hidden information from the features so that the model can find it easy to learn is called feature engineering. We will revisit the same example in the "Feature Engineering Tips and Tricks" section, later in this chapter, where we will try to lift the model accuracy from the current 70 percent to a higher value. In the following section, let us get into the details of some model validation measures.

5.4   MODEL VALIDATION MEASURES: CLASSIFICATION

While working with regression models, we have used R-squared and other deviation-based validation measures. When it comes to binary classification, we have only 0s and 1s in the output. A deviation-based method, that is, actual versus predicted, may not work in the case of classification type of problems. We will create a confusion matrix and derive accuracy from actual and predicted classes.

5.4.1   Confusion Matrix and Accuracy

We have already discussed the confusion matrix depth in Chap. 3. The following table is just a quick recap. For example, we have built a classification model, and Table 5.1 contains both actual and predicted values.

TABLE 5.1   Actual and Predicted Values for a Sample Dataset

S. No.	x_1	x_2	...	x_k	y	$\hat{y}$
1					1	1
2					1	0
3					1	1
4					0	0
5					0	0
6					0	1
7					1	1
8					0	1
9					1	1
10					0	0

Once we have actual and predicted values, we are ready to create a confusion matrix. A confusion matrix is nothing but the cross-tab between actual and predicted classes (Table 5.2).

TABLE 5.2   Confusion Matrix

		Predicted Classes	
		0	**1**
Actual Classes	**0**	Number of times 0 is predicted as 0	Number of times 0 is predicted as 1
	1	Number of times 1 is predicted as 0	Number of times 1 is predicted as 1

Accuracy is derived from the confusion matrix as follows:

$$\text{Accuracy} = \frac{\text{Right Classifications}}{\text{Overall Records}}$$

$$= \frac{\text{cm}[0,0] + \text{cm}[1,1]}{\text{cm}[0,0] + \text{cm}[0,1] + \text{cm}[1,0] + \text{cm}[1,1]}$$

In the actual practice, there also exists a different terminology. Class0 is called positive, while Class1 is termed as negative (Table 5.3); as you will notice, this terminology is much easier to use in the same confusion matrix accuracy formula, as explained above.

TABLE 5.3   Confusion Matrix in Proper Terminology

		Predicted Classes	
		Positive	**Negative**
Actual Classes	**Positive**	True Positives (TP)	False Negatives (FN)
	Negative	False Positives (FP)	True Negatives (TN)

$$\text{Accuracy} = \frac{\text{Right Classifications}}{\text{Overall Records}}$$

$$= \frac{TP + TN}{TP + FN + FP + TN}$$

Table 5.4 presents the confusion matrix and accuracy calculations for our example.

TABLE 5.4   Confusion Matrix for the Example Problem

		Predicted Classes	
		0	**1**
Actual Classes	**0**	3	2
	1	1	4

$$\text{Accuracy} = \frac{3+4}{3+2+1+4}$$
$$= 0.7$$

5.4.2   Measures for Class Imbalance

Let us take an example of a model where we are trying to predict whether there is a bomb in the car or not. If the car stands on a special platform, we collect some specific information using various sensors. Based on this sensor data, we will try to predict if there is a bomb present in the car. We have the test data that has 100,000 records. In this test data, one car has a bomb. We applied the model on this test data, and the model predicts none of the cars has a bomb. The model has predicted all the 100,000 cars as safe. Let us create the confusion matrix. The two classes are 0 and 1 as usual, with 0 as no bomb and 1 as a bomb present in the car (Table 5.5).

TABLE 5.5   Confusion Matrix for Car Bomb Example

		Predicted Classes	
		0 – No Bomb	**1 – Bomb**
Actual Classes	**0 – No Bomb**	99,999	0
	1 – Bomb	1	0

In our test data, there are 99,999 cars without a bomb. The model has predicted all of them as cars without a bomb. Our model could not accurately predict the car with a bomb in it. Overall, there is only one occasion where our model has mispredicted. Let us calculate the accuracy of this model.

$$\text{Accuracy} = \frac{\text{cm}[0,0] + \text{cm}[1,1]}{\text{cm}[0,0] + \text{cm}[0,1] + \text{cm}[1,0] + \text{cm}[1,1]}$$
$$= \frac{99,999 + 0}{100,000}$$
$$= 0.99999$$
$$= 99.999\%$$

The calculated accuracy of the model is 99.999 percent, which sounds perfect. However, is this model solving its purpose? Why did we build this model? Of course, we built this model for prediction of the bomb, but the model has failed here. Is the model useful? This model takes all the input and gives 0 as the output value for 100 percent of the input records. This has happened due to the huge imbalance in the classes. As we observe, one class is not at all comparable to the other class. There is a vast difference between the number of 0s and the number of 1s (99,999 class-0 vs. only one row of class1). In any real-life scenario as well, the car with a bomb may be as rare as one in 1 million.

The car with a bomb is just one example. There are so many such examples where one class is very rare compared to the other.

- *Fraud* versus *non-fraud*: One in 1 million transactions or even less percentage may be a fraud. In these cases also, the two classes are not equally distributed.

- *Responder* versus *non-responder*: If we run a marketing campaign, people who respond to our campaign will be very less in proportion in most cases. People very rarely respond to a marketing mail.

The confusion matrix accuracy gives equal preference to both the classes. In the cases with a huge class imbalance, if we do a good job of prediction in the most occurring class, the overall accuracy with the confusion matrix will automatically look good. In these cases, we need to consider a different validation measure. In all the cases where we have such a class imbalance, we should look at the individual class accuracy rather than the overall accuracy. In the same bomb example, we will try to calculate classwise accuracy. The accuracy of class0 is calculated by dividing the right predictions in class0 to the total of class0 cases (Table 5.6).

TABLE 5.6   Formulas for Calculating Accuracy Using Class Predictions

		Predicted Classes		
		0	**1**	**Classwise Accuracy**
Actual Classes	**0**	Number of times 0 is predicted as 0	Number of times 0 is predicted as 1	$\dfrac{cm[0,0]}{cm[0,0] + cm[0,1]}$
	1	Number of times 1 is predicted as 0	Number of times 1 is predicted as 1	$\dfrac{cm[1,1]}{cm[1,0] + cm[1,1]}$

Here is another way of looking at it (Table 5.7).

TABLE 5.7   Another Way of Looking at the Expressions for Calculating Accuracy Using Class Predictions

		Predicted Classes		
		Positive	**Negative**	**Classwise Accuracy**
Actual Classes	**Positive**	True Positives (TP)	False Negatives (FN)	$\dfrac{TP}{TP + FN}$
	Negative	False Positives (FP)	True Negatives (TN)	$\dfrac{TN}{FP + TN}$

Table 5.8 calculates classwise accuracy for our example.

TABLE 5.8   Calculates Classwise Accuracy for the Car Bomb Example

		Predicted Classes		
		0 – No Bomb	**1 – Bomb**	**Classwise Accuracy**
Actual Classes	**0 – No Bomb**	99,999	0	99,999/99,999
	1 – Bomb	1	0	0/1

Accuracy of class0, which is the no bomb class, is 100 percent.

Accuracy of class1, which is the bomb class, is 0 percent.

In this example, we observe that the overall confusion matrix accuracy is mainly driven by class0. However, we are interested in class1, where the prediction accuracy is 0 percent. This is a classic example where overall accuracy does not give us the right picture, so there is a need to look at the individual classwise accuracy. As an industry standard, the accuracy of class0 is known as sensitivity, while the accuracy of class1 is called specificity.

5.4.2.1   Sensitivity   Sensitivity is the accuracy of the first class, which we usually denote as class0 or positive class as an accepted convention. A model has a high sensitivity when it has correctly predicted many records related to class0.

$$\text{Sensitivity} = \frac{\text{Number of times 0 is predicted as 0}}{\text{Overall occurrences } of \ 0}$$

$$= \frac{\text{cm}[0,0]}{\text{cm}[0,0] + \text{cm}[0,1]}$$

$$= \frac{\text{True Positives}(TP)}{\text{True Positives}(TP) + \text{Flase Negatives}(FN)}$$

There are some cases where sensitivity is of utmost importance. Let us look at the confusion matrix presented in Table 5.9. Here, we are trying to predict whether a customer is good or bad before giving him a personal loan. Bad customers are known as defaulters and good customers are non-defaulters. These models are known as credit risk models.

TABLE 5.9   Sensitivity and Specificity

		Predicted Classes		
		0 – Bad Customer	**1 – Good Customer**	**Classwise Accuracy**
Actual Classes	**0 – Bad Customer**	Model is predicting bad customer as bad	Model is predicting bad customer as good	*Sensitivity*
	1 – Good Customer	Model is predicting good customer as bad	Model is predicting good customer as good	*Specificity*

In the matrix in Table 5.9, what is of interest to us? We are not worried about the diagonal elements. Predicting bad customers as bad and predicting good customers as good are the right predictions. There are two types of errors here. The model predicting bad customers as good is one type of error; the second type of error is predicting good customers as bad. Let us look at the business implications of all these cells in the confusion matrix.

TABLE 5.10   Sensitivity and Specificity in Loan Application Use Case

		Predicted Classes		
		0 – Bad Customer	**1 – Good Customer**	**Classwise Accuracy**
Actual Classes	**0 – Bad Customer**	Reject the loan	Approve the loan	*Sensitivity*
	1 – Good Customer	Reject the loan	Approve the loan	*Specificity*

Whenever the model predicts the customer as a good customer, we go ahead and approve the loan. If the model predicts them as bad customers, we reject the loan application. Out of the two error cases of the confusion matrix, we are wrongly approving a loan in the first case. In the second error, we are wrongly rejecting a loan. Both are serious errors from a business point of view. Both of them should be minimized. However, out of these two errors, the critical error is the first error—wrongly approving a loan (Table 5.10).

The error in $cm[0,1]$ is not the same as the error in $cm[1,0]$. It means approving a loan mistakenly for a bad customer is not the same as rejecting one good customer. By approving one bad loan, the bank may lose the entire principal, while by rejecting a good customer, the bank may lose profit. Normally, in the credit risk industry, we give a higher preference to the class0 or "bad" class in this case. So sensitivity becomes very important here. There are some different names given to sensitivity. It is also known as recall rate or hit rate or true positive rate (TPR).

5.4.2.2 Specificity   The accuracy of the second class, that is, class1, is specificity. Its value depends upon, out of all records in class1, how many times our model has predicted class1 correctly.

$$\text{Specificity} = \frac{\text{Number of times 1 is predicted as 1}}{\text{Overall occurrences of 1}}$$

$$= \frac{cm[1,1]}{cm[1,0] + cm[1,1]}$$

$$= \frac{\text{True Negatives } (TN)}{\text{False Positives } (FP) + \text{True Negatives } (TN)}$$

Sometimes specificity is important for us. For example, we have two classes as authentic versus fraud transactions. Class0 is non-fraud transactions; class1 is a fraud transaction (Table 5.11).

TABLE 5.11   Sensitivity and Specificity in Terms of Good and Fraud Transactions

		Predicted Classes		
		0 – Good transaction	**1** – Fraud transaction	**Classwise Accuracy**
Actual Classes	**0** – Good transaction	Model is predicting a good transaction as good	Model is predicting a good transaction as fraud	*Sensitivity*
	1 – Fraud transaction	Model is predicting a fraud transaction as good	Model is predicting a fraud transaction as fraud	*Specificity*

Just like sensitivity, there are two types of errors here. In the first, we are predicting a good transaction as a fraud. The second error is when the model is predicting fraud transactions as good. Table 5.12 shows the business implications of these predictions.

TABLE 5.12   Transaction Inferences Based on Sensitivity and Specificity

		Predicted Classes		
		0 – Good transaction	**1** – Fraud transaction	**Classwise Accuracy**
Actual Classes	**0** – Good transaction	Accept the transaction	Reject the transaction	*Sensitivity*
	1 – Fraud transaction	Accept the transaction	Reject the transaction	*Specificity*

Mistakenly rejecting a transaction may make the user try the same transaction multiple times, whereas mistakenly accepting a transaction will cause much loss. One error in cm[0,1] is not the same as cm[1,0]—both have different business implications. Technically speaking, both are errors. However, specificity is more serious here. In all fraud detection models, we focus on the fraud class. It is very rare. Specificity is also known as selectivity or true negative rate.

In the case of class imbalance, we do not care about overall accuracy. We focus on the class that is important for us. Sometimes sensitivity is important, and in some cases, specificity takes priority. We need to look at our data and problem statement to identify which class is important. Let us look at an example to calculate accuracy, sensitivity, and specificity.

5.4.2.3   *Case Study: Credit Risk Model*   The following case study is an example of a credit risk model, which is made to find potential defaulters. The credit scores of clients are decided based on these credit risk models. The dataset used here is created for the "Give me some credit" competition on the kaggle.com website. We have to sign up and create a login ID on Kaggle to access this data.

A bank wants to predict if a customer is good or bad. The bank has collected two years of historical data. We will build a model on this historical data and use it to predict defaulters in the new data. The below code is used to import the data and explore it.

```python
import pandas as pd
credit_risk_data = pd.read_csv(r'D:\Chapter5\5. Base Datasets\loans_data\
credit_risk_data_v1.csv')

#Get an idea on the number of rows and columns
print(credit_risk_data.shape)

#Print the column names
print(credit_risk_data.columns)

#Print the column types
print(credit_risk_data.dtypes)
```

The above code gives us the below output.

```
print(credit_risk_data.shape)
(150008, 10)
print(credit_risk_data.columns)
Index(['Cust_num', 'Bad', 'Credit_Limit', 'Late_Payments_Count',
'Card_Utilization_Percent', 'Age', 'Debt_to_income_ratio',
'Monthly_Income',       'Num_loans_personal_loans',       Family_dependents'],
dtype='object')

print(credit_risk_data.dtypes)
Cust_num                    int64
Bad                         int64
Credit_Limit                int64
Late_Payments_Count         int64
Card_Utilization_Percent    float64
Age                         int64
Debt_to_income_ratio        float64
Monthly_Income              int64
Num_loans_personal_loans    int64
Family_dependents           int64
dtype: object
```

From the output, we can see that there are 150,008 records and eight columns. The columns are self-explanatory. All the columns are numerical. Table 5.13 briefly explains the columns.

TABLE 5.13 Columns Description

Column Name	Description
Cust_num	Customer ID or number
Bad	Bad indicator. It is the target variable. Defaulters are denoted with 1
Credit_Limit	The credit limit on their credit card
Late_Payments_Count	Number of times customer was late in paying the bill
Card_Utilization_Percent	Customer credit line's average utilization
Age	Age of the customer
Debt_to_income_ratio	Debt-to-income ratio
Monthly_Income	Monthly income
Num_loans_personal_loans	Number of personal loans
Family_dependents	Number of dependents

Let us look at the basic data summaries.

```
pd.set_option('display.max_columns', None) #This option displays all the
columns
```

```
all_cols_summary=credit_risk_data.describe()
print(round(all_cols_summary,2))
```

The following is the output.

```
print(round(all_cols_summary,2))
            Cust_num         Bad     Credit_Limit      Late_Payments_Count \
count     150008.00    150008.00        150008.00                150008.00
mean       75004.50         0.07          6311.85                     0.26
std        43303.72         0.25          5221.25                     0.74
min            1.00         0.00           100.00                     0.00
25%        37502.75         0.00          4000.00                     0.00
50%        75004.50         0.00          4900.00                     0.00
75%       112506.25         0.00          7400.00                     0.00
max       150008.00         1.00          0000.00                    13.00

          Card_Utilization_Percent         Age   Debt_to_income_ratio \
count                    150008.00   150008.00              150008.00
mean                         30.38       52.30                   0.30
std                          33.41       14.77                   0.20
min                           0.00       21.00                   0.00
25%                           3.00       41.00                   0.18
50%                          15.40       52.00                   0.27
75%                          50.70       63.00                   0.38
max                         100.00      109.00                   1.00

          Monthly_Income    Num_loans_personal_loans    Family_dependents
count         150008.00                   150008.00            150008.00
mean            6324.28                        1.02                 0.74
std             7847.86                        1.11                 1.11
min             1000.00                        0.00                 0.00
25%             4042.00                        0.00                 0.00
50%             4817.00                        1.00                 0.00
75%             7400.00                        2.00                 1.00
max           835040.00                       14.00                10.00
```

Looks like the data is in good shape. No missing values, and no noticeable outliers. We will directly go ahead with model building. Before that, we will create the train and test datasets using the following code.

```python
#Defining X data
X = credit_risk_data[['Credit_Limit', 'Late_Payments_Count',
'Card_Utilization_Percent', 'Age', 'Debt_to_income_ratio',
'Monthly_Income', 'Num_loans_personal_loans', 'Family_dependents']]

y = credit_risk_data['Bad']

from sklearn import model_selection
X_train, X_test, y_train, y_test = model_selection.train_test_split(X, y,
test_size=0.2, random_state=55)

print(X_train.shape)
print(y_train.shape)
print(X_test.shape)
print(y_test.shape)
```

The following is the output table.

```python
print(X_train.shape)
(120006, 8)

print(y_train.shape)
(120006,)

print(X_test.shape)
(30002, 8)

print(y_test.shape)
(30002,)
```

We use the following code to build the model and get its accuracy.

```python
#Building the model
from sklearn.linear_model import LogisticRegression
model_2= LogisticRegression()

model_2.fit(X_train,y_train)

#Coefficients and Intercept
print(model_2.intercept_)
print(model_2.coef_)

##Confusion Matrix Calculation on Train data
from sklearn.metrics import confusion_matrix

y_pred_train=model_2.predict(X_train)
cm1 = confusion_matrix(y_train,y_pred_train)
print(cm1)

##Accuracy on Train data
accuracy1=(cm1[0,0]+cm1[1,1])/(cm1[0,0]+cm1[0,1]+cm1[1,0]+cm1[1,1])
print(accuracy1)
```

The following is the output.

```python
print(model_2.intercept_)
[-0.12910986]
```

```
print(model_2.coef_)
[[-9.88263513e-05  6.44961913e-01  1.07038983e-02  -5.95174153e-02
   -3.06946836e-03  1.23630184e-05  1.22022670e-01  4.07276702e-02]]

print(cm1)
[[111264 650]
 [  7465 627]]

print(accuracy1)
0.932378381080946
```

The overall accuracy on the train data seems to be good. Let us look at the details of the test data. We use the following code to work out the accuracy on test data.

```
##Confusion matrix on test data
y_pred_test=model_2.predict(X_test)
cm2 = confusion_matrix(y_test,y_pred_test)
print(cm2)

#####Accuracy on Test data
accuracy2=(cm2[0,0]+cm2[1,1])/(cm2[0,0]+cm2[0,1]+cm2[1,0]+cm2[1,1])
print(accuracy2)
```

The above code gives us the below output.

```
print(cm2)
[[27920    147]
 [ 1806    129]]

print(accuracy2)
0.934904339710686
```

We can see similar results on the train and test data. Overall accuracy is 93 percent on both the train and test data. If we consider only the accuracy, the model looks good. Is accuracy a sufficient measure here?

As discussed earlier, in the case of credit risk models, overall accuracy may not be the right measure. One bad customer is not the same as one good customer. In this data, more than 90 percent are good customers; less than 10 percent are bad. We can confirm it by looking at the frequency of good customers (0s) and bad customers (1s) in the data.

```
credit_risk_data['Bad'].value_counts()
0    139981
1     10027
Name: Bad, dtype: int64
```

Out of 150,000 records, we can see that the number of 1s is less than 15,000, which means the class1 percentage is less than 10 percent. It indicates a clear class imbalance. Remember that in this example bad customers are denoted with class1. Therefore, we need to focus on class1 accuracy instead of overall accuracy. The following code is used to calculate the sensitivity and specificity of the model.

```
##Sensitivity on train data
Sensitivity1=cm1[0,0]/(cm1[0,0]+cm1[0,1])
print(round(Sensitivity1,3))
0.994

##Specificity on train data
Specificity1=cm1[1,1]/(cm1[1,0]+cm1[1,1])
print(round(Specificity1,3))
0.077
```

```
##Sensitivity on test data
Sensitivity2=cm2[0,0]/(cm2[0,0]+cm2[0,1])
print(round(Sensitivity2,3))
0.995

##Specificity on test data
Specificity2=cm2[1,1]/(cm2[1,0]+cm2[1,1])
print(round(Specificity2,3))
0.067
```

The sensitivity of the train and test data is 99 percent, which is very close to perfect. What matters is the specificity. Specificity on both train and test data is very poor. The train data has a specificity of 7.7 percent, and test data shows the specificity at 6.7 percent. The model needs ample improvement to predict this class correctly.

How do we improve the specificity? We have already built the model. There is a small trick that can help boost the specificity. We can experiment with the threshold. By default, logistic regression gives us the probability as the prediction. We then convert it into a class by taking 0.5 as the threshold. As an accepted convention, if the predicted value is less than 0.5, the predicted class is taken as "0"; otherwise, it is 1 (Fig. 5.1).

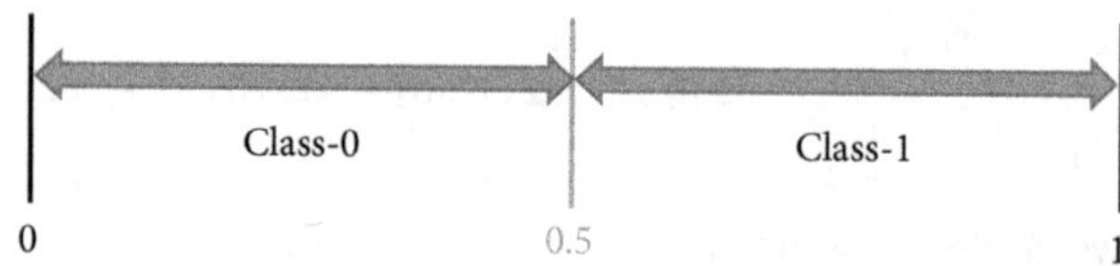

FIGURE 5.1   Default threshold for classification.

If we do not mention any threshold, any predicted value below 0.5 will be considered as class0 in python. Now imagine class1 is the bomb present class and class0 is the no-bomb class. For a car, if the model is giving a predicted value of 0.3 since 0.3 is less than 0.5, can we classify the car as a no-bomb car? Or since having a bomb is a very rare case and a huge class imbalance is present in the data, shall we put the threshold at a lower value, say 0.2, and call any probability above 0.2 as class1? In any case with such a class imbalance, we lower the threshold to 0.2 or 0.25 or 0.3 to increase the chances of finding class1. In the process, we may misclassify some of the data in class0 as class1, which should be okay, as we are interested in class1, so we can afford to misclassify a few records. In our example, it means that to detect the rare event case of bomb cars, we should lower the threshold. As discussed, in that process, we may misclassify some of the non-bomb cars as cars with bombs, but it is still safer than missing a bombed car. That is fine in this scenario since specificity is more important than sensitivity in this business case.

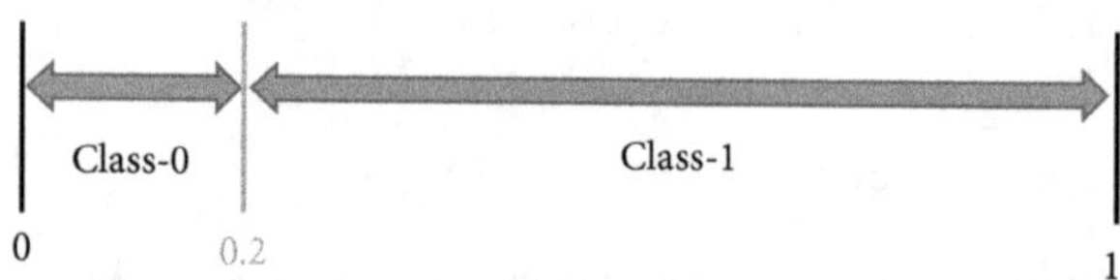

FIGURE 5.2   Lower threshold value.

Figure 5.2 shows a case of a lower threshold value to increase the specificity of the model. Sometimes we may be interested in class0; the aim now is to increase class0 predictions. We increase the threshold from 0.5 to 0.8 to increase the sensitivity.

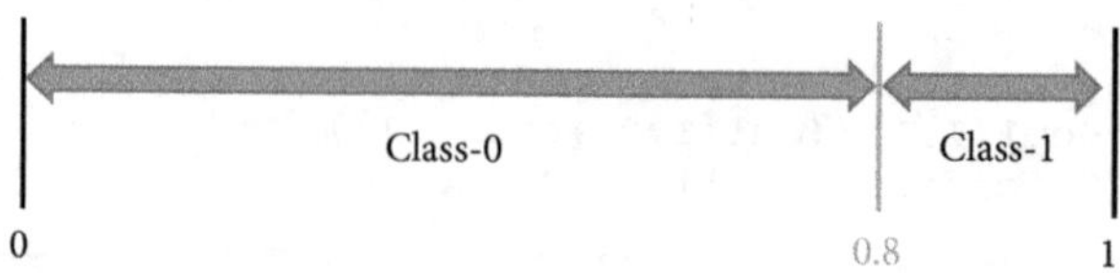

FIGURE 5.3   Higher threshold value.

Let us look at an example of changing the threshold to increase the value of sensitivity and specificity (Fig. 5.3). In previous examples, we have already calculated sensitivity and specificity using the confusion matrix. We did not mention any threshold there. That means the default threshold was considered in those calculations, which stands at 0.5. The following code, explained in Table 5.14, is used to predict the probabilities as the logistic regression output, and later converting it to classes based on different thresholds.

```
y_pred_prob=model_2.predict_proba(X_train)
print(y_pred_prob.shape)
print(y_pred_prob)
print(y_pred_prob[0,])
print(y_pred_prob[0,0])
print(y_pred_prob[0,1])
print(y_pred_prob[0:5,1])
print(y_pred_prob[:,1])

y_pred_prob_1=y_pred_prob[:,1]
## Default Threshold 0.5
threshold=0.5
y_pred_class=y_pred_prob_1*0
y_pred_class[y_pred_prob_1>threshold]=1
```

TABLE 5.14   Code Explanation

Code	Explanation
`y_pred_prob=model_2.predict_proba(X_train)`	Function for predicting the probabilities. This function gives both class-0 and class-1 probabilities for the train data.
`print(y_pred_prob.shape)`	NX2 matrix, where N is the number of rows in train data and two classes.
`print(y_pred_prob)`	Prints a few predicted values.
`print(y_pred_prob[0,])`	Prints the first row of the result.
`print(y_pred_prob[0,0])`	Prints the first row and first column value, which is class-0 probability in the first row.
`print(y_pred_prob[0,1])`	Prints the first row and second column value, which is class-1 probability in the first row.
`print(y_pred_prob[0:5,1])`	Prints class-1 probability for the first five rows.
`print(y_pred_prob[:,1])`	Store all class-1 probabilities in a new variable.
`y_pred_prob_1=y_pred_prob[:,1]`	Number of dependents.
`## Default Threshold of 0.5` `threshold=0.5`	Define threshold.
`y_pred_class=y_pred_prob_1*0`	Create a new class variable with all 0s in it.
`y_pred_class[y_pred_prob_1>threshold]=1`	If the probability is more than 0.5 then fill the classes with one. The rest will be left as zero.

The above code gives us the below output.

```
print(y_pred_prob.shape)
(120006, 2)

print(y_pred_prob)
[[0.96335718 0.03664282]
 [0.93137997 0.06862003]
 [0.74255004 0.25744996]

 ...
 [0.94355759 0.05644241]
 [0.93345083 0.06654917]
 [0.99098599 0.00901401]]
```

```
print(y_pred_prob[0,])
[0.96335718 0.03664282]

print(y_pred_prob[0,0])
0.9633571764052702

print(y_pred_prob[0,1])
0.0366428235947298

print(y_pred_prob[0:5,1])
[0.03664282 0.06862003 0.25744996 0.0425129 0.07394386]

print(y_pred_prob[:,1])
[0.03664282 0.06862003 0.25744996 ... 0.05644241 0.06654917 0.00901401]

y_pred_prob_1=y_pred_prob[:,1]

print(y_pred_class)
[0. 0. 0. ... 0. 0. 0.]
```

Now we will create the confusion matrix and recalculate sensitivity and specificity for different thresholds.

```
##Confusion Matrix and accuracy
cm3 = confusion_matrix(y_train,y_pred_class)
print("confusion Matrix with Threshold ", threshold, "\n",cm3)
accuracy3=(cm3[0,0]+cm3[1,1])/(cm3[0,0]+cm3[0,1]+cm3[1,0]+cm3[1,1])
print("Accuracy is ", round(accuracy3,3))

##Sensitivity and Specificity on Train data
Sensitivity3=cm3[0,0]/(cm3[0,0]+cm3[0,1])
print("Sensitivity is", round(Sensitivity3,3))

Specificity3=cm3[1,1]/(cm3[1,0]+cm3[1,1])
print("Specificity is ", round(Specificity3,3))
```

The above code gives us the below output.

```
confusion Matrix with Threshold 0.5
 [[111264    650]
 [ 7465    627]]
Accuracy is 0.932
Sensitivity is 0.994
Specificity is 0.077
```

We will get the same results as default settings for the threshold 0.5. We can compare these results with the train data results in the previous section, and they are the same. Remember, in this problem statement, class-1 is important for us, so we need to maximize the probability of detecting class-1. We will lower the threshold and try to boost the specificity from less than 10 percent to a higher number. We can compromise a little on sensitivity. We use the same code, and we just need to change the threshold.

```
## New Threshold 0.2
threshold=0.2
y_pred_class=y_pred_prob_1*0
y_pred_class[y_pred_prob_1>threshold]=1

##Confusion Matrix and accuracy
cm3 = confusion_matrix(y_train,y_pred_class)
print("confusion Matrix with Threshold ", threshold, "\n",cm3)
accuracy3=(cm3[0,0]+cm3[1,1])/(cm3[0,0]+cm3[0,1]+cm3[1,0]+cm3[1,1])
print("Accuracy is ", round(accuracy3,3))
```

```
##Sensitivity and Specificity on Train data
Sensitivity3=cm3[0,0]/(cm3[0,0]+cm3[0,1])
print("Sensitivity is", round(Sensitivity3,3))

Specificity3=cm3[1,1]/(cm3[1,0]+cm3[1,1])
print("Specificity is ", round(Specificity3,3))
```

The following is the output.

```
confusion Matrix with Threshold 0.2
 [[104697    7217]
 [ 5389     2703]]
Accuracy is 0.895
Sensitivity is 0.936
Specificity is 0.334
```

By changing the threshold to 0.2, we have lifted specificity from less than 10 percent to 33 percent. Let us further reduce the threshold to 0.1.

```
## New Threshold 0.1
threshold=0.1
y_pred_class=y_pred_prob_1*0
y_pred_class[y_pred_prob_1>threshold]=1

##Confusion Matrix and accuracy
cm3 = confusion_matrix(y_train,y_pred_class)
print("confusion Matrix with Threshold ", threshold, "\n",cm3)
accuracy3=(cm3[0,0]+cm3[1,1])/(cm3[0,0]+cm3[0,1]+cm3[1,0]+cm3[1,1])
print("Accuracy is ", round(accuracy3,3))

##Sensitivity and Specificity on Train data
Sensitivity3=cm3[0,0]/(cm3[0,0]+cm3[0,1])
print("Sensitivity is", round(Sensitivity3,3))

Specificity3=cm3[1,1]/(cm3[1,0]+cm3[1,1])
print("Specificity is ", round(Specificity3,3))
```

The following is the output with a threshold of 0.1.

```
confusion Matrix with Threshold 0.1
 [[89538  22376]
 [ 3294   4798]]
Accuracy is 0.786
Sensitivity is 0.8
Specificity is 0.593
```

By changing the threshold to 0.1, we have further boosted the specificity from 7 to 59 percent. However, sensitivity has decreased from 99 to 80 percent. If we further reduce the threshold, specificity will increase at the cost of sensitivity. We will be losing a considerable business if we keep on classifying good customers as bad customers. There has to be a trade-off between sensitivity and specificity. Next, we will discuss how to choose the optimal threshold when we are satisfied with both sensitivity and specificity.

5.4.2.4   ROC and AUC   In some cases, sensitivity is important, and in some others, specificity takes precedence. We know by now that by lowering the threshold, we can increase the specificity, while increasing the threshold increases the sensitivity. The critical consideration here is that both are interlinked. When we drop the threshold, specificity increases, but at the same time sensitivity decreases. It is called net effect in business terms—while we focus on increasing the accuracy of one class, the other class accuracy decreases. Is there a risk in this whole process?

Let us examine the previous example results again with a threshold of 0.1.

```
confusion Matrix with Threshold 0.1
  [[89538  22376]
   [ 3294   4798]]
Accuracy is 0.786
Sensitivity is 0.8
Specificity is 0.593
```

In the above output, we can see that specificity has increased by reducing the threshold. However, sensitivity has decreased from 99 to 80 percent. If we want to focus only on specificity, why shouldn't we keep the threshold at 0.01? Given below is the result when we follow the threshold at 0.01.

```
confusion Matrix with Threshold 0.01
  [[14646  97268]
   [ 179    7913]]
Accuracy is 0.188
Sensitivity is 0.131
Specificity is 0.978
```

In the above table, the specificity is 97.8 percent, with 0.01 as the threshold. When a customer is bad, we are predicting him or her as a bad customer with high accuracy. However, if the customer is good, we are classifying all of them as bad. Since we are very strict on the threshold, we are classifying every customer as a bad customer. By wrongly classifying a lot of good customers as bad customers, we reject their loan applications as well. If we reject good customer applications, is it not an opportunity loss for us? Imagine a scenario where we kept the threshold at 0; then, we will reject all the loan applications. By rejecting all the applications, we will sanction no bad loans, as we are not giving any loans at all. Note that any business cannot work this way. There has to be a trade-off.

Sensitivity and specificity move in opposite directions; if one increases, the other decreases. We have to set our priorities first. We will try to maximize our gains, but at the same time, we should minimize the associated loss—as both go hand-in-hand. The ROC curve helps us in choosing the desired optimal pair of sensitivity and specificity. ROC stands for the receiver operating characteristic curve.

We plot a ROC curve by taking True-Positive Rate (sensitivities) on the y-axis and False-Positive Rate (1 – Specificity) on the x-axis. If we have got a pair of sensitivity and specificity values with a threshold of 0.5, now by altering the threshold, both sensitivity and specificity will change. The ROC curve is created by considering the possibilities of all the thresholds. We will try all the thresholds between 0 and 1 and note down the corresponding sensitivity and specificity values. Note that the values on the x-axis of the ROC curve are False-Positive Rate (1 – Specificity) (Fig. 5.4).

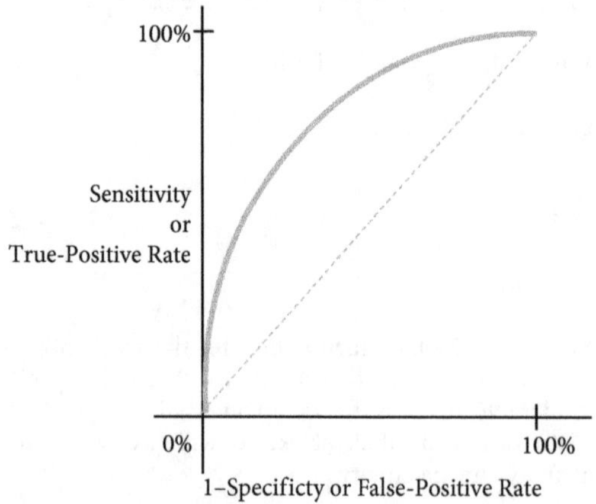

FIGURE 5.4   ROC curve.

The dotted line in Fig. 5.4 is the no-discrimination line. On this line, both sensitivity and (1 – Specificity) are equal. This line is created from a model that gives a 50-50 chance to both the classes.

To understand the ROC curve, we will take an example of a model that predicts the response of a customer for a marketing campaign. In simple terms, it tries to predict whether a customer will respond or not to a given marketing message or mailer. As usual, let class-0 be responders and class-1 be nonresponders. Sensitivity, in this case, would be to predict a responder as a responder accurately, which is the True-Positive Rate (Sensitivity). Specificity is to predict nonresponders as nonresponder accurately, which is a True-Negative Rate. Nevertheless, we want to display (1 – Specificity), i.e., False-Positive Rate, which is nothing but wrongly predicting a nonresponder as a responder. Let us consider Table 5.15.

TABLE 5.15 Revisiting Confusion Matrix

		Predicted Classes		
		Responder	Nonresponder	Classwise Accuracy
Actual Classes	Responder	True Positives (TP)	False Negatives (FN)	$\dfrac{TP}{TP + FN}$
	Nonresponder	False Positives (FP)	True Negatives (TN)	$\dfrac{TN}{FP + TN}$

Table 5.15 stands for a given threshold value. With a change in the threshold, both Sensitivity and (1 – Specificity) will change. A ROC curve helps us to zero in on the most optimal pair of sensitivity and specificity along with the threshold associated with it. We consider three scenarios on the ROC curve: aggressive, defensive, and moderate. Have a look at the ROC curve plot in Fig. 5.5.

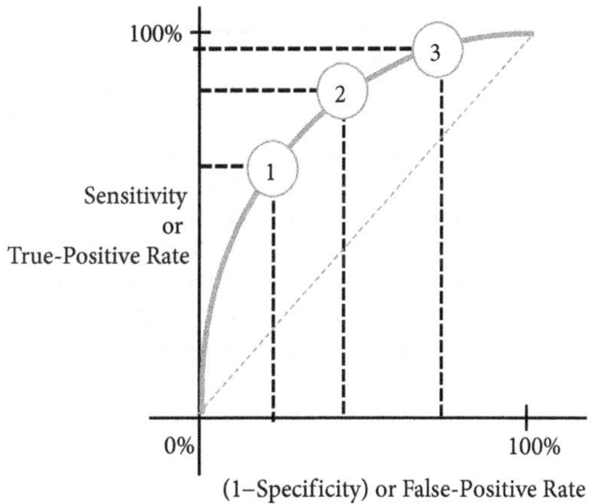

FIGURE 5.5 Aggressive, defensive, and moderate scenarios on ROC curve.

These three scenarios are explained next.

Scenario-1: Defensive

- The focus is on point 1 on the ROC curve of Fig. 5.5. In this scenario, we have a sensitivity of 65 percent, and the corresponding (1 – Specificity) is 25 percent.
- An easy way to understand this is to predict 65 percent of responders correctly; we are making around 25 percent errors. In other words, to get 65 percent of responders correctly, we are wrongly identifying 25 percent of nonresponders as responders.
- We are aiming at capturing fewer responders here; this is a defensive scenario.

Scenario-2: Moderate

- At point 2 on the ROC curve, we have a sensitivity of 80 percent, and the corresponding (1 – Specificity) as 45 percent.

- To capture 80 percent of responders, we are wrongly identifying 45 percent of nonresponders as responders. We are identifying more responders here, but at the same time, false positives are also increasing.
- We are capturing some more responders compared to scenario-1 by making a few more mistakes.

Scenario-3: Aggressive

- At point 3 on the ROC curve, we have a sensitivity of 95 percent and the corresponding (1 – Specificity) as 75 percent.
- To capture 95 percent of responders, we are wrongly identifying 75 percent of nonresponders as responders. False positives are too high in this scenario.
- Though we are capturing maximum responders in this scenario, the mistakes are too many.

All three scenarios are summarized in Table 5.16.

TABLE 5.16   True- and False-Positive Rates

Scenario	Sensitivity True-Positive Rate (% Right Predictions)	(1 – Specificity) False-Positive Rate (% Mistakes)
Defensive	65	25
Moderate	80	45
Aggressive	95	75

We go to the business team with these three scenarios; the business will take judgment based on the factors such as budget, risk factor, and maybe a few others. For example, if we are running an email marketing campaign, we can afford to opt for an aggressive scenario. We do not have any significant loss if we send an email to a nonresponder. But if we are running a telephone marketing campaign, there is some cost associated with every call. We have contact center agents working on these campaigns. In this case, we may not go ahead with the aggressive scenario; there are too many mistakes in the aggressive scenario. We may prefer to go with a defensive or a moderate scenario based on the business conditions and marketing budget. If it is a high budget campaign, we can afford to go ahead with the aggressive scenario. Companies that are new in business and want more focus growth, they generally go with an aggressive scenario. For an established company having adequate customers, the better choice may be a defensive scenario, where the losses are relatively less. It depends on the type of company and the type of business they are into.

A connected measure to ROC is AUC. Look at the two ROC curves in Fig. 5.6.

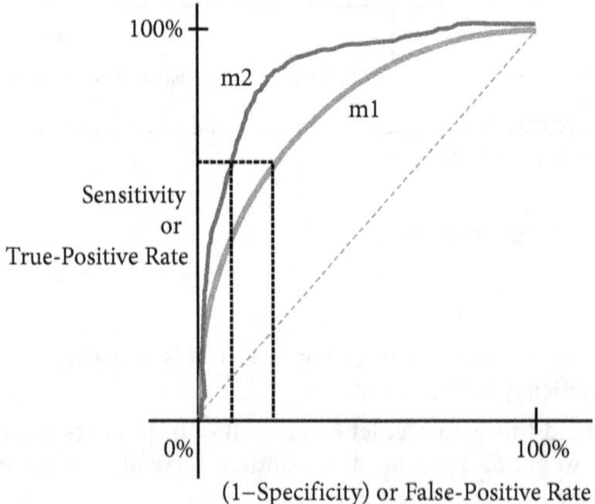

FIGURE 5.6   ROC curves represent two models.

These ROC curves represent two models here—m1 and m2. Which model is better? At any given level of sensitivity, which model is making more mistakes? For example, at 65 percent sensitivity, m1 has 25 percent errors, and m2 has it as 10 percent. Strikingly m2 is performing better than m1 at all levels. The area under the curve for m2 is more than that of m1. Area under the (ROC) curve (AUC) is another validation measure for the model's performance. The plot and its shaded portion in Fig. 5.7 explain it pictorially.

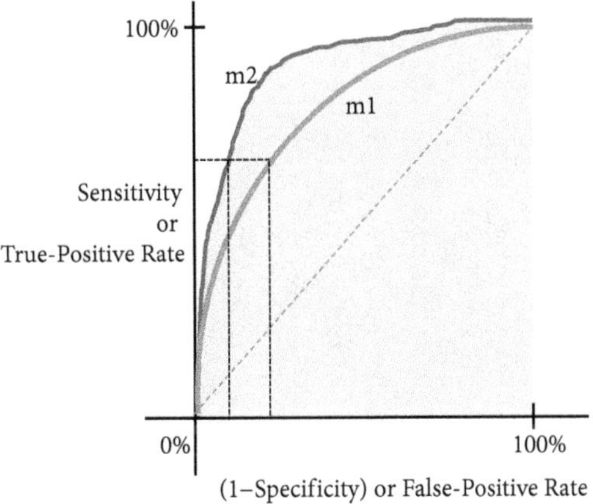

FIGURE 5.7   Area under the curve (AUC).

AUC is a better validation measure in the case of class imbalance. AUC is calculated from the ROC curve, which considers all the threshold values. If the AUC value is approximately 1, the model is considered to be good. Let us now learn to plot the ROC curve and calculate AUC.

```python
from sklearn.metrics import roc_curve, auc
import matplotlib.pyplot as plt

False_positive_rate, True_positive_rate, thresholds = roc_curve(y_train,
y_pred_prob_1)

plt.figure(figsize=(10,10))
plt.title('ROC Curve',fontsize=15)
plt.plot(False_positive_rate, True_positive_ rate)
plt.plot([0,1],[0,1],'r--')
plt.ylabel('True Positive Rate(Sensitivity)',fontsize=15)
plt.xlabel('False Positive Rate(1-Specificity)',fontsize=15)
plt.show()
```

The critical function in the above code is `roc_curve()`. It takes the actual values of y and the corresponding predicted probabilities of y as an input. This function considers all the threshold values and returns a table containing the threshold value, false-positive rate, and true-positive rate. Have a look at the ROC curve output from this code in Fig. 5.8.

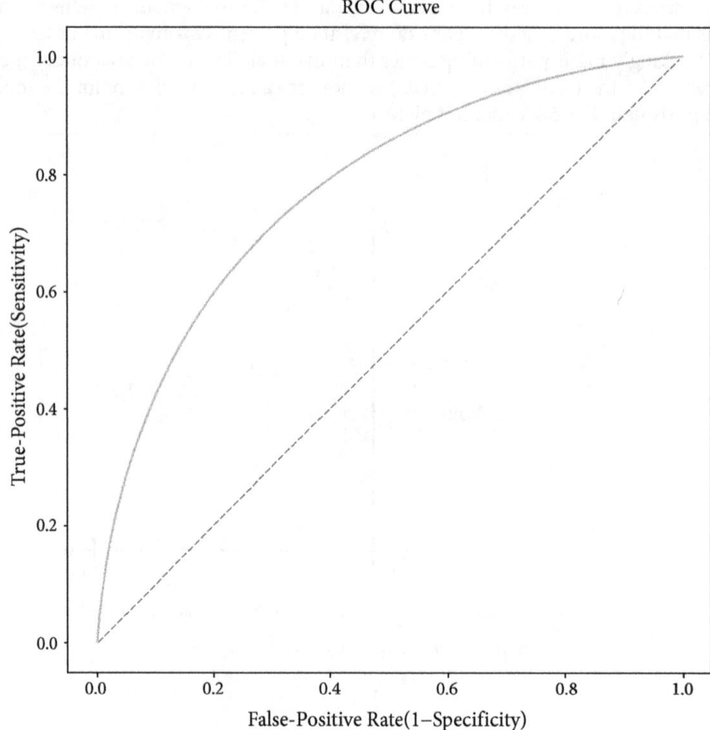

FIGURE 5.8   The ROC curve output from the code.

The following code calculates the area under the curve.

```
###Area under Curve-AUC
auc = auc(False_positive_rate, True_positive_rate)
print(auc)
```

The above code gives us the below output.

```
print(auc)
0.7749199563244183
```

AUC is a good measure for comparison of models. The dotted line in the model is the no-discrimination line. The farther our ROC curve is from this line, the better is our model. There is another validation measure called the Gini coefficient, which is derived from the area between the ROC curve and the nondiscrimination line.

5.4.2.5   F1 Score   Accuracy is a primary validation measure. We discussed measures such as sensitivity and specificity, which are useful for class imbalance cases. There are a few more measures for validating the models. All these measures try to tell us how good the model is. Different industries use different measures.

The F1 score can be considered as an extension of sensitivity and specificity. In sensitivity, we tend to focus on a single class. The F1 score is also calculated for individual classes (Table 5.17).

TABLE 5.17 Classwise Accuracy Formulas to Discuss the F1 Score

		Predicted Classes		
		Positive	**Negative**	**Classwise Accuracy**
Actual Classes	**Positive**	True Positives (TP)	False Negatives (FN)	$\dfrac{TP}{TP+FN}$
	Negative	False Positives (FP)	True Negatives (TN)	$\dfrac{TN}{FP+TN}$

Sensitivity is the True-Positive Rate. It is also known as recall. It tells us, out of all the records in the positive class, how many are predicted correctly. If we are focusing on a single class, sensitivity helps us in predicting its accuracy. There is one more approach to considering a single class. Out of all the predicted values as positive, how many are genuinely positive. It is the accuracy presented in the first column in Table 5.17. This measure is known as precision (Table 5.18).

TABLE 5.18 Recall and Precision

		Predicted Classes		
		Positive	**Negative**	**Classwise Accuracy**
Actual Classes	**Positive**	True Positives (TP)	False Negatives (FN)	$\textbf{Recall} = \dfrac{\textbf{TP}}{\textbf{TP}+\textbf{FN}}$
	Negative	False Positives (FP)	True Negatives (TN)	
		$\textbf{Precision} = \dfrac{\textbf{TP}}{\textbf{TP}+\textbf{FP}}$		

The F1 score is the harmonic mean of recall and precision. The harmonic mean is the inverse of the arithmetic mean. It is preferred when we are dealing with fractions.

$$\text{F1 Score} = \text{Harmonic mean}\left(\text{Recall}, \text{Precision}\right)$$

$$\text{F1 Score} = \dfrac{2}{\dfrac{1}{\text{Recall}} + \dfrac{1}{\text{Precision}}}$$

$$\text{F1 Score} = 2^* \dfrac{\text{Precision}^*\text{Recall}}{\text{Precison} + \text{Recall}}$$

The F1 score is slightly better than sensitivity, as we are adding one more dimension to the formula. While sensitivity and specificity are easy to explain and compare, the F1 score, being a harmonic mean, is not easy to interpret. The formula below is used for the calculation of the F1 score.

```
##F1 Score
from sklearn.metrics import f1_score

## Threshold 0.5
threshold=0.5
y_pred_class=y_pred_prob_1*0
y_pred_class[y_pred_prob_1>threshold]=1
print(f1_score(y_train, y_pred_class))
```

```
## Threshold 0.2
threshold=0.2
y_pred_class=y_pred_prob_1*0
y_pred_class[y_pred_prob_1>threshold]=1
print(f1_score(y_train, y_pred_class))
```

The following is the output of this code. Please note that we get different values of the F1 score for different thresholds.

```
threshold=0.5 f1_score 0.13384566122318287
threshold=0.2 f1_score 0.30013324450366424
```

There are several more validation measures for classification. We can explore them based on the need. We have adequately discussed validation measures to proceed further with bias-variance trade-off.

5.5   BIAS-VARIANCE TRADE-OFF

While building models, we focus on bringing the best out of the data. We want to have the best model having a high accuracy of predictions. While we concentrate a lot on increasing accuracy, we may run into two types of problems: the problem of overfitting and the problem of underfitting. Although we have covered this topic in detail in Chap. 4, we will discuss a few more details here using a different case study.

5.5.1   The Problem of Overfitting: Variance

As a recap, the dataset used for model building is known as train data, and the dataset used for testing the robustness of the model is known as test data. The following are the characteristics of an overfitted model. In the below list, we are trying to give as many different explanations as possible for better understanding.

- An overfitted model works well on the train data and fails on the test data. It means the model has high accuracy on the train data and significantly low accuracy on the test data.
- Instead of learning generic patterns in the train data, an overfitted model tries to learn specific patterns related to only train data. The model almost tries to memorize the train data, rather than learning the generic patterns.
- Small changes in the train data cause a significant change in the model parameters. Since there is a considerable change in the parameters, when we change the data, the overfitted models are also known as models with a lot of variance or high variance models.
- Sometimes a few data points are against the general patterns. Such data points are called the noise in the data. For example, a generic rule (pattern) in the data is "if age <25 then most of the customers are buyers." There will be a few data points where the age <25, but still they fall in the nonbuyers category. If any model tries to memorize these noise points in the data, it will end up as an overfitted model.
- In the case of decision trees, we have overfitted models if the tree is larger than what is necessary. In regression, we will have an overfitted model if we add a lot of polynomial terms.

How to detect overfitted models? Build the model with the train data, and then apply it on the test data; if the accuracy or any other validation measures deteriorate, the model is overfitted. Five percent is the general industry standard of significance. For example, if the train data shows an accuracy level of 90 percent, and the test data on the same model demonstrates less than 85 percent accuracy, the model is concluded as overfitted.

5.5.2   The Problem of Underfitting: Bias

The problem of overfitting frequently occurs in practice. To avoid it, we try to reduce the complexity of the model. Sometimes we tend to settle for a lesser accuracy even if the model has a potential for more. Such models are known as underfitted models. Given below are the characteristics of underfitted models.

- An underfitted model does not work well even on the train data, and definitely, the results on test data will be inferior. If any model exhibits low accuracy on the training data itself, we do not even need to assess it with the test data.

- An underfitted model may be too simple to learn all the generic patterns present in the train data.
- In underfitted models, small changes in the train data will not cause any change in the model.
- The underfitted model's parameters are not decidedly accurate. These models have some inherent bias in their estimates. Underfitted models are also known as the models with a lot of bias or high-bias models.
- In the case of decision trees, we have underfitted models if the tree is smaller than necessary.

How to detect underfitted models? Again, we follow our familiar process—build the model on train data; if it shows accuracy less than the benchmark, then it is an underfitted model. The identification of underfitted models is a little tricky. We should have a base idea on the type of data and the expected accuracy with that data. For some models, 80 percent accuracy is high; for some others, 80 percent accuracy may be less. Unsurprisingly, we try not to settle with underfitted models. High accuracy is always desirable; seldom do we settle for a low accuracy model—never knowingly, however.

5.5.3 Bias-Variance Trade-Off

The model should be neither overfitted nor underfitted; in other words, the model should neither have variance nor bias. The following is an illustration of overfitted and underfitted models. In the three plots in Figs. 5.9 to 5.11, we have taken the target variable on the y-axis, while the input is taken on the x-axis. The dots on all the three plots represent the training data, the actual values of the target variable y for a given input variable x. The three types of model-fit curves are depicted in the three different sketches in Figs. 5.9 to 5.11.

In an overfitted model the regression goes through all the points.

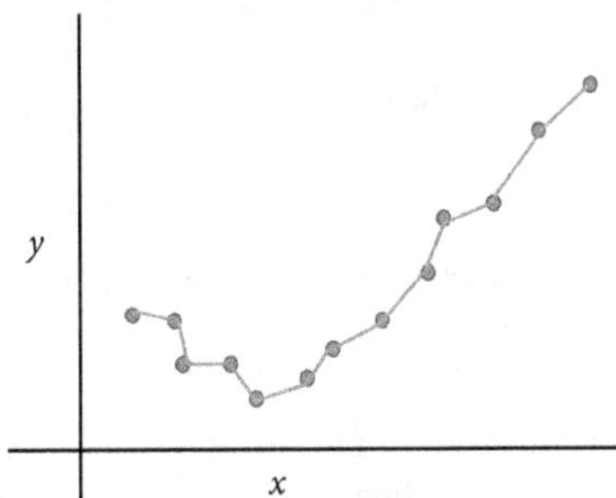

FIGURE 5.9 An overfitted regression model.

In an underfitted model the regression line misses most of the points.

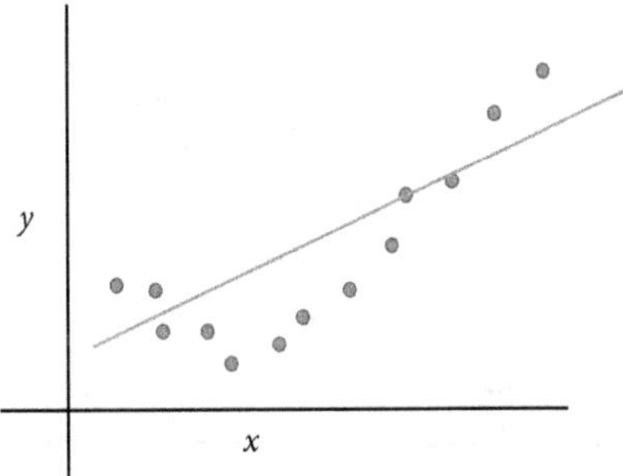

FIGURE 5.10 An underfitted model's regression line.

In an optimal model, regression goes through most of the points and captures the overall pattern in the data.

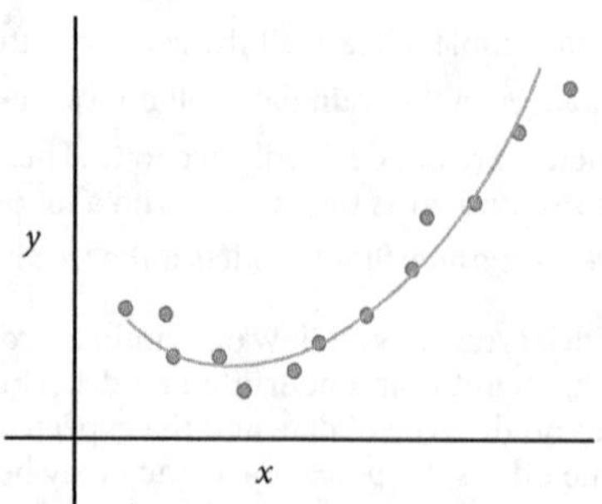

FIGURE 5.11   An optimally fitted curve.

The broad error in a model can be divided into three parts: the irreducible error, bias, and variance. Not every model is 100 percent accurate. There will always be some error inherent in data that cannot be reduced. This error component is known as an irreducible error. Bias component happens due to underfitting, and variance component happens due to overfitting.

$$\text{Overall squared error} = \text{Irreducible error} + \text{Bias}^2 + \text{Variance}$$

The mathematical form of the above equation is

$$\text{Model is } Y = f(X) + \varepsilon$$

$$\text{Irreducible error Var}(\varepsilon) = \sigma^2$$

$$\text{Model error} = E[(Y - \hat{f}(x_0))^2 \mid X = x_0]$$

$$\text{Model error} = \sigma^2 + [E\hat{f}(x_0) - f(x_0)]^2 + E[\hat{f}(x_0) - E\hat{f}(x_0)]^2$$

$$\text{Model error} = \sigma^2 + \text{Bias}^2(\hat{f}(x_0)) + \text{Var}(\hat{f}(x_0))$$

$$\text{Model error} = \text{Irreducible Error} + \text{Bias}^2 + \text{Variance}$$

Bias and variance move in opposite directions. If we increase the complexity of a model, then variance increases and bias decreases. If we decrease the complexity, then we see a decrease in variance, but bias increases. To reduce bias and variance, we need to build models with optimal complexity.

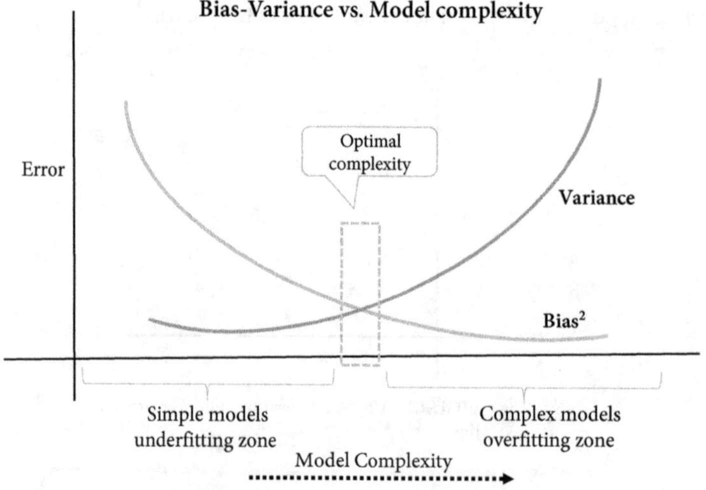

FIGURE 5.12   Bias and variance.

Figure 5.12 depicts bias and variance versus model complexity. As the model complexity increases, the bias reduces, but the variance increases. Variance is low up to a point, but it increases after a specified complexity. We are aware that, in the case of a decision tree, as the tree size grows, the variance increases.

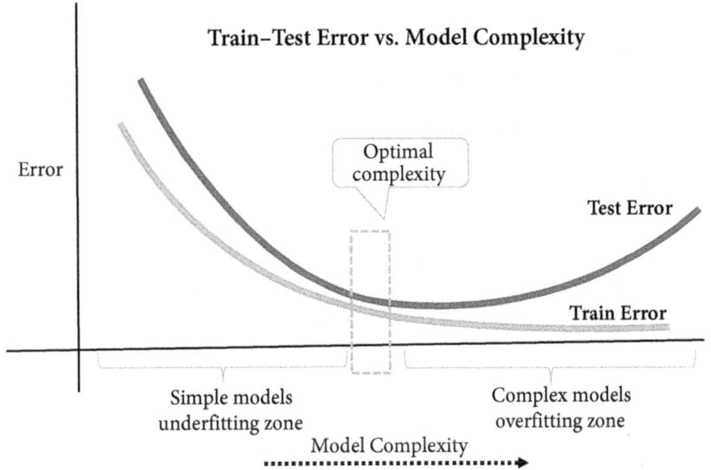

FIGURE 5.13   Error versus model complexity.

Figure 5.13 shows the Train–Test error versus model complexity. As the model complexity increases, the training error reduces. It may even reach a zero-error state. However, as the complexity increases, the test error will increase after a specified complexity.

5.5.4   Case Study: Pima Indians Diabetes Prediction

The dataset used in this study was initially shared by the National Institute of Diabetes and Digestive and Kidney Diseases. The goal is to predict whether a person has diabetes or not, based on several diagnostic measurements. The diagnostic measurements used here are Pregnancies, Glucose, Blood Pressure, SkinThickness, Insulin, BMI, DiabetesPedigreeFunction, and Age. The target variable name is "outcome," and it takes two values: 0 and 1. Class-1 indicates diabetes. Table 5.19 gives us the details of all the diagnostic measures.

TABLE 5.19   Feature Descriptions

Column Name	Details
Pregnancies	Number of times pregnant
Glucose	Plasma glucose concentration at 2 hours in an oral glucose tolerance test
Blood Pressure	Diastolic blood pressure (mm Hg)
SkinThickness	Triceps skinfold thickness (mm)
Insulin	2-hour serum insulin (μ/ml)
BMI	Body mass index (weight in kg/(height in m)^2)
DiabetesPedigreeFunction	Diabetes pedigree function
Age	Age (years)
Outcome	It takes two values 0 and 1. Class-1 indicates diabetes.

After having a glance at the column particulars, we will import this data into Python. We also need some basic statistics on the data before proceeding with the model building. The code below gives us these details.

```
diabetes_data=  pd.read_csv(r'D:\Chapter5\5.  Base  Datasets\pima\diabetes.
csv')
```

```
#Get an idea on the number of rows and columns
print(diabetes_data.shape)

#Print the column names
print(diabetes_data.columns)

#Print the column types
print(diabetes_data.dtypes)
```

The above code gives us the below output.

```
print(diabetes_data.shape)
(768, 9)

print(diabetes_data.columns)
Index(['Pregnancies',    'Glucose',    'BloodPressure',    'SkinThickness',
'Insulin','BMI',      'DiabetesPedigreeFunction',    'Age',    'Outcome'],
dtype='object')

print(diabetes_data.dtypes)
Pregnancies                 int64
Glucose                     int64
BloodPressure               int64
SkinThickness               int64
Insulin                     int64
BMI                         float64
DiabetesPedigreeFunction    float64
Age                         int64
Outcome                     int64
dtype: object
```

There are 768 records in the dataset. All the columns are of the numeric datatype. The following code gives us a summary of all the columns.

```
#Summary
pd.set_option('display.max_columns', None) #This option displays all the
columns

all_cols_summary=diabetes_data.describe()
print(round(all_cols_summary,2))
```

As usual, the output follows.

```
print(round(all_cols_summary,2))
```

	Pregnancies	Glucose	BloodPressure	SkinThickness	Insulin	BMI
count	768.00	768.00	768.00	768.00	768.00	768.00
mean	3.85	120.89	69.11	20.54	79.80	31.99
std	3.37	31.97	19.36	15.95	115.24	7.88
min	0.00	0.00	0.00	0.00	0.00	0.00
25%	1.00	99.00	62.00	0.00	0.00	27.30
50%	3.00	117.00	72.00	23.00	30.50	32.00
75%	6.00	140.25	80.00	32.00	127.25	36.60
max	17.00	199.00	122.00	99.00	846.00	67.10

	DiabetesPedigreeFunction	Age	Outcome
count	768.00	768.00	768.00
mean	0.47	33.24	0.35
std	0.33	11.76	0.48
min	0.08	21.00	0.00
25%	0.24	24.00	0.00

```
50%                      0.37    29.00          0.00
75%                      0.63    41.00          1.00
max                      2.42    81.00          1.00
```

From the summary, we can observe that the data is clean enough to go ahead with model building. The data is already cleaned; naturally, someone had made a conscious effort to it in this state. We now define the train and the test data using the code below.

```python
#Defining X data
X = diabetes_data[['Pregnancies', 'Glucose', 'BloodPressure', 'SkinThickness',
'Insulin', 'BMI', 'DiabetesPedigreeFunction', 'Age']]

y = diabetes_data[['Outcome']]

#Train and Test data creation
from sklearn import model_selection
X_train, X_test, y_train, y_test = model_selection.train_test_split(X, y,
test_size=0.2, random_state=33)

print(X_train.shape)
print(y_train.shape)
print(X_test.shape)
print(y_test.shape)
```

The `random_state` parameter helps us to fetch the same random sample again. This code gives us the following output.

```python
print(X_train.shape)
(614, 8)

print(y_train.shape)
(614, 1)

print(X_test.shape)
(154, 8)

print(y_test.shape)
(154, 1)
```

Let us go ahead and build the decision tree. The code below helps us in building the decision tree.

```python
from sklearn.tree import DecisionTreeClassifier
diabetes_tree1= DecisionTreeClassifier()
diabetes_tree1.fit(X_train, y_train)

#Calculate Accuracy on Train and Test data
print("Max Depth = None")
print("Train data Accuracy", diabetes_tree1.score(X_train, y_train))
print("Test data Accuracy", diabetes_tree1.score(X_test, y_test))
```

The following is the code output; `score()` function helps us in calculating the accuracy value directly. This function internally creates the confusion matrix and gives us the accuracy value.

```
Max Depth = None
Train data Accuracy 1.0
Test data Accuracy 0.70
```

We can make out from the output that this model is overfitted. It is an example of a high-variance model. We will try to decrease the model complexity to get it simplified. In this case, we need to reduce the size of the tree. If the resulting model is underfitted, then we will slightly increase the complexity. The final aim here is to have a model with optimal complexity. Taking into consideration all our iterations, we will use cross-validation to zero in on the optimal model, as explained next.

5.6 CROSS-VALIDATION

We have already discussed the notion of train and test data while working with decision trees. Building the model on train data and validating it on test data is called cross-validation. In the first few attempts of building any model, discovering optimal complexity may be a little tricky and take some time. We need to build and validate many models before deciding on the optimal model complexity level that gives us consistent accuracy—both on train and test data. In some algorithms, there are a few hyperparameters, which can be adjusted to increase or decrease the model complexity. For example, decision trees are very sensitive to the max_depth hyperparameter. Next, we get into a little more detail by taking an example model and altering it using the max_depth parameter.

5.6.1 Cross-Validation: An Example

In this example, as usual, we build the model on train data, find its accuracy, and finally validate it on test data. If the model shows high accuracy on the train data and a lower level of accuracy on the test data, then the model is overfitted. If the model shows low accuracy on the training data itself, we identify it as underfitted. As discussed earlier, detecting underfitting can be a little tricky; we need to look at only train data accuracy to detect an underfitted model.

Given below is the code for building the model with multiple max_depth options. We have already gone through this code multiple times. Here only the max_depth parameter is changing.

```
## Model Building with pruning parameters
from sklearn.tree import DecisionTreeClassifier
diabetes_tree1= DecisionTreeClassifier(max_depth=1)
diabetes_tree1.fit(X_train, y_train)

#Calculate Accuracy on Train and Test data
print("Max Depth = 1")
print("Train data Accuracy", diabetes_tree1.score(X_train, y_train))
print("Test data Accuracy", diabetes_tree1.score(X_test, y_test))
```

Below is the output of the above model.

```
Max Depth = 1
Train data Accuracy 0.74
Test data Accuracy 0.71
```

The above model is not overfitted. There may be a case of underfitting. We will try with a higher value of max_depth.

```
## Model Building with pruning parameters
print("Max Depth = 6")
from sklearn.tree import DecisionTreeClassifier
diabetes_tree1= DecisionTreeClassifier(max_depth=5)
diabetes_tree1.fit(X_train, y_train)

#Calculate Accuracy on Train and Test data
print("Train data Accuracy", diabetes_tree1.score(X_train, y_train))
print("Test data Accuracy", diabetes_tree1.score(X_test, y_test))
```

Below is the output of the above model.

```
Max Depth = 6
Train data Accuracy 0.86
Test data Accuracy 0.68
```

The above model is overfitted. We will try with a lower value of `max_depth`.

```
## Model Building with pruning parameters
print("Max Depth = 3")
from sklearn.tree import DecisionTreeClassifier
diabetes_tree1= DecisionTreeClassifier(max_depth=3)
diabetes_tree1.fit(X_train, y_train)

#Calculate Accuracy on Train and Test data
print("Train data Accuracy", diabetes_tree1.score(X_train, y_train))
print("Test data Accuracy", diabetes_tree1.score(X_test, y_test))
```

The output of the above model is presented below.

```
Max Depth = 3
Train data Accuracy 0.78
Test data Accuracy 0.74
```

This model is neither overfitted nor underfitted. We will see the results of `max_depth` = 2 and finalize the model.

```
## Model Building with pruning parameters
print("Max Depth = 2")
from sklearn.tree import DecisionTreeClassifier
diabetes_tree1= DecisionTreeClassifier(max_depth=2)
diabetes_tree1.fit(X_train, y_train)

#Calculate Accuracy on Train and Test data
print("Train data Accuracy", diabetes_tree1.score(X_train, y_train))
print("Test data Accuracy", diabetes_tree1.score(X_test, y_test))
```

The following is the output of the above model.

```
Max Depth = 2
Train data Accuracy 0.79
Test data Accuracy 0.75
```

From the above output, we can make out that the `max_depth` = 6 tree is overfitted and the `max_depth` = 1 tree is slightly underfitted. Out of these results, we can take either `max_depth` = 3 or `max_depth` = 2 as the final decision tree, since the accuracy on train data is high (no bias) difference in the train and test data accuracy is less than or equal to 5 percent (no variance).

5.6.2 *K*-Fold Cross-Validation

When we approach model validation using train versus test data, some doubts may come to our mind. What if we have a biased test data by any chance? What if we have test data that is not representing the population? What if we have overfitted the model to the test data as well? What if the model works well only with one set of the train and test data? Or, what if the model fails on any other dataset? Most of the time, the train and test data approach works, but as data scientists, we always look for a more robust method that will get us the optimal model accuracy.

Many data scientists prefer the *K*-fold cross-validation method. *K*-fold cross-validation is an extension of the train–test cross-validation method. It generally results in a less biased model when compared to some other methods. The reason being that it ensures that every single observation from the population (original dataset) gets a chance to appear in training and test datasets. It is sometimes called one of the best approaches if the project team has a limited amount of input data (population). The following are the steps in the *K*-fold cross-validation algorithm.

Step 1: Take the whole dataset, and divide it into *K* subsets (*K*-folds). Usually, *K* is taken as a number between 5 and 10. If *K* = 5, then each subset will have 20 percent of the data. If *K* = 10, then each subset will have 10 percent of the data (Fig. 5.14).

Overall Data

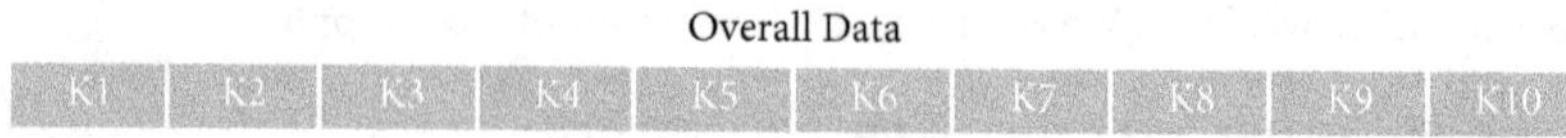

FIGURE 5.14   Data divided into *K* subsets.

Step 2: Build *K* models. While building the first model, take the first K−1 folds (first 9 folds out of the total 10 *K*-folds) as the train data and take the last part (K10) as the test data. Build and fine-tune the model that best fits for this pair of train and test data. Repeat the same by changing the test data. Every fold will be used as the test dataset for one model. For ease of understanding, Figure 5.15 provides the pictorial representation of tenfold cross-validation.

Model1	Train — Test
Model2	Train — Test — Train
Model3	Train — Test — Train
Model4	Train — Test — Train
Model5	Train — Test — Train
Model6	Train — Test — Train
Model7	Train — Test — Train
Model8	Train — Test — Train
Model9	Train — Test — Train
Model10	Test — Train

FIGURE 5.15   Illustration of tenfold cross-validation.

Step 3: Find the accuracy of all models (10 models if you have 10 *K*-folds). Note that all the above *K* models were built and fine-tuned for a combination of train and test data. It is not the case of a single model applied on several datasets. Now we take the average accuracy of all the *K* models (10 models if you have 10 *K* folds) to arrive at the final result. In our example, we are building 10 different models; we will be taking the average accuracy of all the 10 models.

K-fold cross-validation repeats the train–test scenario *K*-times. Since the *K*-fold method is taking the average as the final result, it gives us the optimal value for accuracy. We should see *K*-fold cross-validation as a model validation method, not a model building method. For example, if a *K*-fold cross-validation method yields 80 percent as the final result, then any model that gives us above 80 percent accuracy is overfitted, and any model that gives us below 80 percent accuracy is underfitted.

In the tenfold method, we have to build 10 models. In actual practice, we build more than 10 models. While building individual models, several iterations are tried out, as we may not get the best model in one attempt. We may have to try different values of the hyperparameter(s) to arrive at the most optimal model for that pair of test and train data. For 10 *K*-folds, we repeat this process 10 times.

Given below is the code required for *K*-fold cross-validation.

```
####K-fold cross-validation
diabetes_tree_KF = DecisionTreeClassifier(max_depth=3)
#Simple K-Fold cross-validation. 10 folds.
from sklearn.model_selection import KFold
kfold = KFold(n_splits=10)


## Checking the accuracy of model on 10-folds
from sklearn import model_selection
acc10 = model_selection.cross_val_score(diabetes_tree_KF,X, y,cv=kfold)
print(acc10)
print(acc10.mean())
```

We need to use the function KFold() and mention n_splits that mentions the number of folds. In this example, we are trying a tenfold cross-validation, so we need to mention n_splits=10. Here we are starting with max_depth = 3. In reality, we have to make many iterations and fine-tune each model to arrive at this value. Internally, several depth values are tried by *K*-fold function. The depth may not be 3 for all 10 trees in our final output. The following is the code output.

```
print(acc10)
[0.67532468 0.77922078 0.7012987 0.64935065 0.77922078
 0.81818182 0.83116883 0.83116883 0.67105263 0.71052632]
```

```
print(acc10.mean())
0.7446514012303486
```

From this output, we infer that the optimal accuracy of the model is 74.4 percent. Any model above 74.4 percent accuracy is overfitted, and any model with accuracy below 74.4 percent is an underfitted model.

K-fold cross-validation is computationally expensive. It takes a lot of iterations and much time to arrive at the optimal accuracy value. Next, we discuss one more reliable method that can be used to arrive at optimal accuracy.

5.6.3   Train–Validation–Holdout Cross-Validation Method

In this section, we straightway get into the details of the Train–Validation–Holdout cross-validation method. Divide the data into three parts; call the three parts as train data, validation data, and holdout data. These three data parts are explained below.

- Train data: This dataset is used to build the model. We learn the patterns from this data.
- Validation data: This data is used for validating the model and hyperparameter values. Essentially, we use this dataset to fine-tune parameters and finalize our model. With the validation data, the model and its parameters are finalized (fine-tuned).
- Holdout data: This is like our final test data, which was not used before in fine-tuning the model and its parameters. We will use this data only to test the finalized model. Note that we had already finalized the model and its parameters with validation data in the previous step. We never use holdout data to fine-tune the parameters. It is only for testing the robustness of the finalized model.

The problem with train–test cross-validation was the possibility of overfitting a model both on train and test data. On the other hand, the challenge with the K-fold cross-validation method is that it is computationally expensive. The approach of the Train–Validation–Holdout method can be seen as an acceptable midway compromise. Many professionals call it the train–test1–test2 approach.

Given below is the code for the Train–Validation–Holdout data approach. Here, we split test data again into two parts.

```
#### Train - Validation - Holdout cross-validation
from sklearn import model_selection

## Split overall data into Train and Test split
X_train, X_test, y_train, y_test = model_selection.train_test_split(X, y,
test_size=0.3, random_state=99)

## Split Test data into Validation and Holdout data split
X_val, X_hold, y_val, y_hold =
model_selection.train_test_split(X_test, y_test,test_size=0.5 ,
random_state=11)

print(X_train.shape)
print(y_train.shape)
print(X_val.shape)
print(y_val.shape)
print(X_hold.shape)
print(y_hold.shape)
```

This code first splits the overall data into two parts: 70 percent of the data as the train data and the remaining 30 percent for testing. In the next line of code, the test data is further split into two parts: 50 percent of the test data is stored for validation and the remaining 50 percent as holdout data. This split gives 15 percent of the overall data as the validation data. The following is the resultant output.

```
print(X_train.shape)
(537, 8)

print(y_train.shape)
(537, 1)

print(X_val.shape)
(115, 8)
print(y_val.shape)
(115, 1)

print(X_hold.shape)
(116, 8)

print(y_hold.shape)
(116, 1)
```

We use the code below to build the model with different hyperparameters.

```
## Model Building with pruning parameters
print("Max Depth 6")
from sklearn.tree import DecisionTreeClassifier
diabetes_tree1= DecisionTreeClassifier(max_depth=6)
diabetes_tree1.fit(X_train, y_train)

#Calculate Accuracy on Train and Test data
print("Train data Accuracy", diabetes_tree1.score(X_train, y_train))
print("Validation data Accuracy", diabetes_tree1.score(X_val, y_val))

## Model Building with pruning parameters
print("Max Depth 1")
from sklearn.tree import DecisionTreeClassifier
diabetes_tree1= DecisionTreeClassifier(max_depth=1)
diabetes_tree1.fit(X_train, y_train)

#Calculate Accuracy on Train and Test data
print("Train data Accuracy", diabetes_tree1.score(X_train, y_train))
print("Validation data Accuracy", diabetes_tree1.score(X_val, y_val))

## Model Building with pruning parameters
print("Max Depth 3")
from sklearn.tree import DecisionTreeClassifier
diabetes_tree1= DecisionTreeClassifier(max_depth=3)
diabetes_tree1.fit(X_train, y_train)

#Calculate Accuracy on Train and Test data
print("Train data Accuracy", diabetes_tree1.score(X_train, y_train))
print("Validation data Accuracy", diabetes_tree1.score(X_val, y_val))
```

In this code, we have tried different values of max_depth. The following is the output.

```
Max Depth 6
Train data Accuracy 0.8957169459962756
Validation data Accuracy 0.7130434782608696

Max Depth 1
Train data Accuracy 0.7411545623836127
Validation data Accuracy 0.6869565217391305
```

```
Max Depth 3
Train data Accuracy 0.7746741154562383
Validation data Accuracy 0.7304347826086957
```

We can finalize `max_depth` as three, where the difference in accuracy for train and validation data is less than 5 percent. Finally, we proceed to test the finalized model (with `max_depth` as 3) on holdout data. The following is the code.

```
#Final Model and Result
print("Max Depth 3")
print("Train data Accuracy", diabetes_tree1.score(X_train, y_train))
print("Validation data Accuracy", diabetes_tree1.score(X_val, y_val))
print("Holdout data Accuracy", diabetes_tree1.score(X_hold, y_hold))
```

The resultant output is as given below.

```
Max Depth 3
Train data Accuracy 0.7746741154562383
Validation data Accuracy 0.7304347826086957
Holdout data Accuracy 0.7758620689655172
```

We make it out from this output that the model validates well on the holdout data. Here out of the three datasets, the Holdout data accuracy is the best. However, in practice, this may not always be the case. We look for consistent accuracy levels with all the three sets of data for the model to be recognized as robust. The methods discussed here are the most widely used procedures for cross-validation by the practitioners. However, there are a few more methods for cross-validation such as Leave One Out Cross Validation (LOOCV) and Bootstrap Cross-Validation.

5.7   FEATURE ENGINEERING TIPS AND TRICKS

We have discussed some standard machine learning algorithms for building the models. Regression works for continuous target variables. Logistic regression and decision tree algorithms can help us to solve classification problems. We have also discussed model validation metrics. All the algorithms are already coded and available in various packages. The packages for all the model validation measures are also available. One major question that comes to our mind is, "If all the algorithms and their codes are available for everyone. Then everyone must have the same results." If we supply a dataset and run a data science competition, then everyone gets the same result. What separates the winners from others? Feature engineering is the step that separates the standard models from top-performing models.

We have already reached an accuracy level with the available data. How can we improve the accuracy of the model from here? Can we use the same algorithm and the same dataset but still get better accuracy of the model? How do we make the model learn hidden patterns in the data that are not directly available from the dataset? Can we tweak existing columns and derive new columns from the data and make the model learn from those columns? How do we add or edit or highlight features based on our business knowledge? The answer to all these questions is "feature engineering."

5.7.1   What Is Feature Engineering?

Intentionally adding new features that are derived from the existing features is called feature engineering. Let us take an example; we have a dataset with only two columns—Date and Sales of a book store (Table 5.20).

TABLE 5.20   Date and Sales Data of a Book Store

Date	Sales
27-06-2019	6087
05-07-2019	9489
08-07-2019	9868
13-07-2019	**17461**
13-06-2019	9560
17-05-2019	8696
18-05-2019	**17148**
14-05-2019	7216
05-05-2019	**14110**
29-05-2019	7348
26-05-2019	**18708**
01-06-2019	**13550**
17-06-2019	9247
29-06-2019	**13443**
18-07-2019	7614
25-06-2019	8833
21-06-2019	9206
26-05-2019	**13264**
07-06-2019	8297
28-05-2019	7138

If we feed this data directly to any machine learning model, it may not be able to find the hidden patterns directly. From her business experience, the store manager knows that weekend sales is almost double that of weekdays. The weekend versus weekday sales information is not directly present in the data. Even if we divide the date variable into three parts as date, month, and year, the model will not be able to derive the weekend sales pattern on its own. We now proceed to derive a new feature called Weekend_indicator. We mark Weekend_indicator with 1 for weekends and 0 for weekdays (Table 5.21).

TABLE 5.21   Sales Data with Weekend_indicator

Date	Weekend_indicator	Sales
27-06-2019	0	6087
05-07-2019	0	9489
08-07-2019	0	9868
13-07-2019	1	**17461**
13-06-2019	0	9560
17-05-2019	0	8696
18-05-2019	1	**17148**
14-05-2019	0	7216
05-05-2019	1	**14110**
29-05-2019	0	7348
26-05-2019	1	**18708**
01-06-2019	1	**13550**
17-06-2019	0	9247
29-06-2019	1	**13443**
18-07-2019	0	7614
25-06-2019	0	8833
21-06-2019	0	9206
26-05-2019	1	**13264**
07-06-2019	0	8297
28-05-2019	0	7138

Adding the new field Weekend_indicator gives us some valuable information. If `weekend_ind = 1`, then the bookstore sales value is above 11,000; otherwise, it is less than 10,000. It is an example of feature engineering. This information was already available in the dataset but not in the form that the model can understand; by adding a new feature here, we could identify the underlying pattern in the data. It is an example of feature engineering. It involves adding new features based on our business knowledge. In feature engineering, we derive the features manually and present them to the model in order to get better results. Feature engineering requires the knowledge of statistics as well as the business domain. Sometimes the information is hidden in the modest fields such as dates, latitude and longitude, and region. Feature engineering methods are not standard across the datasets and industries. We need to carefully study the data and apply business experience to create new useful features. We can gain the intuition and knowledge about feature engineering only with practice and experience. In this section, we will discuss some tips and tricks to make feature engineering more effective.

We revisit the case study: House Sales in King County, USA, where we have to predict the prices of the houses based on specific features as inputs. We had built a regression model, which resulted in the R-squared value of 70 percent. We had used the data as it is—without adding any derived features.

Can we use the same data and increase the R-square value using feature engineering techniques? Given below is the code for creation of a basic model and calculating its R-squared value.

```python
#Defining X data
X = kc_house_data[['bedrooms', 'bathrooms', 'sqft_living', 'sqft_lot',
'floors', 'waterfront', 'view', 'condition', 'grade', 'sqft_above', 'sqft_
basement', 'yr_built', 'yr_renovated', 'zipcode', 'lat', 'long', 'sqft_
living15', 'sqft_lot15']]

y = kc_house_data['price']

from sklearn import model_selection
X_train, X_test, y_train, y_test = model_selection.train_test_split(X, y,
test_size=0.2, random_state=55)

print(X_train.shape)
print(y_train.shape)
print(X_test.shape)
print(y_test.shape)

import sklearn
model_1 = sklearn.linear_model.LinearRegression()
model_1.fit(X_train, y_train)

#Coefficients and Intercept
print(model_1.intercept_)
print(model_1.coef_)

#Rsquared Calculation on Train data
from sklearn import metrics
y_pred_train=model_1.predict(X_train)
print(metrics.r2_score(y_train,y_pred_train))

x#Rsquared Calculation on test data
y_pred_test=model_1.predict(X_test)
print(metrics.r2_score(y_test,y_pred_test))

#RMSE
print("RMSE on Train data : ", round(math.sqrt(np.mean(np.abs(y_train -
y_pred_train)**2)),2))
print("RMSE on Test data : ", round(math.sqrt(np.mean(np.abs(y_test -
y_pred_test)**2)),2))
```

Given below is the code output.

```python
print(model_1.intercept_)
8080822.666112712
```

```
print(model_1.coef_)
[-3.76187142e+04    4.39929752e+04    1.11927627e+02    1.12260521e-01
  7.86848634e+03    5.82851207e+05    5.24147307e+04    2.56475517e+04
  9.63780999e+04    7.02315647e+01    4.16960620e+01   -2.66443082e+03
  2.22630357e+01   -5.83768676e+02    6.04058556e+05   -2.04643130e+05
  1.84979337e+01   -3.70481687e-01]

print(metrics.r2_score(y_train,y_pred_train))
0.7004310823997761

print(metrics.r2_score(y_test,y_pred_test))
0.6964362880041228


RMSE on Train data : 202295.52
RMSE on Test data : 196693.42
```

From this output, we can make out that the R-squared value on the given set train and test data is around 70 percent, and the RMSE value approximately stands at 200,000. We will explore this example further in the following section. We will discuss one hot encoding, which sometimes works like magic. Let us explore how it impacts our model.

5.7.2   The Dummy Variable Creation or One Hot Encoding

Dummy variable creation is one of the most basic and also the easiest way to extract hidden information. Dummy variable creation is used for non-numerical as well as categorical variables. We had already discussed dummy variable creation earlier in Chap. 2. As a very first step, we identify and list all the categorical variables. In some datasets, a few columns take numerical values, but still, we need to convert them to dummy variables. In this example, we have "zipcode" as one of the numeric variables. However, the zip code number does not have any directly visible pattern.

The code below creates boxplots for all the discrete and categorical variables versus the target variable. We have discussed boxplots in detail in earlier chapters (Chap. 2).

Bedrooms versus price:

```
import matplotlib.pyplot as plt
import seaborn as sns
plt.figure(figsize=(10,10))
sns.boxplot( x=kc_house_data["bedrooms"],y=kc_house_data["price"])
plt.title('Bedrooms vs House Price', fontsize=20)
```

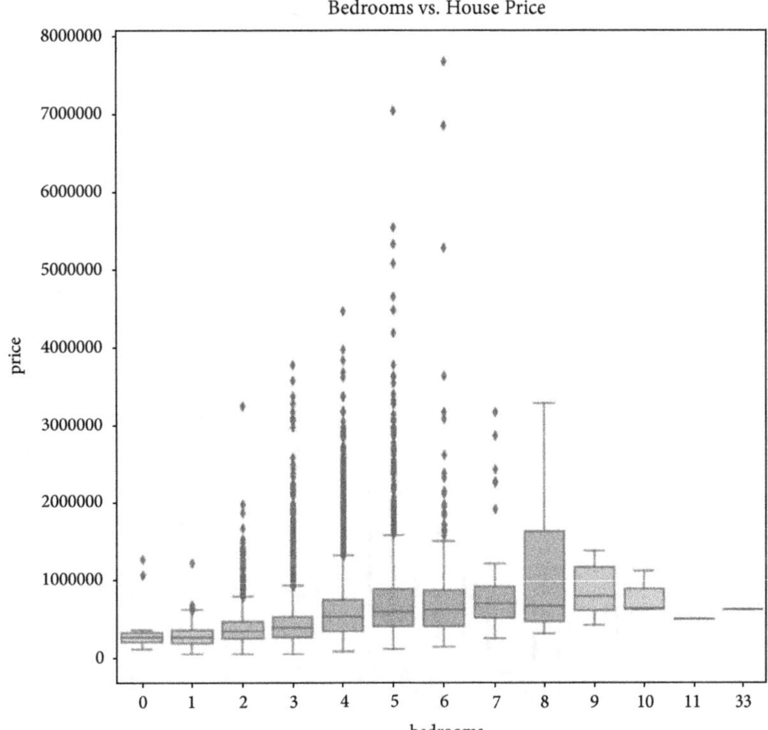

FIGURE 5.16   Bedroom versus house price boxplot.

Inference from the boxplot in Fig. 5.16—as the number of bedrooms increases, the price of the house also increases.

Bathrooms versus price:

```
plt.figure(figsize=(10,10))
sns.boxplot( x=kc_house_data["bathrooms"],y=kc_house_data["price"])
plt.title('Bathrooms vs House Price', fontsize=20)
```

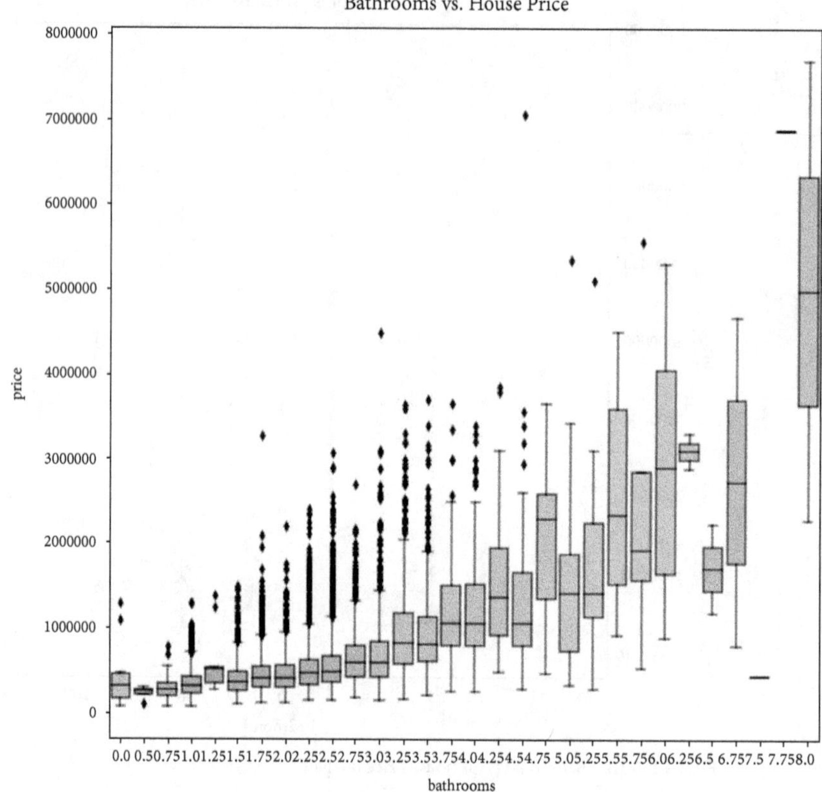

FIGURE 5.17   Bathrooms versus house price boxplot.

Inference from the graph in Fig. 5.17—as the number of bathrooms increases, the price of the house also increases.

Number of floors versus price:

```
plt.figure(figsize=(10,10))
sns.boxplot( x=kc_house_data["floors"],y=kc_house_data["price"])
plt.title('Floors vs House Price', fontsize=20)
```

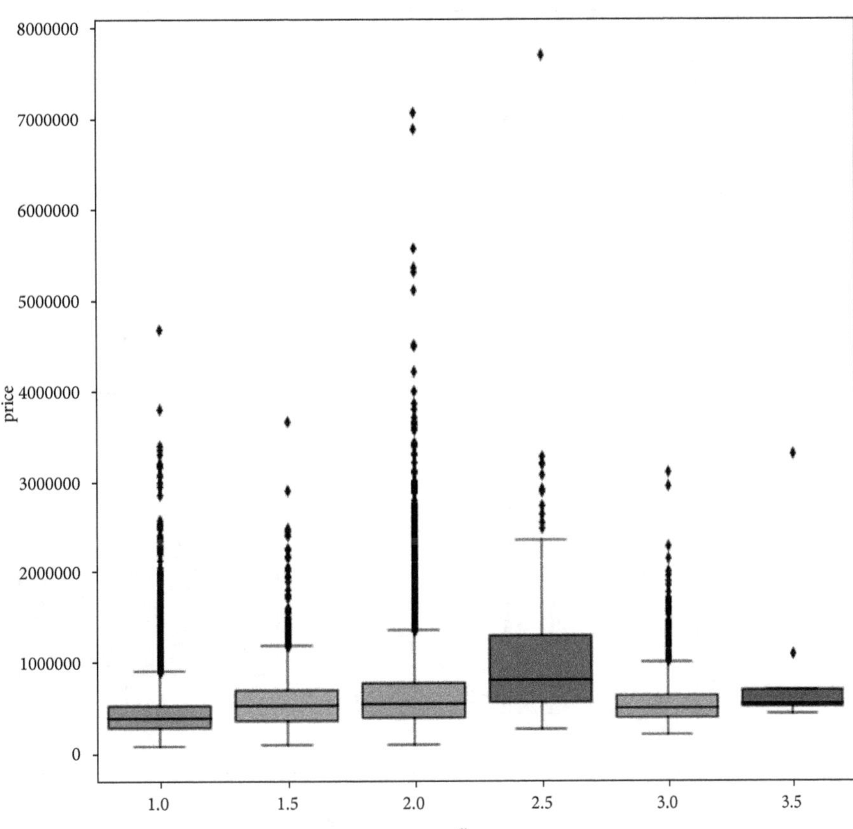

FIGURE 5.18   Floors versus house price boxplot.

Inference from the graph in Fig. 5.18—the number of floors does not have a direct relation with the price. One hot encoding may help here.

Waterfront versus price:

```python
plt.figure(figsize=(10,10))
sns.boxplot( x=kc_house_data["waterfront"],y=kc_house_data["price"])
plt.title('Waterfront vs House Price', fontsize=20)
```

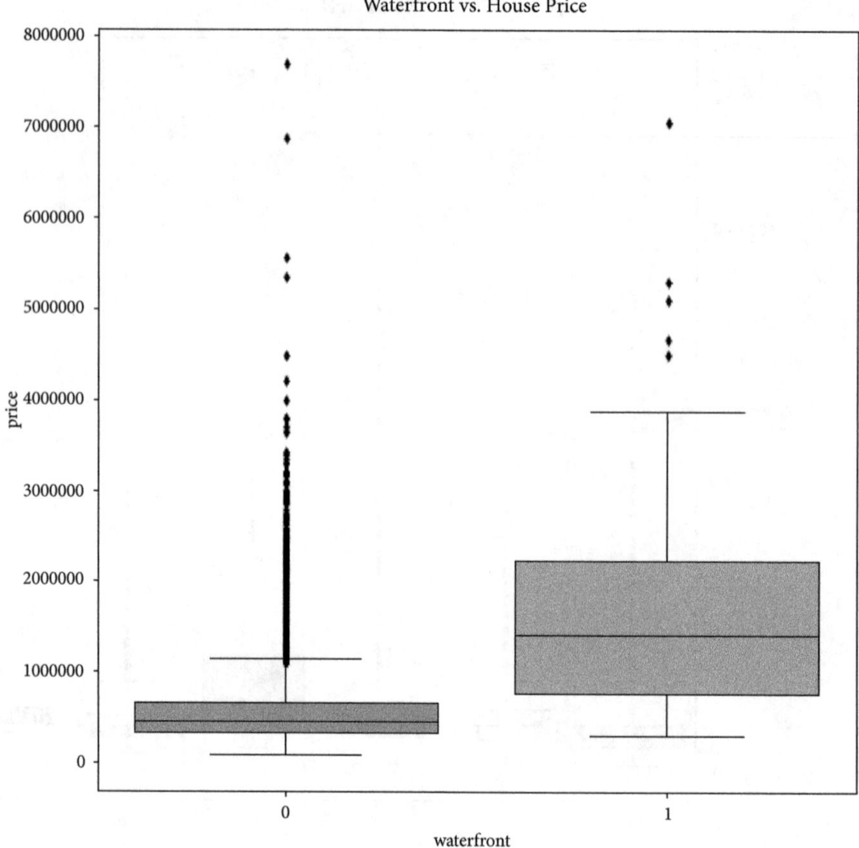

FIGURE 5.19 Waterfront versus house price boxplot.

Inference from the graph in Fig. 5.19—the waterfront indicator has a strong relationship with the house price.

View versus price:

```
plt.figure(figsize=(10,10))
sns.boxplot( x=kc_house_data["view"],y=kc_house_data["price"])
plt.title('View vs House Price', fontsize=20)
```

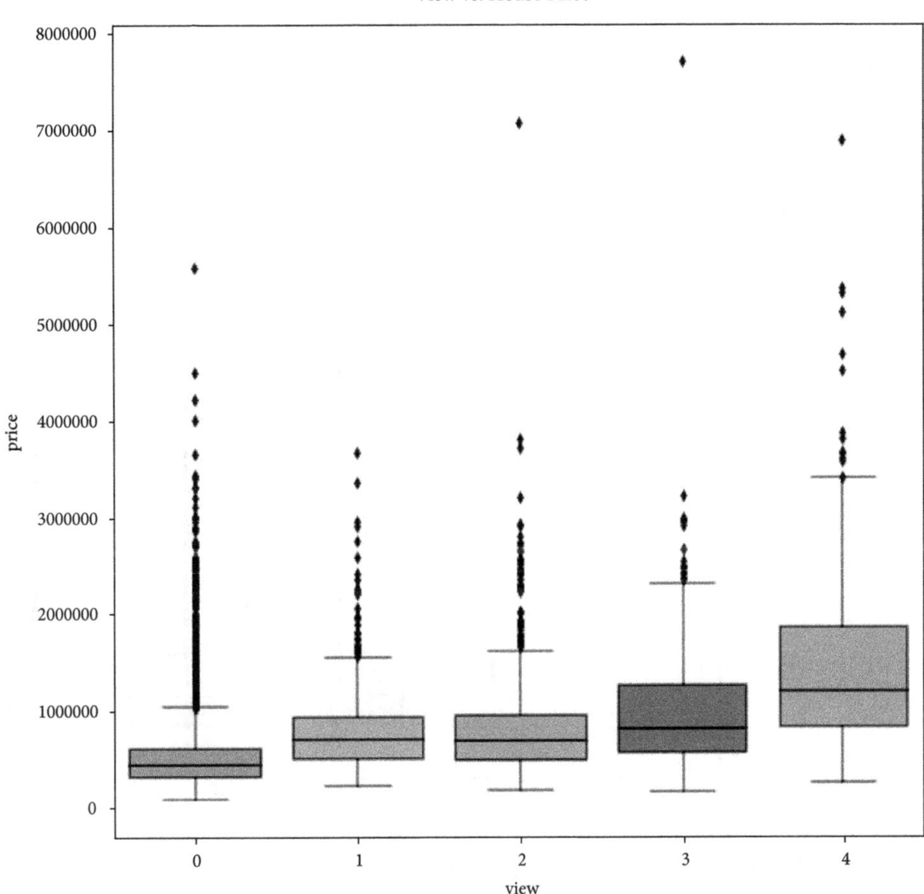

FIGURE 5.20 View versus house price boxplot.

Inference from the graph in Fig. 5.20—house price increases as view increases.

Condition versus price:

```
plt.figure(figsize=(10,10))
sns.boxplot( x=kc_house_data["condition"],y=kc_house_data["price"])
plt.title('Condition vs House Price', fontsize=20)
```

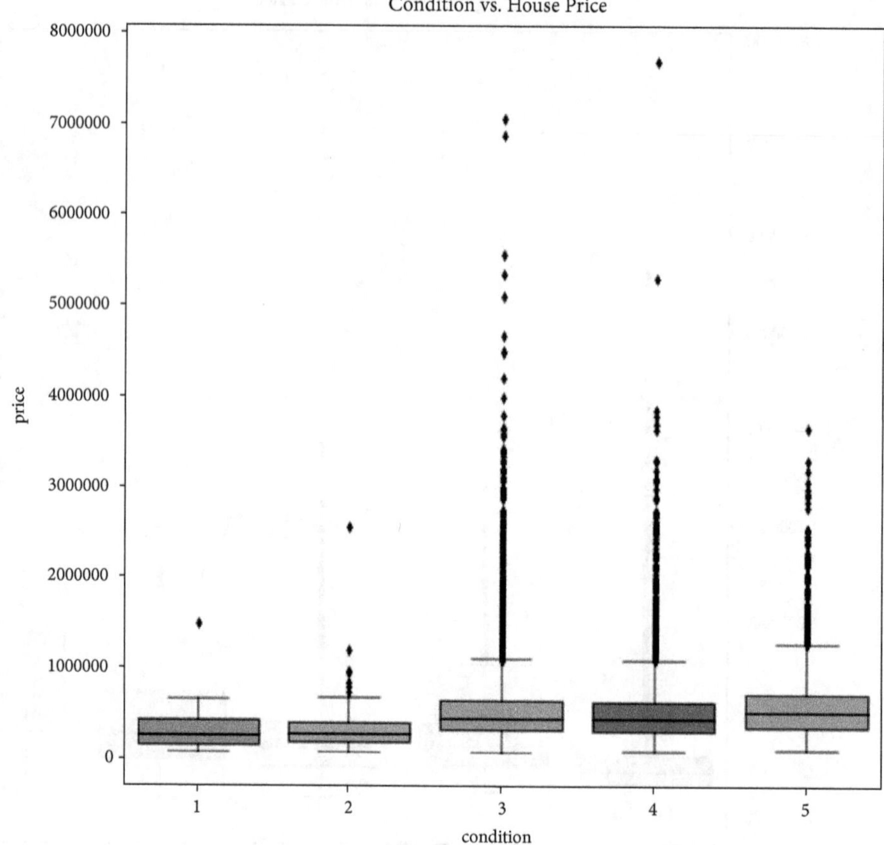

FIGURE 5.21   Condition versus house price boxplot.

Inference from the graph in Fig. 5.21—the condition does not show a direct strong relation with house price.

Grade versus price:

```
plt.figure(figsize=(10,10))
sns.boxplot( x=kc_house_data["grade"],y=kc_house_data["price"])
plt.title('Grade vs House Price', fontsize=20)
```

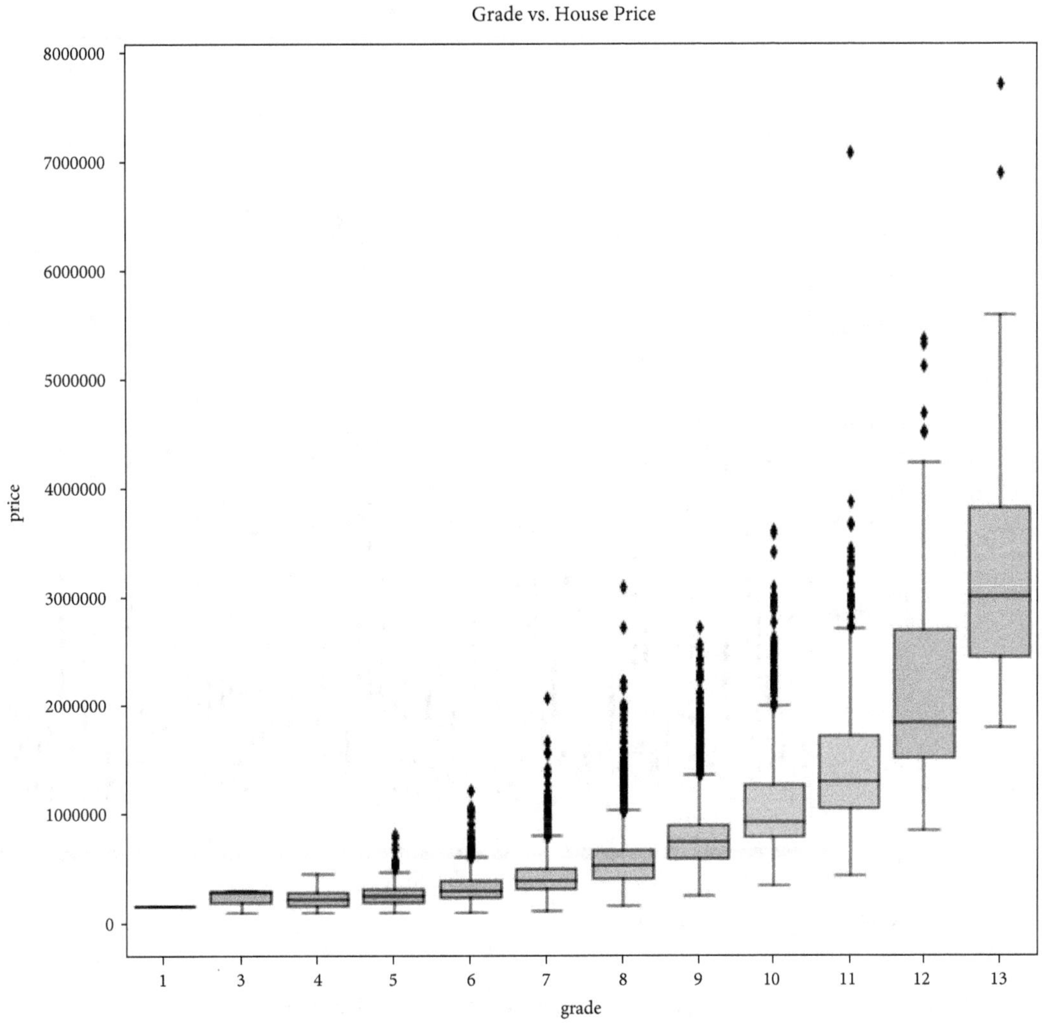

FIGURE 5.22   Grade versus house price boxplot.

Inference from the graph in Fig. 5.22—house price increases as grade increases.

Zipcode versus price:

```
plt.figure(figsize=(10,10))
sns.boxplot( x=kc_house_data["zipcode"],y=kc_house_data["price"])
plt.title('Zipcode vs House Price', fontsize=20)
```

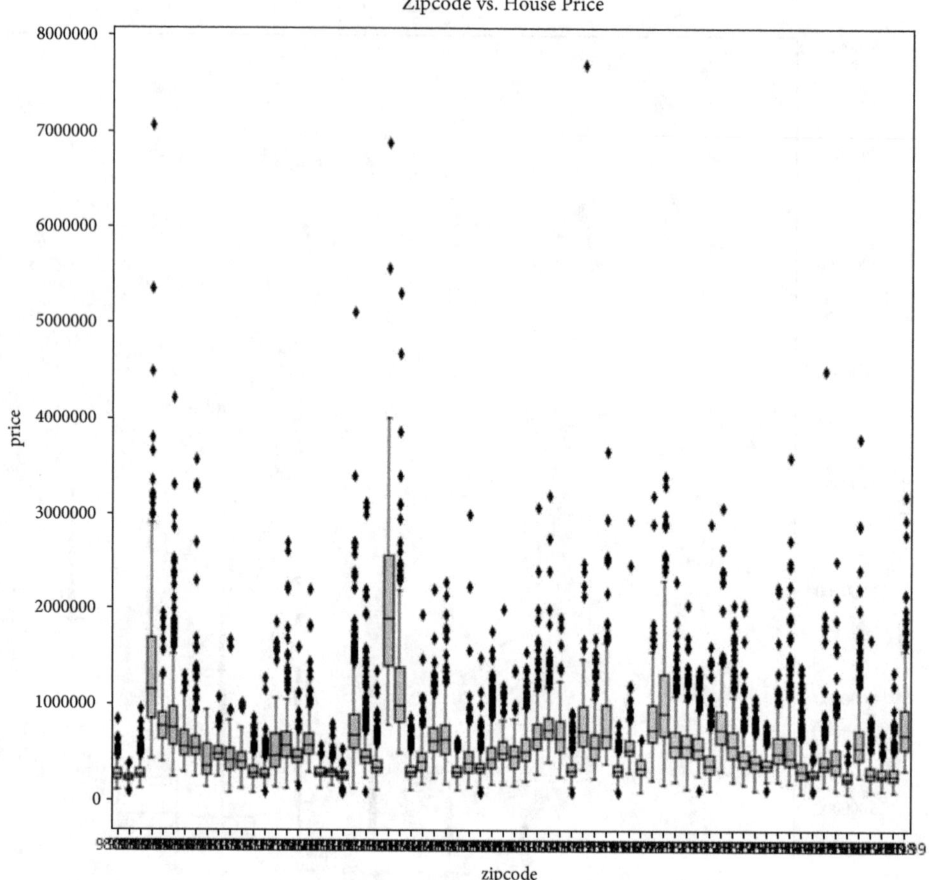

FIGURE 5.23 Zipcode versus house price boxplot.

Inference from the graph in Fig. 5.23—house price has no seeming relation with zip code.

So far, we have used all these variables directly in the model—without any variable alteration. If we reason that by converting these variables into dummy variables, we may get some more patterns. We can use the following code to create new dummy variables. It is always a good idea to create dummy variables and check the impact of variables on the model.

```
from sklearn.preprocessing import OneHotEncoder

kc_house_data.shape
categorical_vars=['bedrooms', 'bathrooms', 'floors', 'waterfront', 'view',
'condition', 'grade', 'zipcode']

encoding=OneHotEncoder()
encoding.fit(kc_house_data[categorical_vars])
onehotlabels = encoding.transform(kc_house_data[categorical_vars]).toarray()
onehotlabels_data=pd.DataFrame(onehotlabels)

print(kc_house_data.shape)

kc_house_data1 = kc_house_data.drop(categorical_vars,axis = 1)
print(kc_house_data1.shape)
```

```
kc_house_data_onehot=kc_house_data1.join(onehotlabels_data)
print(kc_house_data_onehot.shape)
```

The `OneHotEncoder()` function converts the categorical columns to one hot encoded columns. In the `fit()` function, we need to mention the column names. The transform function will transform the columns to one hot encoded columns. The number of one hot encoded columns will depend on the number of unique values in the variable. We then drop actual columns and update the dataset with one hot encoded column. This code gives us the following output.

```
print(kc_house_data.shape)
(21613, 21)

print(kc_house_data1.shape)
(21613, 13)

print(kc_house_data_onehot.shape)
(21613, 132)
```

Now we will use this updated dataset to build the regression line. The code below creates the train and test data.

```
col_names = kc_house_data_onehot.columns.values
print(col_names)

x_col_names=col_names[3:]
print(x_col_names)

X = kc_house_data_onehot[x_col_names]
y = kc_house_data_onehot['price']

from sklearn import model_selection
X_train, X_test, y_train, y_test = model_selection.train_test_split(X, y,
test_size=0.2, random_state=55)

print(X_train.shape)
print(y_train.shape)
print(X_test.shape)
print(y_test.shape)
```

Following is the resultant output of this code.

```
col_names = kc_house_data_onehot.columns.values
print(col_names)
['id' 'date' 'price' 'sqft_living' 'sqft_lot' 'sqft_above' 'sqft_basement'
 'yr_built' 'yr_renovated' 'lat' 'long' 'sqft_living15' 'sqft_lot15' 0 1 2 3
 4 5 6 7 8 9 10 11 12 13 14 15 16 17 18 19 20 21 22 23 24 25 26 27 28 29 30 31
 32 33 34 35 36 37 38 39 40 41 42 43 44 45 46 47 48 49 50 51 52 53 54 55 56 57
 58 59 60 61 62 63 64 65 66 67 68 69 70 71 72 73 74 75 76 77 78 79 80 81 82 83
 84 85 86 87 88 89 90 91 92 93 94 95 96 97 98 99 100 101 102 103 104 105 106
 107 108 109 110 111 112 113 114 115 116 117 118]

x_col_names=col_names[3:]
print(x_col_names)
['sqft_living' 'sqft_lot' 'sqft_above' 'sqft_basement' 'yr_built'
 'yr_renovated' 'lat' 'long' 'sqft_living15' 'sqft_lot15' 0 1 2 3 4 5 6 7 8 9
 10 11 12 13 14 15 16 17 18 19 20 21 22 23 24 25 26 27 28 29 30 31 32 33 34 35
 36 37 38 39 40 41 42 43 44 45 46 47 48 49 50 51 52 53 54 55 56 57 58 59 60 61
 62 63 64 65 66 67 68 69 70 71 72 73 74 75 76 77 78 79 80 81 82 83 84 85 86 87
 88 89 90 91 92 93 94 95 96 97 98 99 100 101 102 103 104 105 106 107 108 109
 110 111 112 113 114 115 116 117 118]
```

```
print(X_train.shape)
(17290, 129)

print(y_train.shape)
(17290,)

print(X_test.shape)
(4323, 129)

print(y_test.shape)
(4323,)
```

Now we are ready to build the model. Remember that for the previous model, we had the R-square value as 70 percent and the RMSE value was around 200,000.

```
import sklearn
model_1 = sklearn.linear_model.LinearRegression()
model_1.fit(X_train, y_train)

#Coefficients and Intercept
print(model_1.intercept_)
print(model_1.coef_)

#Rsquared Calculation on Train data
from sklearn import metrics
y_pred_train=model_1.predict(X_train)
print(metrics.r2_score(y_train,y_pred_train))

#Rsquared Calculation on test data
y_pred_test=model_1.predict(X_test)
print(metrics.r2_score(y_test,y_pred_test))

#RMSE
print("RMSE on Train data : ", round(math.sqrt(np.mean(np.abs(y_train -
y_pred_train)**2)),2))
print("RMSE on Test data : ", round(math.sqrt(np.mean(np.abs(y_test -
y_pred_test)**2)),2))
```

The following are the results from this code.

```
Train data R-Squared : 0.8430167450266046
Test data R-Squared : 0.8244800765868934
RMSE on Train data : 146441.63
RMSE on Test data : 149564.36
```

We have a piece of great news. The R-squared has jumped from 70 to 84 percent, and RMSE dropped from 200,000 to 145,000. It is a very significant improvement. This finding may raise doubts on our methodology—better, we revalidate to be doubly sure. The only fishy variable here is zipcode. It has too many distinct values.

A second look at the entire process reassures that everything is fine. Even if we consider an additional check of test data R-square, the model passes in the flying colors. By using the same data and same model building technique, we have achieved far better accuracy. All the credit goes to feature engineering.

There are a few more checks available to validate models, which we have not discussed in this book. You can explore them as an exercise. In the following sections, we will discuss how to explore a few specific variables such as longitude and latitude and dates along with some other topics.

5.7.3 Handling Longitude and Latitude

From our everyday experience, we know that house prices vary depending upon the location. The location of the house can be most scientifically captured in terms of longitude and latitude. A model may not be able to learn directly from the numerical values of longitude and latitude. We can do it better by deriving a couple of new features from longitude and latitude. First, we will try to grasp the relationship between price and longitude and latitude values. The following code does it for us.

```
###' House Price versus Longitude and Latitude'
bubble_col= kc_house_data["price"] > kc_house_data["price"].quantile(0.7)

import matplotlib.pyplot as plt
plt.figure(figsize=(12,12))
plt.scatter(kc_house_data["long"],kc_house_data["lat"], c=bubble_
col,cmap="RdYlGn",s=10)
plt.title('House Price vs Longitude and Latitude', fontsize=20)
plt.xlabel('Longitude', fontsize=15)
plt.ylabel('Latitude', fontsize=15)
plt.show()
```

This code is trying to draw a scatter plot between longitude and latitude. The house prices are color-coded in the plot. The bubble color is filled with green (black in the grayscale graph) if the house price is in the top 30 percentile. The code above gives us the output sketch presented in Fig. 5.24.

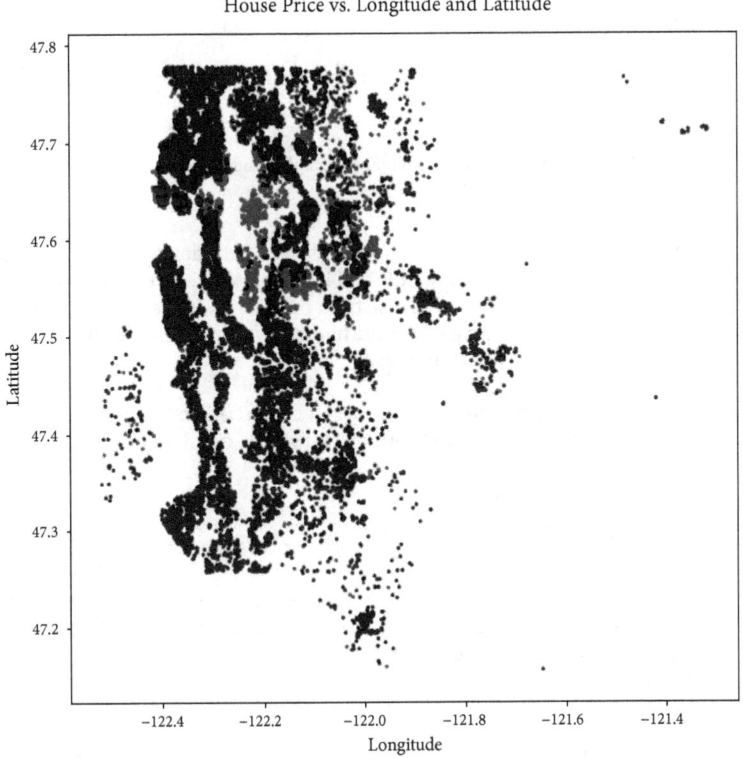

FIGURE 5.24   House price versus longitude and latitude.

The same graph is presented in grayscale in Fig. 5.25. Use the option `cmap="Greys"` to produce this graph.

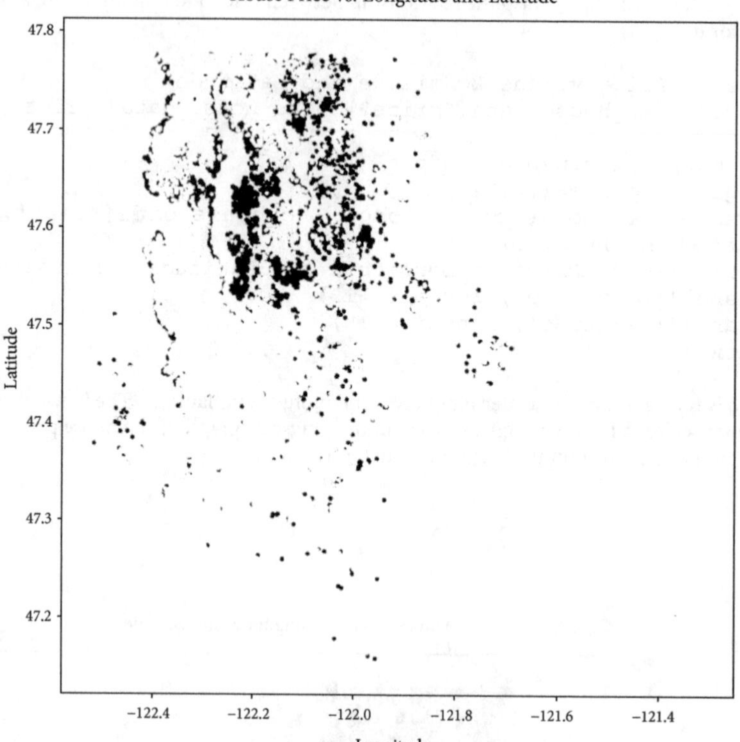

FIGURE 5.25   House price versus longitude and latitude (grayscale).

It looks like there is no significant impact on longitude and latitude. We will now create an additional feature to extract the maximum information out of longitude and latitude. We will create a center value for high-priced houses, and then we will calculate the distance of each house from that center. It's based on an intuition that the houses near to the center are high priced; the other way round for the houses away from this newly created center. The following code finds the geographical center of high-priced houses.

```python
high_long_mean=kc_house_data["long"][bubble_col].mean()
high_lat_mean=kc_house_data["lat"][bubble_col].mean()

import matplotlib.pyplot as plt
plt.figure(figsize=(12,12))
plt.scatter(kc_house_data["long"],kc_house_data["lat"], c=bubble_col,
cmap="RdYlGn",s=10)
plt.scatter(high_long_mean,high_lat_mean, c="balck", s=1000)
plt.title('House Price vs Longitude and Latitude', fontsize=20)
plt.xlabel('Longitude', fontsize=15)
plt.ylabel('Latitude', fontsize=15)
plt.show()
```

The point with the coordinate values of high_long_mean on the x-axis and high_lat_mean on the y-axis is the center of high-priced houses. This code gives us the output plot shown in Fig. 5.26.

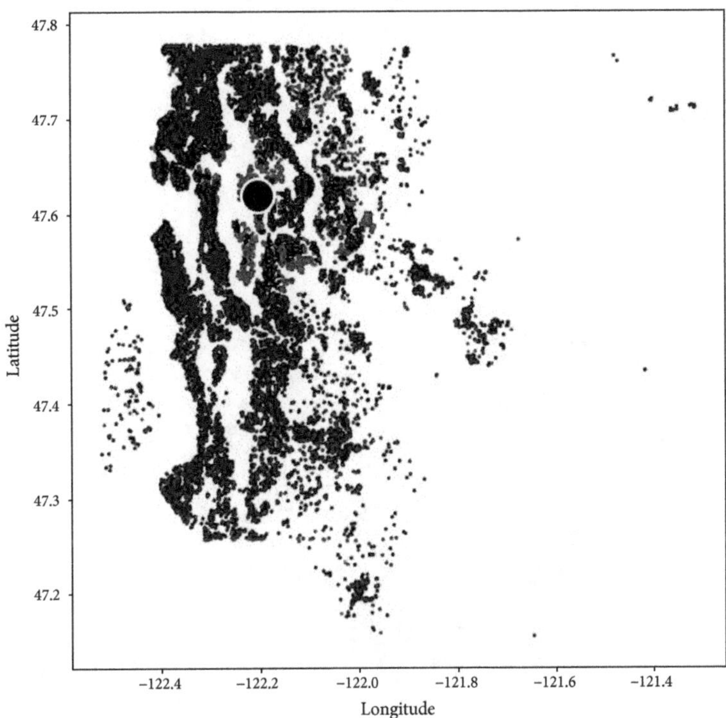

FIGURE 5.26   House price versus longitude and latitude with center of high-priced houses.

We can see the center of the high-priced houses plotted in Fig. 5.26. One more column is to be created to include the distance from this center of high-priced houses. This we will use in model building later. We will see if the distance variable can boost accuracy. We will also see the relationship between house price and this new distance variable by drawing a scatter plot. In the code below, we have taken a log (logarithm) of the housing prices to visualize it better.

```
##Distance from center to every house
kc_house_data["High_cen_distance"]=np.sqrt((kc_house_data["long"]  -  high_
long_mean) ** 2 + (kc_house_data["lat"] - high_lat_mean) ** 2)

plt.figure(figsize=(15,15))
plt.scatter(kc_house_data["High_cen_distance"],np.log(kc_house_
data["price"]))
plt.title('House Price vs Distance from center', fontsize=20)
plt.xlabel('Distance from center', fontsize=15)
plt.ylabel('log(house price)', fontsize=15)
```

This code gives us the output presented in Fig. 5.27.

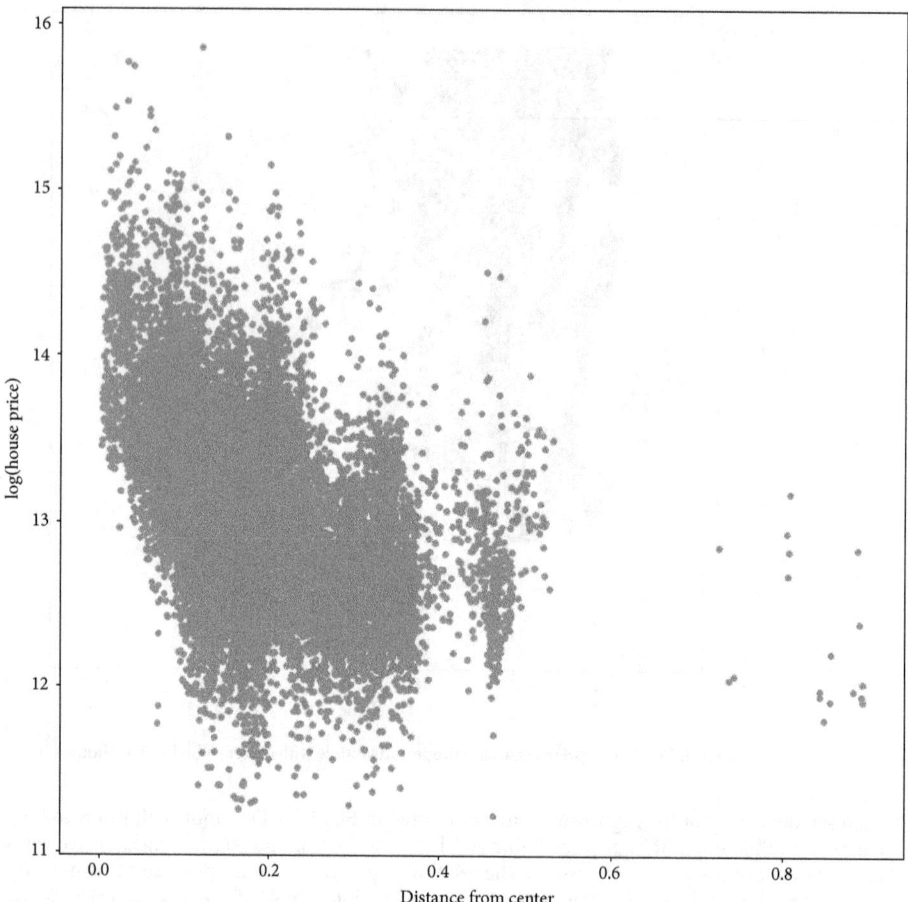

FIGURE 5.27 Scatter plot of house price versus distance from center.

From the output shown in Fig. 5.27, we can see that as the x variable (distance from the center) increases, the y variable (price) goes down. We cannot call it a strong pattern, but it is a pattern, nevertheless. We will now use this distance variable in our initial standard model. The following code does it for us.

```
#Defining X data
col_names = kc_house_data.columns.values
print(col_names)

x_col_names=col_names[3:]
print(x_col_names)

X = kc_house_data[x_col_names]
y = kc_house_data['price']

from sklearn import model_selection
X_train, X_test, y_train, y_test = model_selection.train_test_split(X, y,
test_size=0.2, random_state=55)

print(X_train.shape)
print(y_train.shape)
print(X_test.shape)
print(y_test.shape)
```

```python
import sklearn
model_1 = sklearn.linear_model.LinearRegression()
model_1.fit(X_train, y_train)

#Coefficients and Intercept
print(model_1.intercept_)
print(model_1.coef_)

#Rsquared Calculation on Train data
from sklearn import metrics
y_pred_train=model_1.predict(X_train)
print("Train data R-Squared : ", metrics.r2_score(y_train,y_pred_train))

#Rsquared Calculation on test data
y_pred_test=model_1.predict(X_test)
print("Test data R-Squared : " , metrics.r2_score(y_test,y_pred_test))

#RMSE
print("RMSE  on  Train  data  :  ",  round(math.sqrt(np.mean(np.abs(y_train  -
y_pred_train)**2)),2))
print("RMSE  on  Test  data  :  ",  round(math.sqrt(np.mean(np.abs(y_test  -
y_pred_test)**2)),2))
```

The above code gives us the results as given below.

```python
print(x_col_names)
['bedrooms' 'bathrooms' 'sqft_living' 'sqft_lot' 'floors' 'waterfront' 'view'
 'condition' 'grade' 'sqft_above' 'sqft_basement' 'yr_built' 'yr_renovated'
 'zipcode' 'lat' 'long' 'sqft_living15' 'sqft_lot15' 'High_cen_distance']

print(X_train.shape)
(17290, 19)

print(y_train.shape)
(17290,)

print(X_test.shape)
(4323, 19)

print(y_test.shape)
(4323,)

Train data R-Squared : 0.7148088489464938
Test data R-Squared : 0.7090610925035006
RMSE on Train data : 197381.26
RMSE on Test data : 192559.88
```

The R-squared value increased just by 1 percent, and there is a reduction in the RMSE value by 5000. The improvement in this case is not substantial by any standards. Sometimes, such derived features add significant value—the key is to try out as many iterations with new derived variables as the time and resources permit. Our added familiarity with King County can help us to add extra feature(s) to explore if you can improve the accuracy any further. For instance, we can derive one more feature as top_city_indicator; the feature name is self-explanatory. Figure 5.28 shows the published map of King County.

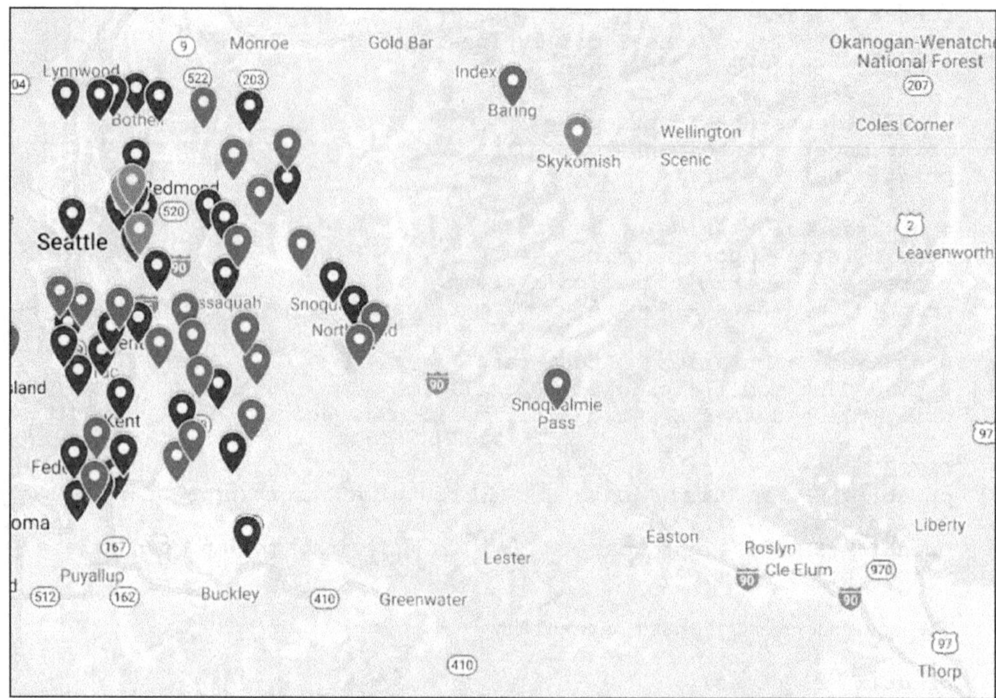

FIGURE 5.28   Published map of King County.

We can handpick cities such as "Seattle" and "Redmond," and give them more weight. There are some more towns and unincorporated areas in King County. The house prices are different in all these cities and towns. If we do an in-depth study, then we may be able to discover more and more hidden patterns, but all of them are not easy. Feature engineering requires much patience and time. It may sometimes be resource-intensive. Next, we will explore how to handle the date variable.

5.7.4   Handling Date Variables

The date variables should always be considered for feature engineering. Date and DateTime variables have fixed sets of formats like DD-MM-YY-HH-MM-SS, among others. The DD-MM-YY-HH-MM-SS format makes it very difficult for the model to learn any pattern(s). We can follow the below steps to handle the date variables more productively.

- Divide the single DateTime column into multiple components. For example, a column in the DD-MM-YY-HH-MM-SS format can be separated into six distinct columns—the day of the month, month, year, hour, minute, and second. Typically, minutes and seconds will not have any considerable impact.

- We can create the new derived variables such as the day of the week, weekend, month-start, month-end, quarter, half-year, holiday, week of the year, year-start, year-end, and so on. We can use these variables as is, or we can perform one hot encoding on these variables as well.

- If we are aware of any seasonal effects, then we can include a season index. For example, if sales are high in winter, then we can have winter_index as a new column.

- Knowing the date, sometimes it may be possible to derive the age or interval duration. For example, if we have the account opening date, then it is possible to derive the age of the account.

As mentioned below, in our example, there are three date-related variables.

```
print(kc_house_data.columns)
date_vars = ['date', 'yr_built', 'yr_renovated']
kc_house_dates=kc_house_data[date_vars]
kc_house_dates.head()
```

The code above gives us the following output.

```
          date yr_built yr_renovated
0 20141013T000000     1910         1987
1 20140611T000000     1940         2001
2 20140919T000000     2001            0
3 20140804T000000     2001            0
4 20150413T000000     2009            0
```

We have sale date variable by the name date. We can use it to derive the year of sales, the month of sales, and the day of sales. We can also derive the age of construction using the year_built variable. We will derive a new indicator variable renovation_ind, which will indicate all the houses that are renovated. The code below is used to create all these variables.

```
kc_house_dates['sale_year'] = np.int64([d[0:4] for d in
kc_house_dates["date"]])

kc_house_dates['sale_month'] = np.int64([d[4:6] for d in
kc_house_dates["date"]])

kc_house_dates['day_sold'] = np.int64([d[6:8] for d in
kc_house_dates["date"]])

kc_house_dates['age_of_house'] = kc_house_dates['sale_year']
- kc_house_dates['yr_built']

kc_house_dates['Ind_renovated'] = kc_house_dates['yr_renovated']>0
```

We will plot separate graphs (Figs. 5.29 to 5.33) to see how all the new columns are related to the price variable

Sales year versus price:

```
plt.figure(figsize=(10,10))
sns.boxplot( x=kc_house_dates['sale_year'],y=kc_house_data["price"])
plt.title('sale_year vs House Price', fontsize=20)
```

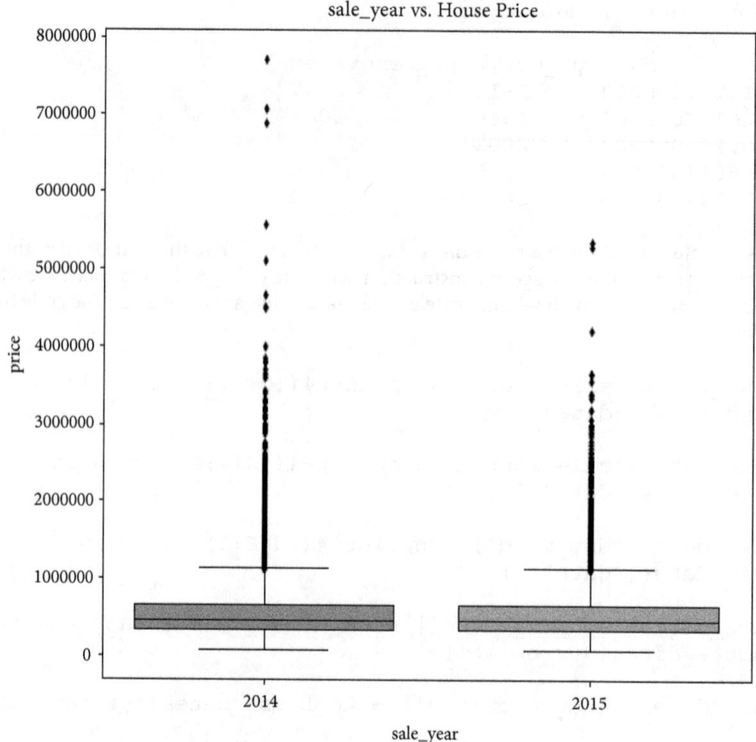

FIGURE 5.29   Price versus sale_year boxplots.

Inference from the graph in Fig. 5.29 is that sale year has no direct relation with house prices.

Sales month versus price:

```
plt.figure(figsize=(10,10))
sns.boxplot( x=kc_house_dates['sale_month'],y=kc_house_data["price"])
plt.title('sale_month vs House Price', fontsize=20)
```

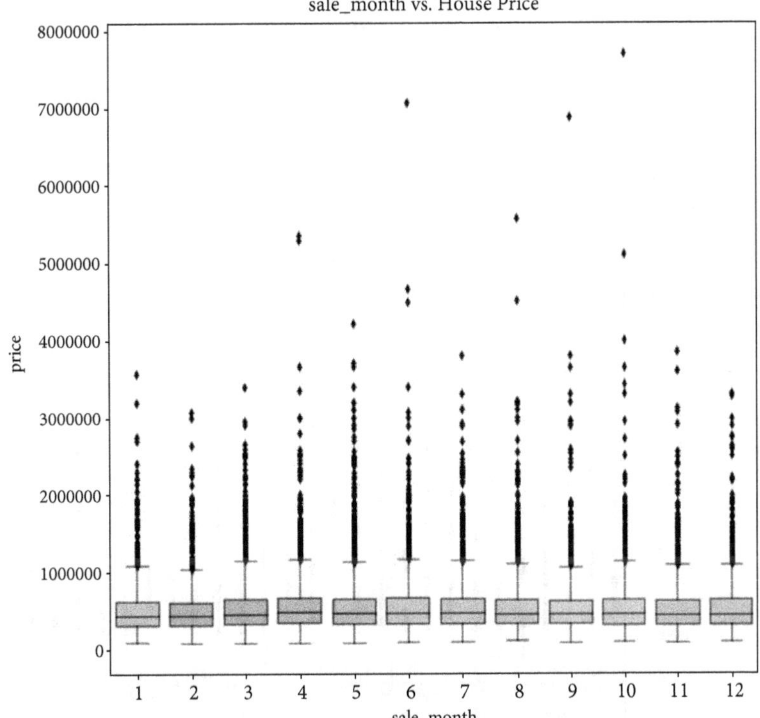

FIGURE 5.30   Price versus sale_month boxplots.

Inference from the graph in Fig. 5.30 is that sale month has no direct relation with house prices.

Day sold versus price:

```
plt.figure(figsize=(10,10))
sns.boxplot( x=kc_house_dates['day_sold'],y=kc_house_data["price"])
plt.title('day_sold vs House Price', fontsize=20)
```

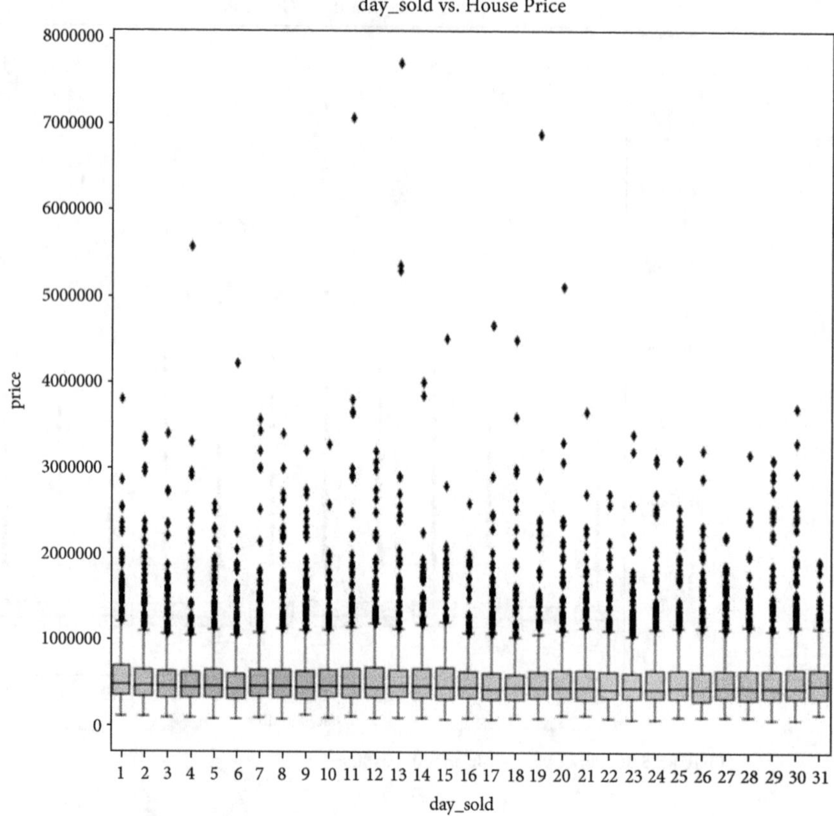

FIGURE 5.31   Day_sold versus house price boxplots.

Inference from the graph in Fig. 5.31—sale day has no direct relation with house prices.

Age of house versus price:

```
plt.figure(figsize=(10,10))
plt.scatter(kc_house_dates["age_of_house"],kc_house_data["price"])
plt.title('age_of_house vs House Price', fontsize=20)
```

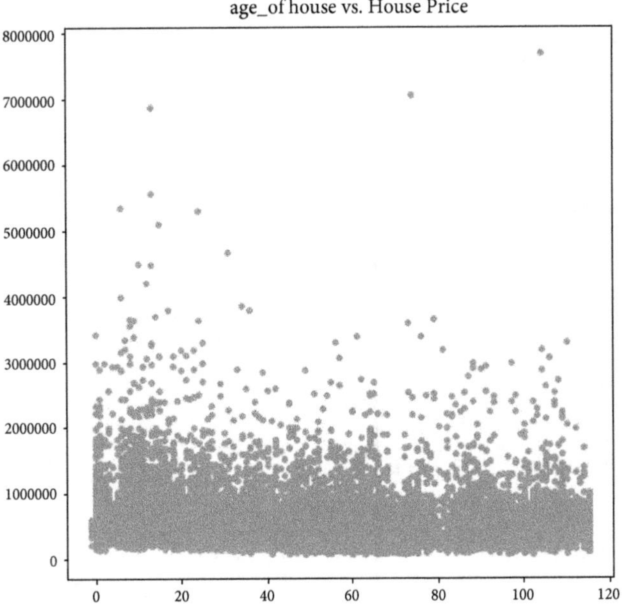

FIGURE 5.32   Age_of_house versus house price.

Inference from the graph in Fig. 5.32—age of the construction has no direct relation with house prices. This result is surprising.

Ind_rennovated versus price:

```
plt.figure(figsize=(10,10))
sns.boxplot( x=kc_house_dates['Ind_renovated'],y=kc_house_data["price"])
plt.title('Ind_renovated vs House Price', fontsize=20)
```

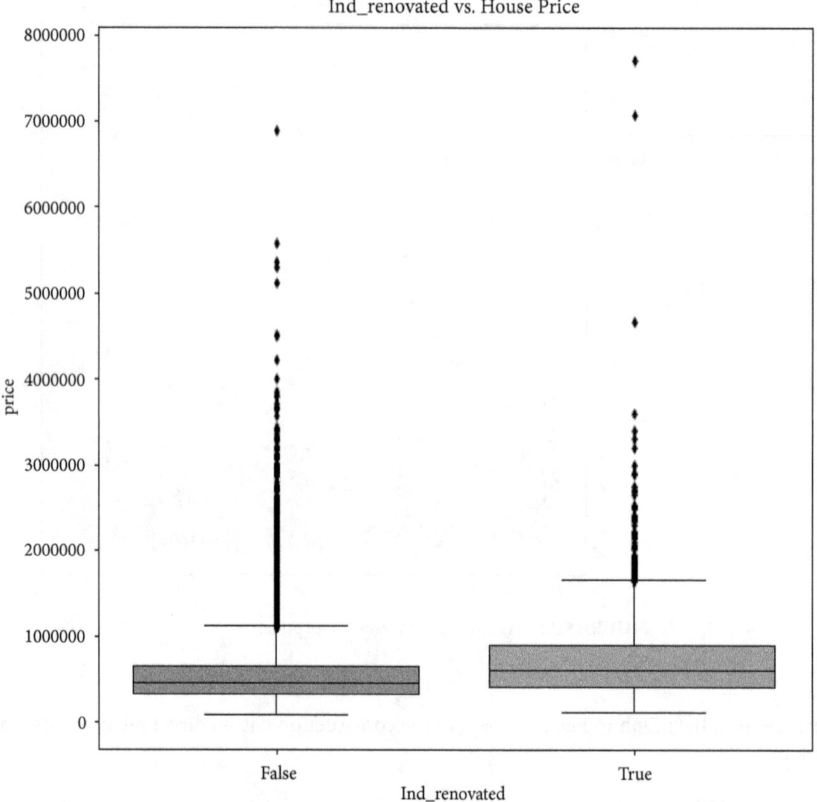

FIGURE 5.33   Ind_renovated versus house price boxplots.

Inference from the graph in Fig. 5.33—renovation has an impact on house prices. This result is in line with our intuition and the business experience

The code below is used for building the model with the new variables. We add all these variables to check if we can boost the accuracy any further.

```
##Model building with date variables
kc_house_dates1=kc_house_dates.drop(date_vars,  axis=1)  #keep  only  newly
derived variables
kc_house_with_dates=kc_house_data.join(kc_house_dates1)
print(kc_house_with_dates.shape)

###Model building with date variables

#Defining X data
col_names = kc_house_with_dates.columns.values
print(col_names)

x_col_names=col_names[3:]
print(x_col_names)

X = kc_house_with_dates[x_col_names]
y = kc_house_with_dates['price']
```

```python
from sklearn import model_selection
X_train, X_test, y_train, y_test = model_selection.train_test_split(X, y,
test_size=0.2, random_state=55)

print(X_train.shape)
print(y_train.shape)
print(X_test.shape)
print(y_test.shape)
```

Presented below is the result of the above code.

```python
print(x_col_names)
['bedrooms' 'bathrooms' 'sqft_living' 'sqft_lot' 'floors' 'waterfront' 'view'
'condition' 'grade' 'sqft_above' 'sqft_basement' 'yr_built' 'yr_renovated'
'zipcode' 'lat' 'long' 'sqft_living15' 'sqft_lot15' 'High_cen_distance'
'sale_year' 'sale_month''day_sold' 'age_of_house' 'Ind_renovated']

print(X_train.shape)
(17290,24)

print(y_train.shape)
(17290,)

print(X_test.shape)
(4323,24)

print(y_test.shape)
(4323,)
```

We will now build the model with these newly created variables.

```python
import sklearn
model_1 = sklearn.linear_model.LinearRegression()
model_1.fit(X_train, y_train)

#Coefficients and Intercept
print(model_1.intercept_)
print(model_1.coef_)

#Rsquared Calculation on Train data
from sklearn import metrics
y_pred_train=model_1.predict(X_train)
print("Train data R-Squared : ", metrics.r2_score(y_train,y_pred_train))

#Rsquared Calculation on test data
y_pred_test=model_1.predict(X_test)
print("Test data R-Squared : " , metrics.r2_score(y_test,y_pred_test))

#RMSE
print("RMSE on Train data : ", round(math.sqrt(np.mean(np.abs(y_train -
y_pred_train)**2)),2))
print("RMSE on Test data : ", round(math.sqrt(np.mean(np.abs(y_test -
y_pred_test)**2)),2))
```

The above code gives us the below output.

```
Train data R-Squared : 0.7172127242618249
Test data R-Squared : 0.710839122114219
RMSE on Train data : 196547.63
RMSE on Test data : 191970.57
```

Again, we see only a marginal improvement—the R-squared value is increased by 1 percent, and RMSE reduced by 5000. This improvement is also not significant. If you wish you can try it with a few more derived features, based on intuition and the domain knowledge, such as a quarter day of the week. Next, we will discuss how we can make numeric variables more productive using transformations.

5.7.5 Transformations

We have discussed how one hot encoding technique can be applied to categorical variables. We have also discussed how we can get more out of date variables. What about the continuous variables? We can apply transformations to these variables. If some variables take exponential values, then we can use log transformation for better predictions. In some other cases, if the data is skewed, then log transformation can also be used to normalize it. We must ensure that there is no negative value in the column before we apply log or square-root transformation. Alternatively, we can try some other situation-based transformations. A couple of more transformations are discussed here.

Occasionally, we can derive polynomial terms from the existing data. We can handle the outliers by replacing them with the mean or median. We can apply transformation even to the target column if that helps us in improving the accuracy.

Taking the same housing price example, the code below helps with the scatter plots for the continuous variables against the target variable.

```
grid_plot1= sns.PairGrid(kc_house_data, y_vars=["price"], x_vars=["sqft_liv-
ing", "sqft_lot"], height=5)
grid_plot1.map(sns.regplot)

grid_plot2 = sns.PairGrid(kc_house_data, y_vars=["price"], x_vars=["sqft_
above", "sqft_basement"], height=5)
grid_plot2.map(sns.regplot)

grid_plot3 = sns.PairGrid(kc_house_data, y_vars=["price"], x_vars=["sqft_
living15","sqft_lot15"], height=5)
grid_plot3.map(sns.regplot)
```

In the above code, the parameter `sns.regplot` tries to draw the scatter plot and build the regression line. This plot will help us to get an idea of the relation between the two variables. The above code gives us the output presented in Fig. 5.34.

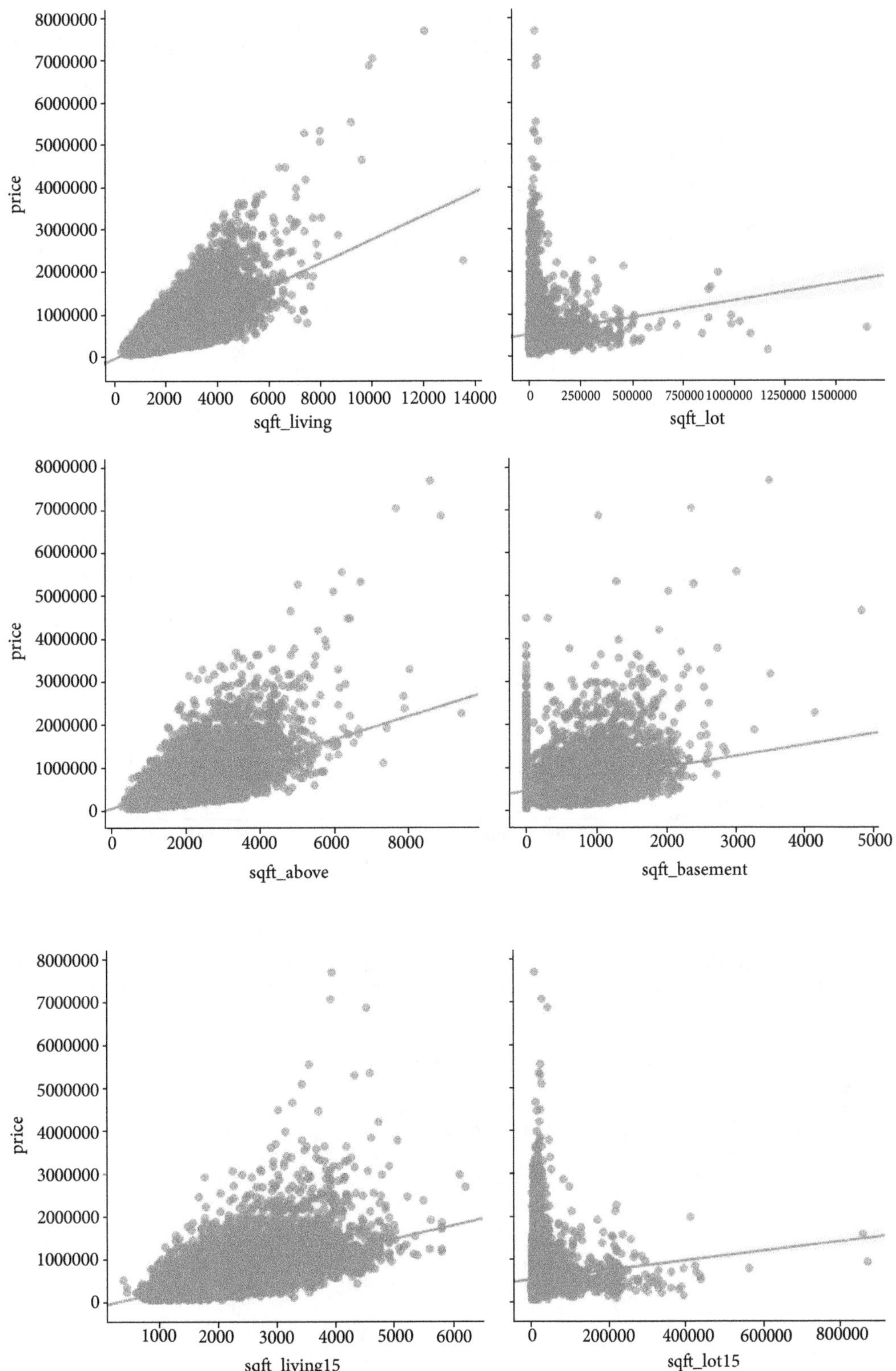

FIGURE 5.34 Scatter plots for the continuous variables against the target variable.

From plots in Fig. 5.34, we can see that some variables such as sqft_living, sqft_above, and sqft_living15 have a direct relationship with the price variable. The price itself has a few extremes. Let us plot the distribution of the price variable.

```
#Histogram on target variable
plt.figure(figsize=(10,10))
sns.distplot(kc_house_data["price"])
plt.title('House Price distribution', fontsize=20)
```

The output is plotted in Fig. 5.35.

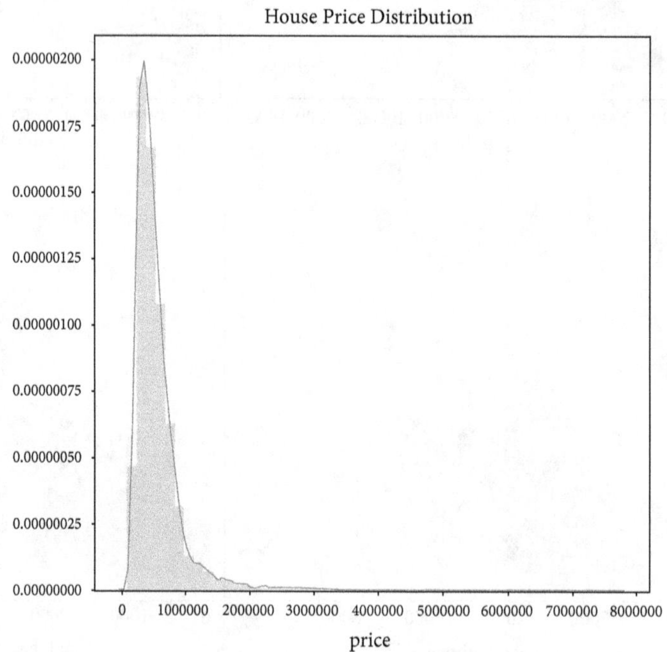

FIGURE 5.35   House price distribution.

We can see the distribution is skewed. We need to make an outlier treatment or apply log transformation on this data. The code below creates the `log_price` variable and draws the distribution chart for the transformed variable.

```
#Log transformation
kc_house_data["log_price"]=np.log(kc_house_data["price"])
plt.figure(figsize=(10,10))
sns.distplot(kc_house_data["log_price"])
plt.title('log(House Price) distribution', fontsize=20)
```

We depict the code output in the plot in Fig. 5.36.

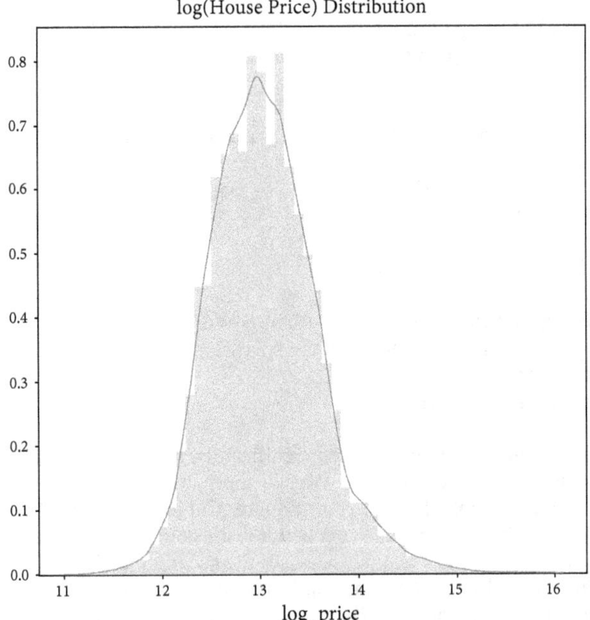

FIGURE 5.36   Log distribution of house price.

The graph in Fig. 5.36 yields negligible skewness. We will try to build the model by taking this log transformation on the target variable. As usual, we will compare these results with our original model. The code below is used for building the model after log transformation.

```
###Model building after Transformations
#Defining X data
X = kc_house_data[['bedrooms', 'bathrooms', 'sqft_living', 'sqft_lot',
'floors', 'waterfront', 'view', 'condition', 'grade', 'sqft_above', 'sqft_
basement', 'yr_built', 'yr_renovated', 'zipcode', 'lat', 'long', 'sqft_liv-
ing15', 'sqft_lot15']]

y = kc_house_data['log_price']

from sklearn import model_selection
X_train, X_test, y_train, y_test = model_selection.train_test_split(X, y,
test_size=0.2, random_state=55)

import sklearn
model_1 = sklearn.linear_model.LinearRegression()
model_1.fit(X_train, y_train)

#Coefficients and Intercept
print(model_1.intercept_)
print(model_1.coef_)

#Rsquared Calculation on Train data
from sklearn import metrics
y_pred_train=model_1.predict(X_train)
print("Train data R-Squared : ", metrics.r2_score(y_train,y_pred_train))
```

```
#Rsquared Calculation on test data
y_pred_test=model_1.predict(X_test)
print("Test data R-Squared : " , metrics.r2_score(y_test,y_pred_test))

#RMSE
print("RMSE  on  Train  data  :  ",  round(math.sqrt(np.mean(np.abs(y_train  -
y_pred_train)**2)),2))
print("RMSE  on  Test  data  :  ",  round(math.sqrt(np.mean(np.abs(y_test  -
y_pred_test)**2)),2))
```

The following is the output.

```
Train data R-Squared : 0.7718064095694628
Test data R-Squared : 0.7644327349746673
RMSE on Train data : 0.25
RMSE on Test data : 0.25
```

We can see that the R-squared value increased significantly. It has gone up to 77 percent. As we did the log transformation, we cannot compare the RMSE value. We can apply a transformation on predictor variables as well. The log transformation is just one example. Square root, Square, Cube root, Cube, Inverse, and Binning are some other examples of transformation that we can use to get better results out of a given dataset.

There is no promise that every new feature increases the accuracy of the model. In our example, one hot encoding and log transformation worked well. Nevertheless, feature engineering on the date, longitude, and latitude variables did not show much improvement. For a specific type of datasets, a particular class of feature engineering tricks works. What works best for our dataset in hand needs to be discovered manually; knowledge of the business domain, a good grasp on data, practice, and experience would help.

In the next section we discuss about dealing with class imbalance. We also discuss adjustments that we need to do before building the model so that the model can learn the patterns related to rare events.

5.8   DEALING WITH CLASS IMBALANCE

Let us start with a bit of recap. While discussing sensitivity and specificity, we have discussed class imbalance. In some classification problems, the classes in the target have this problem of class imbalance. In some cases, the overall accuracy is primarily driven by a single class. If we are not interested in that class, then overall accuracy looks good, but the model fails to fulfill its fundamental objective(s). We then concentrate on individual class accuracy—we call it sensitivity and specificity. In the same section, we have discussed the model validation measures in the case of class imbalance. In this section, we discuss the adjustments that we need to do before building the model so that the model can learn the patterns related to rare events.

We will revisit the credit risk data and try to work on the problem of class imbalance. Given below are the confusion matrix, accuracy, sensitivity, and specificity of the credit risk model.

```
print(cm1)
[[111264    650]
 [ 7465    627]]

print(accuracy1)
0.932

print(round(Sensitivity1,3))
0.994

print(round(Specificity1,3))
0.077
```

Accuracy stands at 93 percent, which is primarily driven by class-0. Sensitivity is 99 percent, but specificity is 7.7 percent. We have built this model to identify the likely defaulters before giving a loan. Class-1 indicates "bad" in this data with a massive class imbalance. Class-1 is less than 10 percent of all the data. We aim to increase the specificity by some means. In the previous sections, we tried changing the threshold to increase specificity. Next, we will learn a few more techniques to handle the data having class imbalance.

5.8.1 Oversampling and Undersampling

In practice, we usually are more interested in a specific class—looking for a fraud transaction or detecting a loan defaulter are a few examples. Unfortunately, as often may be the case, our class of interest has a very less proportion of records compared to the overall data. Therefore, the model is not able to learn or find it difficult to grasp the patterns associated with the minor portion of the data. Can we create a new sample by taking very few records from the dominating (but unimportant) class and take several records from the significant class (the class of interest)? If class-0 is unimportant for us, then we will reduce the records from class-0 and increase the records from class-1 in our data sample. For example, if Fraud and Non-Fraud are two classes, then we will take a subset of Non-Fraud records and take the full set of "Fraud" records. By introducing some duplication, we can even increase the number of records in the "Fraud" class.

Taking a subset of the majority class is known as undersampling. Taking duplicate copies of the minority class is known as oversampling. We will try to create a balanced dataset from the imbalanced dataset. We expect the model to pick the patterns associated with the minority class present in the balanced data. In a way, we are sending skewed data to the model so that it can focus on minority class. Since we are focusing on minority class, this balanced data creation works well for us (Fig. 5.37).

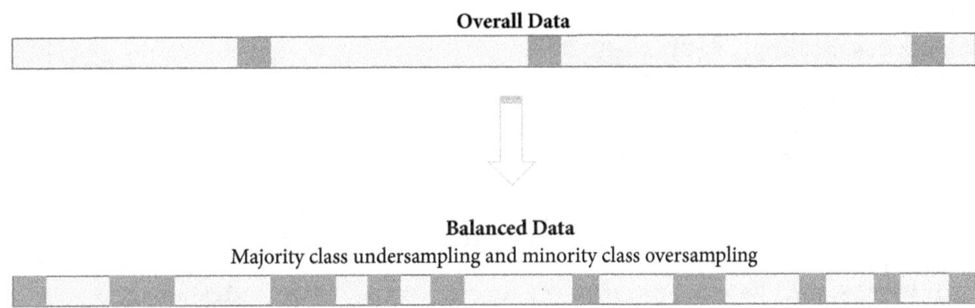

FIGURE 5.37   Oversampling and undersampling.

We will try to use these undersampling and oversampling techniques to solve our credit risk problem. The code below gives us a balanced sample.

```python
import pandas as pd
credit_risk_data  =  pd.read_csv(r'D:\Chapter5\5. Base  Datasets\loans_data\
credit_risk_data_v1.csv')

print("Actual Data :", credit_risk_data.shape)

#Frequency count on target column
freq=credit_risk_data['Bad'].value_counts()
print(freq)
print((freq/freq.sum())*100)

#Classwise data
credit_risk_class0 = credit_risk_data[credit_risk_data['Bad'] == 0]
credit_risk_class1 = credit_risk_data[credit_risk_data['Bad'] == 1]

print("Class0 Actual :", credit_risk_class0.shape)
print("Class1 Actual :", credit_risk_class1.shape)
```

The above code gives us the below output.

```
Actual Data : (150008, 10)
Overall Data
0    139981
1     10027

print((freq/freq.sum())*100)
0    93.31569
1     6.68431
Name: Bad, dtype: float64

Class0 Actual : (139981, 10)
Class1 Actual : (10027, 10)
```

We can see from the output that the proportion of class-1 is less than 10 percent. We will now perform oversampling and undersampling.

```
##Undersampling of class-0
## Consider half of class-0
credit_risk_class0_under = credit_risk_class0.
sample(int(0.5*len(credit_risk_class0)))
print("Class0 Undersample :", credit_risk_class0_under.shape)

##Oversampling of Class-1
# Lets increase the size by four times
credit_risk_class1_over = credit_risk_class1.sample(4*len(credit_risk_class1),
replace=True)
print("Class1 Oversample :", credit_risk_class1_over.shape)

#Concatenate to create the final balanced data
credit_risk_balanced=pd.concat([credit_risk_class0_under,  credit_risk_class1_
over])
print("Final Balanced Data :", credit_risk_balanced.shape)

#Frequency count on target column in the balanced data
freq=credit_risk_balanced['Bad'].value_counts()
print(freq)
print((freq/freq.sum())*100)
```

In the code above, we used the `sample()` function to fetch a sample from the data. For the undersample, we chose 50 percent of the records from class0. In the case of oversample, we increased the records four times. We need to use `replace=True` option for oversampling. The following is the code output.

```
Class0 Undersample : (69990, 10)
Class1 Oversample : (40108, 10)

Final Balanced Data : (110098, 10)

Balanced Data
0    69990
1    40108

print((freq/freq.sum())*100)
0    63.570637
1    36.429363
```

We can see from the output that class1 was just 6 percent in the overall data (the original data sample). In the balanced data, the same class-1 is 36 percent. We will now build a model with balanced data. We expect the updated model to demonstrate a better specificity. The following code builds the model for us.

```python
X = credit_risk_balanced[['Credit_Limit', 'Late_Payments_Count',
'Card_Utilization_Percent', 'Age', 'Debt_to_income_ratio',
'Monthly_Income', 'Num_loans_personal_loans', 'Family_dependents']]

y = credit_risk_balanced['Bad']

from sklearn import model_selection
X_train, X_test, y_train, y_test = model_selection.train_test_split(X, y,
test_size=0.2, random_state=55)

print(X_train.shape)
print(y_train.shape)
print(X_test.shape)
print(y_test.shape)

#Building the model
from sklearn.linear_model import LogisticRegression
model_2= LogisticRegression()
###fitting logistic regression for active customer on rest of the
variables#######
model_2.fit(X_train,y_train)

#Coefficients and Intercept
print(model_2.intercept_)
print(model_2.coef_)

##Confusion Matrix Calculation on Train data
from sklearn.metrics import confusion_matrix

y_pred_train=model_2.predict(X_train)
cm1 = confusion_matrix(y_train,y_pred_train)
print(cm1)

##Accuracy on Train data
accuracy1=(cm1[0,0]+cm1[1,1])/(cm1[0,0]+cm1[0,1]+cm1[1,0]+cm1[1,1])
print(accuracy1)

##Confusion matrix on test data
y_pred_test=model_2.predict(X_test)
cm2 = confusion_matrix(y_test,y_pred_test)
print(cm2)

#####Accuracy on Test data
accuracy2=(cm2[0,0]+cm2[1,1])/(cm2[0,0]+cm2[0,1]+cm2[1,0]+cm2[1,1])
print(accuracy2)

#Frequency of target variable
credit_risk_data['Bad'].value_counts()

#Sensitivity and Specificity on Train data
Sensitivity1=cm1[0,0]/(cm1[0,0]+cm1[0,1])
print(round(Sensitivity1,3))

Specificity1=cm1[1,1]/(cm1[1,0]+cm1[1,1])
print(round(Specificity1,3))
```

```python
#Sensitivity and Specificity on Test data
Sensitivity2=cm2[0,0]/(cm2[0,0]+cm2[0,1])
print(round(Sensitivity2,3))

#Specificity
Specificity2=cm2[1,1]/(cm2[1,0]+cm2[1,1])
print(round(Specificity2,3))
```

Given below are the results after building the model on the balanced data.

```
Confusion Matrix on Train Data
[[47911   8062]
 [13850  18255]]
Accuracy on Train data 0.7512205090942119
Sensitivity Train data 0.856
Specificity Train data 0.569

Confusion Matrix on Test Data
[[11955 2062]
 [ 3414 4589]]
Accuracy on Test data 0.7513169845594914
Sensitivity Test data 0.853
Specificity Test data 0.573
```

In all cases, we look for a model with high specificity. By creating the balanced data, we boosted the specificity of the model from 7 to 57 percent. Oversampling and undersampling is one method of handling class imbalance. There are a few other methods such as synthetic sampling and cluster centroids. You can explore them if the technique discussed here does not work well on your data sample.

5.9 CONCLUSION

In this chapter, we have comprehensively discussed various model validation measures and metrics for regression and classification type of problems. We have also discussed the merits and demerits of each of these measures. We have covered overfitting and underfitting and that most models suffer from overfitting. We need to fine-tune hyperparameters carefully. The discussion has also been on some of the tips and tricks of feature engineering. The feature engineering process involves a lot of creativity and business knowledge. The concepts discussed in this chapter are fundamental in solving practical business problems. The entire chapter is an essential part of the model building life cycle. We pick and choose techniques on case-by-case basis.

In later chapters, we will explore more sophisticated machine learning algorithms that have several hyperparameters; we need to fine-tune each hyperparameter carefully to avoid overfitting or underfitting. Stay tuned.

5.10 PRACTICE PROBLEMS

1. Download the New York taxi fare prediction data.
   - Import the data. Complete the necessary exploration and sanitization of the data.
   - Build a machine learning model to predict the right fare amount.
   - Perform the model validation and measure the accuracy of the model.
   - Check for the innovative ways to improve the accuracy of the model.
   - Add the feature engineering tricks on the date and location columns to improve the accuracy of the model.

   Dataset credits—https://console.cloud.google.com/bigquery and https://www.kaggle.com/c/new-york-city-taxi-fare-prediction/data. The dataset in this link is a large dataset. Take a random sample of one million rows.

2. Download the "default of credit card clients Data Set."

- Import the data. Complete the necessary exploration and sanitization of the data.
- Build a machine learning model to predict the defaulters.
- Perform the model validation and measure the accuracy of the model.
- Check for innovative ways to improve the accuracy of the model.

Dataset credits—from the UCI website: https://archive.ics.uci.edu/ml/datasets/default+of+credit+card+clients. Source: I-Cheng Yeh: (1) Department of Information Management, Chung Hua University, Taiwan. icyeh @chu.edu.tw; (2) Department of Civil Engineering, Tamkang University, Taiwan. 140910@mail.tku.edu.tw. Phone: 886-2-26215656, ext. 3181.

5.11   REFERENCES

1. Dataset: House sales in King County, USA. License CC0: Public Domain. Retrieved from https://www.kaggle.com/harlfoxem/ housesalesprediction. Accessed on August 3, 2020.

2. Dataset: "Give me some credit" competition from Kaggle.com. credit_risk_data_v1 used in specificity and sensitivity. Retrieved from https://www.kaggle.com/c/GiveMeSomeCredit. Accessed on August 3, 2020.

3. Dataset: Pima Indian Diabetes dataset used in Bias variance trade-off. Retrieved from https://archive.ics.uci.edu/ml/support/diabetes and https://raw.githubusercontent.com/jbrownlee/Datasets/master/pima-indians-diabetes.names. Accessed on August 3, 2020.

4. Resampling strategies for imbalanced datasets. Retrieved from https://www.kaggle.com/rafjaa/resampling-strategies-for-imbalanced-datasets. Accessed on August 3, 2020.

CHAPTER 6
CLUSTER ANALYSIS

In previous chapters, we have discussed concepts such as regression, logistic regression, and decision trees. These algorithms are essential to comprehend advanced machine learning algorithms. The models such as regression, logistic regression, and decision trees are simple and easy to interpret. If we look at the dataset of all the three algorithms, we have a target variable, and we have a list of predictor variables. If we do not have any target variable, then none of these algorithms will work. For example, a decision tree requires the target variable for calculation of entropy and information gain to perform classification. In linear regression as well, the information on the target variable y is an absolute must. Even in classification models like logistic regression, we need the target variable y. The machine learning algorithms that work on the data with a target variable are known as "supervised learning" algorithms. In other words, building a model on a training dataset that has input and output pairs is known as supervised learning. This data is also known as labeled data. Supervised algorithms work only with labeled data.

In some cases, we do not have labeled data, as the target variable may not be present in every case. What if we are in a situation where we have independent x variables only. The machine learning algorithms that work on unlabeled data are known as "unsupervised learning" algorithms. The following are a few examples of unsupervised learning datasets.

- Training data contains only the transaction details such as "amount," "time," "location," and "product type," but we do not have any information if the transaction is genuine or fraudulent.

- Historical data contains transaction details such as "number of loans," "monthly income," and "average utilization of the card," but we do not have any information if the customer is a defaulter or not.

- In a customer segmentation case, we have customer profile details such as income, age, region, and spending. Still we have no clue if the customer falls under the "Buying" or "Not buying" category.

You may have many questions, such as the following: In what situations do we have the data without any target variable information? With unlabeled data, what is the aim of analysis? How do we analyze the data that has no target variable? Which algorithm works with this kind of unlabeled data? We will answer all these questions with adequate details in this chapter. We will explore one significant unsupervised learning algorithm.

6.1   UNSUPERVISED LEARNING

Machine learning algorithms are often classified into two major categories: supervised learning and unsupervised learning. If the data is labeled, we apply a supervised learning algorithm; for unlabeled data, we use unsupervised learning algorithms. Have a look at the pictorial representations in Table 6.1.

TABLE 6.1   Supervised and Unsupervised Learning Datasets

Supervised Learning Dataset						Unsupervised Learning Dataset				
x_1	x_2	x_3	..	x_p	y	x_1	x_2	x_3	..	x_p

There are two significant algorithms of unsupervised learning. The first algorithm deals with the columns, and it is known as principal component analysis (PCA). In the PCA, we find a linear combination of the columns with maximum information. The second algorithm is cluster analysis. Cluster analysis is used for customer segmentation. Cluster analysis divides the whole dataset into smaller subsets in such a way that customers inside a subset are very similar to each other. Cluster analysis sounds very similar to decision trees. In a decision tree algorithm also, we try to divide the whole dataset into smaller subsets. Both cluster analysis and decision trees are used for customer segmentation; the only difference is data. The decision tree is a supervised learning algorithm, and cluster analysis is categorized as an unsupervised learning algorithm. It may be interesting to note that clustering is used in computer vision applications to divide the geospatial images into distinct regions and detect borders and objects. Let us get into the details of cluster analysis next.

6.1.1   Cluster Analysis

Dividing the data into clusters (subsets) in such a way that within a specific cluster all the elements are similar to each other and they are dissimilar from elements in other clusters. In market research, cluster analysis is used for grouping people with similar properties to run targeted market campaigns. In the insurance industry, cluster analysis is used for identifying groups of people with similar properties so that the premium rate can be fixed. You can appreciate a simple case in the pictorial presentation in Fig. 6.1.

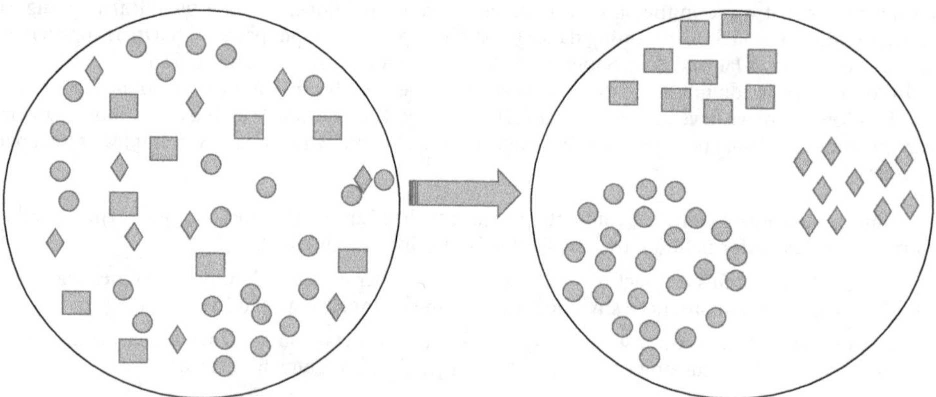

FIGURE 6.1   Cluster analysis—dividing the population into subsets.

Since we do not have any target variables in data, we will use all the input variables for spotting the similarities to divide the population into smaller subsets called clusters. While forming clusters, we aim to keep intracluster distance at a minimum and at the same time, intercluster distance is maximized. Here distance means dissimilarity. Next, we take an example case study that will be solved using cluster analysis.

6.1.2   Case Study: Customer Segmentation Wholesale Customers Data

A supermarket wants to send some promotional offers and coupons. An irrational option is to select a few customers and give them promotional offers. As one can expect, this approach may not yield any significant returns. How can we best make these promotional offers to increase the revenue and see customers happy? One size does not fit all; applying a uniform strategy to the entire population will not be effective. We will divide the whole population into subsets and apply segment-specific cross-selling and upselling strategies. Let us have a look now at the data to collect some vital details.

6.1.2.1   Data   The dataset for this example is "Wholesale Customers Data Set." It is publicly available data, available at https://archive.ics.uci.edu/ml/datasets/wholesale+customers. This dataset contains annual spending of customers on the product categories such as milk, groceries, frozen items, fresh items, detergents, and paper. Each row in the dataset represents one customer. Each column represents the type of product. The values in the data are annually aggregated spending in monetary units. The code below is used for importing the data and getting some descriptive statistics.

```
# Load the wholesale customers dataset
cust_data = pd.read_csv(r"D:\Chapter6\Datasets\Wholesale\Wholesale_customers_
data.csv")

#Rows and Columns
print(cust_data.shape)

#Dataset Information
cust_data.info()

#Sample
pd.set_option('display.max_columns', None) #This option displays all the columns
cust_data.sample(n=5, random_state=77)
```

The code above gives us the output as shown below.

```
print(cust_data.shape)
(440, 9)

print(cust_data.columns.values)
['Cust_id' 'Channel' 'Region' 'Fresh' 'Milk' 'Grocery' 'Frozen'
 'Detergents_Paper' 'Delicatessen']

cust_data.info()
<class 'pandas.core.frame.DataFrame'>
RangeIndex: 440 entries, 0 to 439
Data columns (total 9 columns):
Cust_id             440 non-null int64
Channel             440 non-null int64
Region              440 non-null int64
Fresh               440 non-null int64
Milk                440 non-null int64
Grocery             440 non-null int64
Frozen              440 non-null int64
Detergents_Paper    440 non-null int64
Delicatessen        440 non-null int64
dtypes: int64(9)
memory usage: 31.1 KB
```

The dataset has 440 rows and nine columns. The first three column names are Cust_id, Channel, and Region. Other columns represent various product types. Channel represents the type of customer—retail and commercial customers; commercial customers are populated as HoReCa (Hotel-Restaurants-Café). The region, as the name suggests, is the customer region. We will look into the individual variable descriptions to get a better understanding of the data.

```
pd.set_option('display.max_columns', None) #This option displays all the
columns
cust_data.sample(n=5, random_state=77)
     Cust_id  Channel  Region   Fresh   Milk  Grocery  Frozen  \
45        46        2       3    5181  22044    21531    1740
223      224        2       1    2790   2527     5265    5612
64        65        1       3    4760   1227     3250    3724
366      367        1       3    9561   2217     1664    1173
288      289        1       3   16260    594     1296     848

     Detergents_Paper  Delicatessen
45               7353          4985
223               788          1360
64               1247          1145
366               222           447
288               445            25
```

The output above shows a random sample of five records. We can use the `random_state` parameter to regenerate the same sample. `Channel` takes two values: 1 for HoReCa (Hotel-Restaurants-Café) while 2 represents retail. The region takes three values: 1-Lisbon, 2-Oportom, and 3-Other Region. The rest of the columns show the annual spending for different product types. The code below is used for getting the desired basic descriptive statistics.

```
#Frequency Counts
cust_data["Channel"].value_counts()
cust_data["Region"].value_counts()
```

The code above gives us the following output.

```
cust_data["Channel"].value_counts()
1    298
2    142
Name: Channel, dtype: int64

cust_data["Region"].value_counts()
3    316
1     77
2     47
Name: Region, dtype: int64
```

From the output, we can verify that `Channel` has two values: 1-HoReCa (Hotel-Restaurants-Café) and 2-retail. Channel-1 is the highest frequency. The `region` takes three values: 1-Lisbon, 2-Oporto, and 3-Other. The "other" region is the most occurring value in this variable.

Let us check the summary of the variables.

```
round(cust_data.describe(),2)
```

	Cust_id	Channel	Region	Fresh	Milk	Grocery	Frozen
count	440.00	440.00	440.00	440.00	440.00	440.00	440.00
mean	220.50	1.32	2.54	12000.30	5796.27	7951.28	3071.93
std	127.16	0.47	0.77	12647.33	7380.38	9503.16	4854.67
min	1.00	1.00	1.00	3.00	55.00	3.00	25.00
25%	110.75	1.00	2.00	3127.75	1533.00	2153.00	742.25
50%	220.50	1.00	3.00	8504.00	3627.00	4755.50	1526.00
75%	330.25	2.00	3.00	16933.75	7190.25	10655.75	3554.25
max	440.00	2.00	3.00	112151.00	73498.00	92780.00	60869.00

	Detergents_Paper	Delicatessen
count	440.00	440.00
mean	2881.49	1524.87
std	4767.85	2820.11
min	3.00	3.00
25%	256.75	408.25
50%	816.50	965.50
75%	3922.00	1820.25
max	40827.00	47943.00

In the above output, we ignore the Channel and Region for the time being. We will focus on the rest of the numerical columns. The product category "fresh" is seen as the most dominant, the average spending on fresh products is the highest when compared to any other product. This observation is evident from the average and percentile values. Similarly, the least bought product is of type delicatessen. There are some outliers present in each of these columns, which we will take up later. For now, we will boxplot all these six variables to get a visual representation (Fig. 6.2).

```
#Box Plots
plt.figure(figsize=(10,10))
plt.title("All the variables box plots", size=20)
sns.boxplot(x="variable",    y="value",    data=pd.melt(cust_data[['Fresh',
'Milk', 'Grocery','Frozen', 'Detergents_Paper', 'Delicatessen']]))
plt.show()
```

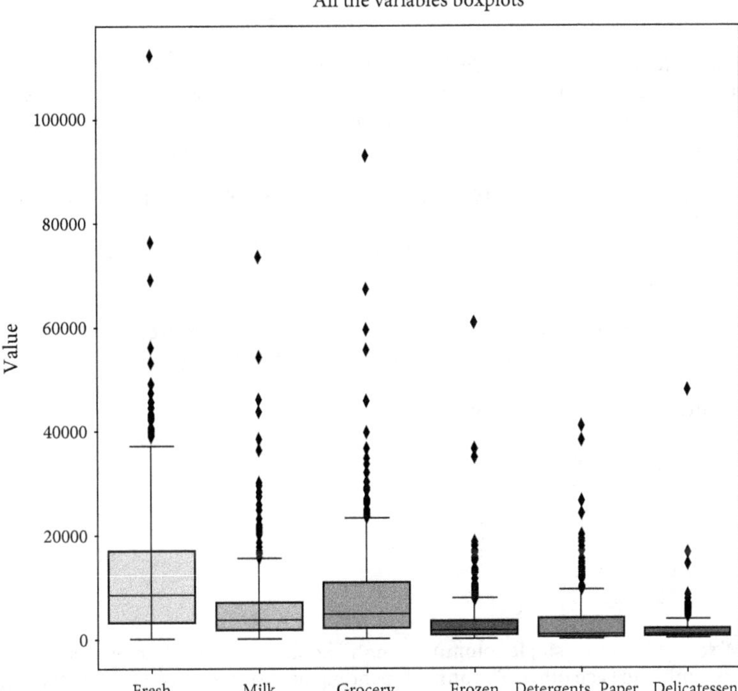

FIGURE 6.2   Boxplot for all variables.

From the output, we can make out that Fresh items are the most sold, followed by grocery and milk. Delicatessen is the least sold item. We got some basic ideas on the data. Now we proceed with the objective of this case study.

6.1.2.2   Business Problem and Objective   We aim to send one or the other promotional offer to almost all customers. We know that sending a single offer to every customer in the population will not be helpful. We will prefer to divide the whole population into subsets and send subset-specific offers. These offers are unique to each subset. With these offer letters, we hope to promote cross-selling and upselling tactics.

If you are not yet initiated in this kind of sales terminology, cross-selling is a sales trick used to inspire a customer to buy other items related to the product that is already being bought. Upselling is to influence a customer to buy more or buy something related and expensive. For example, if we order a burger in a restaurant, we usually get two offers. Offer1: Buy two burgers and get 15% off. Offer2: Add French fries and soft drink to your burger, make it a meal, and get 20% off. Offer1 is upselling. Offer2 is cross-selling.

Given below are the complete objectives of our case study.

1. Fresh items are the most sold items in the population. Is there a subset in the population where fresh items are sold less than other items such as grocery or milk? If yes, we would like to identify such segments and send them the promo offers on "Fresh items."

2. We want to find a group of customers with high spending on Fresh and low spending on Frozen. We will try cross-selling frozen items to such groups.

3. We aim to identify a customer segment that spends very less on groceries and send them promotional offers related to "Grocery."

We are going to use a cluster analysis algorithm to solve the above problem. Let us get into the details, starting with the following section.

6.2   *DISTANCE MEASURE*

The key idea of cluster analysis is to divide the whole population into subsets. The subsets should be homogenous. The records (customers) inside a subset should be similar to each other. We need first to understand what is similarity and dissimilarity. Let us try to understand dissimilarity with the help of an example. Take a sample (only five records) from the customer spending data that we have; consider the following two scenarios.

- Scenario-1: Consider only a single column, "Fresh." Looking at the output, find the two customers who are very similar to each other.
- Scenario-2: Take two columns, "Fresh" and "Grocery." Looking at these two column values, find which customers are similar to each other.

```
cust_data_sample=cust_data.sample(n=5, random_state=11)
cust_data_sample[["Cust_id", "Fresh", "Grocery"]]
```

The code above gives us the following output.

```
Cust_id   Fresh   Grocery
     46    5181     21531
    224    2790      5265
     65    4760      3250
    367    9561      1664
    289   16260      1296
```

Scenario-1: We consider only a single column, "Fresh." Looking at the output, we make out that customers with ID 46 and 65 are very similar to each other. The annual spending of customer 46 on the Fresh category is 5181 and the closest spending in this sample is 4760, and it is from customer ID 65.

Scenario-2: Consider two columns, "Fresh" and "Grocery." In Scenario-1 it was easy for us to choose the similar customers. Now we need to consider two columns to do the same. The Cust_id 46 and 65 are not closer to each other; grocery has a high value for customer 46. Proceed further and consider these five customers as five points in the space. Plot these points by taking Fresh on the *x*-axis and Grocery on the *y*-axis. The code given below helps us in drawing the graph.

```
plt.figure(figsize=(10,10))
plt.title("Fresh and Grocery spending plot", size=20)
plot=sns.scatterplot(x="Fresh",y="Grocery", data=cust_data_sample, s=500)
for i in list(cust_data_sample.index):
    plot.text(cust_data_sample.Fresh[i],
    cust_data_sample.Grocery[i],
    cust_data_sample.Cust_id[i], size=20)
```

The above code gives us the graph in Fig. 6.3.

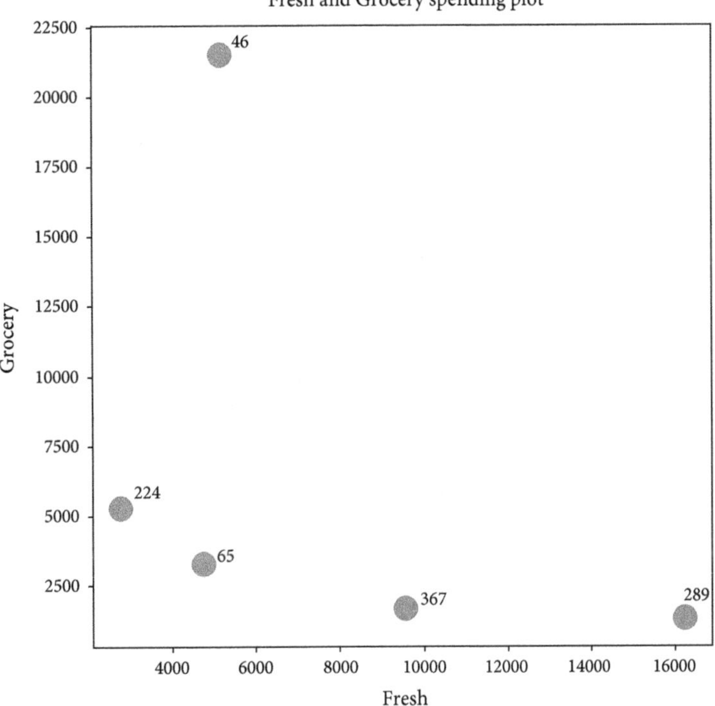

FIGURE 6.3   Fresh versus Grocery spending plot.

We have taken spending on Fresh on the *x*-axis and Grocery spending on the *y*-axis. Customer IDs are shown as labels. From this plot, we can see that customers 65 and 224 are very close to each other. We can conclude that these two customers are similar to each other. We can confirm the same by looking at the data for customers 65 and 224. Since there are only five points here, we can easily make a plot to visualize them. In practice, datasets are large. So, we need a metric that can quantify the similarity and dissimilarity between any given two points. How do we find the distance between two points in traditional coordinate geometry? Using the Euclidean distance formula? Right!

6.2.1 Euclidean Distance

Let (x_1, y_1) and (x_2, y_2) be two points in the two-dimensional space. The distance between them is measured using the formula

$$D = \sqrt{(x_2 - x_1)^2 + (y_2 - y_1)^2}$$

This formula is used to measure the standard distance between two points, known as Euclidean distance. You can also visualize it as the length of the straight line drawn between two points. We will use this distance as our measure of dissimilarity. We can consider each customer as a point in the space and find the distance between the customers— one pair at a time. We can use this Euclidean distance formula to quantify the dissimilarity between any given pair of customers. This distance formula can be extended to multiple dimensions. In a three-dimensional space, the distance between two points (x_1, y_1, z_1) and (x_2, y_2, z_2) can be calculated as

$$D = \sqrt{(x_2 - x_1)^2 + (y_2 - y_1)^2 + (z_2 - z_1)^2}$$

We extend it to multiple dimensions as follows:

$$D = \sqrt{(x_2 - x_1)^2 + (y_2 - y_1)^2 + \cdots + (k_2 - k_1)^2}$$

Let us take the example of two customers, IDs 224 and 65, and the distance between them.

```
Cust_id   Fresh   Grocery
     46    5181     21531
    224    2790      5265
     65    4760      3250
    367    9561      1664
    289   16260      1296
```

Customer ID 224 is our first data point, which takes values (2790,5265), and the second data point, customer ID 65, takes the values (4760,3250). We go ahead and calculate the Euclidean distance as follows:

$$(x_1, y_1) = (2790, 5265)$$

$$(x_2, y_2) = (4760, 3250)$$

$$D = \sqrt{(x_2 - x_1)^2 + (y_2 - y_1)^2}$$

$$= \sqrt{(4760 - 2790)^2 + (3250 - 5265)^2}$$

$$= 2818.$$

In the next section, let us proceed further with the other customer IDs in our data sample of five rows.

6.2.2   Distance Matrix

In our sample dataset, we have five customers. We need to find the distance of each customer from every other customer; the result will look like a matrix. The matrix that shows the distance of every customer in the population to every other customer is known as the distance matrix. In our case, our distance matrix will be of size 5×5. If there are N customers, then the distance matrix will be of size $N \times N$. The distance matrix size is independent of the number of columns in the input dataset. All the columns are used in the calculation of distance. If a data matrix has a size of $N \times K$, then the size of the distance matrix will be $N \times N$. Have a look at the pictorial presentation of the matrix in Table 6.2.

TABLE 6.2   Data Matrix and Distance Matrix

Data Matrix						**Distance Matrix**					
	x_1	x_2	x_3	..	x_K		**1**	**2**	**3**	..	..	**N**
1						**1**	0	$d(1,2)$				
2						**2**	$d(2,1)$	0				
3						**3**			0			
..						..				0		
..						..					0	
N						**N**						0

The diagonal elements in the distance matrix are zeros. $d(1,2)$ represents the distance between customer-1 and customer-2, which is the same as $d(2,1)$. The distance matrix is symmetrical with all the entities as 0s in its principal diagonal. Let us create the distance matrix for our sample dataset.

```python
def distance_cal(data_frame):
 distance_matrix=np.zeros((data_frame.shape[0],data_frame.shape[0]))
    for i in range(0 , data_frame.shape[0]):
        for j in range(0 , data_frame.shape[0]):
            distance_matrix[i,j]=
 round(np.sqrt(sum((data_frame.iloc[i] - data_frame.iloc[j])**2)))
     return(distance_matrix)

distance_matrix=distance_cal(cust_data_sample[["Fresh", "Grocery"]])
print(distance_matrix)
print(distance_matrix[0,0])
print(distance_matrix[1,0])
print(distance_matrix[2,1])
```

In the above code, we wrote a function that navigates all the rows and calculates the distance of one row to every other row. Finally, the function returns the distance matrix as a two-dimensional array. The code above gives us the following result.

```
print(distance_matrix)
[[    0. 16441. 18286. 20344.  23069.]
 [16441.     0.  2818.  7669.  14043.]
 [18286.  2818.     0.  5056.  11665.]
 [20344.  7669.  5056.     0.   6709.]
 [23069. 14043. 11665.  6709.      0.]]

print(distance_matrix[0,0])
0.0

print(distance_matrix[1,0])
16441.0

print(distance_matrix[2,1])
2818.0
```

The customer IDs are 46, 224, 65, 367, and 289. In the matrix, their respective indexes are 0, 1, 2, 3, and 4. The distance matrix will return distances in the same order.

- distance_matrix[1,0] is the distance between 224 and 46 → 16,441
- distance_matrix[2,1] is distance between 65 and 224 → 2,818

From the distance matrix, we can see that the customer pair (65,224) is the closest; they are the utmost similar customers. Customers (289,46) (distance_matrix[0,4]) are the farthest from each other, so they are as dissimilar as possible. We can verify it by looking at the plot shown in Fig. 6.3. The distance measure is the most crucial metric in cluster analysis. This metric will help us quantify the dissimilarity. Using this matrix, we can now make clusters.

Euclidean distance is one of the most widely used distance measures. There are other distance measures available. Manhattan distance is another distance measure with a slightly different formula:

$$\text{Manhattan distance} = \sum |x_2 - x_1| + |y_2 - y_1| + \cdots$$

The next section discusses how to make clusters based on the similarity of customers. The most popular clustering algorithm is the K-means clustering algorithm.

6.3   *K-MEANS CLUSTERING ALGORITHM*

A clustering algorithm takes the population as input and gives the customer segments as output. "*K*-means" refers to the clustering algorithm. It gives us *K*-clusters as output. We need to mention *K* while building clusters. If we mention *K* = 10, then 10 clusters will be formed. Given below are the steps followed in a *K*-means clustering algorithm.

6.3.1   Steps in Clustering Algorithm

1. *Initialization step:* Fix the number of clusters *K*. Randomly select *K* points from the data and call them cluster centers.
2. *Assignment step:* Assign each data point to its closest cluster center.
3. *Update step:* Recompute cluster centers by taking the average of all the records in a cluster. Update the cluster centers.
4. *Reassignment step:* If the cluster center changes then go back to the assignment step.
5. *Stopping criteria:* Repeat steps 2, 3, and 4 until there are no further changes in the cluster centers.

Assign-Update-Resign are the three significant steps in this algorithm. We will try to understand the algorithm using an illustration. Let us take an example of the data points, as shown in the plot in Fig. 6.4. We will use it to understand the cluster analysis steps (Table 6.3).

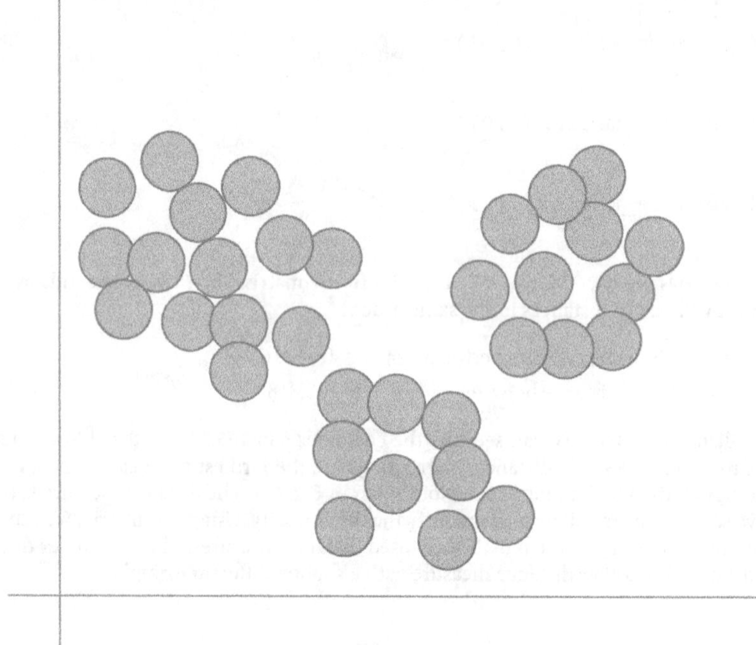

FIGURE 6.4   Input data.

TABLE 6.3   Steps in Creating Clusters

Algorithm Step	Visualization
Step 1: Initialization: Fix the number of clusters K. Randomly select K points from the data; call them cluster centers. We can explicitly define this initialization of centers if we have prior information on the data. In this data, we will build three clusters. We will randomly take three points and call them cluster centers.	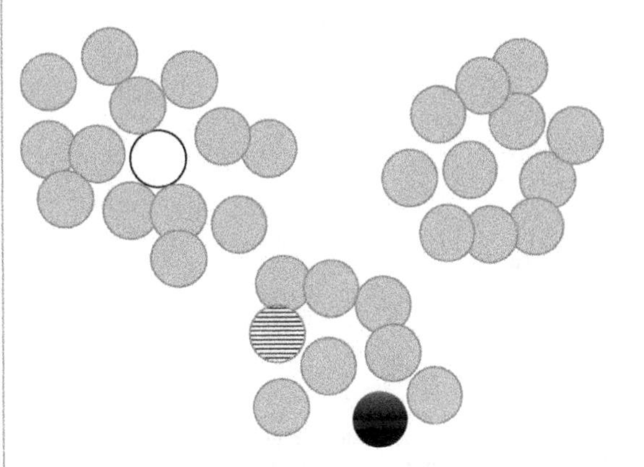
Step 2: Assignment step: Assign each data point to its closest cluster center. We have three center points. We can calculate the distance of each point to all three cluster centers. We will add the point to the closest cluster based on the minimum distance criteria.	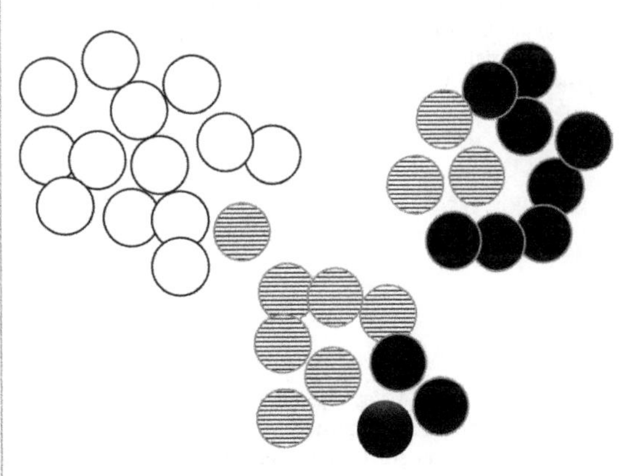

Step 3: Update step: Recompute cluster centers by taking the average of all the records in a cluster. Update the cluster centers. For example, if two customers are falling in a cluster, customer 1 has income 10,000, and customer 2 has income 12,000, then the cluster center will be at income 11,000. In the graph, we can see the updated cluster centers denoted by stars.

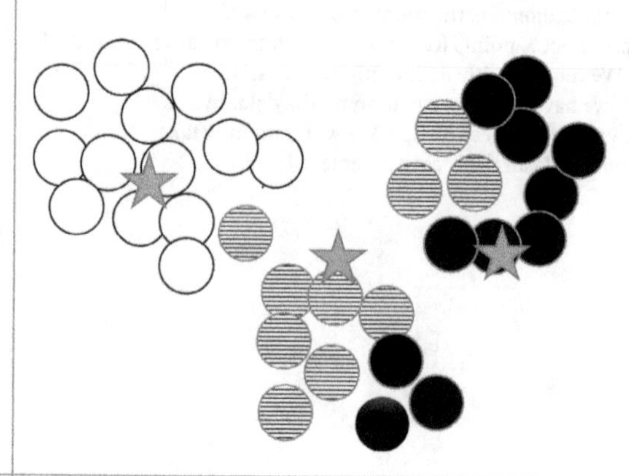

Step 4: Reassignment step: If the cluster center changes, then go back to the assignment step and reassign all the points to their nearest clusters based on distance from the cluster center. In our example, the cluster centers have changed. We need to reassign all the points to their nearest cluster centers. Some points may change their cluster in this step. In our example, a few points on the top right have changed their cluster. Many points at the bottom of the graph also changed their cluster in the reassignment step. This reassignment is a crucial step; it makes sure that the right points are assigned to the closest cluster.

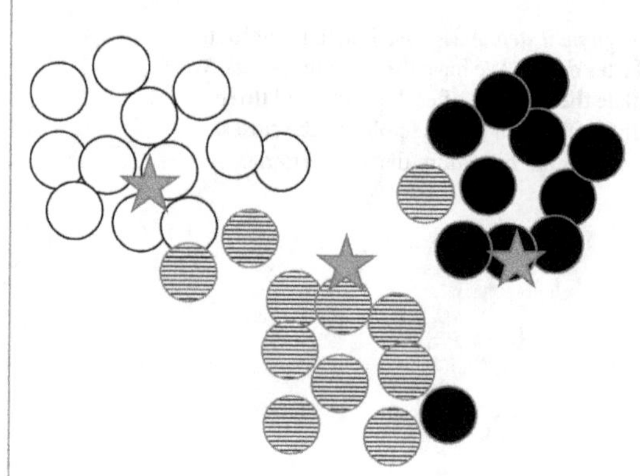

Step 5: Stopping criteria: Repeat steps 2, 3, and 4 until there are no further changes in the cluster centers. We have seen only one iteration here. We keep on repeating the assigning step followed by center updating, followed by the reassignment step. Finally, when all the points are added to their nearest clusters, there will be no change or update in the cluster centers. We can stop the training process here and return the final clusters.

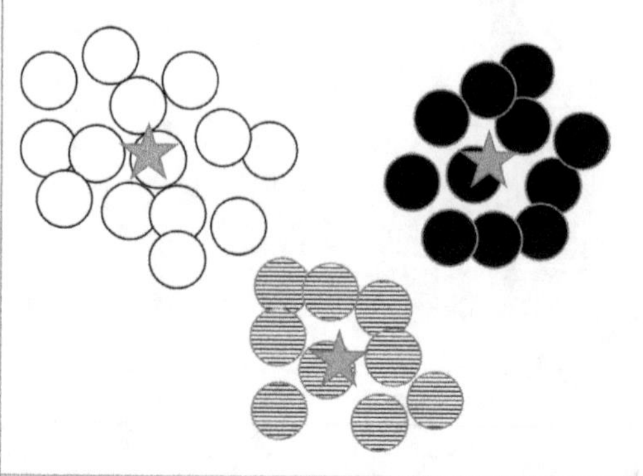

6.3.2   *K*-Means Clustering Algorithm: Illustration

We will take another small sample of 30 records from the data and try to grasp the algorithm iterations. The plots in Fig. 6.5 show the state of the clusters at each iteration. While solving real-world business problems, we need not create these visualizations.

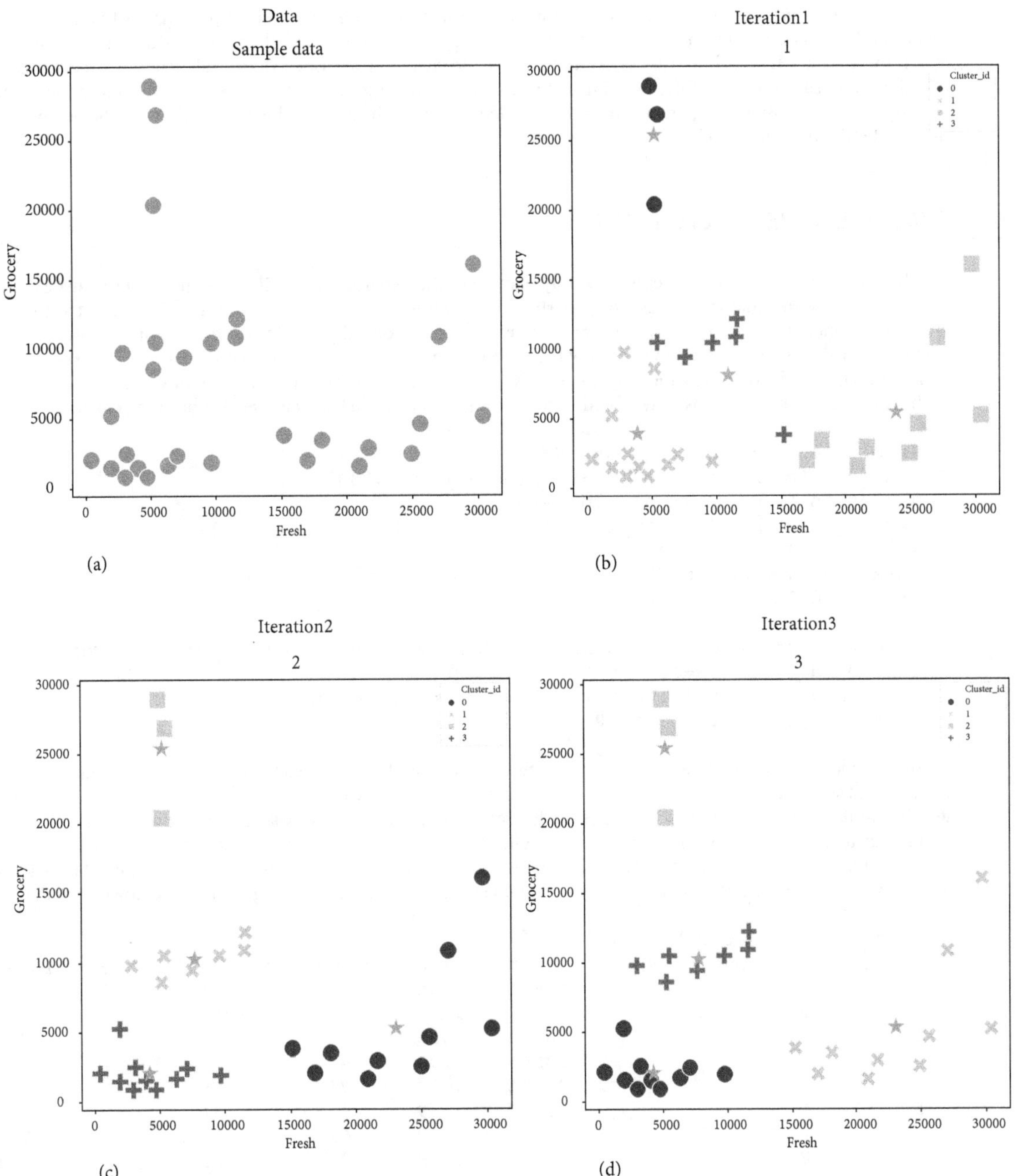

FIGURE 6.5   (*a*) Data points. (*b*) Iteration 1 clustering. (*c*) Iteration 2 clustering. (*d*) Iteration 3 clustering.

In Fig. 6.5, the cluster IDs are changing amid the graphs. Colors and shapes change based on the cluster IDs. Ignore the color code and shapes; focus only on the four clusters in the graphs in Fig. 6.5.

6.3.3 *K*-Means Clustering Algorithm: Output

What is the final output of the *K*-means clustering algorithm? In the end, we will get the clusters that tell about each record and to which cluster does it belong to. *K*-means algorithm gives us the cluster IDs as the final result; we then attach these cluster IDs back to the dataset and analyze the cluster properties. We can calculate the cluster means and distribution of each column in the individual clusters. After studying the cluster properties, we choose the cluster for the target market campaigning or any other strategy depending on the data. In the next section, we will get into further details of building *K*-means clusters.

6.4 BUILDING K-MEANS CLUSTERS

Building *K*-means clusters is very easy. The only parameter that we need to specify is the number of clusters. Usually, based on the business that we are handling, we should know how many clusters are optimal. For market segmentation and sales segmentation, four of five clusters may be sufficient. Normally, we do not find 50 or 60 different types of customers in a population. If we do not know the optimal value of *K*, then we start with a number, observe the output, and rebuild the model by increasing or decreasing *K*. We will revisit the case study of previous sections. In our dataset, we have annual spending of customers on different products. The code below is used for building clusters.

```python
from sklearn.cluster import KMeans
kmeans = KMeans(n_clusters=5, random_state=333)
# Mention the Number of clusters
X=cust_data.drop(['Cust_id', 'Channel', 'Region'],axis=1)
# Custid is not needed
kmeans = kmeans.fit(X)
#Model building

# Getting the cluster labels and attaching them to the original data
cust_data_clusters=cust_data
cust_data_clusters["Cluster_id"]= kmeans.predict(X)
cust_data_clusters.head(10)
```

In the code above, we used the `kmeans.fit()` function for building the model. We dropped irrelevant columns from the data and supplied it to the `fit()` function. We have mentioned five clusters using the `Cluster_id` parameter. We use the `random_state` parameter to regenerate the same output again. Otherwise, there may be minor changes in the output due to the random initialization. Once we build the model, we get predictions for each record by using the `predict()` function, which returns the `Cluster_id`. We attach this `Cluster_id` to the dataset by adding it to all the rows. This `Cluster_id` will be common to all the rows in one specific cluster. This code gives us the following results.

```
cust_data_clusters.head(10)
```

	Cust_id	Channel	Region	Fresh	Milk	Grocery	Frozen	Detergents_Paper	\
0	1	2	3	12669	9656	7561	214	2674	
1	2	2	3	7057	9810	9568	1762	3293	
2	3	2	3	6353	8808	7684	2405	3516	
3	4	1	3	13265	1196	4221	6404	507	
4	5	2	3	22615	5410	7198	3915	1777	
5	6	2	3	9413	8259	5126	666	1795	
6	7	2	3	12126	3199	6975	480	3140	
7	8	2	3	7579	4956	9426	1669	3321	
8	9	1	3	5963	3648	6192	425	1716	
9	10	2	3	6006	11093	18881	1159	7425	

```
     Delicatessen  Cluster_id
0            1338            0
1            1776            0
2            7844            0
3            1788            4
4            5185            4
5            1451            0
6             545            0
7            2566            0
8             750            0
9            2098            2
```

With the output of *K*-means, we have added `Cluster_id` to the data. Now we will analyze each cluster. We will look at the cluster counts and the cluster means. The following code gives us cluster-wise count and mean.

```python
cluster_counts = cust_data_clusters['Cluster_id'].value_counts(sort=False)

cluster_means = cust_data_clusters.groupby(['Cluster_id']).mean()

print(cluster_counts)
print(cluster_means)
```

The following is the output.

```
print(cluster_counts)
0    217
1     27
2     82
3      7
4    107
```

From this output, we can see that five clusters are formed. Cluster IDs are 0, 1, 2, 3, and 4. Cluster ID 0 has 217 records, cluster ID 4 has 107, and cluster ID 2 has 82 records. These three clusters have more than 50 records each. Cluster ID 3 has just seven records. When we apply the clustering algorithm on the real-time datasets; this is how clusters are formed. We *cannot* expect an equal distribution of cluster counts. There will always be one or two clusters with quite a lot of records and a few others with negligible records. For example, when we think of typical customer segments, people with extreme spending will be very less. Customers with near to average spending will be more, and they will always form a cluster. We will now look at the cluster means—the average value of every field in each cluster.

```
            Cust_id  Channel  Region     Fresh     Milk  Grocery   Frozen  \
Cluster_id
0             230.0      1.2     2.5    5834.0   3322.4   4096.3   2635.2
1             225.1      1.1     2.7   46916.6   7033.6   6205.3   9757.0
2             207.1      1.9     2.5    5057.0  12105.1  18414.1   1580.7
3             127.9      2.0     2.6   20031.3  38084.0  56126.1   2564.6
4             216.4      1.2     2.5   20490.7   3554.1   5040.1   3446.8

            Detergents_Paper  Delicatessen
Cluster_id
0                     1234.7         995.5
1                      936.4        4199.3
2                     8092.0        1828.7
3                    27644.6        2548.1
4                     1099.0        1623.9
```

From the output, we can see the mean value in each cluster. Cluster ID 3 seems to be an outlier with just seven records. Milk, Grocery, and Detergents_Paper have the highest mean in this cluster. Even if we build four clusters, this outlier cluster will appear in those four clusters. To get a better understanding of these clusters, we need to look at the distribution of the variables in each cluster rather than just looking at the mean value. The following code will boxplot all the variables within each cluster.

```
df_melt = pd.melt(cust_data_clusters.drop(['Cust_id', 'Channel', 'Region'],axis=1),
"Cluster_id", var_name="Prod_type", value_name="Spend")

plt.figure(figsize=(20,10))
sns.boxplot(x='Cluster_id', hue="Prod_type", y="Spend", data=df_melt)
plt.title("Cluster wise Spending", size=20)
```

In this code the `pd.melt()` function transforms the data into three columns; it will be used in sketching the graph later. The code above gives us the boxplots in Fig. 6.6. We have cluster_id on the *x*-axis and spending on the *y*-axis

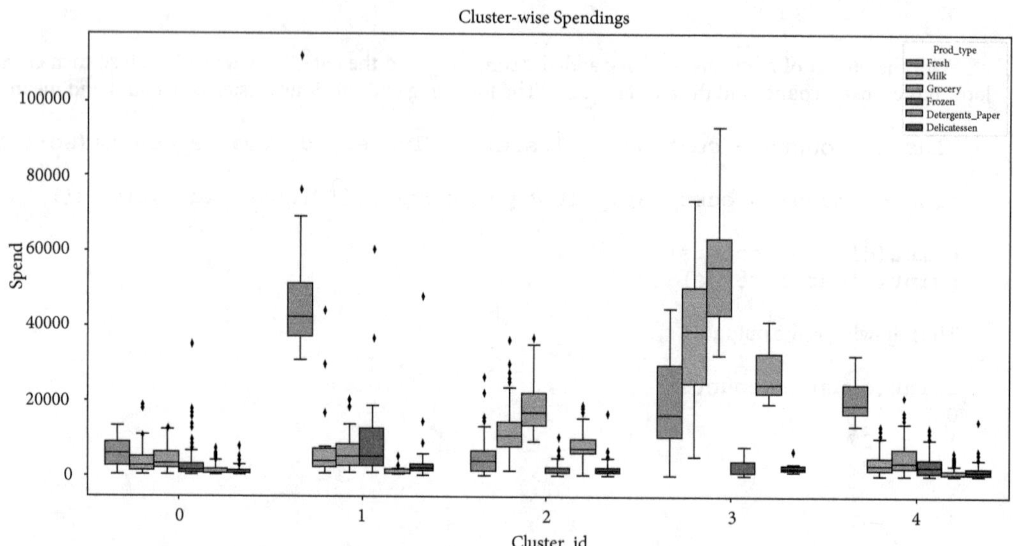

FIGURE 6.6   Cluster-wise spendings resulting from the code.

The following code provides the same boxplots in five different presentations for better visualization.

```
##Cluster 0
Cluster=df_melt[(df_melt['Cluster_id']==0)] #Change this from 0 to 4
plt.figure(figsize=(7,7))
sns.boxplot(x='Cluster_id', hue="Prod_type", y="Spend", data=Cluster)
plt.title("Cluster0", size=20)
```

This code gets a subset for each cluster and draws the boxplot individually for each cluster. The output is presented in Fig. 6.7.

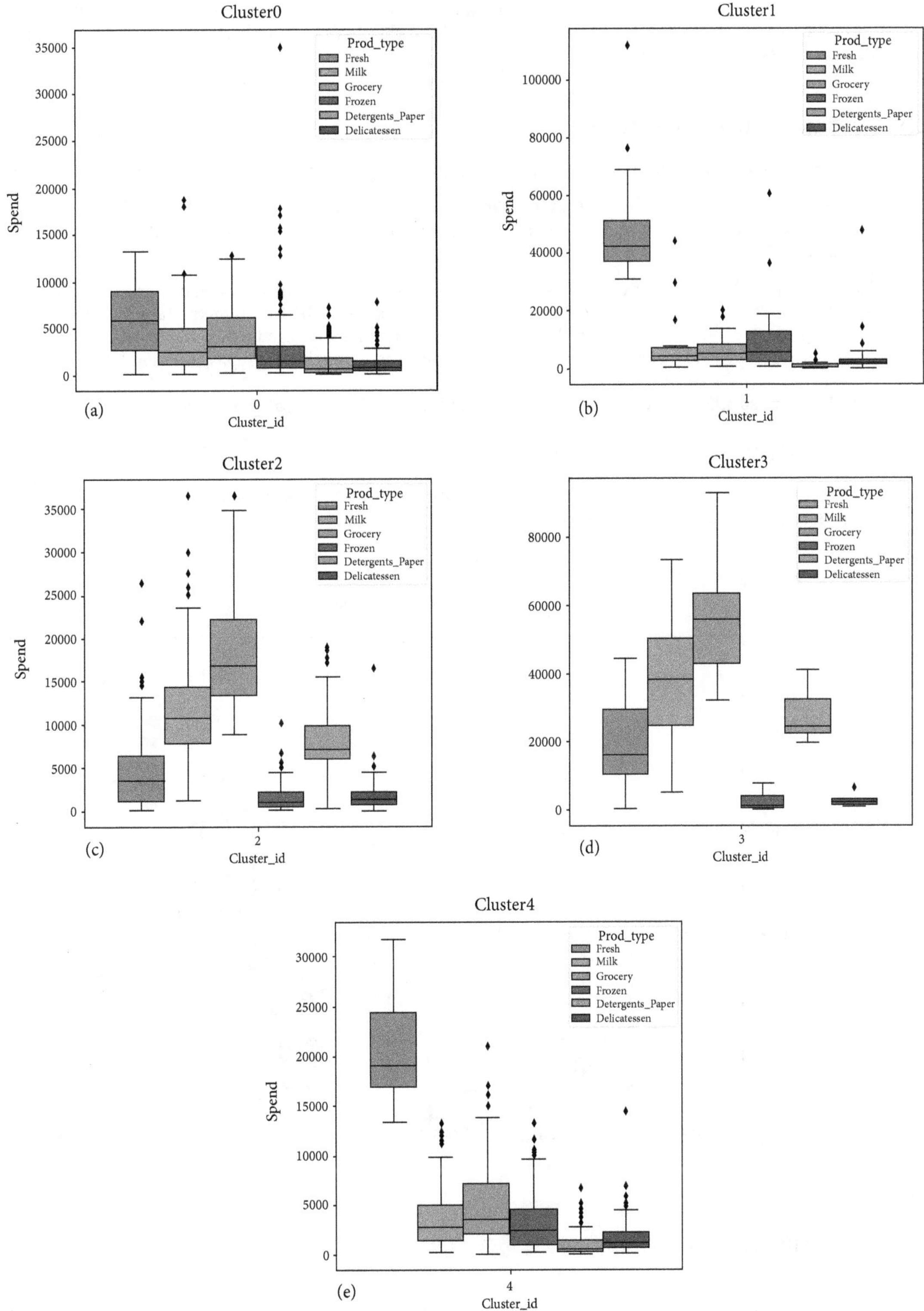

FIGURE 6.7   (*a*) Output cluster 0 boxplots. (*b*) Output cluster 1 boxplots. (*c*) Output cluster 2 boxplots. (*d*) Output cluster 3 boxplots. (*e*) Output cluster 4 boxplots.

The output in Fig. 6.7 shows the spending in individual clusters. We are now all set to explain the wholesale data case study. In the following section, we will revisit all the objectives previously stated in the problem statement and discuss the respective solution.

6.4.1   Wholesale Data Case Study Final Result

We started this case study to run a targeted customer marketing campaign. We have three inclusive objectives.

Objective 1: Fresh items are the most sold items in the population. Is there a subset in the population where fresh items are sold less than other items such as grocery or milk? We want to identify such segments and send them offers on "Fresh items."

Solution: Cluster IDs 2 and 3 are the two clusters where fresh items are sold less when compared to other items. There are a total of 82 + 7 = 89 customers in these two clusters. We can send these customers a targeted promo offer that will encourage them to buy products from the fresh category. If required, we can further divide them based on region and customer type and send separate offers. Let us redraw only these two cluster graphs, as in Fig. 6.8.

```
Cluster_2and3=df_melt[(df_melt['Cluster_id']==2)|(df_melt['Cluster_id']==3)]

plt.figure(figsize=(10,7))

sns.boxplot(x='Cluster_id', hue="Prod_type", y="Spend", data=Cluster_2and3)
plt.title("Cluster 2 and 3 only", size=20)
```

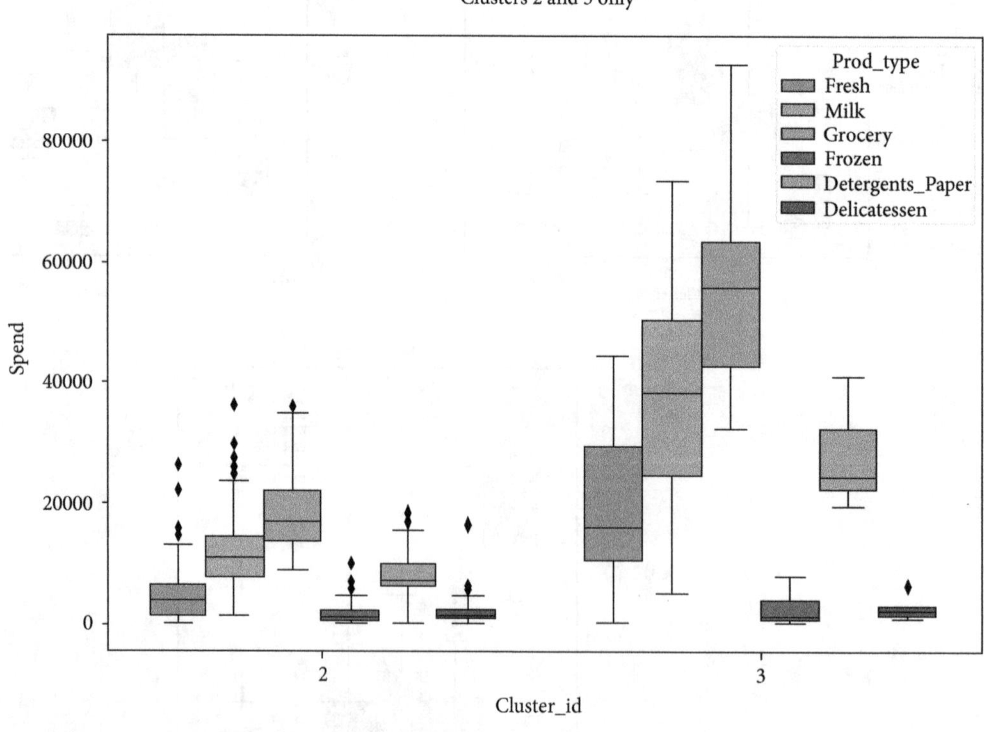

FIGURE 6.8   Spending for clusters 2 and 3.

Inference from the graph in Fig. 6.6 and Fig. 6.8: In both these clusters, the spending on "fresh" is less than other items such as milk, grocery, and detergents. The below code lists the customer IDs from these two segments.

```
obj1_data= cust_data_clusters[(cust_data_clusters['Cluster_id']==2)| (cust_data_
clusters ['Cluster_id']==3)]

print(list(obj1_data["Cust_id"]))
[10, 11, 17, 24, 29, 38, 39, 43, 44, 46, 47, 48, 50, 54, 57, 58, 62, 64, 66,
78, 82, 83, 86, 87, 93, 95, 101, 102, 107, 108, 110, 112, 146, 156, 157, 160,
164, 166, 171, 172, 174, 176, 183, 189, 190, 194, 201, 202, 206, 210, 212,
215, 216, 217, 219, 246, 252, 265, 266, 267, 269, 294, 302, 304, 305, 306,
307, 310, 313, 316, 320, 332, 334, 344, 347, 350, 352, 354, 358, 377, 385,
397, 408, 417, 419, 421, 427, 431, 438]
```

Objective 2: Find the group of customers with high spending on Fresh and low spending on Frozen, and try cross-selling frozen for that group.

Solution: In cluster ID 4, there is a massive gap between fresh and frozen items. Our target population for cross-selling frozen items is cluster ID 4. We have not chosen cluster ID 1; frozen items have maximum sales in that group. We can always talk to marketing team and choose additional clusters to apply this cross-selling strategy. As of now, based on the elementary logic, we are choosing cluster 4.

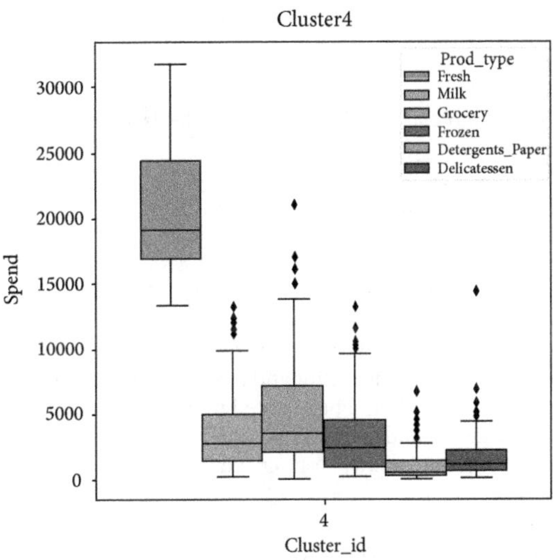

FIGURE 6.9   Spending for cluster 4.

Inference from the graph in Fig. 6.6 and Fig. 6.9: Spending on Frozen is much less than Fresh items. It is less than Milk and Grocery in this cluster. Customer IDs from this cluster are printed below.

```
obj2_data= cust_data_clusters[cust_data_clusters['Cluster_id']==4]
print(list(obj2_data["Cust_id"]))
[4, 5, 13, 14, 15, 19, 21, 23, 25, 26, 28, 31, 33, 34, 37, 41, 42, 55, 59,
68, 71, 72, 74, 76, 84, 90, 105, 106, 113, 114, 115, 119, 121, 127, 128, 133,
139, 141, 142, 145, 150, 151, 153, 158, 163, 191, 192, 196, 203, 211, 218,
221, 227, 233, 235, 238, 241, 242, 243, 248, 249, 254, 256, 263, 268, 270,
277, 280, 284, 288, 289, 295, 297, 301, 308, 312, 323, 324, 325, 329, 333,
335, 336, 337, 348, 355, 357, 361, 369, 372, 374, 381, 382, 388, 394, 402,
403, 404, 405, 407, 422, 423, 424, 425, 433, 435, 436]
```

Objective 3: Find a customer segment that spends very less on groceries and send those customers promotional offers on "Grocery."

Solution: Cluster ID 1 satisfies this criterion. We choose this cluster for sending the promotional offers on Grocery. Cluster_id 0 also has fewer sales of grocery, but the rest of the product's sales is also very less in that cluster.

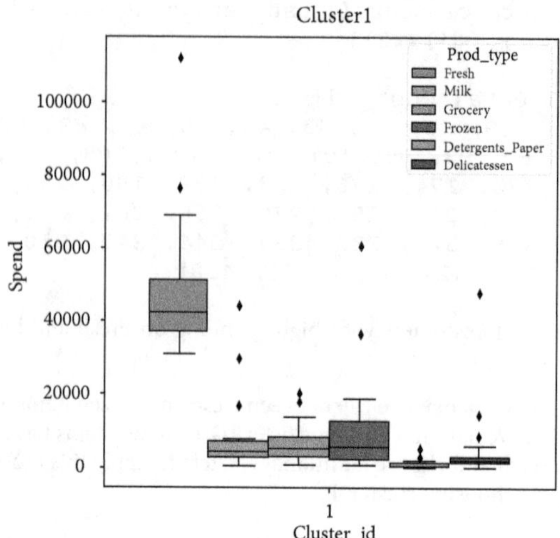

FIGURE 6.10   Spending for cluster 1.

Inference from the graph in Fig. 6.6 and Fig. 6.10: Grocery sales is less in this cluster as compared to other items. Customer IDs are printed below.

```
obj3_data= cust_data_clusters[cust_data_clusters['Cluster_id']==1]
print(list(obj3_data["Cust_id"]))
[30, 40, 53, 88, 104, 125, 126, 130, 143, 177, 182, 184, 197, 240, 259, 260,
274, 283, 285, 286, 290, 326, 371, 378, 383, 428, 437]
```

The above mentioned cross-selling and upselling strategies are just a few examples. The cluster numbers might change from user to user. However, the final result and customer IDs in the result of each objective will be the same. In reality, we study each segment and apply the best-suited marketing strategy in each segment. The choice of strategies depends on the budget for marketing and promotional offers. Marketing professionals aim to maximize the returns (in terms of sales revenue or maybe brand equity) for a given budget.

6.5   DECIDING THE NUMBER OF CLUSTERS

While building the model, we can choose the number of clusters based on our business knowledge. Alternatively, we can follow a technique called the "elbow method" to get an idea of the optimal number of clusters for a given dataset. The elbow method, as discussed in the following section, gives us a basic estimate on the number of clusters. We can use this information to fine-tune and finalize the number of clusters.

6.5.1   Elbow Method

While structuring clusters, we have to make sure that the intracluster distance is at a minimum. Intracluster distance is the distance of every point from its cluster center. We build several cluster sets by taking multiple values of K. For each of these models, we calculate the cluster-wise within-cluster squared distance and sum it up. This value is known as inertia. Intuitively, inertia is the sum of squares of errors (SSE). It is represented by the following equation:

$$\text{Inertia} = \sum_{i=1}^{n_1}(x_i - \mu_1)^2 + \sum_{i=1}^{n_2}(x_i - \mu_2)^2 + \sum_{i=1}^{n_3}(x_i - \mu_3)^2 ..... + \sum_{i=1}^{n_k}(x_i - \mu_k)^2$$

where $\mu_1\,\mu_2\,\mu_3 ...$ are cluster centers

In the elbow method, we plot a graph and decide the optimal value for K. The graph is plotted by taking the number of clusters K on the x-axis and inertia on the y-axis. The graph looks like an elbow (Fig. 6.11).

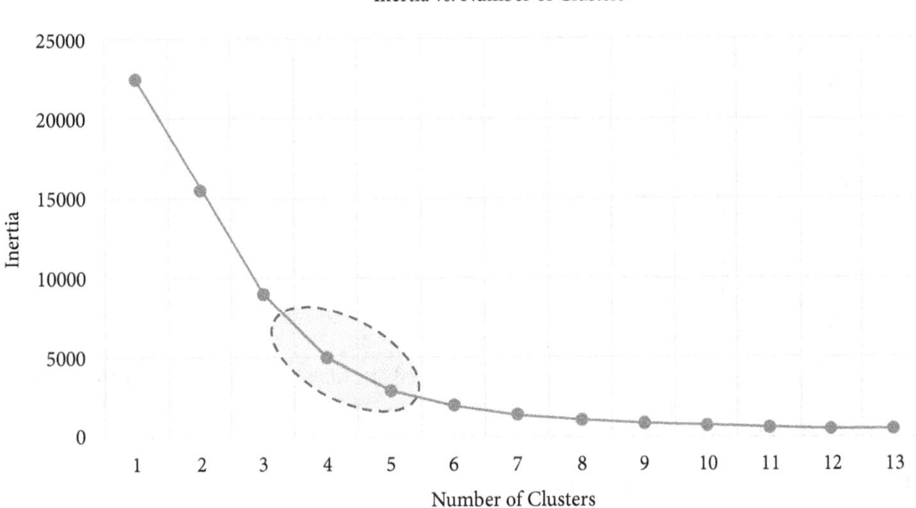

FIGURE 6.11   Inertia versus number of clusters.

As the number of clusters K increases, each cluster size becomes smaller and smaller. As a result, within-cluster distance decreases, so the inertia also decreases. If K is too high, $K = N$ (the number of records), as an extreme, then each cluster will have one record. In such a case, inertia becomes zero. However, so many clusters are not desired. On the other hand, if the value of K is too low, say 1, then inertia is at the maximum. Naturally, for $K = 1$, the global cluster center is not close to every point in the dataset when compared to several individual cluster centers. Inertia is at the maximum in case of lower values of K. As K increases, inertia reduces rapidly; after a specific value of K, inertia saturates. The saturation point where inertia does not reduce much can be picked as the optimal value of K. The graph looks like an elbow. Therefore, we need to pick the elbow point as our optimal value of clusters.

In the previous section, we built five clusters; we can check the value of inertia using the command "kmeans.inertia_" **`print(kmeans.inertia_)`**.

We will now build several models by changing the number of clusters K. We will start with $K = 1$ and then increase the number of clusters iteratively to 15. We consider all the integer values of K between 1 and 15, store the values of inertia, and create the elbow plot.

```
elbow_data=pd.DataFrame()

for i in range(1,15):
    kmeans_m2 = KMeans(n_clusters=i, random_state=333)
    X=cust_data.drop(['Cust_id', 'Channel', 'Region'],axis=1)
    model= kmeans_m2.fit(X)
    elbow_data.at[i,"K"]=i
    elbow_data.at[i,"Inertia"]=round(model.inertia_)/10000000

print(elbow_data)
```

The code above builds 15 models and stores the values of inertia in elbow_data. Note that the values of inertia are in billions; we have scaled them so that we can plot the elbow curve. The following is the output.

```
print(elbow_data)
         K        Inertia
1      1.0   15759.585837
2      2.0   11321.752961
3      3.0    8034.216798
4      4.0    6485.574107
5      5.0    5313.877813
6      6.0    4727.387665
7      7.0    4131.258478
8      8.0    3627.142213
9      9.0    3281.378566
10    10.0    3006.729241
11    11.0    2837.224803
12    12.0    2659.655791
13    13.0    2455.059454
14    14.0    2258.887159
15    15.0    2118.656948
```

We can see from the output that the inertia has reduced at a higher rate until the fourth or fifth cluster. After that, the rate of decrement is low. We will draw the elbow curve to verify and finalize the optimal value of K. The code below draws the graph of K versus inertia, as in Fig. 6.12.

```
plt.figure(figsize=(15,8))
plt.title("Elbow Plot", size=20)
plt.plot(elbow_data["K"],elbow_data["Inertia"],'--bD')
plt.xticks(elbow_data["K"])
plt.xlabel("K", size=15)
plt.ylabel("Inertia", size=15)
```

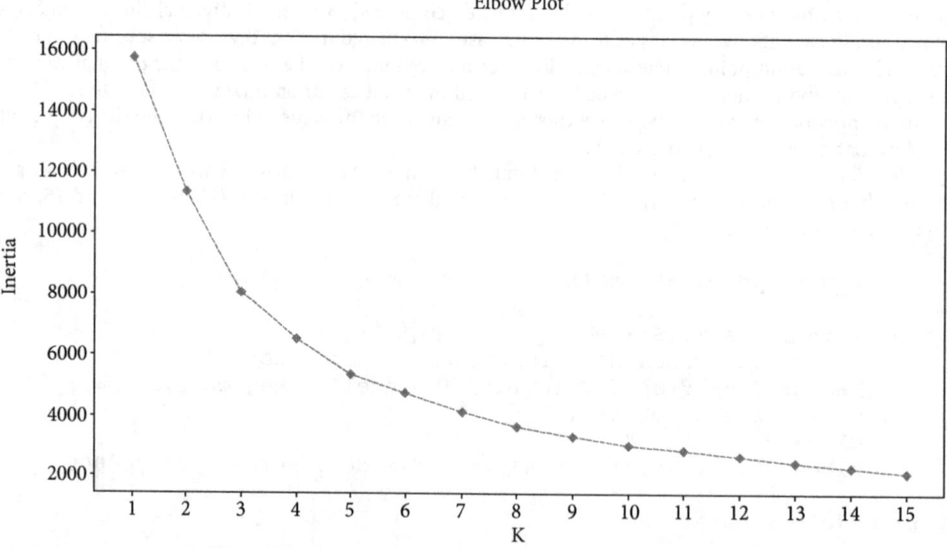

FIGURE 6.12   Inertia versus clusters–elbow plot.

From the elbow plot in Fig. 6.12, we can see the elbow at $K = 5$. We can choose $K = 4$ or 5 or 6 as the optimal number of clusters for this data. This elbow estimation is just a suggested number for K. Depending on the business problem, we can choose a different number. We can choose the best number near the suggested optimal elbow K value based on our business knowledge.

The elbow method does not give us the best result in every scenario. While solving practical problems, we should rely more on business knowledge to choose clusters. In our case study, you need to look at the cluster output and fine-tune K accordingly. For example, if we choose a very high value of K, then several clusters will display the same properties in the output. We need to merge the similar clusters under one cluster, reducing the K value. If K is too less, then each cluster will have a massive size. For such a model, when we look at the distribution of variables inside a specific cluster, we will find a significant variance. We need to split such massive clusters into two or three, increasing the K value. As a general practice, we take a basic estimate from the elbow method and fine-tune further. In most cases, we finalize the K value from our knowledge of the business domain.

6.6   CONCLUSION

K-means cluster analysis is one of the most widely used methods of unsupervised learning. K-means is relatively fast and easy to scale up. Interpretation of the results is straightforward in the K-means clustering algorithm. The validation of the unsupervised algorithms is a challenge. We do have some measures to validate the models, but they are not as strong as the supervised learning validation metrics. We can also use the clustering method as a feature reengineering techniques in supervised learning. We can divide the data into clusters and assign IDs; this assigned ID variable can give us some boost in the accuracy. Here we have taken up a numeric data example only. For non-numeric data, we need to use different distance measures. There are other types of cluster analysis algorithms which you can explore when needed.

6.7   PRACTICE PROBLEMS

1. Download the World Happiness Report data (years 2018 and 2019).
   - Import the data. Complete the necessary exploration and sanitization of the data.
   - Build a machine learning model to divide the data into homogenous segments based on the attributes.
   - Check for the innovative ways to improve the model.

   Dataset credits: Login to Kaggle website at https://www.kaggle.com/unsdsn/world-happiness for world happiness report 2019. License CC0: Public Domain.

2. Download the Young People Survey dataset.
   - Import the data. Complete the necessary exploration and sanitization of the data.
   - Build a machine learning model to divide the participants into homogenous segments based on the behavioral attributes.
   - Check for the innovative ways to improve the model.

   Dataset credits: Data collected by students of the statistics class at FSEV UK, 2013. https://www.kaggle.com/miroslavsabo/young-people-survey. License CC0: Public Domain.

6.8   REFERENCES

1. Dataset: "Wholesale Customers Data Set." It is publicly available data at https://archive.ics.uci.edu/ml/datasets/wholesale+customers. Original source: Margarida G. M. S. Cardoso, margarida.cardoso@iscte.pt, ISCTE-IUL, Lisbon, Portugal.

2. Elbow method: https://www.coursera.org/learn/machine-learning.

RANDOM FORESTS AND BOOSTING

Algorithms such as regression, logistic regression, and decision trees are simple and easy to interpret. They are good at identifying simple patterns in the data. These models are easy to build, and they take less training time. Now we are going to start with some advanced machine learning algorithms, also known as black-box methods. Random forest, boosting, and artificial neural networks (ANNs) are a few examples of advanced machine learning algorithms. These algorithms can identify intricate patterns in the data. Also, these black-box methods are more accurate compared to the simple models, discussed earlier in this book. As a flip side, these black-box methods are not as straightforward or easy to interpret as basic machine learning models.

In general, the black-box methods do an excellent job of identifying the interaction effects of the input variable and hidden patterns. These models take higher training time, and we need a high-performance machine to execute these algorithms on more massive datasets. In case we are looking only for high accuracy without worrying about the impact of individual variable and model interpretability, then we choose to go ahead with black-box methods. If model interpretation and variable impact are also crucial along with accuracy, then we opt for simpler models such as logistic regression.

Trying to predict an accident based on sensor data is a case for choosing black-box methods. In the banking and finance industry, all credit risk models need to be submitted to the regulatory authorities for approval before the model can be applied to real-time customers. The authorities need to understand each variable and its impact on the target. In such cases, we need to go for a simple model, even if it means a little compromise on accuracy.

We have logically divided this work into four parts. The first part is about the basics of Python and statistics. In the second part, we will discuss so-called simple algorithms, such as linear and logistic regression. In the third part, we will discuss black-box methods such as random forest, boosting, and ANNs. The fourth and final part covers the deep learning algorithms such as convolutional neural networks (CNNs), recurrent neural networks (RNNs), and long short-term memories (LSTMs). We will start getting into more details after introducing a few essential terms, such as *ensemble models* and *wisdom of crowds* in the following section.

7.1 ENSEMBLE MODELS

Building several models instead of one model falls under ensemble model building techniques. Ensemble models are analogous to a concept in psychology called the wisdom of crowds. If we are unsure and not able to decide the right option, then we can follow the most liked option by the crowd, given that everyone in the crowd is smart, and everyone has independent opinions.

7.1.1 Wisdom of Crowds

Suppose we want to buy a product online—a product such as a mobile phone or laptop. We are unsure about which is the best one. What do we do in such a situation? We read all the reviews. We do not consider all the reviews, but we consider only smart and constructive reviews. We tend to choose the product that has received the most positive reviews or ratings. This approach is nothing but the wisdom of crowds. Aristotle was the first person to write about the wisdom of the crowds in his work titled *Politics*. He writes:

> It is possible that the many, though not individually good men, yet when they come together may be better, not individually but collectively, than those who are so, just as public dinners to which many contribute are better than those supplied at one man's cost.

This wisdom of crowds concept is popularized by the author and journalist James Surowiecki. In his book titled *Wisdom of Crowds*, he writes:

> One should not expend energy trying to identify an expert within a group but instead rely on the group's collective wisdom, however, make sure that Opinions must be independent and some knowledge of the truth must reside with some group members.

You can always follow the crowd's common choice, given that everyone in the crowd is intelligent, and everyone is thinking independently. The wisdom of crowds approach fails when the crowd consists of fools. The wisdom of crowds also does not work if people are not thinking independently. Just one person making a decision and everyone else following him will not satisfy the underlying assumption of the wisdom of crowds.

Take the example of a model-building exercise where we want to build a model that predicts the average expenses of a family in a city. We gave this problem statement to one eminent professor, who is also a renowned data scientist. He built a model and gave a predicted value. The same problem statement later was given to 200 assistant professors—each one a good data scientist. Each one of them built their models and shared their version of the predicted value of the average expense. For the same city now, we have 200 predictions from assistant professors. The results are as shown in Fig. 7.1.

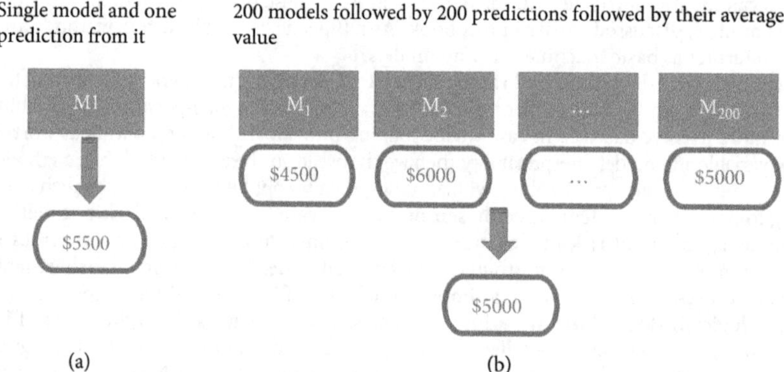

FIGURE 7.1   (*a*) Prediction using a single model. (*b*) Average prediction from 200 models.

Given the two scenarios in Fig. 7.1, which value, $5500 or $5000, has a higher chance to be near to the actual value? We feel the average value of 200 models will have higher accuracy, as just one model may not have the flexibility to learn all the patterns in the data. A few patterns may still be left out. If we build multiple independent models, there is less chance to miss any pattern in the data. The wisdom of crowds tells the same. Multiple independent and moderately good models always beat the accuracy of a single model. Building multiple models and collating their predictions is known as the ensemble model-building technique. When compared to a single model, ensemble models give us slightly better accuracy. We will discuss more on ensemble models next.

7.1.2   Ensemble Models Approach

We have discussed many model-building algorithms until now. The process was to take a dataset and to build a model on it. Each model got us a single prediction for every new data point. Let us try ensemble models in this chapter, where we build several moderately good models. For every new data point, we will get multiple predictions from these individual models. All these predictions will be collated to give a single value as the final prediction. An ensemble model has two steps: the first step involves building several models, and the second step aggregates the results. The aggregation happens at the time of prediction. If we are solving a regression problem, then predictions are aggregated using the average method. If we are solving a classification problem, then the most frequent class is the final result. For example, if we build 200 regression models, then every new data point will have 200 predicted values. We will consider their average as a single prediction. If we have built 200 classification models for a new data point, 160 models predict as class-0, and the remaining 40 predict the result as class-1, then the predicted value from the ensemble model is the most frequent or most voted class. In this case, it is class-0 (Fig. 7.2).

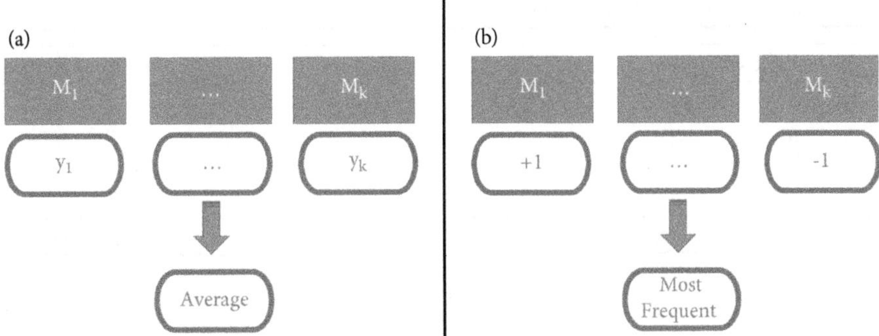

FIGURE 7.2   (*a*) A regression ensemble. (*b*) A classification ensemble.

Ensemble models do an excellent job of identifying the interaction effect between the features. If one model is making a mistake, then other models may cover up for it. Since every model is moderately good, we do not expect several models making the same mistake for the same data point. If we have a single model with 90% accuracy, then the remaining 10% of results will be errors. Assume we have 100 models with 80% accuracy. Then for a new data point, a few models may predict wrongly. However, several other models (out of a total of 100) might predict correctly at the same time. It will eventually improve the overall accuracy of the prediction. Since we are working with multiple models, we need better computation power in the hardware being used when compared to working with a classic single model. With the recent advances in computer hardware in terms of storage and computational power, we can comfortably work with ensemble models. There are two standard ensemble methods—bagging and boosting. We will now start with bagging.

7.2   BAGGING

Bagging stands for bootstrap aggregating (**b**ootstrap **agg**regat**ing**). Bagging has two steps: bootstrap sampling and aggregating the classification results. First, we will talk about bootstrap sampling in the following section. We know what a simple random sample is. For example, if we have 10 records in the population, and we draw a simple random sample of 10 records, then the sample will be the same as the population. Bootstrap sampling works in a slightly different manner.

7.2.1   Bootstrap Sampling

In the example discussed in the previous section, while drawing the bootstrap sample of 10 records, we do not sample all the records in one shot; we draw only one sample record at a time. For example, out of 10 records, we draw one record randomly; it may be record 7. We draw one more sample, between 1 and 10; by pure chance, it may be the record 4 this time. We now have two samples. Let us draw one more sample; it may be record 7 again. We repeat this process 10 times to create a sample of 10 records. The critical point to note here is that while picking the second record, we consider all 10 records of the population; this may lead to duplication of some of the data points. In a bootstrap sample, a few records are repeated multiple times, and some others never get picked. This example creates a bootstrap sample1. We go on to repeat the same process and create bootstrap sample2 and more (Table 7.1).

TABLE 7.1   Original Population with Different Bootstrap Sample Sets

Original Population			
1	2	3	4
5	6	7	8
	9	10	

Bootstrap Sample1			
10	6	8	8
6	7	3	6
	9	4	

Bootstrap Sample2			
1	4	10	6
7	10	9	3
	4	5	

Bootstrap Sample3			
6	4	9	2
3	2	10	5
	7	3	

Bootstrap Sample4			
6	1	9	4
5	8	7	9
	5	2	

Bootstrap Sample5			
4	8	4	4
2	2	1	6
	4	9	

A bootstrap sample is also known as a sample with replacement. While we are drawing the sample records individually one by one, we replace the previously drawn sample point in the dataset. We then consider all the points again in the data for drawing the sample point randomly. This process is known as a sample with replacement. Is there any possibility of picking only one record in the complete bootstrap sample set (say record 4 each time)? Probably yes, but this situation is almost impossible. On average, every bootstrap sample will have 63% unique records and the remaining 37% are duplicates. In our example, we can see six to seven unique numbers in each of the bootstrap samples. By default, the size of the bootstrap sample is the same as the size of the population. Each bootstrap sample has N records. N is the count records of the data. However, the bootstrap sample is not a complete replica of the population.

7.2.2 Bagging Algorithm

The first step in the bagging algorithm is the bootstrap sampling step. Once we understand the bootstrap sample, the rest of the steps are straightforward. Given below is the bagging algorithm.

1. Draw K (number of) bootstrap samples; a higher value of K is preferred.
2. Build a model on each bootstrap sample set. (There will be finally K models.)
3. Collate the results for the new data points based on the average for regression, maximum votes for classification models.

Figure 7.3 is the diagrammatic representation of the bootstrap algorithm. B_1 to B_k are bootstrap samples. M_1 to M_k are individual models. The final result will be an average of predictions for regression, the class with the maximum votes for classification models.

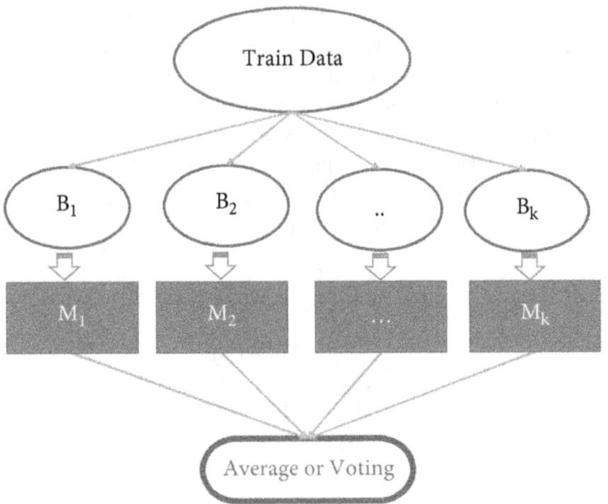

FIGURE 7.3   Representation of a bootstrap algorithm.

Here the models M$_1$ to M$_k$ can be logistic regression models, decision trees, or regression models. If we are repeating to build the same models, then what is it changing from one model to the other? As you notice, the data is changing from model to model. Model1 is built on bootstrap sample1, model2 is built on sample2, and so on. What we discuss in the following section is a specific case of bagging called a random forest.

7.3   RANDOM FOREST

In the bagging algorithm if we explicitly build all decision trees, then it is called a random forest. Several trees put together form a forest (in real life); several decision trees form a random forest (in machine learning). The random forest model most often outperforms a single decision tree model. If we have good computing power available, then random forests are the first choice to have an accurate model. There is a bit of a problem with the standard bagging algorithm. While building individual decision trees, if we use all the (predictor) variables, then almost all the trees give us the same rules. Which means we are building the same model several times and aggregating its output. There is some randomness in the data that varies from model to model. However, it is not sufficient. One of the points in the wisdom of crowds theory was independence. If we are building and counting the same model 100 times, it may not improve the overall accuracy. We need to add one more course of randomness that will make all these models independent. The random forest algorithm has one additional tweak in the original bagging algorithm. The following are the details of this algorithm.

7.3.1   Random Forest Algorithm

Random forest is a particular case of bagging. Given below is the random forest algorithm.

1. Draw K bootstrap samples; a higher value of K is preferred.
2. On each of the bootstrap samples, build a decision tree model. While building this model, do not consider all the variables (in any single model) for splitting.

    a. Consider only randomly selected p-variables while splitting every node. If there are t-variables in the data, then $p \ll t$.

    b. Use the variable with the highest information gain out of these p-variables to split the node into child nodes.

    c. Go to each child node and repeat the above two steps of selecting p-variables randomly and using the best variable to split the node.

    d. Grow the tree as long as possible without pruning.
3. Collate the results for the new data points based on the class with the maximum votes.

In a random forest, there are two sources of randomness: the randomness in the data and the randomness in the features. These two random factors contribute to the independence between the models. While we are building the trees, we consider only p-columns in each cut. Where p is much smaller than the overall number of columns t in the training data, note that we consider p-columns every time we are splitting in the tree. For example, take the first tree. While making two child nodes from the root node, we consider a random set of p-columns. In this subset, we use the variable with maximum information gain for splitting. We go to child node-1 and again consider a random sample set of p-columns and choose the best variable for splitting. We repeat this step and grow the tree without any pruning. Figure 7.4 is the diagrammatic representation of the random forest. B_1 to B_k are bootstrap samples. D_1 to D_k are individual models.

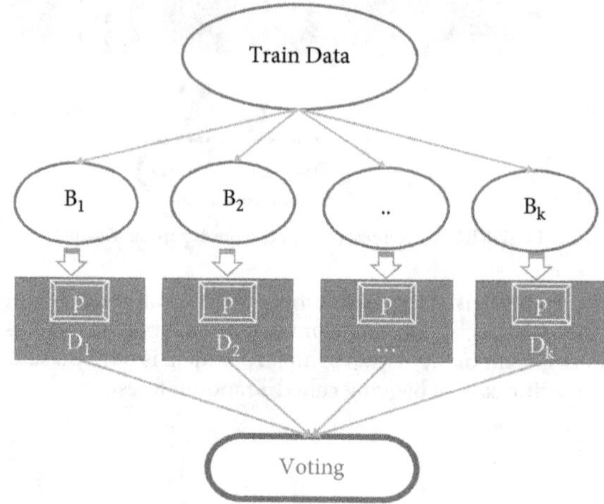

FIGURE 7.4   Representation of a random forest algorithm.

Let us look at one more illustration to understand random forests in depth. The p features are sampled before every split; some variables might get repeated. For example, if there are 30 columns in the training data and we choose $p = 5$, while building the first decision tree we take a random sample of five columns at its root node. We go to each child node again and take a random sample of 5 out of 30 columns. We repeat this step to grow the tree. We build all the trees in the same manner. The diagram in Fig. 7.5 shows the difference between a traditional decision tree and its comparison with a tree in the random forest.

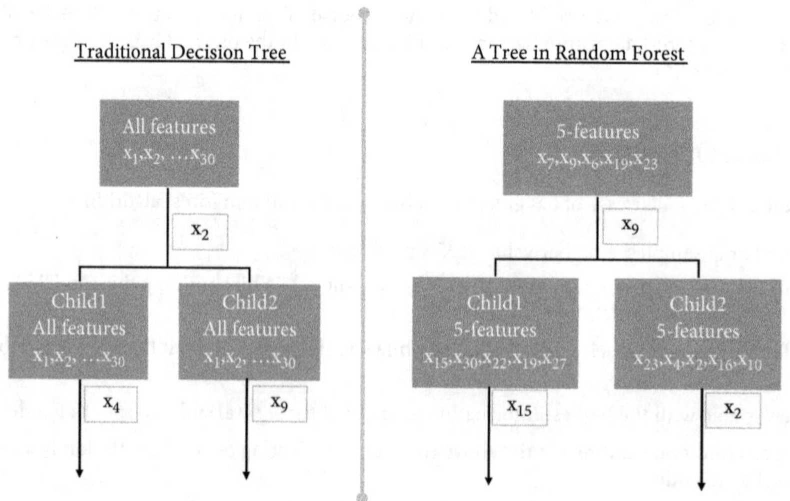

FIGURE 7.5   Standalone classic decision tree and a tree in random forest.

Let us get into more details in the following section.

7.3.2 Hyperparameters in Random Forest

The p has to be small compared to t (number of columns in the full data). If $p = t$, then almost all the trees will be the same. We will be compromising on the independence of the trees. As a general rule of thumb, we can consider p as $\sqrt{t}$. We can try to search for the optimal value of p around this number $\sqrt{t}$.

- The number of features p is the first hyperparameter. It has to be as low as possible. Start with $\sqrt{t}$; search for the optimal value around it. If p is too large, then trees are not independent.

If the value of p is very less, then we cannot afford to build fewer trees. We need to make sure that all the variables are getting sufficient chances to be picked up. In case p is less and the number of trees is also less, then we will end up with an underfitted model.

- The number of trees k is the second hyperparameter. The number of trees has to be high. The default value is set as 500 trees in some tools. We can search between 50 and 500 to find the optimal value. There is hardly any real risk if the number of trees is larger than the optimal value.

While building the random forest model, we need to fine-tune these two hyperparameters. These parameters assume the default values if we do not specify their values. Therefore, we change the default values and search for optimal ones. In theory, we should not prune the decision trees in the random forest; each tree is grown as long as possible. In practice, we mention `max_depth` to avoid overfitting and to reduce the computation time.

- The third hyperparameter is `max_depth`. If `max_depth` is too less, then we will end up with an underfitted model. We mention `max_depth` as a slightly higher number and work with the rest of the two hyperparameters.

7.4 CASE STUDY: CAR ACCIDENTS PREDICTION

This case study is inspired by a data science competition run by Ford Motor Company. This data is derived from the original competition on Kaggle.com. You need to sign up and create a login ID to download this data. The challenge name is "Stay Alert! The Ford Challenge." The Source data link is https://www.kaggle.com/c/stayalert/data. In this case study, we are NOT considering the original data as it is; we will take a random sample from the source, clean it, and change column names for better readability.

7.4.1 Background and Objective

In this case study, we will try to predict the chance of an accident based on the sensor data. If we predict the accident a few seconds early, then we can take some preventive measures to reduce the fatality of the accident. There are multiple causes of an accident. Some factors are vehicle-related, vehicle speed, and wheel alignment. Some other factors are related to the driver, such as attending a phone call and eating while driving. The weather may be another reason, such as snowy, rainy, or foggy weather conditions. Based on these (maybe some more) factors, we want to predict an accident using an appropriately constructed model. We can deploy this model in cars and try to reduce the fatality of the accidents by adjusting the car parameters such as reducing the car speed, opening the airbags, or changing the car suspension. Ford has installed a different type of sensors in their cars. These sensors collect weather, surroundings, vehicular, and driver's physiological data. The actual names and measurement units of these sensors are not disclosed by Ford.

7.4.2 Data Exploration

The dataset has data from 22 sensors and one target variable. Below are some of the basic details of the data.

```
#Importing dataset
car_train=pd.read_csv(r"D:\Chapter7\5.Datasets\car_accidents\car_sensors.
csv")
```

```python
#Get an idea on the number of rows and columns
print(car_train.shape)

#Print the column names
print(car_train.columns)

#Print the column types
print(car_train.info())
```

The following are the code results.

```
print(car_train.shape)
(33239, 23)

print(car_train.columns)
Index(['safe', 'S1', 'S2', 'S3', 'S4', 'S5', 'S6', 'S7', 'S8', 'S9', 'S10',
'S11', 'S12', 'S13', 'S14', 'S15', 'S16', 'S17', 'S18', 'S19', 'S20', 'S21',
'S22'],
      dtype='object')

print(car_train.info())
<class 'pandas.core.frame.DataFrame'>
RangeIndex: 33239 entries, 0 to 33238
Data columns (total 23 columns):
safe     33239 non-null int64
S1       33239 non-null float64
S2       33239 non-null float64
S3       33239 non-null float64
S4       33239 non-null int64
S5       33239 non-null float64
S6       33239 non-null float64
S7       33239 non-null float64
S8       33239 non-null int64
S9       33239 non-null int64
S10      33239 non-null float64
S11      33239 non-null int64
S12      33239 non-null int64
S13      33239 non-null int64
S14      33239 non-null int64
S15      33239 non-null float64
S16      33239 non-null float64
S17      33239 non-null int64
S18      33239 non-null float64
S19      33239 non-null int64
S20      33239 non-null float64
S21      33239 non-null int64
S22      33239 non-null float64
dtypes: float64(12), int64(11)
memory usage: 5.8 MB
None
```

From this output, we can see that there are 33,239 records in the dataset. There are 23 variables in the data. The target variable name is "safe"; other variables are representing the data collected from 22 sensors. All the columns are numerical. As usual, we will perform basic data exploration on the predictor and target variables.

```
##Data Exploration
#Summary
all_cols_summary=car_train.describe()
print(round(all_cols_summary,2))

##Target variable
print(car_train['safe'].value_counts())
```

The code output is as given below.

```
all_cols_summary=car_train.describe()
print(round(all_cols_summary,2))
            safe          S1          S2          S3           S4          S5          S6  \
count   33239.00    33239.00    33239.00    33239.00     33239.00    33239.00    33239.00
mean        0.58       35.46       12.04        0.18       837.14       77.98       10.44
std         0.49        7.27        3.75        0.33      2187.28       18.95       13.96
min         0.00      -22.16      -45.61        0.04       136.00        0.26        0.00
25%         0.00       31.79        9.92        0.09       668.00       66.67        0.00
50%         1.00       34.16       11.43        0.11       800.00       75.00        0.00
75%         1.00       37.37       13.72        0.14       900.00       89.82       28.18
max         1.00      101.34       71.15       11.72    228812.00      441.18       96.84

              S7          S8          S9         S10          S11         S12         S13  \
count   33239.00    33239.00    33239.00    33239.00     33239.00    33239.00    33239.00
mean      103.32        0.28       -4.05        0.02       358.82        1.37        0.88
std       127.53        0.99       35.90        0.00        27.26        1.58        0.33
min         0.00        0.00     -250.00        0.01       260.00        0.00        0.00
25%         0.00        0.00       -8.00        0.02       348.00        0.00        1.00
50%         0.00        0.00        0.00        0.02       365.00        1.00        1.00
75%       213.52        0.00        6.00        0.02       367.00        2.00        1.00
max       359.96        4.00      250.00        0.02       513.00        9.00        1.00

             S14         S15         S16         S17          S18         S19         S20  \
count   33239.00    33239.00    33239.00    33239.00     33239.00    33239.00    33239.00
mean       63.35        1.39       -0.04      572.37        20.01        0.18       12.62
std        18.94        5.40        0.40      297.92        63.43        0.38       11.55
min         0.00        0.00       -4.30      240.00         0.00        0.00        0.00
25%        52.00        0.00       -0.18      255.00         1.49        0.00        0.00
50%        67.00        0.00        0.00      511.00         3.02        0.00       12.80
75%        73.00        0.00        0.07      767.00         7.48        0.00       21.90
max       126.00       52.00        3.60     1023.00       481.51        1.00       71.50

             S21         S22
count   33239.00    33239.00
mean        3.31       11.60
std         1.25        8.98
min         1.00        1.68
25%         3.00        7.97
50%         4.00       10.77
75%         4.00       15.25
max         7.00      262.45

print(car_train['safe'].value_counts())
1    19139
0    14100
```

The target variable takes two values: 1-Safe, 0-Not safe. If we know the complete details about these sensors, then we can perform some feature engineering tasks. First, we will go ahead with the model building.

7.4.3   Model Building and Validation

We will build a decision tree model first. We will then compare the results of the decision tree with the random forest model. We expect a slightly better accuracy from the random forest model.

The code below is used for creating train and test datasets.

```
##Defining train and test data
features=car_train.columns.values[1:]
print(features)
X = car_train[features]
y = car_train['safe']

X_train, X_test, y_train, y_test = model_selection.train_test_split(X, y,
test_size=0.2, random_state=55)

print("X_train Shape ",X_train.shape)
print("y_train Shape ", y_train.shape)
print("X_test Shape ",X_test.shape)
print("y_test Shape ", y_test.shape)
```

The above code gives us the below output.

```
X_train Shape   (26591, 22)
y_train Shape   (26591,)
X_test Shape   (6648, 22)
y_test Shape   (6648,)
```

Below is the code for building a decision tree model.

```
###Building Decision tree on the training data ####
D_tree = tree.DecisionTreeClassifier(max_depth=7)
D_tree.fit(X_train,y_train)

#####Accuracy on train data ####
tree_predict1=D_tree.predict(X_train)
cm1 = confusion_matrix(y_train,tree_predict1)
accuracy_train=(cm1[0,0]+cm1[1,1])/sum(sum(cm1))
print("Decision Tree Accuracy on Train data = ", accuracy_train )

#####Accuracy on test data ####
tree_predict2=D_tree.predict(X_test)
cm2 = confusion_matrix(y_test,tree_predict2)
accuracy_test=(cm2[0,0]+cm2[1,1])/sum(sum(cm2))
print("Decision Tree Accuracy on Test data = ", accuracy_test )

##AUC on Train data
false_positive_rate, true_positive_rate, thresholds = roc_curve(y_train,
tree_predict1)
auc_train = auc(false_positive_rate, true_positive_rate)
print("Decision Tree AUC on Train data = ", auc_train )

##AUC on Test data
false_positive_rate, true_positive_rate, thresholds = roc_curve(y_test,
tree_predict2)
auc_test = auc(false_positive_rate, true_positive_rate)
print("Decision Tree AUC on Test data = ", auc_test )
```

After many trials, we will arrive at the optimal `max_depth` for this data, which is `max_depth = 7`. The following is the code output.

```
Decision Tree Accuracy on Train data =  0.88
Decision Tree Accuracy on Test data =  0.88
Decision Tree AUC on Train data =  0.87
Decision Tree AUC on Test data =  0.87
```

The best decision tree gives us 88% accuracy and AUC of 87%. If we try a higher `max_depth`, then we will be getting into the overfitting zone. We will now build a random forest model. The two important hyperparameters are the number of trees and the number of features. We can set the `max_depth` to a fixed number, say 10 in this case.

```
####Building Random Forest Model
R_forest=RandomForestClassifier(n_estimators=300, max_features=4, max_depth=10)
R_forest.fit(X_train,y_train)

#####Accuracy on train data ####
forest_predict1=R_forest.predict(X_train)
cm1 = confusion_matrix(y_train,forest_predict1)
accuracy_train=(cm1[0,0]+cm1[1,1])/sum(sum(cm1))
print("Random Forest Accuracy on Train data = ", round(accuracy_train,2) )

####Accuracy on test data ####
forest_predict2=R_forest.predict(X_test)
cm2 = confusion_matrix(y_test,forest_predict2)
accuracy_test=(cm2[0,0]+cm2[1,1])/sum(sum(cm2))
print("Random Forest Accuracy on Test data = ", round(accuracy_test,2) )

##AUC on Train data
false_positive_rate, true_positive_rate, thresholds = roc_curve(y_train,
forest_predict1)
auc_train = auc(false_positive_rate, true_positive_rate)
print("Random Forest AUC on Train data =  ", round(auc_train,2) )

##AUC on Test data
false_positive_rate, true_positive_rate, thresholds = roc_curve(y_test,
forest_predict2)
auc_test= auc(false_positive_rate, true_positive_rate)
print("Random Forest AUC on Test data =  ", round(auc_test,2) )
```

Let us go through the random forest code.

```
R_forest=RandomForestClassifier(n_estimators=300, max_features=4, max_depth=10)
```

- `n_estimators = 300` as we are building 300 trees here. A higher number is preferred. If the dataset size is large, then we can use a smaller number. We can try between 100 and 500 and choose the optimal value.
- `max_features=4`, the number of features randomly chosen. A lower number is preferred. We can try 3, 4, or 5.
- `max_depth=10`. We can fix this parameter at a slightly higher value as compared to a standard decision tree. If we try a lower value for this, we will get minimal accuracy on train and test data.

The above code gives us the below output.

```
Random Forest Accuracy on Train data =  0.92
Random Forest Accuracy on Test data =  0.91
Random Forest AUC on Train data =  0.91
Random Forest AUC on Test data =  0.9
```

The random forest gives us 91% accuracy and AUC of 90% on the test data. When compared to a single decision tree, we can see an improvement of 3%. With a random forest model, we generally observe an improvement of 1% to 5% over a single decision tree.

There is an impressive webpage on random forests containing all the relevant and useful information on the subject. It is maintained by Leo Breiman and Adele Cutler (and team). This website has the most accurate and comprehensive information on random forests: https://www.stat.berkeley.edu/~breiman/RandomForests/cc_home.htm.

There is one more ensemble method that works very differently from random forests. The method is called boosting. We will discuss in detail the boosting algorithm in the following sections.

7.5   BOOSTING

In bagging, we built multiple models in parallel. In boosting we build various weak models and combine them to form a robust model. Nevertheless, in boosting, we build models sequentially, which is the main difference between bagging and boosting. If we think bagging as the wisdom of crowds, then boosting is the wisdom of crowds with some weight given to individuals based on their skills.

7.6   ADABOOSTING ALGORITHM

The first boosting algorithm is the AdaBoosting algorithm. This algorithm uses a small trick on top of decision trees, and that trick gives us a lift in the accuracy. The name comes from the phrase adaptive boosting. Given below are the four critical steps of the AdaBoosting algorithm.

1. *Step 1: Data and Weak Classifier:* Start with the whole training data. We build a weak classifier model. What is a weak classifier? Any model that beats random guessing or a coin-tossing model is called a weak classifier. For example, in the car accident case study, if we predict the accidents by tossing a coin, then for getting heads, we give the prediction as "Accident." This model will have approximately 50% accuracy. A weak classifier is a model that can beat 50% accuracy.

2. *Step 2: Error Calculation and Weighted Sample:* After building the weak classifier model, as discussed in step 1, take note of the error. There will be some records that are wrongly classified by the model. We go on to re-create the new weighted sample. In this weighted sample, we will give more weight to incorrectly classified records by the previous model. Just to explain the weight concept, we take an example for now. If there are 10 records in the data, then the probability of picking up a record is 0.1. We give 0.5 as the weight to record number 6. Now, if we draw a sample of 10 records from the data, record 6 will appear five times in the new weighted sample.

3. *Step 3: Rebuild and Repeat:* We will take the weighted sample explained in the previous step and rebuild a new weak classifier on this data. Since we gave more weight to wrongly classified records, this model is bound to learn those patterns that the previous model had failed to learn. This new model may not be perfect. Again we take note of the error rate for this model. There will be some records that are wrongly classified by this model as well. We will create a new weighted sample. We give more weight to records that are wrongly classified by this model. Again, we go ahead and rebuild a new model. We repeat this step of creating a weighted sample and building a classifier on them.

4. *Step 4: Stopping Criteria:* Building a model is one iteration. We will stop this process when we reach zero error or maximum iterations criteria. Strict zero error may lead to overfitting. We need to mention the number of iterations, which is a hyperparameter. In every iteration, the error reduces. Too many iterations will lead to overfitting; too few will lead us to underfitting. All models are not treated equally. The final class will be decided based on all model predictions and the respective model weights, which will be given based on their accuracy. Model-1 can see many data points. Model-50 (for example) might see only a part of the data. The final class will be decided not on simple voting but weighted voting.

The philosophy behind doing all this is whichever pattern or record, the model has not learned (mispredicted), we give it more chance by giving it a higher weight.

Figure 7.6 is a diagrammatic representation of the above-mentioned boosting process.

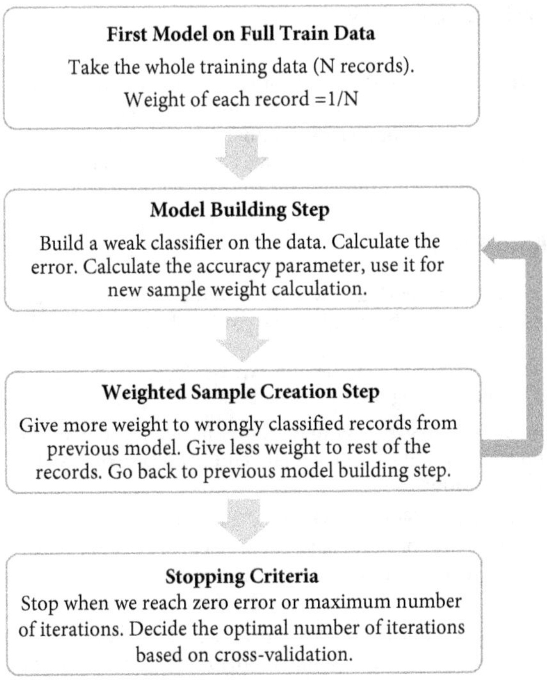

First Model on Full Train Data

Take the whole training data (N records).

Weight of each record =1/N

Model Building Step

Build a weak classifier on the data. Calculate the error. Calculate the accuracy parameter, use it for new sample weight calculation.

Weighted Sample Creation Step

Give more weight to wrongly classified records from previous model. Give less weight to rest of the records. Go back to previous model building step.

Stopping Criteria

Stop when we reach zero error or maximum number of iterations. Decide the optimal number of iterations based on cross-validation.

FIGURE 7.6   Boosting process.

The sample weights will be decided based on the accuracy of the previous model. There is a formula for calculation of this weight. We will now get into the actual theoretical steps involved in the AdaBoosting algorithm.

1. Consider the full training data with N records. Calculate the weight of each sample w_i. For the first iteration $w_1 = 1/N$.

2. Build a model M and calculate the error factor and accuracy factor using the below formulas. Error is the sum of misclassified weights/overall weight. The accuracy factor is analogous to the accuracy of the model. It will be used later in updating the weights.

   a. Error factor $e_M = \dfrac{\sum_{i=1}^{N} w_i I\left(y_i \neq \hat{y}_i\right)}{\sum_{i=1}^{N} w_i}$

   b. Accuracy factor $\alpha_M = \log\left(\dfrac{1 - e_M}{e_M}\right)$

3. First, identify all the misclassified records. Update the weights using the formula below.

   a. Updated weights $w_{i+1} = w_i e^{\alpha_M I}(y_i \neq \hat{y}_i)$.

   b. Normalize weights to make their sum equal to 1.

4. Repeat the model-building and weight updating processes until we have zero error.

5. The final collation is done based on individual votes from the model, and each vote is given a weight of α_m

   a. Predicted class $= \text{sign}\left(\sum \alpha_m \hat{y}_m\right)$

The AdaBoosting algorithm is one of the early versions of boosting. The gradient boosting algorithm, discussed in the following sections, is the latest and most widely used boosting technique. In the AdaBoosting algorithm, we give more weight to previously misclassified records and build models iteratively to reduce the error. In gradient boosting, we deal with errors in a slightly different manner.

7.7   *GRADIENT BOOSTING ALGORITHM*

The gradient boosting algorithm can be easily understood by making an illustration with regression. We take the whole training data. We build the first regression model. This model may not be perfect. We will take predictions from this model to calculate the errors. We will build a new model that can exclusively learn these errors. We will recalculate the errors and repeat this process. Let us look into the steps of the gradient boosting algorithm in detail.

7.7.1   Gradient Boosting Algorithm

To understand the gradient boosting algorithm, we will take the example of regression. Regression gives us a better intuition to follow each step easily. Given below are the steps involved in the gradient boosting algorithm.

1. *Step 1: Initial Model:* Take the training data having (x_1, y_1), (x_2, y_2), (x_3, y_3) ... (x_N, y_N) as the data points. Build a regression model on the whole training data. Let that be $y = F(x)$. The predicted values of y are $\hat{y}_i = F(x_i)$.

2. *Step 2: Residual Calculation:* The first model may not be perfect. Calculate the errors at each point: $y_i = \hat{y}_i + e1_i$. These errors are also known as residuals. The actual values can be expressed as $y = F(x) + e1$. Here term $e1$ is the error made by the model in iteration1.

3. *Step 3: Build Model on the Residuals:* Now, we will build a new model $h(x)$ on the residuals. For this new model, the target is not the actual y-values but the residuals. The model $h(x)$ is built on $(x_1, e1_1)$, $(x_2, e1_2)$, $(x_3, e1_3)$ ... $(x_N, e1_N)$ data points. If there are any left-out patterns from the previous model, then we are giving a chance for this model to pick them.

4. *Step 4: Update the Residuals and Update the Model:* The model $h(x)$ will give us new residuals while predicting errors in $e1$. Updated residuals $e1_i = \widehat{e1}_i + e2_i$. The residual model is $e1 = h(x) + e2$. The original model is $y = F(x) + e1$. We can now express the original model as $y = F(x) + h(x) + e2$. This equation is slightly modified to $y = F(x) + \rho *$ $h(x) + e2$. Instead of considering $h(x)$ values directly, we can multiply them with a shrinkage parameter ρ. It is the weight given to the predictions from the new model $h(x)$. $\rho = 1$ that indicates the full weight.

5. *Step 5: Stopping Criterion:* We can repeat the steps of residual calculation followed by model building on residuals, followed by updating the overall model until we see zero error. One model building is one iteration; we can mention the parameter "maximum number of iterations" to avoid the problem of overfitting. Too few iterations may lead to underfitting; too many iterations lead to overfitting. As we know, the number of iterations is a hyperparameter. We need to fine-tune it by comparing validation metrics on training and test data. If ρ is very less, then the number of iterations needed would be high. If ρ is small, then the model gives less preference to predictions from $h(x)$. Usually, ρ is set to a number between 0.01 and 1. This parameter slows down the learning process. If shrinkage is less then the model takes more iterations to overfit on the data. The slow learning helps us in stopping at the optimal iterations with more precision.

The same algorithm can be applied to trees. We have already discussed classification trees in the decision trees chapter. Now we will discuss how to work with regression trees and how gradient boosting works for classification.

7.7.2   Gradient Boosting on Trees

In Chap. 4, we have discussed classification trees in detail. In the classification tree, we try to divide the whole dataset in such a way that the subsets are pure. If a subset contains only records from one class, then it is pure. We have also discussed entropy and the Gini index as measures of impurity. Entropy is low for pure segments. Imagine building a regression tree where we want to divide the data with a continuous target variable into pure subsets. The only difference between classification and regression is the measure of impurity. We have used entropy in classification. Here in regression trees, the sum of squares of deviations from the mean is the measure of impurity. Impurity of a segment in the regression tree is $\sum (y_i - \bar{y})^2$.

- If all the records in a subset have the same y value, then the segment is pure, and the value of squared deviations from the mean will be zero: $\sum (y_i - \bar{y})^2 = 0$.
- If a few records are different and most of the records have the same value, then the segment is slightly impure.
- If the target variable values are very distinct, then the segment is impure. The value of squared deviations from the mean will be very high.

In the regression tree, we follow the same steps as in classification trees. We start with the overall population. We choose the variable that does the best job splitting the data into segments. The best variable is chosen based on the highest reduction in impurity. The average of impurities in the child nodes should be the least. We continue this process until we achieve maximum purity. We can use train and test data validation to find the optimal pruning parameters. For any new data point, the final prediction will be made based on the leaf node that it falls in. The prediction will be the average value of the target, i.e., $\bar{y}$.

Now we will use the above regression tree and apply the gradient boosting technique on top of it to solve a classification problem. Take the classification data. Let the first tree be a simple weak learner with the probability of the class as the prediction. If there are two classes, then we can take 0.5 as the initial probability, or we can consider the probability based on their frequency as an initial prediction. This probability value will be our first model $y = F(x)$. The target value y takes 0s and 1s. The $F(x)$ results are a number between 0 and 1. We can calculate the values of errors: $y_i = F(x_i) + e1_i$ The first set of residuals will be the difference between actual values and probability: $e1_i = y_i - F(x_i)$. We can now build a regression tree model $h(x)$ for $\left(x_1, e1_1\right), \left(x_2, e1_2\right) \ldots \left(x_N, e1_N\right)$. The updated model will be $y = F(x) + \rho * h(x) + e2$.

We repeat the process of building the model on residuals and updating the model. We stop this learning process at the optimal number of iterations.

7.7.3 Hyperparameters in Boosting

There are several hyperparameters in boosting. We will discuss the essential hyperparameters that will have a high impact on the final boosted model.

- *Number of iterations (n):* Boosting is an iterative algorithm. The error reduces in each iteration. The number of iterations is the first hyperparameter in boosting. A large number will lead to overfitting, and a minimal number will lead to underfitting. The number of iterations needs to be fine-tuned based on train and test data validation.

- *Shrinkage or learning rate(ρ):* Instead of considering the predictions as it is from the residual models, we shrink them by factor ρ. The shrinkage or learning rate parameter makes the whole learning process very slow. The error reduction will be very less in each iteration. Let us take an example to understand the importance of shrinkage. We have a dataset, and we have the prior information that we can build a machine learning model and reduce the error to 10%. Now we start building the boosted model with $\rho = 1$. Our boosted model has 40% error in iteration1, 20% error in iteration2, and 0% error in iteration3. When we validate this model on test data, we find that with three iterations, the model is overfitting. With two iterations, the model gives a 20% error on train data. However, we know there is a scope to reduce the error to 10%. How do we build a decision tree and stop at a 10% error? Here in this example, our boosted model has exited after three iterations. It has learned and modeled the error too fast. How do we slow down this process? We introduce the shrinkage parameter ρ in the equation of the model $F(x) + \rho * h(x)$. If shrinkage = 1, then we take predictions from $h(x)$ as it is. If shrinkage = 0.1, then we suppress predictions from $h(x)$ by a factor ρ. There will be much error even after the first iteration. Roughly we can understand that if $\rho = 0.1$ then it will take 10 steps to reduce the error that was reduced in one step when $\rho = 1$. In the above example, to reach the zero error, our model took three iterations, and if we set $\rho = 0.1$ then it may take 30 iterations to get to zero error. By the end of iteration 20, the error will be 20%. Then we can set the number of iterations at about 25 or 26 to reach the error 10%. Similarly, if we set $\rho = 0.01$ then it may take 300 iterations to reach zero error. Then our optimal number of iterations may be about 240 to 260. A few tools have 0.01 as the default value of shrinkage. Usually, we set the value of ρ between [0.001 and 1] and choose the optimal number of iterations. The optimal number of iterations is chosen based on the model performance on train and test data. If both ρ and number of iterations are very less, then the final model will be underfitted.

- *Size of the tree:* While solving practical problems, it is always preferable to make the trees learn slowly. The real power of the ensemble is in building weak models and collating them. To make the individual trees as weak learners, we need to set the `max_depth` of each tree as a small value. Usually, we choose the `max_depth` of trees as a number between [1 and 5]. It can change based on the data.

There are several other optional parameters in gradient boosting. A few parameters are alternatives to the above three parameters. In general, weak learners (small size trees), slow learning process ($\rho = 0.01$), and a large number of trees ($n > 100$) yield a good boosted tree.

7.7.4   **Gradient Boosting Illustration**

We will use a dataset to understand the boosting and learning rate parameter in depth. Pet adoption data has two columns. One is the age of the customer and another is the target column. The target column has two classes: 0s and 1s: 0-Not adopted the pet and 1-Adopted the pet. We will try using the GBM (gradient boosting method) model to predict the customer's likelihood of adopting a pet based on his or her age (customer age). The below code is trying to build boosted trees. We will try with learning rate = 1, and below are the results from the first 20 iterations.

```python
import pandas as pd
pets_data = pd.read_csv(r"D:\Chapter7\Pet_adoption\\adoption.csv")
pets_data.columns.values
pets_data.head(10)


X=pets_data[["cust_age"]]
y=h=pets_data['adopted_pet']

from sklearn.ensemble import GradientBoostingClassifier
from sklearn.metrics import f1_score
import matplotlib.pyplot as plt

for i in range (1,21):

    #Model and predictions
    boost_model=GradientBoostingClassifier(n_estimators=i,learning_rate=1,
    max_depth=1)
    boost_model.fit(X,y)
    pets_data["iteration_result"]=boost_model.predict_proba(X)[:,1]
    boost_predict= boost_model.predict(X)

    #Graph
    fig = plt.figure()
    plt.rcParams["figure.figsize"] = (7,5)
    plt.title(['Iteration :', i ], fontsize=20)
    ax1 = fig.add_subplot(111)
    ax1.scatter(pets_data["cust_age"],pets_data["adopted_pet"],
    s=50, c='b', marker="x")
    ax1.scatter(pets_data["cust_age"],pets_data["iteration_result"],
    s=50, c='r', marker="o")
    ax1.set_xlabel('cust_age')
    ax1.set_ylabel('adopted_pet')

    #SSE and Accuracy
    print("SSE : ", sum((pets_data["iteration_result"] - y)**2))
    accuracy=f1_score(y, boost_predict, average='micro')
    print("Accuracy : ", accuracy)
```

Figure 7.7 shows the results from the above code.

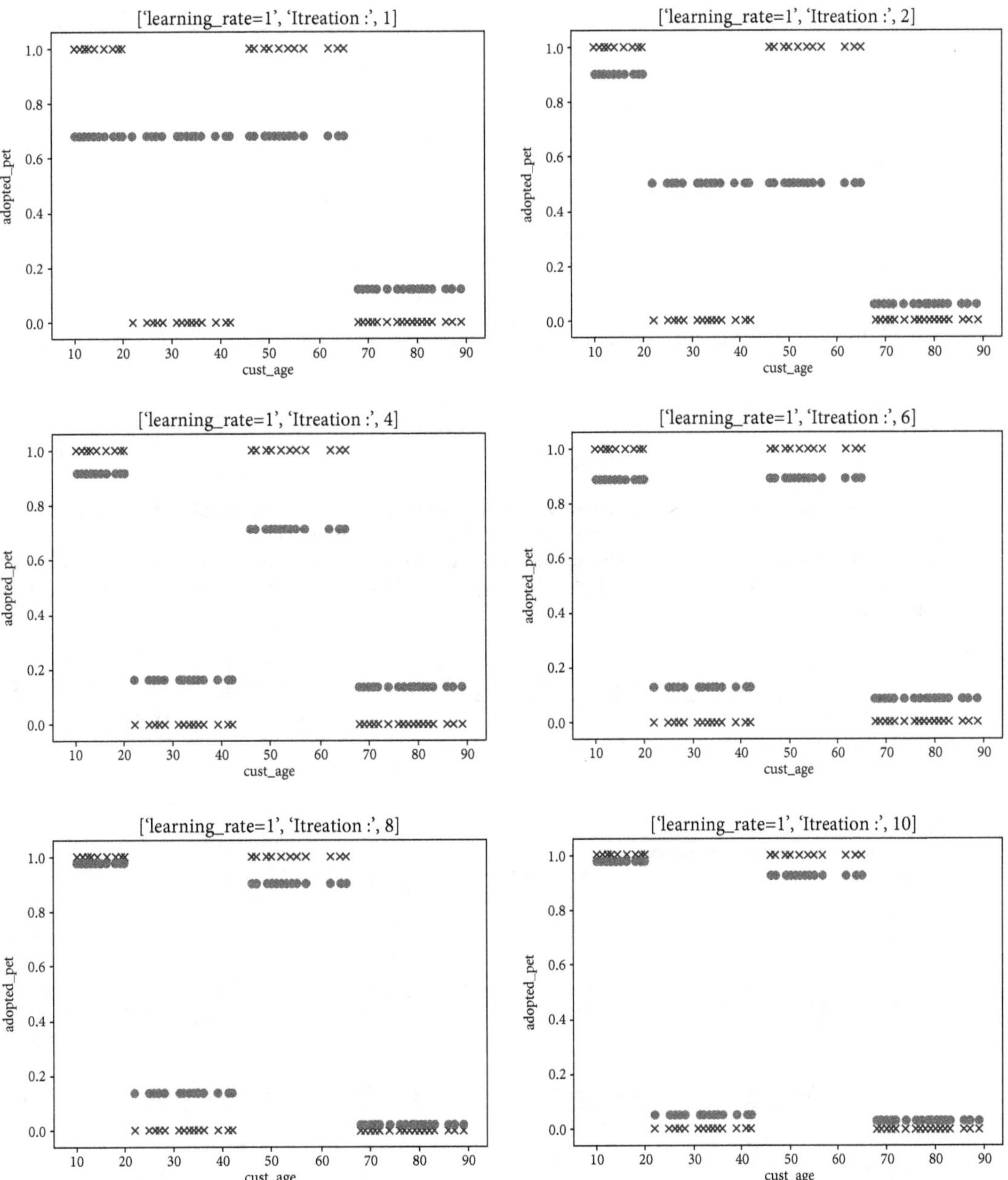

FIGURE 7.7   Boosting iterations.

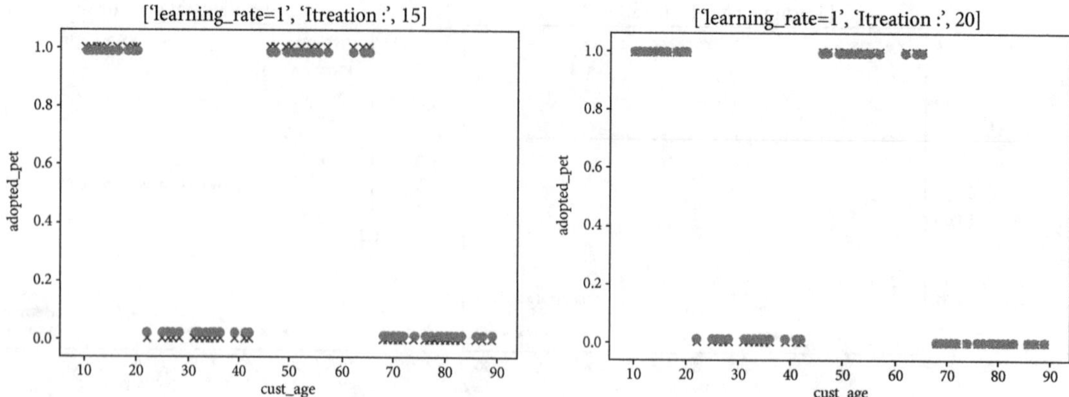

FIGURE 7.7   (*Continued*)

In the output in Fig. 7.7, we can see the actual values are shown as "x" and predicted values are shown with "●." In the first iteration, there was much error. Slowly by the end of the 20th iteration, the predictions have moved almost on top of the actual values. If we try a lower learning rate, then by the end of the 20th iteration, we will still have much error left. We can reuse the same code by just changing the learning rate. Figure 7.8 shows the results for learning rate = 0.1.

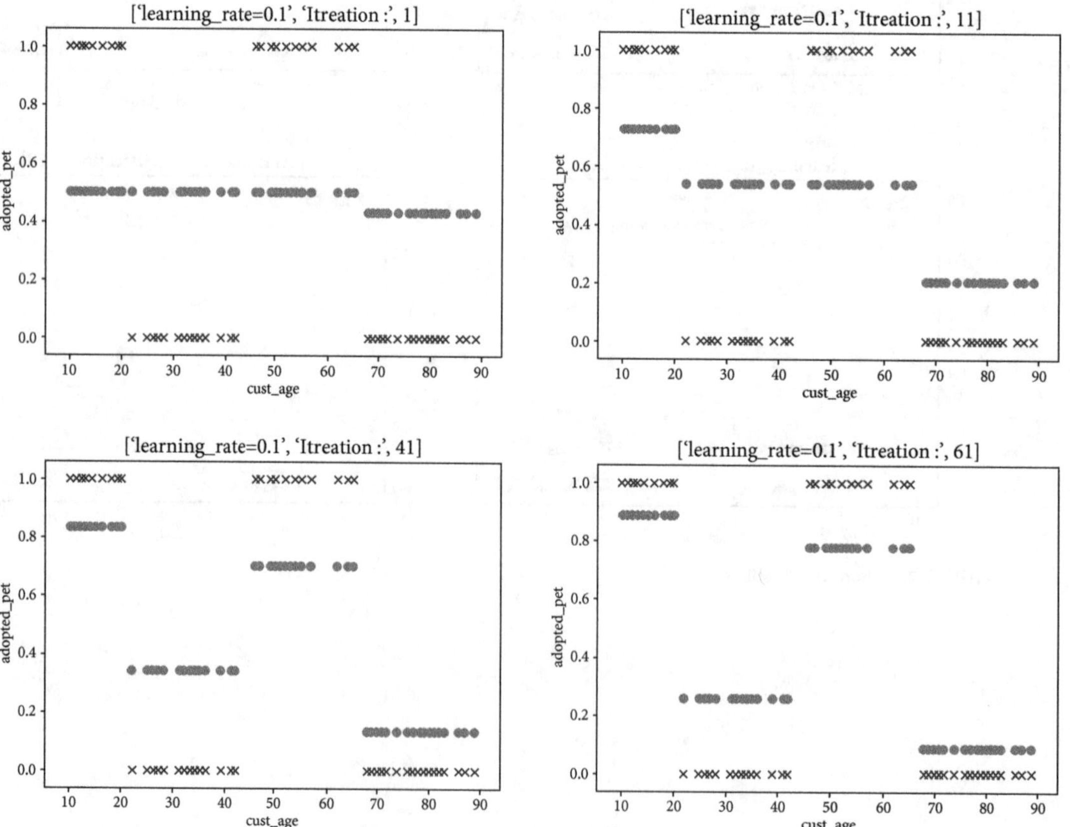

FIGURE 7.8   Boosting iterations with learning rate = 0.1.

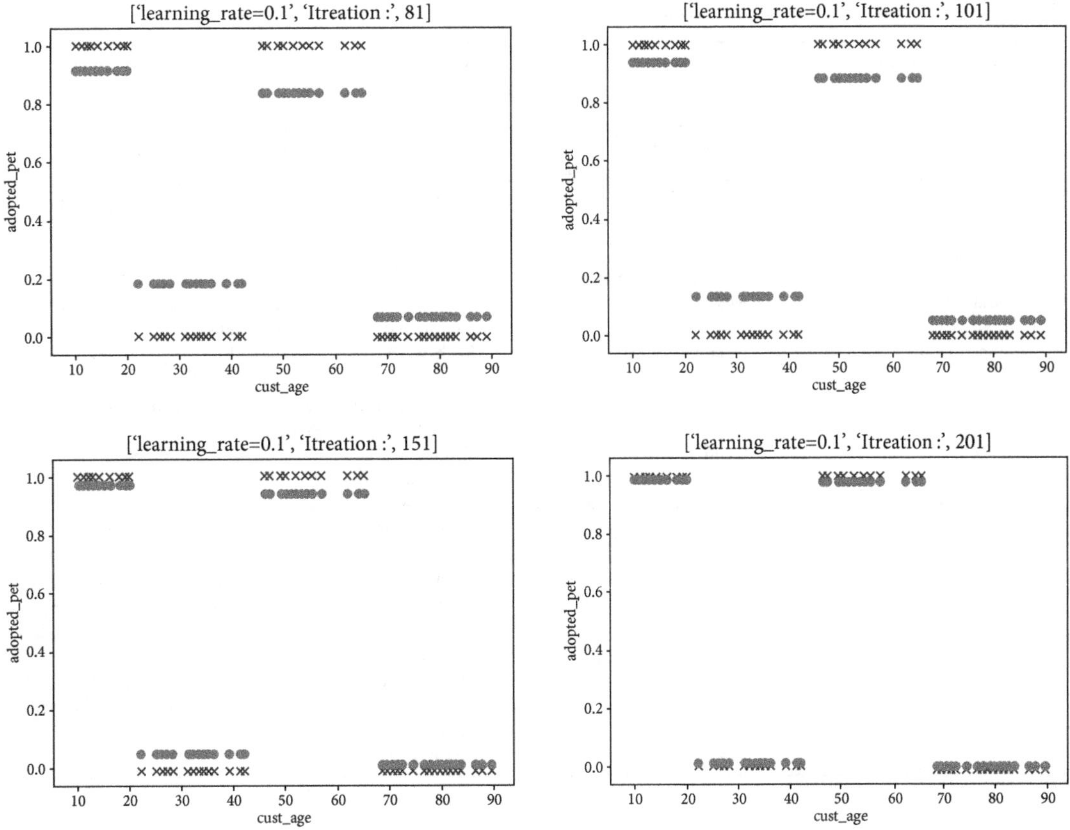

FIGURE 7.8   (*Continued*)

From the above output, we can see the impact of the learning rate parameter. The error reduction that we could achieve in 10 steps with a learning rate of 1 almost took 100 iterations with a learning rate of 0.1. In general, a slow learning model with a lower learning rate (shrinkage) is preferred.

7.8   *CASE STUDY: INCOME PREDICTION FROM CENSUS DATA*

In this case study, we will consider data from the 1994 US Census database. This dataset is publicly available under CC0: Public Domain license. We have cleaned the original dataset and are considering a subset for this case study.

7.8.1   Background and Objective

We need to study this dataset and predict whether a person can earn more than 50,000 dollars per year. Before making a sales pitch or making a marketing call to a customer, it is good to get an idea of the expected income of a customer. This dataset has customer profile details along with the income class. Education, occupation, marital status, and hours-per-week are a few columns in the data. The target column is income, which takes two values: ≤50K and >50K. The objective is to predict the customer income class based on customer details. Table 7.2 is the list of columns and their values.

TABLE 7.2 Column Names, Descriptions, and Values

S. No.	Column Name	Description	Values
1	age	Age of the individual	Continuous
2	workclass	Employment type of an individual	Private, Self-emp-not-inc, Self-emp-inc, Federal-gov, Local-gov, State-gov, Without-pay, Never-worked
3	education	Educational qualification	Bachelors, Some-college, 11th, HS-grad, Prof-school, Assoc-acdm, Assoc-voc, 9th, 7th-8th, 12th, Masters, 1st-4th, 10th, Doctorate, 5th-6th, Preschool
4	education-num	Educational qualification converted to number	Continuous
5	marital-status	Marital status	Married-civ-spouse, Divorced, Never-married, Separated, Widowed, Married-spouse-absent, Married-AF-spouse
6	occupation	Occupation category	Tech-support, Craft-repair, Other-service, Sales, Exec-managerial, Prof-specialty, Handlers-cleaners, Machine-op-inspect, Adm-clerical, Farming-fishing, Transport-moving, Priv-house-serv, Protective-serv, Armed-Forces
7	sex	Gender	Female, Male
8	capital-gain	Capital gains by the individual	Continuous
9	capital-loss	Capital losses by the individual	Continuous
10	hours-per-week	Working hours per week	Continuous
11	native-country	Country of origin	United-States, Cambodia, England, Puerto-Rico, Canada, Germany, Outlying-US(Guam-USVI-etc), India, Japan, Greece, South, China, Cuba, Iran, Honduras, Philippines, Italy, Poland, Jamaica, Vietnam, Mexico, Portugal, Ireland, France, Dominican-Republic, Laos, Ecuador, Taiwan, Haiti, Columbia, Hungary, Guatemala, Nicaragua, Scotland, Thailand, Yugoslavia, El-Salvador, Trinidad&Tobago, Peru, Hong, Holland-Netherlands.
12	Income	Target column	$\leq$50K and >50K

The objective is to use the first 11 profile-related variables and predict the income band of a customer.

7.8.2 Data Exploration

There are a few continuous variables and a few categorical variables. We will now perform a few necessary data exploration steps. The below code imports the data and gives us the necessary details on the data.

```
income = pd.read_csv(r"D:\
Chapter7\5.Datasets\Adult_Census_Income\Adult_Income.csv")

#Get an idea on the number of rows and columns
print(income.shape)

#Print the column names
print(income.columns)

#Print the column types
print(income.info())

##Data Exploration
#Summary
all_cols_summary=income.describe()
print(round(all_cols_summary,2))
```

The above code gives us the below output.

```
print(income.shape)
(32561, 12)

print(income.columns)
Index(['age',     'workclass',     'education',     'education.num',     'marital.
status','occupation',  'sex',  'capital.gain',  'capital.loss',  'hours.per.
week','native.country', 'income'],
      dtype='object')

print(income.info())
<class 'pandas.core.frame.DataFrame'>
RangeIndex: 32561 entries, 0 to 32560
Data columns (total 12 columns):
age                32561 non-null  int64
workclass          32561 non-null  object
education          32561 non-null  object
education.num      32561 non-null  int64
marital.status     32561 non-null  object
occupation         32561 non-null  object
sex                32561 non-null  object
capital.gain       32561 non-null  int64
capital.loss       32561 non-null  int64
hours.per.week     32561 non-null  int64
native.country     32561 non-null  object
income             32561 non-null  object
dtypes: int64(5), object(7)
memory usage: 3.0+ MB
None

all_cols_summary=income.describe()
print(round(all_cols_summary,2))
          age  education.num  capital.gain  capital.loss  hours.per.week
count  32561.00       32561.00      32561.00      32561.00        32561.00
mean      38.58          10.08       1077.65         87.30           40.44
std       13.64           2.57       7385.29        402.96           12.35
min       17.00           1.00          0.00          0.00            1.00
25%       28.00           9.00          0.00          0.00           40.00
50%       37.00          10.00          0.00          0.00           40.00
75%       48.00          12.00          0.00          0.00           45.00
max       90.00          16.00      99999.00       4356.00           99.00
```

The `describe()` function prints the summary for the numerical columns only. From the output, we can see that a few numerical columns, such as capital gain and capital loss, have outliers. The rest of the columns are fine. We can clean the outliers data or use it as it is to build our first version of the model. We will now study the categorical variables. The below code extracts all the categorical columns and prints their frequency counts.

```
##Categorical Variables Exploration
categorical_vars=income.select_dtypes(include=['object']).columns
print(categorical_vars)

##Frequency tables for all the categorical columns
for col in categorical_vars:
    print("Frequency Table for the column ", col )
    print(income[col].value_counts())
```

The above code gives us the below output.

```
print(categorical_vars)

Index(['workclass', 'education', 'marital.status', 'occupation', 'sex',
'native.country', 'income'], dtype='object')
##Frequency Table for the column  workclass
Private              22696
Self-emp-not-inc      2541
Local-gov             2093
?                     1836
State-gov             1298
Self-emp-inc          1116
Federal-gov            960
Without-pay             14
Never-worked             7
Name: workclass, dtype: int64

##Frequency Table for the column  education
HS-grad           10501
Some-college       7291
Bachelors          5355
Masters            1723
Assoc-voc          1382
11th               1175
Assoc-acdm         1067
10th                933
7th-8th             646
Prof-school         576
9th                 514
12th                433
Doctorate           413
5th-6th             333
1st-4th             168
Preschool            51
Name: education, dtype: int64

##Frequency Table for the column  marital.status
Married-civ-spouse        14976
Never-married             10683
Divorced                   4443
Separated                  1025
Widowed                     993
Married-spouse-absent       418
Married-AF-spouse            23
Name: marital.status, dtype: int64

##Frequency Table for the column  occupation
Prof-specialty        4140
Craft-repair          4099
Exec-managerial       4066
Adm-clerical          3770
Sales                 3650
Other-service         3295
Machine-op-inspct     2002
?                     1843
Transport-moving      1597
Handlers-cleaners     1370
Farming-fishing        994
Tech-support           928
Protective-serv        649
```

```
Priv-house-serv        149
Armed-Forces             9
Name: occupation, dtype: int64

##Frequency Table for the column  sex
Male     21790
Female   10771
Name: sex, dtype: int64

##Frequency Table for the column  native.country
United-States               29170
Mexico                        643
?                             583
Philippines                   198
Germany                       137
Canada                        121
Puerto-Rico                   114
El-Salvador                   106
India                         100
Cuba                           95
England                        90
Jamaica                        81
South                          80
China                          75
Italy                          73
Dominican-Republic             70
Vietnam                        67
Guatemala                      64
Japan                          62
Poland                         60
Columbia                       59
Taiwan                         51
Haiti                          44
Iran                           43
Portugal                       37
Nicaragua                      34
Peru                           31
France                         29
Greece                         29
Ecuador                        28
Ireland                        24
Hong                           20
Cambodia                       19
Trinidad&Tobago                19
Thailand                       18
Laos                           18
Yugoslavia                     16
Outlying-US(Guam-USVI-etc)     14
Honduras                       13
Hungary                        13
Scotland                       12
Holland-Netherlands             1
Name: native.country, dtype: int64

##Frequency Table for the column  income
<=50K    24720
>50K      7841
Name: income, dtype: int64
```

From the above output, we can see that a few columns have missing values denoted by "?" In a few variables, a few classes have an almost negligible frequency. For example, pre-school frequency is just 51 out of 32,561 records. There are just nine individuals with armed forces occupation. We will clean this data by combining a couple of lesser frequency classes as others. Then we will create one-hot encoded columns from each of these categorical variables. The target column is also categorical; we need to convert it into 0s and 1s.

7.8.3   Data Cleaning and Feature Engineering

We will perform some necessary data cleaning and feature engineering steps before building the model. We take the non-numeric variables and combine the less frequent classes to the "other" category using the below code.

```
income["workclass"] = income["workclass"].replace(['?','Never-worked','Without-
pay'], 'Other')
print(income["workclass"] .value_counts())
Private          22696
Self-emp-not-inc  2541
Local-gov         2093
Other             1857
State-gov         1298
Self-emp-inc      1116
Federal-gov        960
Name: workclass, dtype: int64

income["marital.status"] = income["marital.status"].replace(['Never-married
','Divorced','Separated','Widowed'], 'Not-married')
print(income["marital.status"] .value_counts())
Not-married          17144
Married-civ-spouse   14976
Married-spouse-absent   418
Married-AF-spouse        23
Name: marital.status, dtype: int64

income["occupation"] = income["occupation"].replace(['?'], 'Other-service')
print(income["occupation"] .value_counts())
Other-service      5138
Prof-specialty     4140
Craft-repair       4099
Exec-managerial    4066
Adm-clerical       3770
Sales              3650
Machine-op-inspct  2002
Transport-moving   1597
Handlers-cleaners  1370
Farming-fishing     994
Tech-support        928
Protective-serv     649
Priv-house-serv     149
Armed-Forces          9
Name: occupation, dtype: int64

freq_country=income["native.country"].value_counts()
less_frequent= freq_country[freq_country <100].index
print(less_frequent)
Index(['Cuba', 'England', 'Jamaica', 'South', 'China', 'Italy',
'Dominican-Republic', 'Vietnam', 'Guatemala', 'Japan', 'Poland',    'Colum-
bia', 'Taiwan', 'Haiti', 'Iran', 'Portugal', 'Nicaragua', 'Peru', 'France',
'Greece', 'Ecuador', 'Ireland', 'Hong', 'Cambodia',    'Trinidad&Tobago',
'Thailand', 'Laos', 'Yugoslavia', 'Outlying-US(Guam-USVI-etc)', 'Honduras',
'Hungary', 'Scotland', 'Holland-Netherlands'], dtype='object')
```

```
income["native.country"]=income["native.country"].replace([less_frequent],
'Other')
income["native.country"] = income["native.country"].replace(['?'], 'Other')
print(income["native.country"].value_counts())
United-States    29170
Other             1972
Mexico             643
Philippines        198
Germany            137
Canada             121
Puerto-Rico        114
El-Salvador        106
India              100
Name: native.country, dtype: int64
```

We will now convert the binary classes into numbers and multiclass variables into the one-hot encoding format. The get_dummies() function in Pandas creates the one-hot encoding variables.

```
print(income["sex"].value_counts())
income['sex']=income['sex'].map({'Male': 0, 'Female': 1})
Male       21790
Female     10771
Name: sex, dtype: int64

print(income["income"].value_counts())
income['income']=income['income'].map({'<=50K': 0, '>50K': 1})
<=50K     24720
>50K       7841
Name: income, dtype: int64

one_hot_cols=['workclass','marital.status','occupation','native.country']
one_hot_data = pd.get_dummies(income[one_hot_cols])
print(one_hot_data.shape)
(32561, 34)

print(one_hot_data.columns.values)
['workclass_Federal-gov' 'workclass_Local-gov' 'workclass_Other''workclass_
Private' 'workclass_Self-emp-inc' 'workclass_Self-emp-not-inc' 'workclass_
State-gov' 'marital.status_Married-AF-spouse''marital.status_Married-civ-spouse'
 'marital.status_Married-spouse-absent' 'marital.status_Not-married' 'occu-
pation_Adm-clerical' 'occupation_Armed-Forces' 'occupation_Craft-repair'
 'occupation_Exec-managerial' 'occupation_Farming-fishing' 'occupation_Han-
dlers-cleaners' 'occupation_Machine-op-inspct' 'occupation_Other-service' 'occu-
pation_Priv-house-serv' 'occupation_Prof-specialty' 'occupation_Protective-serv'
 'occupation_Sales' 'occupation_Tech-support''occupation_Transport-moving'
 'native.country_Canada''native.country_El-Salvador' 'native.country_Germany'
 'native.country_India' 'native.country_Mexico' 'native.country_ Other' 'native.
country_Philippines' 'native.country_Puerto-Rico' 'native.country_
United-States']
```

We have created 34 one-hot encoded columns from 4 categorical columns. We will merge this data with numerical data. The below code creates the final data with all numeric and one-hot encoded variables.

```
##Final Data
print(income.shape)
income_final = pd.concat([income, one_hot_data], axis=1)
print(income_final.shape)
print(income_final.info())
```

```
##Features
one_hot_features=list(one_hot_data.columns.values)
numerical_features=['age',   'education.num', 'sex', 'capital.gain', 'capital.
loss', 'hours.per.week']
all_features=one_hot_features+numerical_features

##Data
X=income_final[all_features]
y=income_final['income']
X_train, X_test, y_train, y_test = train_test_split(X, y, test_size=0.2)

print(X_train.shape)
print(y_train.shape)
print(X_test.shape)
print(y_test.shape)
```

The above code gives us the below output.

```
print(income.shape)
(32561, 12)


income_final = pd.concat([income, one_hot_data], axis=1)
print(income_final.shape)
(32561, 46)


print(all_features)
['workclass_Federal-gov',   'workclass_Local-gov',   'workclass_Other',   'work-
class_Private', 'workclass_Self-emp-inc', 'workclass_Self-emp-not-inc', 'work-
class_State-gov','marital.status_Married-AF-spouse','marital.status_Married-civ-
spouse', 'marital.status_Married-spouse-absent', 'marital.status_Not-married',
'occupation_Adm-clerical',      'occupation_Armed-Forces',      'occupation_Craft-
repair', 'occupation_Exec-managerial', 'occupation_Farming-fishing', 'occupa-
tion_Handlers-cleaners',   'occupation_Machine-op-inspct',   'occupation_Other-
service', 'occupation_Priv-house-serv', 'occupation_Prof-specialty', 'occupa-
tion_Protective-serv', 'occupation_Sales', 'occupation_Tech-support', occupa-
tion_Transport-moving', 'native.country_Canada', 'native.country_El-Salvador',
'native.country_Germany',   'native.country_India',   'native.country_Mexico',
'native.country_Other', 'native.country_Philippines', 'native.country_Puerto-
Rico', 'native.country_United-States', 'age', 'education.num', 'sex', 'capital.
gain', 'capital.loss', 'hours.per.week']


print(X_train.shape)
(26048, 40)


print(y_train.shape)
(26048,)


print(X_test.shape)
(6513, 40)


print(y_test.shape)
(6513,)
```

7.8.4   Model Building and Validation

We are now ready to build the model. We have to configure three parameters: `learning_rate`, `max_depth`, and
`n_estimators`. The suggested range for the `learning_rate` is [0.1 to 0.01], `max_depth` in the range [1 to 5], and

the n_estimators range is [10 to 1000]. We will start our first model with a slightly low learning_rate of 0.01, max_depth of 4, and n_estimators as 10. We expect this model to be an underfitted one, largely due to the fewer number of iterations and minor learning_rate.

```
##Model Building
gbm_model1  =  GradientBoostingClassifier(learning_rate=0.01,  max_depth=4,
n_estimators=10, verbose=1)
gbm_model1.fit(X_train, y_train)

##Validation on the train and test data

#Train data
predictions=gbm_model1.predict(X_train)
actuals=y_train
cm = confusion_matrix(actuals,predictions)
print(cm)
accuracy=(cm[0,0]+cm[1,1])/(sum(sum(cm)))
print(accuracy)
#Test data
predictions=gbm_model1.predict(X_test)
actuals=y_test
cm = confusion_matrix(actuals,predictions)
print(cm)
accuracy=(cm[0,0]+cm[1,1])/(sum(sum(cm)))
print(accuracy)
```

The above model gives us the results below.

```
      Iter       Train Loss   Remaining Time
         1           1.0911           0.73s
         2           1.0834           0.59s
         3           1.0759           0.50s
         4           1.0686           0.42s
         5           1.0615           0.35s
         6           1.0546           0.28s
         7           1.0479           0.21s
         8           1.0413           0.14s
         9           1.0349           0.07s
        10           1.0287           0.00s
Confusion Matrix on Train data
 [[19832     0]
 [ 6216     0]]
Train Accuracy 0.7613636363636364
Confusion Matrix on Test data
 [[4888    0]
 [1625    0]]
Test Accuracy 0.7504990019960079
```

We will now increase the number of iterations and give a better chance for the algorithm to learn. When we increase the number of iterations to 100, we get the below result.

```
      Iter       Train Loss   Remaining Time
         1           1.0911           6.81s
         2           1.0834           6.50s
         3           1.0759           6.54s
         4           1.0686           6.43s
         5           1.0615           6.34s
         6           1.0546           6.52s
         7           1.0479           6.59s
```

```
         8              1.0413              6.49s
         9              1.0349              6.39s
        10              1.0287              6.33s
        20              0.9737              5.46s
        30              0.9290              4.79s
        40              0.8918              4.10s
        50              0.8597              3.41s
        60              0.8323              2.92s
        70              0.8085              2.17s
        80              0.7880              1.43s
        90              0.7701              0.71s
       100              0.7545              0.00s
Confusion Matrix on Train data
 [[19276    556]
 [ 3450   2766]]
Train Accuracy 0.8462070024570024
Confusion Matrix on Test data
 [[4741   147]
 [ 890   735]]
Test Accuracy 0.8407799785045295
```

We have fixed the low learning rate and tree size. We can write a loop on the number of iterations and see the result on train data and test data. The below code builds 20 models with different numbers of iterations.

```python
##Loop for different iterations
for i in range(5,1000, 50):
    gbm_model1 = GradientBoostingClassifier(
    learning_rate=0.01, max_depth=4,  n_estimators=i)
    gbm_model1.fit(X_train, y_train)

    print("N_estimators=" , i)
    #Train data
    predictions=gbm_model1.predict(X_train)
    actuals=y_train
    cm = confusion_matrix(actuals,predictions)
    accuracy=(cm[0,0]+cm[1,1])/(sum(sum(cm)))
    print("Train Accuracy", accuracy)

    #Test data
    predictions=gbm_model1.predict(X_test)
    actuals=y_test
    cm = confusion_matrix(actuals,predictions)
    accuracy=(cm[0,0]+cm[1,1])/(sum(sum(cm)))
    print("Test Accuracy", accuracy)
```

The above code gives us the result presented in Table 7.3.

TABLE 7.3 The Train and Test Accuracy Output from the Code

n_estimators	Train Accuracy	Test Accuracy
5	76.2%	74.9%
55	81.7%	80.7%
105	84.9%	84.1%
155	85.5%	84.6%
205	86.1%	85.1%
255	86.3%	85.3%
305	86.4%	85.4%
355	86.5%	85.5%
405	86.6%	85.6%
455	86.7%	85.7%
505	86.8%	85.8%
555	86.9%	85.8%
605	87.0%	85.8%
655	87.2%	85.8%
705	87.3%	86.0%
755	87.4%	86.2%
805	87.5%	86.2%
855	87.6%	86.3%
905	87.7%	86.4%
955	87.8%	86.4%

From Table 7.3, we can see that the accuracy of the test data gets saturated at the 85-86% range. Fewer than 200 iterations will lead to underfitting, and more than 1000 iterations may overfit. We can finalize the learning rate as 0.01 and the number of iterations as 200. We can further try to change the learning rate and check results from the combination of different learning rates and different values of the number of iterations. We can do a lot of feature engineering and improve model accuracy. Since there is a slight class imbalance, we can even consider focusing on a single class while validating these models. As suggested, always try to construct a boosted tree with weak learners and with a slow learning rate.

7.9 CONCLUSION

The ensemble models are powerful machine learning algorithms. When prediction accuracy is the high priority and we are not worried about the variable impact, then ensemble models are the first choice. If each variable impact and model interpretation are also very important, then simple models such as regression and trees are preferred. Building and fine-tuning ensemble models are complicated compared to other simple models. Nevertheless, the accuracy levels of optimally built ensemble models are high. There is a sudden increased use of these ensemble models in recent years. Nowadays, the computation power is high, even in personal computers and laptops. A data scientist can effortlessly build these ensemble models within the stipulated time.

7.10 PRACTICE PROBLEMS

1. Download heart disease data.
   - Import the data. Complete the necessary exploration and sanitization of the data.
   - Build a machine learning model to predict the heart disease based on the attributes.
   - Perform the model validation and measure the accuracy of the model.
   - Check for innovative ways to improve the accuracy of the model.

   Dataset credits: https://archive.ics.uci.edu/ml/datasets/Heart+Disease. Creators: 1. Hungarian Institute of Cardiology. Budapest: Andras Janosi, MD. 2. University Hospital, Zurich, Switzerland: William Steinbrunn, MD. 3. University Hospital, Basel, Switzerland: Matthias Pfisterer, MD. 4. V.A. Medical Center, Long Beach and Cleveland Clinic Foundation: Robert Detrano, MD, PhD.

2. Download New York taxi fare prediction data.
   - Import the data. Complete the necessary exploration and sanitization of the data.
   - Build a machine learning model to predict the right fare amount.
   - Perform the model validation and measure the accuracy of the model.
   - Check for innovative ways to improve the accuracy of the model.
   - Add the feature engineering tricks on date and location columns to improve the accuracy of the model.

   Dataset credits: https://console.cloud.google.com/bigquery and https://www.kaggle.com/c/new-york-city-taxi-fare-prediction/data. It is a large dataset. Take a random sample of one million rows.

7.11   REFERENCES

1. James Surowiecki. (2004). *The Wisdom of Crowds: Why the Many Are Smarter Than the Few and How Collective Wisdom Shapes Business, Economies, Societies, and Nations.* Doubleday; Anchor.
2. Case Study: Car Accidents Prediction—Ford data: https://www.kaggle.com/c/stayalert/data.
3. Random forest algorithm: https://www.stat.berkeley.edu/~breiman/RandomForests/cc_home.htm#overview.
4. Case Study: Income prediction from census data: https://archive.ics.uci.edu/ml/datasets/Census+Income.

CHAPTER 8
ARTIFICIAL NEURAL NETWORKS

Regression, logistic regression, and decision trees are relatively less complicated machine learning algorithms. However, you need to develop a good understanding of these algorithms to comprehend advanced machine learning algorithms. Regression and tree algorithms are simple; the results from these algorithms are easy to interpret and explain. These models are good at identifying the simple patterns in the data and take very little execution time. However, these basic algorithms fail at identifying nonlinear patterns in the data. We have more sophisticated algorithms that can find the interaction effects and identify nonlinear hidden patterns automatically. Neural networks, also known as universal function approximators, are one of the most sophisticated machine learning algorithms. A neural network model has the flexibility and complexity to learn almost any nonlinear patterns in the data. In the following section, we will start with logistic regression to get the instinct behind neural networks.

8.1   NETWORK DIAGRAM FOR LOGISTIC REGRESSION

We discussed logistic regression in previous chapters. The terminology of the neural networks is different from the terminology of logistic regression. We used Greek letters such as β for coefficients. We called the variables as independent and dependent variables. Until now, we used typical mathematical or statistical terminology. Here in neural networks, we represent the same logistic regression in computer science terminology. We call the coefficients as weights. Variables will be called input and output. The model equation will be represented as a network diagram. We first need to understand the vocabulary used in neural networks. Given below is the standard logistic regression line equation with two predictor variables.

$$y = \frac{e^{\beta_0 + \beta_1 x_1 + \beta_2 x_2}}{1 + e^{\beta_0 + \beta_1 x_1 + \beta_2 x_2}}$$

The curve equation above, now will be represented in a very different format. As given below, the coefficients are characterized as weights (w) instead of β.

$$y = \frac{e^{w_0 + w_1 x_1 + w_2 x_2}}{1 + e^{w_0 + w_1 x_1 + w_2 x_2}}$$

The logistic function $g(x)$ is also known as a sigmoid activation function

$$g(x) = \frac{e^x}{1 + e^x}$$

The same function is usually rewritten in the following equivalent format:

$$g(x) = \frac{1}{1 + e^{-x}}$$

Let us rewrite the logistic regression equation:

$$y = g\left(w_0 + w_1 x_1 + w_2 x_2\right)$$

The above equation is further rewritten as follows:

$$y = g\left(\sum w_i x_i\right)$$

Finally, we represent it as a network diagram, as shown in Fig. 8.1.

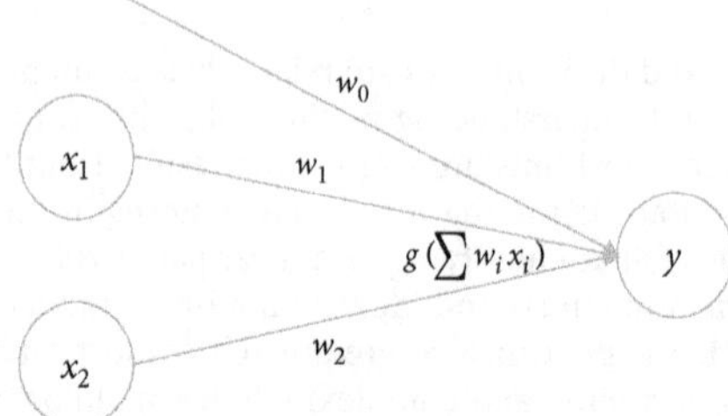

FIGURE 8.1   A primitive network illustration of logistic regression.

In the diagram in Fig. 8.1, the weights (coefficients) are represented as edges. The nodes are the input and output variables. Usually, the function $g(x)$ is not shown in the diagram. It is understood that the logistic or sigmoid function is applied to the final summation. Figure 8.2 is the final representation of the diagram.

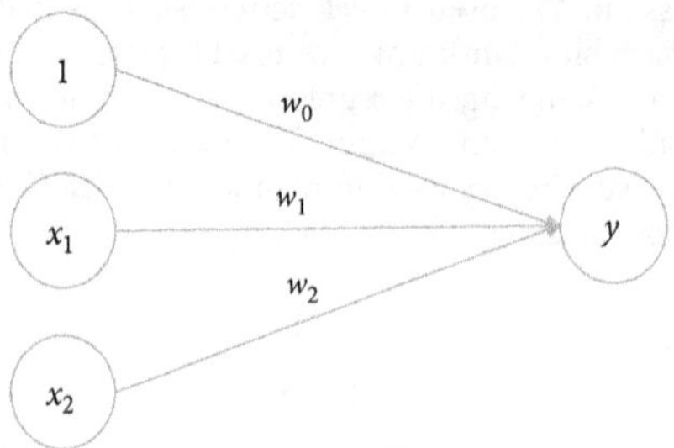

FIGURE 8.2   Refined network illustration of logistic regression.

In the diagram in Fig. 8.2, we have connected the intercept-related weight to the constant node, which always takes the value 1. This node is known as the bias term. First the summation $(w_0 * 1) + \left(w_1 * x_1\right) + (w_2 * x_2)$ will be calculated, which is nothing but $\sum w_i x_i$. Later the sigmoid function $g(x)$ will be applied to it. Finally, the logistic regression equation is $y = g\left(\sum w_i x_i\right)$. This network diagram is our new representation of the logistic regression line.

In this and the upcoming chapters, we will see the equations in this network diagram format only. There is a solid reason behind it, which we will discuss in the later sections.

Here the function $g(x)$ is known as the activation function. Sigmoid activation is one type of activation function. If the activation is a linear function, then the equation becomes the linear regression equation. The same network representation can be made for linear regression as well. The only difference is the activation function $g(x)$. We used a sigmoid activation function in the logistic regression line. Instead, if we use a linear function $g(x) = x$, then $y = g\left(\sum w_i x_i\right)$ becomes a linear regression line. The same diagrammatic representation is used for different activation functions—with a linear $g(x)$ function, it represents the simple linear regression, and when we replace $g(x)$ by a logistic regression function, the same sketch represents the logistic regression. Until now, we have discussed a different type of representation of logistic regression. Table 8.1 summarizes the comparison between logistic regression old and new terminologies.

TABLE 8.1   Comparison of Logistic Regression and New Terminologies

Logistic Regression Terminology	New Terminology
x_1 and x_2 are predictor variables y is the target variable	x_1 and x_2 are inputs y is the output
β_0, β_1, and β_2 are coefficients	w_0, w_1, and w_2 are weights
β_0 is the intercept	w_0 is the weight of the bias
The logistic function $g(x) = \dfrac{e^x}{1 + e^x}$	The sigmoid activation function $g(x) = \dfrac{1}{1 + e^{-x}}$
$y = \dfrac{e^{\beta_0 + \beta_1 x_1 + \beta_2 x_2}}{1 + e^{\beta_0 + \beta_1 x_1 + \beta_2 x_2}}$	$y = \dfrac{e^{w_0 + w_1 x_1 + w_2 x_2}}{1 + e^{w_0 + w_1 x_1 + w_2 x_2}}$
$y = \dfrac{e^{\beta_0 + \beta_1 x_1 + \beta_2 x_2}}{1 + e^{\beta_0 + \beta_1 x_1 + \beta_2 x_2}}$	$y = g\left(\sum w_i x_i\right)$
$y = \dfrac{e^{\beta_0 + \beta_1 x_1 + \beta_2 x_2}}{1 + e^{\beta_0 + \beta_1 x_1 + \beta_2 x_2}}$	

We will revisit this network representation multiple times in the later sections of this chapter and even in the upcoming chapters.

8.2   CONCEPT OF DECISION BOUNDARY

The logistic regression line gives us the predicted values between 0 and 1 as the final output. For a new point, if the predicted value is 0.95, then we can consider the predicted class as Class-1; if the predicted value is 0.05, then we consider the predicted class as Class-0. Usually, we set a threshold at 0.5. If the predicted values are below 0.5, then we classify them as Class-0 and the rest are classified as Class-1. The logistic regression line looks like an "S"-shaped curve. When it comes to decision-making, we use it as a decision boundary that separates Class-0 and Class-1 (Fig. 8.3).

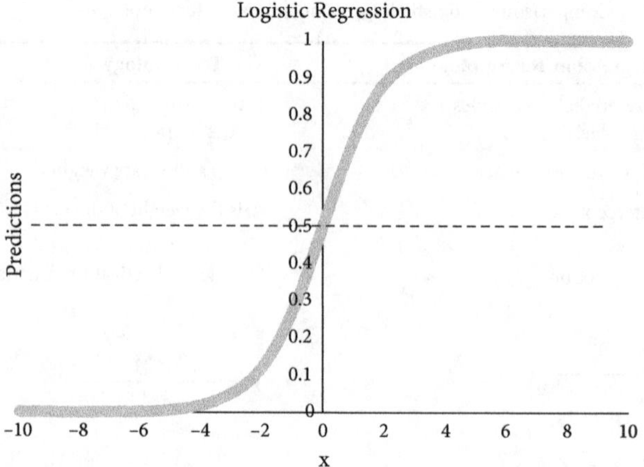

FIGURE 8.3   "S" curve of logistic regression.

We do not see the decision boundary at 0.5 in the logistic regression output. However, we finally perform the classification based on this decision boundary only. The same is the case for multiple predictor variables. We will take the example of two predictor variables and one target variable—x_1, x_2, and y. Figure 8.4 shows the actual data graph for x_1, x_2 vs. y.

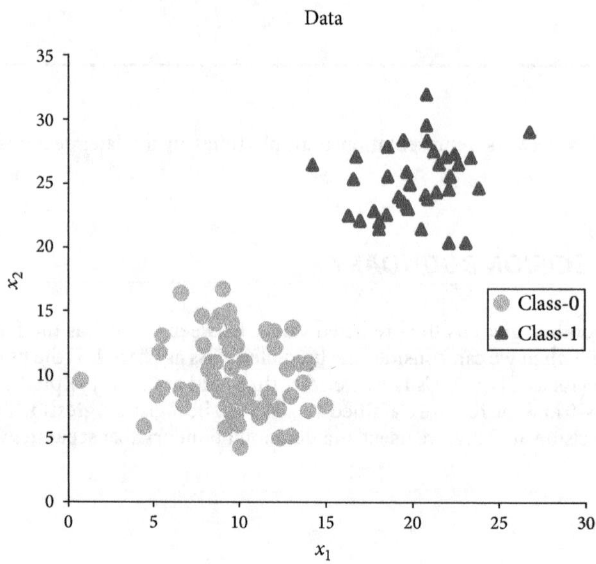

FIGURE 8.4   Actual data points of Class-0 and Class-1.

In the graph in Fig. 8.4, we are trying to show the relation between x_1, x_2, and y. The y variable has two classes: Class-0 and Class-1. If we build a logistic regression line for this data, the decision boundary will appear right between the two classes. The resultant decision boundary acts as a separator between Class-0 and Class-1. Given below is the derivation of that decision boundary.

Logistic Regression Line

$$y = \frac{e^{\beta_0 + \beta_1 x_1 + \beta_2 x_2}}{1 + e^{\beta_0 + \beta_1 x_1 + \beta_2 x_2}}$$

The same equation can be rewritten as follows:

$$\log\left(\frac{y}{1 - y}\right) = \beta_0 + \beta_1 x_1 + \beta_2 x_2$$

Let us substitute $y = 0.5$ in place of y. That will give us the decision boundary line equation:

$$\log\left(\frac{0.5}{0.5}\right) = \beta_0 + \beta_1 x_1 + \beta_2 x_2$$

$$0 = \beta_0 + \beta_1 x_1 + \beta_2 x_2$$

$$-\beta_2 x_2 = \beta_0 + \beta_1 x_1$$

$$\beta_2 x_2 = -\beta_0 - \beta_1 x_1$$

$$x_2 = -\frac{\beta_0}{\beta_2} - \frac{\beta_1}{\beta_2} x_1$$

The above resulting equation is for the decision boundary line. We can build a logistic regression line and derive the decision boundary by using the above formula. We can overlay the decision boundary on top of the original data distribution. Drawing this decision boundary is possible only with two variables—as we cannot visualize beyond three dimensions (Fig. 8.5).

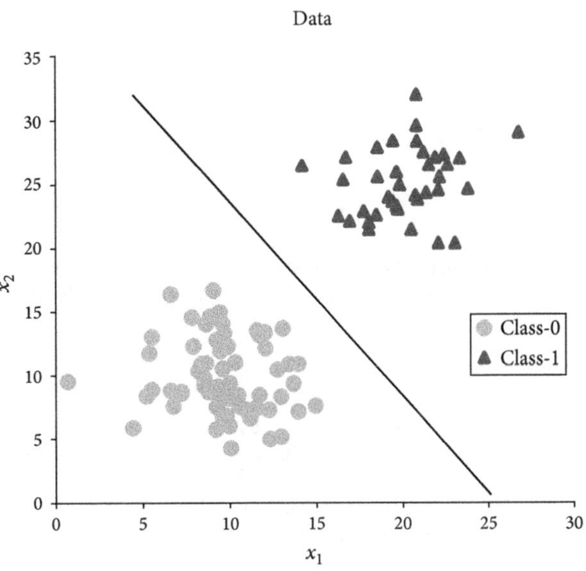

FIGURE 8.5   Data and decision boundary made by logistic regression.

The graph in Fig. 8.5 shows the data along with the decision boundary. The logistic regression line is an "S"-shaped curve; the decision boundary made by the logistic regression is always a straight line that separates the two classes.

8.2.1 Decision Boundary: Code

For this example, the employee purchase data has three columns: employee age, work experience, and the representation of whether they have purchased the product or not. The product here is related to insurance. The objective is to predict the target variable "purchase" by using Age and Experience as predictor variables.

We will build a logistic regression line. However, this time, we are interested in the decision boundary, which we will get after the creation of logistic regression. For this demo, we will use a subset of the data. We will use the complete data in the later sections.

The following code helps us in importing the data and displaying the necessary details of the data.

```
Emp_Purchase_raw  =  pd.read_csv(r"D:\Chapter8  ANN\5.Datasets\Emp_Purchase\
Emp_Purchase.csv")

#### Filter the data and take a subset from the dataset. The filter condition
is Sample_Set<3
Emp_Purchase1=Emp_Purchase_raw[Emp_Purchase_raw.Sample_Set<3]
print(Emp_Purchase1.shape))
print(Emp_Purchase1.columns.values)
print(Emp_Purchase1.head(10))
```

The above code gives us the below output.

```
print(Emp_Purchase1.shape)
(74, 4)


print(Emp_Purchase1.columns.values)
['Age' 'Experience' 'Purchase' 'Sample_Set']


print(Emp_Purchase1.head(10))
     Age  Experience  Purchase  Sample_Set
0   20.0         2.3         0           1
1   16.2         2.2         0           1
2   20.2         1.8         0           1
3   18.8         1.4         0           1
4   18.9         3.2         0           1
5   16.7         3.9         0           1
6   16.3         1.4         0           1
7   20.0         1.4         0           1
8   18.0         3.6         0           1
9   21.2         4.3         0           1
```

There are 74 records in this subset. We will use Age and Experience to predict purchase. We will now plot the data that shows the relation between the predictor and target variables. The code below helps us in plotting the data of all three columns.

```
fig = plt.figure()
ax1 = fig.add_subplot(111)
plt.rcParams["figure.figsize"] = (8,6)
plt.title('Age, Experience  vs Purchase', fontsize=20)

ax1.scatter(Emp_Purchase1.Age[Emp_Purchase1.Purchase==0],Emp_Purchase1.
Experience[Emp_Purchase1.Purchase==0], s=100, c='b', marker="o",
label='Purchase 0')
ax1.scatter(Emp_Purchase1.Age[Emp_Purchase1.Purchase==1],Emp_Purchase1.
Experience[Emp_Purchase1.Purchase==1], s=100, c='r', marker="x", label=
'Purchase 1')
```

```
ax1.set_xlabel('Age',fontsize=15)
ax1.set_ylabel('Experience',fontsize=15)

plt.xlim(min(Emp_Purchase1.Age), max(Emp_Purchase1.Age))
plt.ylim(min(Emp_Purchase1.Experience), max(Emp_Purchase1.Experience))
plt.legend(loc='upper left');

plt.show()
```

Please note that we need not plot these graphs while solving the actual problems. Here we are using them to get a better visual intuition. Figure 8.6 shows the code output.

FIGURE 8.6   Purchase Class-0 and Class-1 data points plotted for different Age and Experience.

In the plot in Fig. 8.6, we observe both the classes of the output: Purchase = 0 and Purchase = 1. We will now build the logistic regression curve, derive the decision boundary, and then draw the decision boundary on top of this plot. We can already guess how the decision boundary would appear. The code below gives us the logistic regression line, and the plot in Fig. 8.7 provides the decision boundary.

```
###Logistic Regression model1
model1 = sm.logit(formula='Purchase ~ Age+Experience', data=Emp_Purchase1)
fitted1 = model1.fit()
fitted1.summary2()

######Accuracy and error of the model1
#Create the confusion matrix
predicted_values=fitted1.predict(Emp_Purchase1[["Age"]+["Experience"]])
predicted_values[1:10]
threshold=0.5

import numpy as np
predicted_class=np.zeros(predicted_values.shape)
predicted_class[predicted_values>threshold]=1

predicted_class
```

```python
from sklearn.metrics import confusion_matrix as cm
ConfusionMatrix = cm(Emp_Purchase1[['Purchase']],predicted_class)
print(ConfusionMatrix)
accuracy=(ConfusionMatrix[0,0]+ConfusionMatrix[1,1])/sum(sum(Confusion
Matrix))
print('Accuracy : ',accuracy)
error=1-accuracy
print('Error: ',error)


#coefficients
slope1=fitted1.params[1]/(-fitted1.params[2])
intercept1=fitted1.params[0]/(-fitted1.params[2])

#Finally draw the decision boundary for this logistic regression model

fig = plt.figure()
ax1 = fig.add_subplot(111)
plt.rcParams["figure.figsize"] = (8,6)
plt.title('Decision Boundary', fontsize=20)

ax1.scatter(Emp_Purchase1.Age[Emp_Purchase1.Purchase==0],Emp_Purchase1.
Experience[Emp_Purchase1.Purchase==0], s=100, c='b', marker="o", label=
'Purchase 0')
ax1.scatter(Emp_Purchase1.Age[Emp_Purchase1.Purchase==1],Emp_Purchase1.
Experience[Emp_Purchase1.Purchase==1], s=100, c='r', marker="x", label=
'Purchase 1')
ax1.set_xlabel('Age',fontsize=15)
ax1.set_ylabel('Experience',fontsize=15)

plt.xlim(min(Emp_Purchase1.Age), max(Emp_Purchase1.Age))
plt.ylim(min(Emp_Purchase1.Experience), max(Emp_Purchase1.Experience))
plt.legend(loc='upper left');

x_min, x_max = ax1.get_xlim()
ax1.plot([0, x_max], [intercept1, x_max*slope1+intercept1])

plt.show()
```

Figure 8.7 shows the output generated from the code.

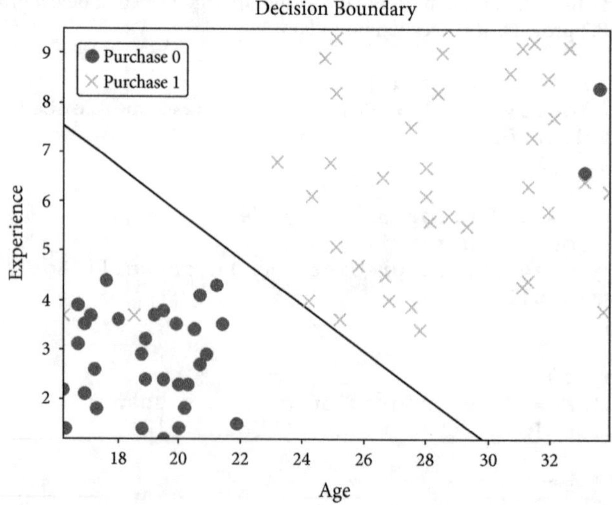

FIGURE 8.7   The decision boundary for classes Purchase 0 and Purchase 1.

From the output in Fig. 8.7, we can see the decision boundary created by the logistic regression line. As expected, the decision boundary is between the two classes. That concludes this section. The takeaway from the discussions in this section is that every logistic regression line creates a decision boundary, which looks like a straight line between the two classes.

8.3   *MULTIPLE DECISION BOUNDARIES PROBLEM*

The creation of the decision boundary works correctly when the two classes in the target variable are separable with a straight line. Not every dataset has a clear separating boundary. Take a look at the dataset in Fig. 8.8. There are two input variables and two classes in the output. It looks like the usual case of the logistic regression line. However, we know that logistic regression can give us a single straight-line decision boundary. For the data shown in Fig. 8.8, we need two decision boundaries to separate the two classes.

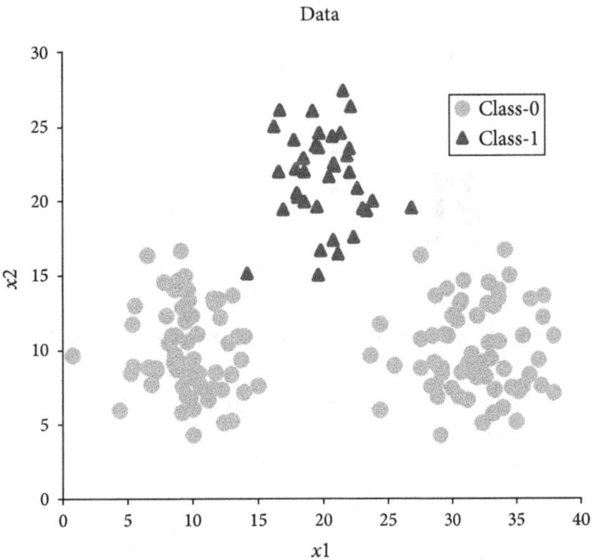

FIGURE 8.8   Multiple decision boundaries case.

A single decision boundary cannot separate the two classes effectively. A logistic regression line fails in this case. In all the cases where the separating boundary is nonlinear or when we need more than one decision boundary, logistic regression fails. In the example in Fig. 8.7 (code), we considered only a subset of the data. We will now look at the full data. The following code helps us in plotting the data.

```
##plotting the overall data

fig = plt.figure()
ax = fig.add_subplot(111)
plt.rcParams["figure.figsize"] = (8,6)
plt.title('Age, Experience  vs Purchase - Overall Data', fontsize=20)

ax.scatter(Emp_Purchase_raw.Age[Emp_Purchase_raw.Purchase==0],Emp_Purchase_
raw.Experience[Emp_Purchase_raw.Purchase==0],  s=100,  c='b',  marker="o",
label='Purchase 0')
ax.scatter(Emp_Purchase_raw.Age[Emp_Purchase_raw.Purchase==1],Emp_Purchase_
raw.Experience[Emp_Purchase_raw.Purchase==1],  s=100,  c='r',  marker="x",
label='Purchase 1')
ax1.set_xlabel('Age',fontsize=15)
ax1.set_ylabel('Experience',fontsize=15)
```

```python
plt.xlim(min(Emp_Purchase_raw.Age), max(Emp_Purchase_raw.Age))
plt.ylim(min(Emp_Purchase_raw.Experience), max(Emp_Purchase_raw.Experience))
plt.legend(loc='upper left');
plt.show()
```

Figure 8.9 presents the code output.

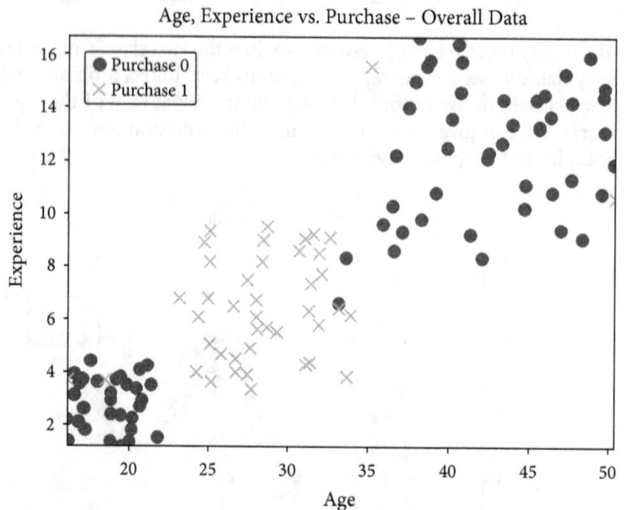

FIGURE 8.9   The code output for the classes Purchase 0 and Purchase 1.

We will now force-fit a logistic regression line to this data and try to sketch the decision boundary. In the earlier case, we could easily guess where the decision boundary will end up, but now we cannot guess it. Logistic regression does not work here. We will try to force-fit one and see the results. The code below builds a logistic regression line and plots the decision boundary on top of the data.

```python
model = sm.logit(formula='Purchase ~ Age+Experience', data=Emp_Purchase_raw)
fitted = model.fit()
fitted.summary2()

# getting slope and intercept of the line
slope=fitted.params[1]/(-fitted.params[2])
intercept=fitted.params[0]/(-fitted.params[2])

##Accuracy and error of the model1
predicted_values=fitted.predict(Emp_Purchase_raw[["Age"]+["Experience"]])
predicted_values[1:10]

#Lets convert them to classes using a threshold
threshold=0.5
threshold

predicted_class=np.zeros(predicted_values.shape)
predicted_class[predicted_values>threshold]=1

#Predicted Classes
predicted_class[1:10]
```

```python
from sklearn.metrics import confusion_matrix as cm
ConfusionMatrix = cm(Emp_Purchase_raw[['Purchase']],predicted_class)
print(ConfusionMatrix)
accuracy=(ConfusionMatrix[0,0]+ConfusionMatrix[1,1])/sum(sum(Confusion
Matrix))
print(accuracy)

error=1-accuracy
error

fig = plt.figure()
ax = fig.add_subplot(111)
plt.rcParams["figure.figsize"] = (8,7)
plt.title('Decision Boundary - Overall Data', fontsize=20)

ax.scatter(Emp_Purchase_raw.Age[Emp_Purchase_raw.Purchase==0],Emp_Purchase_
raw.Experience[Emp_Purchase_raw.Purchase==0],  s=100,  c='b',  marker="o",
label='Purchase 0')
ax.scatter(Emp_Purchase_raw.Age[Emp_Purchase_raw.Purchase==1],Emp_Purchase_
raw.Experience[Emp_Purchase_raw.Purchase==1], s=100, c='r', marker="x", label=
'Purchase 1')
plt.xlim(min(Emp_Purchase_raw.Age), max(Emp_Purchase_raw.Age))
plt.ylim(min(Emp_Purchase_raw.Experience), max(Emp_Purchase_raw.Experience))
plt.legend(loc='upper left');

x_min, x_max = ax.get_xlim()
ax.plot([0, x_max], [intercept, x_max*slope+intercept],linewidth=5)
plt.show()
```

Figure 8.10 shows the code output.

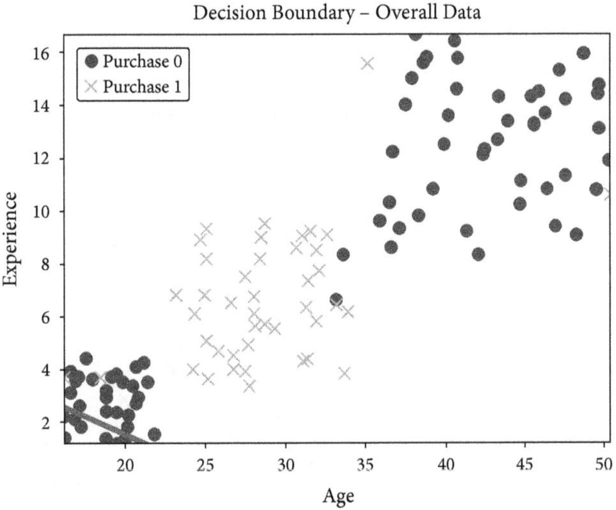

FIGURE 8.10   The decision boundary can be seen near the origin.

In the output presented in Fig. 8.10, you can find the decision boundary on the bottom-left side, near to the origin. It is visibly apparent that this decision boundary cannot separate the classes. We conclude this section with a takeaway: a single logistic regression line does not work in datasets with multiple decision boundaries.

8.4 *MULTIPLE DECISION BOUNDARIES SOLUTION*

If there are multiple decision boundaries in data, then directly predicting the target with input variables does not work. We are sure we need a better classification model. In the plot with full data (Figs. 8.11 and 8.12), if we consider only region-1 (R1), then logistic regression works perfectly to separate the two classes. In the same way, if we consider region-2 (R2), then also logistic regression does the best job of separating both the classes. We will build a logistic regression model-1 for region-1 to get the predicted values. These predicted values will be our intermediate output h_1. Similarly, we will build another model for region-2; this will be our second intermediate output h_2. Finally, we will use these intermediate outputs h_1 and h_2 for the prediction of y. Instead of building one logistic regression line x_1, x_2 versus y, we are now building three logistic regression lines. They are x_1, x_2 versus h_1; x_1, x_2 versus h_2; and finally h_1, h_2 versus y. Since we are changing the region, the values of x_1 and x_2 are different in the intermediate models for h_1 and h_2.

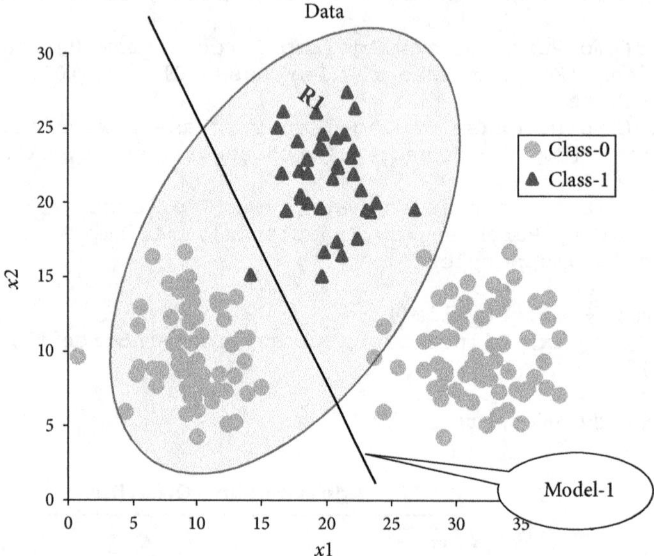

FIGURE 8.11   Model-1 decision boundary in region-1.

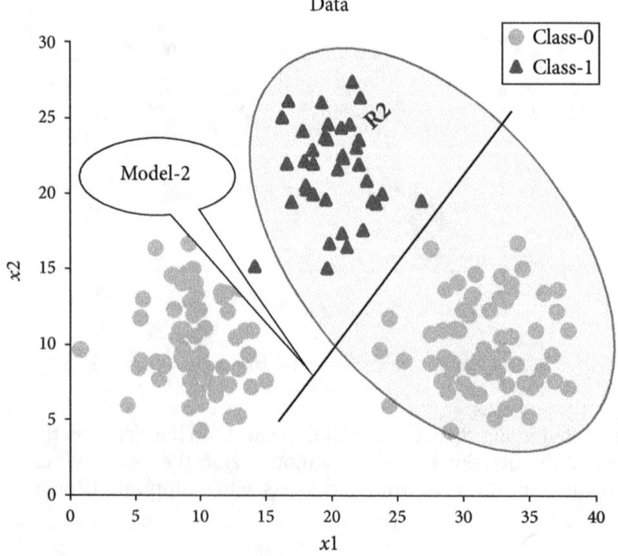

FIGURE 8.12   Model-2 decision boundary in region-2.

Given below are the three models that we are building.
Model-1 in region-1 gives intermediate output-1:

$$h_1 = \frac{e^{w_{01}+w_{11}x_1+w_{21}x_2}}{1+e^{w_{01}+w_{11}x_1+w_{21}x_2}}$$

Model-2 in region-2 gives intermediate output-2:

$$h_2 = \frac{e^{w_{02}+w_{12}x_1+w_{22}x_2}}{1+e^{w_{02}+w_{12}x_1+w_{22}x_2}}$$

Model-3 uses the intermediate outputs to predict the final target:

$$y = \frac{e^{w_0+w_1h_1+w_2h_2}}{1+e^{w_0+w_1h_1+w_2h_2}}$$

Figure 8.13 shows the network representation of the above model equations.

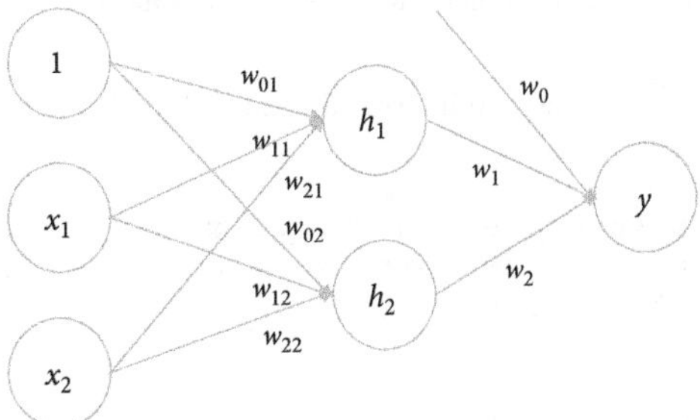

FIGURE 8.13   Network representation of the model equations.

If we were to build a single logistic regression line to predict the target variable, then we needed to find only three weights. However, that model did not work due to nonlinearity in the data. We now have to find inclusive nine weights—three weights associated with h_1, three weights for h_2, and finally three more from h_1,h_2 versus y.

8.4.1 Building Intermediate Output Models

We will get back to our dataset in the exercise. In this data also, we can consider region-1 and region-2 to build the two intermediate models. There is a column sample_set in the dataset; this column will help us in separating these two regions. Previously, in our first exercise of the decision boundary, we had already worked on region-1 and created a decision boundary (Fig. 8.7).

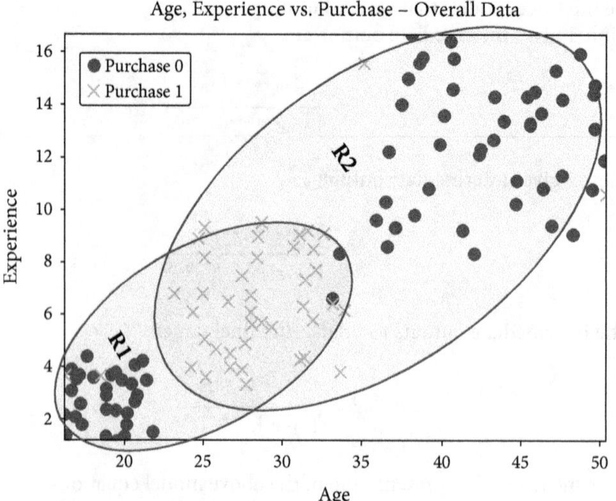

FIGURE 8.14   Regions to build intermediate models.

The code below helps us in creating two intermediate models and creates the two intermediate outputs *h*1 and *h*2.

```
## h1 model
Emp_Purchase1=Emp_Purchase_raw[Emp_Purchase_raw.Sample_Set<3]
model1 = sm.logit(formula='Purchase ~ Age+Experience', data=Emp_Purchase1)
fitted1 = model1.fit(method="bfgs")

##Predictions
Emp_Purchase_raw['h1']=fitted1.predict(Emp_Purchase_raw[["Age"]+
["Experience"]])

## h2 model
Emp_Purchase2=Emp_Purchase_raw[Emp_Purchase_raw.Sample_Set>1]
model2 = sm.logit(formula='Purchase ~ Age+Experience', data=Emp_Purchase2)
fitted2 = model2.fit(method="bfgs")

##Predictions
Emp_Purchase_raw['h2']=fitted2.predict(Emp_Purchase_raw[["Age"]+
["Experience"]])

##h1 and h2 in the data
print(Emp_Purchase_raw[['Age', 'Experience','h1','h2','Purchase']])
```

The code above gives us the output as follows.

```
print(Emp_Purchase_raw[['Age', 'Experience','h1','h2','Purchase']])
```

```
      Age   Experience         h1         h2   Purchase
0    20.0          2.3   0.114232   0.999578          0
1    16.2          2.2   0.040805   0.999910          0
2    20.2          1.8   0.092027   0.999595          0
3    18.8          1.4   0.051521   0.999790          0
4    18.9          3.2   0.139552   0.999661          0
..    ...          ...        ...        ...        ...
114  48.8         15.9   0.999999   0.000865          0
115  48.3          9.1   0.999944   0.005522          0
116  45.4         14.3   0.999994   0.004948          0
117  40.8         15.8   0.999992   0.021197          0
118  49.5         10.8   0.999985   0.002266          0
```

In this output, we can see the values of $h1$ and $h2$. These are the predictions made from the logistic regression lines. Before we go ahead with model building, we will plot $h1$, $h2$ versus the target variable. The code below helps us in plotting the target against $h1$ and $h2$.

```
fig = plt.figure()
ax = fig.add_subplot(111)
plt.rcParams["figure.figsize"] = (8,6)
plt.title('h1, h2 vs target ', fontsize=20)

ax.scatter(Emp_Purchase_raw.h1[Emp_Purchase_raw.Purchase==0],Emp_Purchase_raw.
h2[Emp_Purchase_raw.Purchase==0], s=100, c='b', marker="o", label='Purchase 0')
ax.scatter(Emp_Purchase_raw.h1[Emp_Purchase_raw.Purchase==1],Emp_Purchase_raw.
h2[Emp_Purchase_raw.Purchase==1], s=100, c='r', marker="x", label='Purchase 1')
ax.set_xlabel('h1',fontsize=15)
ax.set_ylabel('h2',fontsize=15)

plt.xlim(min(Emp_Purchase_raw.h1), max(Emp_Purchase_raw.h1)+0.2)
plt.ylim(min(Emp_Purchase_raw.h2), max(Emp_Purchase_raw.h2)+0.2)

plt.legend(loc='lower left');
plt.show()
```

Figure 8.15 is the output.

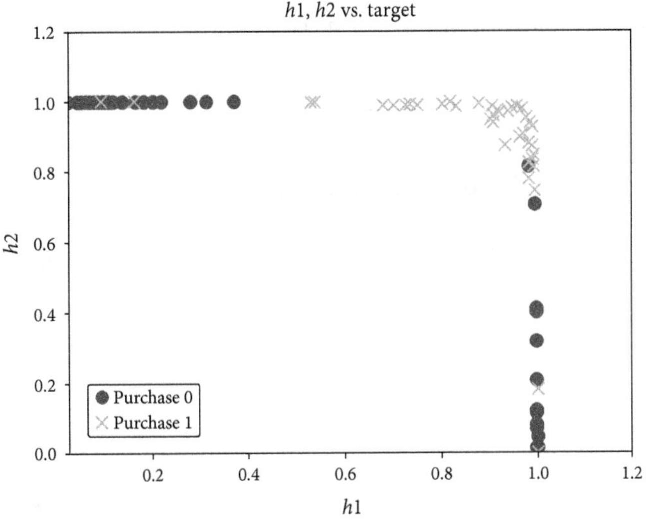

FIGURE 8.15   The target plotted for different values of $h1$ and $h2$.

Look at the output. Observe it and try to answer this question. Can we separate the classes now? Can we now build a logistic regression line that gives us a decision boundary between two classes? In the plot, we observe that the input variables $h1$ and $h2$ can classify the target variable y. We can draw one straight line that separates Class-0 and Class-1. If we build a logistic regression model, it may appear on the top-right corner like a diagonal line. Let us draw the decision boundary.

```
###Logistic Regression model with Intermediate outputs as input
model_combined = sm.logit(formula='Purchase ~ h1+h2', data=Emp_Purchase_raw)
fitted_combined = model_combined.fit(method="bfgs")
fitted_combined.summary()

# getting slope and intercept of the line
slope_combined=fitted_combined.params[1]/(-fitted_combined.params[2])
intercept_combined=fitted_combined.params[0]/(-fitted_combined.params[2])
```

```
##Finally, draw the decision boundary for this logistic regression model
fig = plt.figure()
ax2 = fig.add_subplot(111)
plt.rcParams["figure.figsize"] = (8,7)
plt.title('h1, h2 vs target ', fontsize=20)

ax2.scatter(Emp_Purchase_raw.h1[Emp_Purchase_raw.Purchase==0],Emp_Purchase_
raw.h2[Emp_Purchase_raw.Purchase==0],  s=100,  c='b',  marker="o",  label=
'Purchase 0')
ax2.scatter(Emp_Purchase_raw.h1[Emp_Purchase_raw.Purchase==1],Emp_Purchase_raw.
h2[Emp_Purchase_raw.Purchase==1], s=100, c='r', marker="x", label='Purchase 1')
ax2.set_xlabel('h1',fontsize=15)
ax2.set_ylabel('h2',fontsize=15)

plt.xlim(min(Emp_Purchase_raw.h1), max(Emp_Purchase_raw.h1)+0.2)
plt.ylim(min(Emp_Purchase_raw.h2), max(Emp_Purchase_raw.h2)+0.2)

plt.legend(loc='lower left');

x_min, x_max = ax2.get_xlim()
y_min,y_max=ax2.get_ylim()
ax2.plot([x_min, x_max], [x_min*slope_combined+intercept_combined, x_max*slope_
combined+intercept_combined],linewidth=4)
plt.show()
```

The code above gives us the output, as shown in Fig. 8.16.

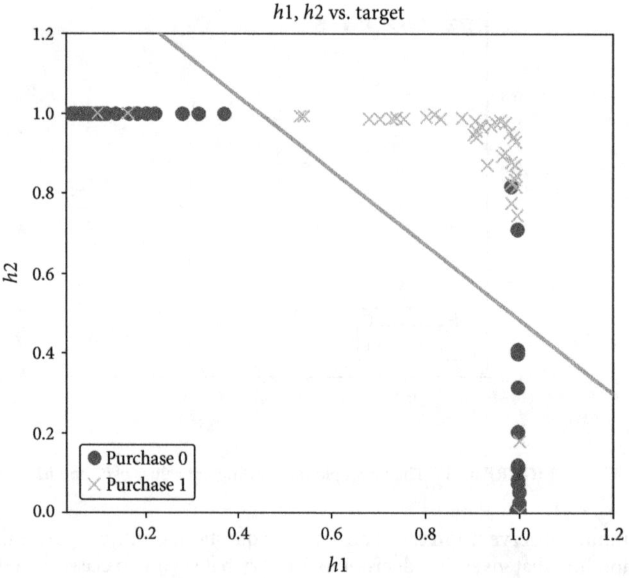

FIGURE 8.16 Decision boundary for the final logistic regression model.

The below code gives us the accuracy of the model shown in Fig. 8.16.

```
#######Accuracy and error of the final model
predicted_values=fitted_combined.predict(Emp_Purchase_raw[["h1"]+["h2"]])
```

```
#Lets convert them to classes using a threshold
threshold=0.5
threshold

predicted_class=np.zeros(predicted_values.shape)
predicted_class[predicted_values>threshold]=1

#ConfusionMatrix
from sklearn.metrics import confusion_matrix as cm
ConfusionMatrix = cm(Emp_Purchase_raw[['Purchase']],predicted_class)
print(ConfusionMatrix)
accuracy=(ConfusionMatrix[0,0]+ConfusionMatrix[1,1])/sum(sum(Confusion
Matrix))
print(accuracy)
```

The code above gives us the following output.

```
ConfusionMatrix
 [[74  2]
 [ 4 39]]
accuracy
 0.9495798319327731
```

The decision boundary appears as expected. We have now solved the problem of multiple decision boundaries using the intermediate output models. We have transformed the x_1, x_2 versus y data into a different space of h_1, h_2 versus y. In the first case, x_1, x_2 versus y, we could not separate the classes with a straight-line decision boundary. When these input variables are transformed into intermediate variables, we can now find the separating boundary.

To solve our problem of nonlinear or multiple decision boundaries, we used a layered logistic regression line approach. The first layer of logistic regressions is from input to an intermediate output. The second layer is from intermediate output to the final output.

The final target is a function of intermediate outputs h_1 and h_2:

$$y = g\left(\sum w_i h_i\right)$$

But h_1 and h_2 are a function of x_1 and x_2:

$$h_i = g\left(\sum w_{ji} x_i\right)$$

Which means y is indeed a function of x_1 and x_2:

$$y = g\left(\sum w_i \left(g\left(\sum w_{ji} x_i\right)\right)\right)$$

Figure 8.17 shows the network representation of the above equation.

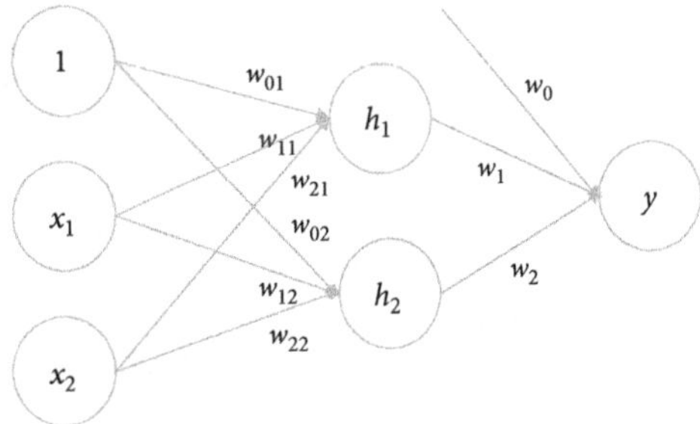

FIGURE 8.17   The network representation.

We have finally classified y with high accuracy. We could achieve more than 90 percent accuracy with three models. If there is a high degree of nonlinearity, then we may need more of these intermediate models in the first layer. We needed two decision boundaries for the classification of this data. The code below helps us in visualizing the decision boundaries from the two intermediate output models h_1 and h_2.

```
##The two decision boundaries
slope1=fitted1.params[1]/(-fitted1.params[2])
intercept1=fitted1.params[0]/(-fitted1.params[2])

slope2=fitted2.params[1]/(-fitted2.params[2])
intercept2=fitted2.params[0]/(-fitted2.params[2])

fig = plt.figure()
ax1 = fig.add_subplot(111)
plt.rcParams["figure.figsize"] = (8,6)
plt.title('Age, Experience  vs Purchase - Overall Data', fontsize=20)

ax1.scatter(Emp_Purchase_raw.Age[Emp_Purchase_raw.Purchase==0],Emp_Purchase_
raw.Experience[Emp_Purchase_raw.Purchase==0], s=100, c='b', marker="o", label=
'Purchase 0')
ax1.scatter(Emp_Purchase_raw.Age[Emp_Purchase_raw.Purchase==1],Emp_Purchase_
raw.Experience[Emp_Purchase_raw.Purchase==1], s=100, c='r', marker="x", label=
'Purchase 1')
ax1.set_xlabel('Age',fontsize=15)
ax1.set_ylabel('Experience',fontsize=15)

plt.xlim(min(Emp_Purchase_raw.Age), max(Emp_Purchase_raw.Age))
plt.ylim(min(Emp_Purchase_raw.Experience),                max(Emp_Purchase_raw.
Experience))

x_min, x_max = ax1.get_xlim()
ax1.plot([0, x_max], [intercept1, x_max*slope1+intercept1],linewidth=4)
ax1.plot([0, x_max], [intercept2, x_max*slope2+intercept2],linewidth=4)

plt.legend(loc='upper left');
plt.show()
```

This code gives us the output shown in Fig. 8.18.

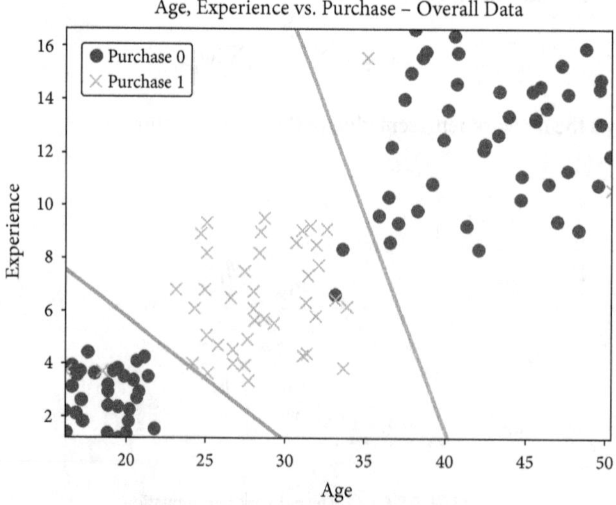

FIGURE 8.18   The code output: multiple boundaries between the two classes.

From the output, we observe that the two models in layer-1 do an excellent job of finding the overall boundaries between the two classes. This is the point we wanted to make. In essence, we used a layered modeling approach and created intermediate outputs to solve the problem of nonlinear data or a dataset with multiple decision boundaries. By increasing the number of intermediate outputs, we can solve any level of intricate nonlinear patterns.

A critical question that comes to our mind: how can we divide the data in a practical scenario? In real-life datasets, we CANNOT draw the data like this. Here we have just two variables; in real-life datasets, we have multiple variables. We cannot plot the graph in a two-dimensional space with multiple independent variables. Also, we cannot handpick the regions and the intermediate models. How do we find all the weights associated with the first layer and second layer in such a case? It is impossible to plot the multidimensional data and guess the number of decision boundaries. We use a neural network to solve this problem. A neural network automatically finds all the weights in one shot without us manually building models one by one. Until now, whatever we have discussed in this chapter cannot be categorized under neural networks. However, we are going to use these insights to understand the concepts explained hereafter. We are now going to get into the details of the neural network models.

8.5   NEURAL NETWORK INTUITION

A neural network has three primary layers: an input layer, a hidden layer, and an output layer. Figure 8.19 is a simple neural network diagram with two input variables and one hidden layer.

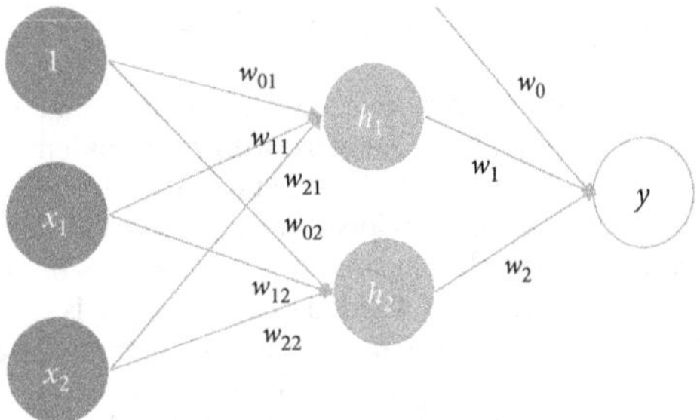

FIGURE 8.19   A simple neural network diagram.

The input layer contains input variables; the number of nodes in the input layer depends on the number of input variables. The input layer is connected to a hidden layer. We can relate it to the intermediate output in our previous example. The number of nodes in the hidden layer depends on the nonlinearity or the complexity in the data. If there is a clear separation between the classes, then we need fewer hidden nodes. If the data is very involved with many decision boundaries, then we need many hidden nodes. The output layer has the target variable. All these connections from the input layer to the hidden layer and hidden layer to the output layer are made using an activation function $g(x)$. The activation function is sigmoid for the classification type of problems and a linear function for regression problems. There are some other activation functions as well. We will discuss them later.

Building a neural network model is nothing but finding the weights. Once we have the weights, we can use the weights to construct the model equation (as given below) to get predictions. Since there can be multiple nodes in the input layer and the hidden layer, it is difficult to write an equation with all the weights. Neural network models are represented using the network diagrams. We do not write model equations like linear and logistic regression.

$$y = g\left(\sum w_i \left(g\left(\sum w_{ji} x_i\right)\right)\right)$$

8.5.1   Hidden Layers and Hidden Nodes

Imagine a dataset where there is a clear separation between Class-0 and Class-1. Then how many hidden nodes will we require for the dataset depicted in the plot in Fig. 8.20?

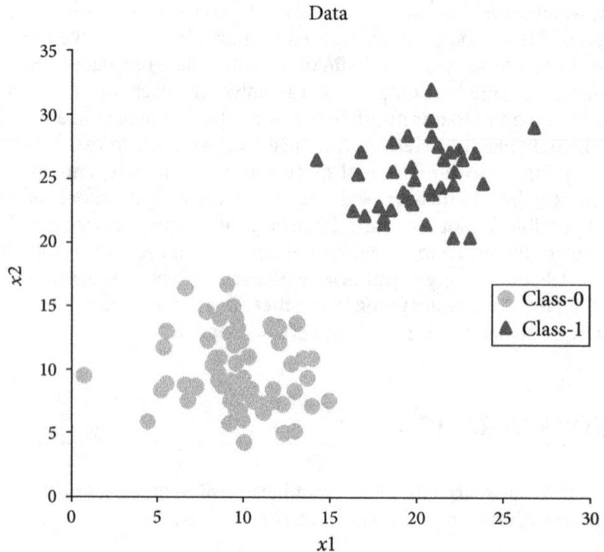

FIGURE 8.20   Class-0 and Class-1 clusters for an imaginary dataset.

In this case, we do not need any hidden nodes; we can use a standard logistic regression to make our predictions directly. A logistic regression line is a neural network with zero hidden layers. While building neural network models, apart from supplying the data on X and Y, we also need to configure the neural network by mentioning the number of hidden layers and hidden nodes. If the classes form two separable clusters, as shown in the plot in Fig. 8.20, then we do not need any hidden layer; it is the simplest possible case. If there is a bit more complexity in the data, then we probably need only one hidden layer with a couple of hidden nodes. If there is much complexity in the data, then we may need more than one hidden layer with several hidden nodes in each of them. Image processing and computer vision problems take pixel values as input and perform, for example, the final classification of object detection or facial recognition. In such cases, we cannot work with simple architecture; we need multiple hidden layers with several hidden nodes.

8.5.1.1 *Fine-tuning Hidden Nodes*   For the sake of discussion, let us first imagine all our neural networks have a single hidden layer and multiple hidden nodes. How do we decide the number of nodes? Now, imagine that a particular dataset requires six hidden nodes for an optimal solution. If we build that model with 50 hidden nodes, then the model will be overfitted. We can always check the model accuracy with train and test data to detect overfitting. However, if we build the same model (requiring six nodes for an optimal solution) with just two hidden nodes, then the model will underfit. We cannot decide the right network configuration in one attempt. We need to build one model and fine-tune the number of hidden nodes using the binary search approach. For the same example, where the optimal hidden nodes are six, we follow the steps as given in Table 8.2 to arrive at it.

TABLE 8.2   Fine-tuning Hidden Nodes

Model	Hidden Nodes	Result	Comment
M1	0	Train Accuracy—50% Test Accuracy—50%	M1 is underfitted; build model M2 by considering large number of hidden nodes
M2	50	Train Accuracy—100% Test Accuracy -55%	M2 is overfitted; build model M3 with number of hidden nodes = 25
M3	25	Train Accuracy—97% Test Accuracy—54%	M3 is still overfitted; build model M4 with a number of hidden nodes = 12
M4	12	Train Accuracy—92% Test Accuracy—76%	M4 is overfitted; build model M5 with a number of hidden nodes = 6
M5	6	Train Accuracy—85% Test Accuracy—84%	The M5 model looks perfect. To be on the safer side, we can also check with hidden nodes = 7 or 8

It took five attempts for us to build the final model with the optimal number of hidden nodes. Next time when we build a model with similar data, we can directly start with six or seven hidden nodes. We need to start by fixing a number for the hidden nodes, build a neural network model, and then fine-tune the number of hidden nodes to arrive at the most optimal model.

8.5.1.2 *Hidden Layer vs. Hidden Nodes*   Until now, we have primarily discussed a neural network with a single hidden layer. Simple logistic regression solves many real-life business problems. It is a neural network with zero hidden layers. If we add one more layer and many nodes, we can solve more complex problems. We can fit a model to any data by adding more and more hidden nodes to a single hidden layer. Alternatively, we can add one more hidden layer. Generally speaking, adding one more layer is more effective than adding more nodes in a single layer. For example, what we can achieve with a single layer with 15 nodes can be achieved with two layers of 4 nodes. To achieve the same accuracy with one hidden layer, we may need to add an exponentially higher number of nodes in that single hidden layer.

Theoretically, we can model any data by using a single hidden layer, but it requires an exponentially higher number of nodes. It is better to build a neural network with multiple layers for highly complex problems. However, fine-tuning these layers and nodes may take much effort. In later chapters, we will introduce some better methods of fine-tuning neural network models. Banking, finance, insurance, and many other customer-related data in real-life business scenarios may not require multiple layers. However, computer vision, self-driving cars, object detection, and many others may require multiple layers. In this chapter, we will use a neural network with a single layer. In upcoming chapters, we will deal with neural networks with multiple hidden layers (Fig. 8.21).

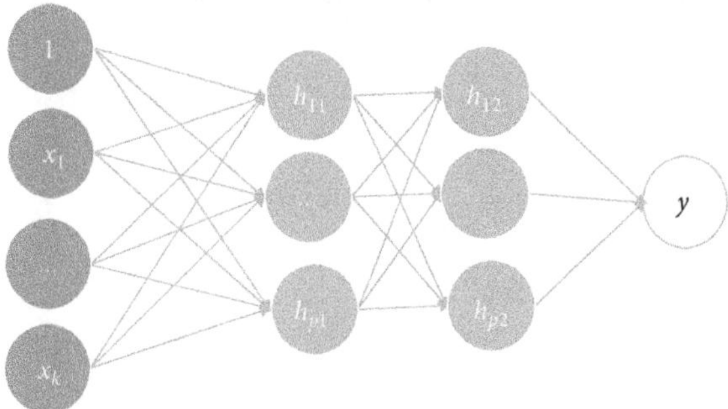

FIGURE 8.21   Neural networks with hidden layers.

The number of hidden layers and hidden nodes should be decided based on the nonlinearity in the data. The number of hidden nodes does not depend on the input variables. In some datasets, there are hundreds of variables, but there is a clear separation in the target variable. In such cases, simple logistic regression is sufficient, meaning a neural network with zero hidden nodes. In some datasets, we have a couple of variables, but there is much nonlinearity in the target. In such scenarios, we need several hidden nodes. To conclude, the number of hidden nodes is not a function of input variables, but it is a function of nonlinearity in the data.

8.6   NEURAL NETWORK ALGORITHM

Building a neural network model is all about finding the weights. The number of weights depends on the network configuration. If we are building a model with the configuration of [2,2,1]➔[2-input, 2-Hidden, 1-Output] nodes, then we need to calculate an inclusive nine weights. If we add one more node in the hidden layer, i.e., for the configuration [2,3,1], we need to calculate 13 weights. As shown earlier, we can get an idea of the number of weights by counting the edges in that network. Alternatively, you can use the below formula.

Single hidden layer formula:

$$Input\ nodes - n;\ Hidden\ nodes - k;\ Output\ nodes - r$$

$$Weights\ from\ input\ to\ hidden\ layer = (n+1) * k$$

$$\text{Weights from hidden layer to output layer} = (k+1) * r$$

$$\text{Overall weights} = (n+1) * k + (k+1) * r$$

Multiple hidden layers formula:

$$\text{Input nodes} - n;\ \text{Hidden nodes in layer } i - k_i;\ \text{Output nodes} - r$$

$$\text{Weights from input to hidden layer } 1 = (n+1) * k_1$$

$$\text{Weights from hidden layer 1 to hidden layer 2} = (k_1 + 1) * k_2$$

$$\text{Weights from hidden layer i to hidden layer } i+1 = (k_i + 1) * k_{i+1}$$

$$\text{Weights from final hidden layer to output layer} = (k_l + 1) * r$$

$$\text{Overall Weights} = (n+1) * k_1 + (k_1 + 1) * k_2 + \cdots + (k_i + 1) * k_{i+1} + \cdots (k_l + 1) * r$$

Table 8.3 shows us a few examples of networks and the number of weights.

TABLE 8.3   Examples of Networks and the Number of Weights

Input Nodes	Hidden Layers	Hidden Nodes List	Output Nodes	Weights Calculation	Weights
2	1	[2]	1	$((2+1)*2)$ $+((2+1)*1)$	9
2	1	[3]	1	$((2+1)*3)$ $+((3+1)*1)$	13
3	1	[2]	1	$((3+1)*2)$ $+((2+1)*1)$	11
3	1	[5]	1	$((3+1)*5)$ $+((5+1)*1)$	26
5	2	[3,3]	1	$((5+1)*3)$ $+((3+1)*3)$ $+((3+1)*1)$	34
10	3	[4,4,3]	1	$((10+1)*4)$ $+((4+1)*4)$ $+((4+1)*3)$ $+((3+1)*1)$	83

8.6.1 Neural Network Algorithm: Nontechnical

We need to supply data on X and Y along with the information on hidden layers. Once we configure the network, the neural network tries to find the weights using a backpropagation algorithm. The algorithm has five significant steps. At the end of this algorithm, we will have the optimal value of all the weights.

We will take a simple network to visualize these steps. Figure 8.22 shows our simple neural network.

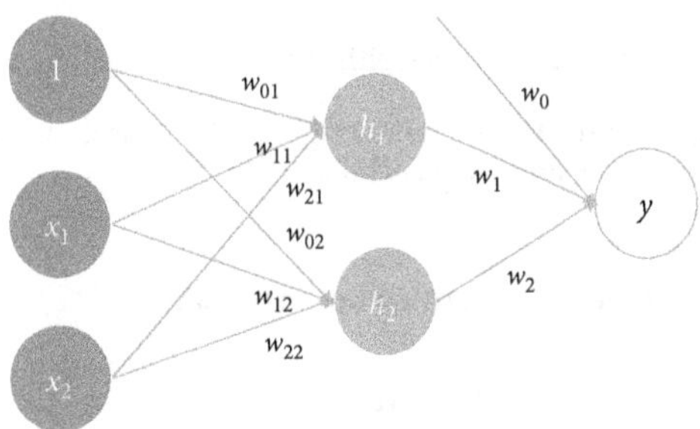

FIGURE 8.22   A simple neural network showing weights.

- *Step 1: Random initialization*
  - Initialize the weights by randomly selecting the values for the weights.
  - We can smartly initialize the weights if we have some prior information. In that case, the algorithm will converge faster.
  - Most of the time, the initialization is randomly chosen values from any standard distribution.
  - In our example, we initialize all nine weights w_{ij} and w_j.
- *Step 2: Activation and feed forward*
  - We take the data and use the current weights. We multiply weights with the x values, find the summation, and apply the activation function on them. The activation function is represented as $h_j = \left(\sum w_{ij} x_i\right)$.
  - We can now calculate the values of hidden nodes and further calculate the values of y from there.
  - We will have the predicted values of y at the end of this step.

$$\hat{y} = g\left(\sum w_j\left(g\left(\sum w_{ij} x_i\right)\right)\right)$$

- *Step 3: Error calculation and backpropagation*
  - By now, we have actual values and predicted values. We can calculate the error.
  - The error at the output layer is not entirely due to the wrong weights connected to that layer. We need to correct those weights first, and we also need to calculate the error contributions by hidden layers.
  - The target y is a function of $h1$ and $h2$. If there is some error at y, then we can find the error contribution by $h1$ and $h2$. For example, if $y = 80h1 + 20h2$, and if we know the error at y, then we can backtrack and find the error at $h1$ and $h2$. The output y is not always a simple linear function. We use the chain rule and partial derivatives to find the error contributions.
  - Finding the error at the output layer and propagating backward to find the error contribution at the hidden layers is known as the backpropagation step. Neural network algorithms are popularly known as the backpropagation algorithms.
- *Step 4: Weight updating*
  - Now we have the error at every hidden node. We will try to adjust the weights so that the error at hidden nodes decreases, which will, in turn, reduce the total error.
  - We will move the weights in that direction where the error reduces. If we have to increase the weights, we will increase; if we have to decrease the weights, we will decrease the weights.
  - This weight updating will be done using the gradient descent method: $w_{new} = w_{old} + \Delta w$.
  - At the end of this step, we will have a set of updated weights. From random weights, we have now moved to these new weights.

- *Step 5: Stopping criteria*
  - A full cycle of sending the whole data in the feed-forward step followed by error calculation and backpropagation followed by weight updating is one epoch.
  - At the end of epoch-1, we will have a new set of weights and slightly reduced errors. Nevertheless, we may not have the smallest total error yet.
  - We will again send the whole data and perform feedforward, error calculation, backpropagation, and weight upgradation steps.
  - We will repeat these epochs until we reach either zero error or when weights stop updating.

The calculation of error, backpropagation, and weight updating is done using some mathematical formulas, which we will take up for discussion in the following sections.

8.6.2   Neural Network Algorithm: Mathematical Formulas

$x_1, x_2, x_3 \ldots x_k$	Input variables
$h_1, h_2, h_3 \ldots h_j$	Hidden nodes (a single hidden layer)
y	Target variable
$w_{11}, w_{21} .. w_{ij}, w_1, w_2 .. w_j$	Weights

- *Step 1: Random initialization*
  - We can use uniform distribution or truncated normal distribution. Or, let them be any random values.
- *Step 2: Activation and feed forward*
  - $\hat{y} = g\left(\sum w_j h_j\right)$ where $h_j = g\left(\sum w_{ij} x_i\right)$
  - $\hat{y} = g\left(\sum w_j \left(g\left(\sum w_{ij} x_i\right)\right)\right)$
- *Step 3: Error calculation and backpropagation*
  - Consider squared error or cross-entropy error. Overall squared error $E = \dfrac{1}{2}\sum (y - \hat{y})^2$
  - Error gradient at the output layer $\delta = \hat{y}(1 - \hat{y})(y - \hat{y})$
  - Error gradient at the hidden layer $\delta_j = \hat{y}(1 - \hat{y}) w_j \delta$
- *Step 4: Weight updating*
  - Weight correction at the output layer $\Delta w_j = \eta h_j \delta$
  - Updated weights at the output layer $w_j := w_j + \Delta w_j$
  - Weight correction at the hidden layer $\Delta w_{ij} = \eta x_i \delta_j$
  - Updated weights at the hidden layer $w_{ij} := w_{ij} + \Delta w_{ij}$
- *Step 5: Stopping criteria*
  - Minimum error threshold criteria met
  - Error not reducing or weights not changing
  - Maximum epochs criteria reached

η is the learning rate parameter. We will discuss this in detail later.

8.6.3   Neural Network Algorithm: A Worked-Out Example

Let us consider a simple case of a neural network example and solve it using the formulas in Sec. 8.6.2 to get a better understanding of this algorithm. Table 8.4 provides the dataset and its visualization.

TABLE 8.4   The Dataset

x1	x2	Y
1	1	0
0	1	1
1	0	1
0	0	0

Figure 8.23 is the plot for the dataset in Table 8.4.

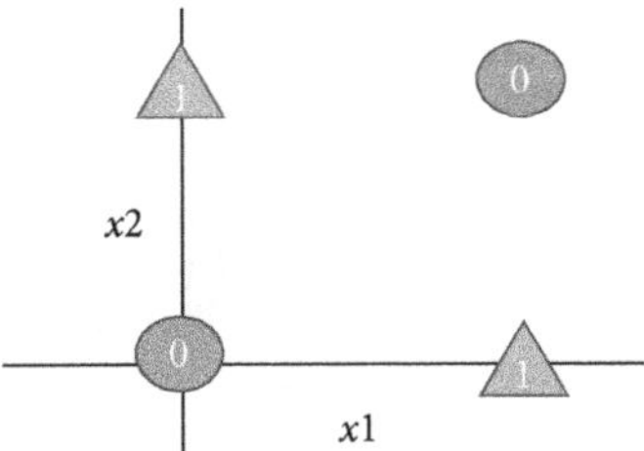

FIGURE 8.23   Plot of the dataset.

There are just four records and two input variables. We need two hidden nodes to separate the classes. As discussed earlier, the number of hidden nodes only depends on the nonlinearity in the data, not the number of input variables or the records. Figure 8.24 is the network diagram to solve the above neural network problem.

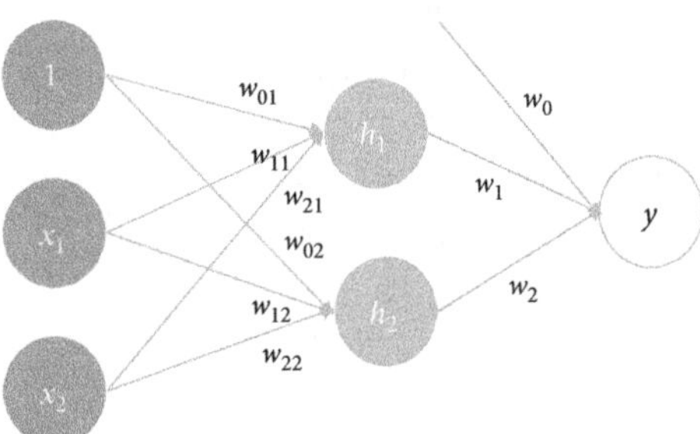

FIGURE 8.24   The network diagram to solve the neural network problem.

Let us follow the below steps and apply the formulas one by one.

- *Step 1: Random initialization*
  - We can use uniform distribution or truncated normal distribution. Alternatively, let them be any random values (Fig. 8.25).
  - $w_{01} = 1$, $w_{11} = 0.5$, $w_{21} = 0.5$, $w_{02} = 1$, $w_{12} = 1$, $w_{22} = -1$
  - $w_0 = 1, w_1 = -1, w_2 = 1$

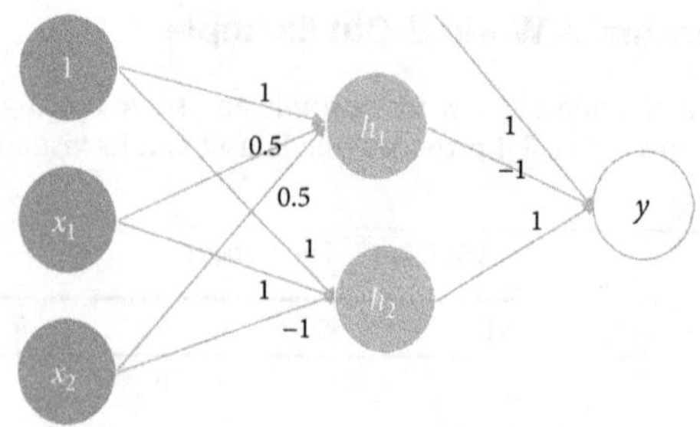

FIGURE 8.25   Random initialization of weights.

- *Step 2: Activation and feed forward*
  - Hidden layer activation is $h_j = g\left(\sum w_{ij} x_i\right)$.
  - Output layer activation is $\hat{y} = g\left(\sum w_j h_j\right)$.
  - The first data point is [1,1,0] (Fig. 8.26).
  - $x_1 = 1, \, x_2 = 1, \, y = 0.$
  - $h_1 = 0.8808, \, h_2 = 0.7311, \, \hat{y} = 0.7006.$

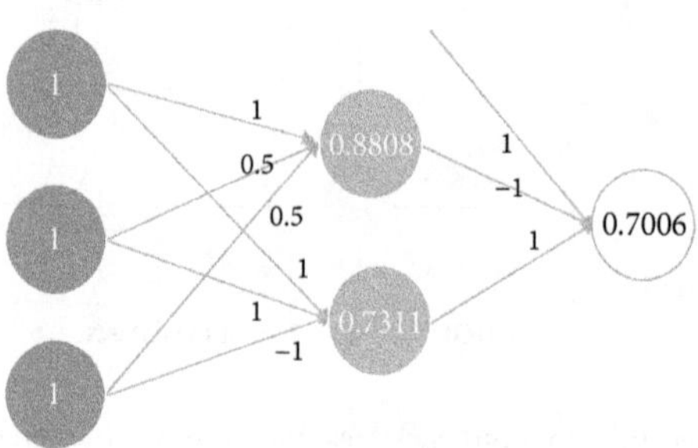

FIGURE 8.26   Activation and feed-forward step.

- *Step 3: Error calculation and backpropagation*
  - Error gradient at the output layer $\delta = \hat{y}(1 - \hat{y})(y - \hat{y})$
  - Weight correction at the output layer $\Delta w_j = \eta h_j \delta.$
  - Updated weights at the other layers $w_j := w_j + \Delta w_j.$
  - Error gradient at the other layers $\delta_j = \hat{y}(1 - \hat{y}) w_j \delta.$ (Fig. 8.27)
  - $\delta = -0.1470, \, \eta = 0.5$
  - $\Delta w_0 = -0.0735, \, \Delta w_1 = -0.0647, \, \Delta w_2 = -0.0537$
  - Updated weights $w_0 = 0.9265, \, w_1 = -1.0647, \, w_2 = 0.9463$
  - $\delta_1 = 0.0164, \, \delta_2 = -0.0273$

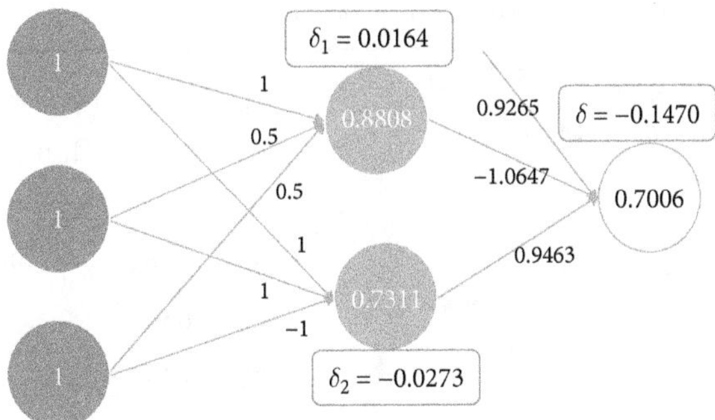

FIGURE 8.27   Network diagram resulting with error gradients updated from Step 3.

- *Step 4: Weight updating*
  - Weight correction at the output layer $\Delta w_{ij} = \eta x_i \delta_j$
  - Updated weights at the output layer $w_{ij} := w_{ij} + \Delta w_{ij}$ (Fig. 8.28).
  - $\Delta w_{01} = 0.0082,\ \Delta w_{11} = 0.0082,\ \Delta w_{21} = 0.0082,\ \Delta w_{02} = -0.0137,\ \Delta w_{12} = -0.0137,\ \Delta w_{22} = -0.0137$
  - $w_{01} = 1.0082,\ w_{11} = 0.5082,\ w_{21} = 0.5082,\ w_{02} = 0.9863,\ w_{12} = 0.9863,\ w_{22} = -1.0137$

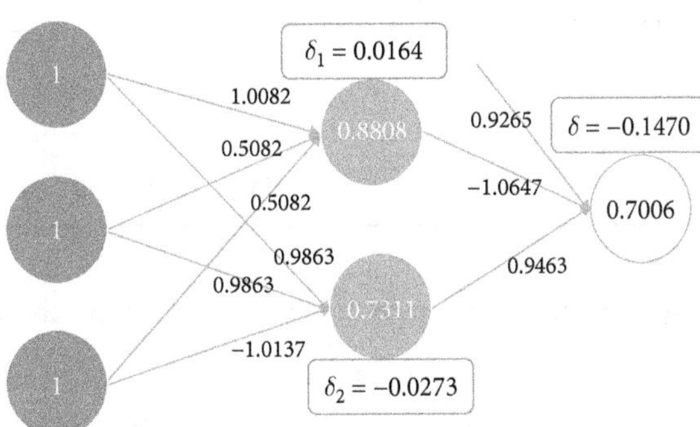

FIGURE 8.28   Network diagram resulting in updated weights from Step 4.

- *Step 5: Stopping criteria*
  - Minimum error threshold criteria.
  - Error not reducing or weights not changing.
  - Maximum epochs criteria.

TABLE 8.5   Multiple Iterations in Backpropagation

Epoch-0	Random initialization	$w_{01} = 1,\ w_{11} = 0.5,\ w_{21} = 0.5$ $w_{02} = 1,\ w_{12} = 1,\ w_{22} = -1,$ $w_0 = 1,\ w_1 = -1,\ w_2 = 1$
Epoch-1	Iteration-1 [1,1,0] (shown above)	$w_{01} = 1.0082,\ w_{11} = 0.5082,\ w_{21} = 0.5082$ $w_{02} = 0.9863,\ w_{12} = 0.9863,\ w_{22} = -1.0137$ $w_0 = 0.9265,\ w_1 = -1.0647,\ w_{02} = 0.9463$
	Iteration-2 [1,0,1]	$w_{01} = 1.0036,\ w_{11} = 0.5036,\ w_{21} = 0.5082$ $w_{02} = 0.9895,\ w_{12} = 0.9895,\ w_{22} = -1.0137$ $w_0 = 0.9545,\ w_1 = -1.0399,\ w_2 = 0.9728$
	Iteration-3 [0,1,1]	$w_{01} = 0.9998,\ w_{11} = 0.5036,\ w_{21} = 0.5045$ $w_{02} = 0.9956,\ w_{12} = 0.9895,\ w_{22} = -1.0076$ $w_0 = 0.9781,\ w_1 = -1.0197,\ w_2 = 0.9850$
	Iteration-4 [0,0,0]	....
Epoch-2	Iteration-1	.....
...	....	....
Epoch 1000	Iteration-4	$w_{01} = -2.1562,\ w_{11} = 4.3476,\ w_{21} = -4.778$ $w_{02} = 4.2160,\ w_{12} = 6.6021,\ w_{22} = -6.9031$ $w_0 = 3.2884,\ w_1 = 6.8628,\ w_2 = -6.8066$ $(y - \hat{y}) = 0$ and $\delta = 0$

Figure 8.29 is the classification diagram, with the final weights from Table 8.5 being used in creating decision boundaries.

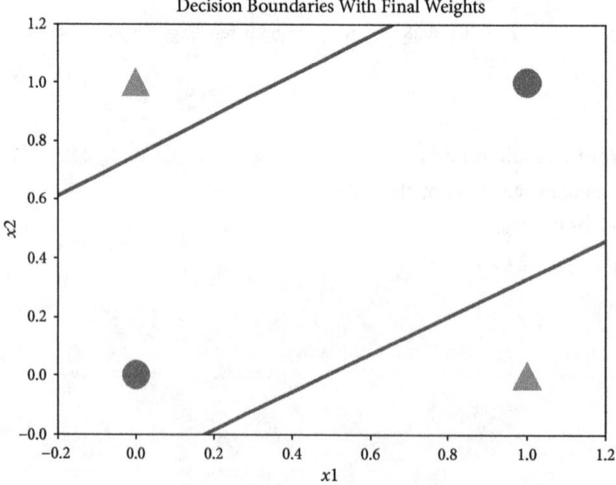

FIGURE 8.29   Classification diagram using final weights.

8.7   *THE CONCEPT OF GRADIENT DESCENT*

Look at the neural network algorithm as an optimization problem and try to find the optimal weights that will reduce the overall error.

$$\text{Minimize } E = \frac{1}{2}\sum(y - \hat{y})^2$$

$$\text{Mminimize } E = \frac{1}{2}\sum\left(y - g\left(\sum w_j h_j\right)\right)^2$$

$$\text{Find } w\text{'s that minimize } E = \frac{1}{2}\sum\left(y - g\left(\sum w_j\left(g\left(\sum w_{ij}x_i\right)\right)\right)\right)^2$$

We use the gradient descent method to solve the above optimization problem. We start with random weights and move the weights in that direction where the overall error reduces. The negative gradient is the direction in which the error reduces. You can imagine the error surface as the mountain range. With random initialization, we have started somewhere in that mountain range. We have to reach the ground. We have to move in the direction of the steepest descent, which is nothing but a negative gradient. The neural network algorithm is often called a gradient descent algorithm.

$$W := W + \Delta W$$

$$\Delta W \propto -\frac{\partial E}{\partial W}$$

The diagram in Fig. 8.30 shows an error surface and how weights are moving in the direction of the steepest descent.

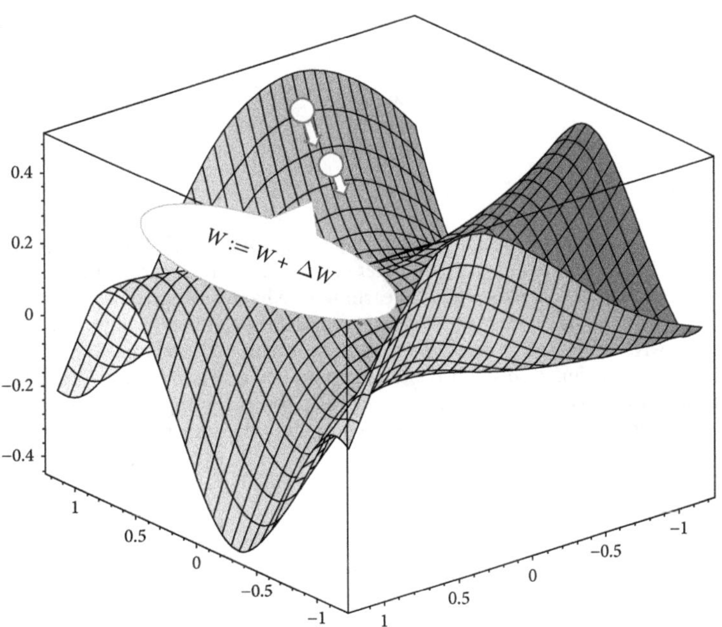

FIGURE 8.30   The error surface.

8.7.1   Gradient Descent for Regression

To understand and visualize the gradient, we will use a regression model example.

$$y = w_0 + w_1 x$$

$$\text{Minimize } E = \frac{1}{2}\sum(y - \hat{y})^2$$

$$\text{Minimize } E = \frac{1}{2}\sum\left(y - (w_0 + w_1 x)\right)^2$$

Calculate negative gradients with respect to weights:

$$\Delta W \propto -\frac{\partial E}{\partial W}$$

$$-\frac{\partial E}{\partial w_0} = -\sum\left(y - (w_0 + w_1 x)\right)$$

$$-\frac{\partial E}{\partial w_1} = -\sum x * \left(y - (w_0 + w_1 x)\right)$$

Weight updating formula:

$$W := W + \Delta W$$

$$w_0 := w_0 + \left(-\frac{\partial E}{\partial w_0}\right)$$

$$w_1 := w_1 + \left(-\frac{\partial E}{\partial w_1}\right)$$

8.7.2   Learning Rate

We make a small adjustment to the above weight updating formula. Instead of directly adding the weight correction ΔW, we multiply it with a learning rate parameter η. If $\eta = 1$ then we will get the same formula as above. We can pick and choose the learning rate. By keeping the learning rate handy, we can dictate how much weights should move in each iteration. A higher value of learning rate will make the weights take bigger jumps, and a lower value makes the weights move slowly. The learning should be neither too small nor too large. We have to fine-tune it. Here regression error function is a simple function; choosing the learning rate is relatively easy. When the error surface is complicated, like in neural networks, an optimal learning rate will help us in avoiding the local minima. Some tools assume a default value of the learning rate as 0.1 or 0.01. So the new weight updating formula is

$$W := W + \eta * \Delta W$$

$$w_0 := w_0 + \eta * \left(-\frac{\partial E}{\partial w_0}\right)$$

$$w_1 := w_1 + \eta * \left(-\frac{\partial E}{\partial w_1}\right)$$

We will repeat these weight updating steps until the error reaches zero.

8.7.3 Code of Gradient Descent for Regression

Let us take a sample dataset and solve it using the gradient descent method to find the weights. Below is the gradient descent function.

```python
def lr_gd(X, y, w1, w0, learning_rate, epochs):
    for i in range(epochs):
        y_pred = (w1 * X) + w0
        error = sum([k**2 for k in (y-y_pred)])

        ##Gradients
        w0_gradient = -sum(y - y_pred)
        w1_gradient = -sum(X * (y - y_pred))

        ##Weight Updating
        w0 = w0 - (learning_rate * w0_gradient)
        w1 = w1 - (learning_rate * w1_gradient)

    return error, w0, w1
```

We will use the above function to solve this regression problem, as presented below. We will generate some random data using the formula $y = 10 + 20x$. This means that after solving this regression line, we should get the weights as $[w_0 = 10$ and $w_1 = 20]$ as the output. Below is the code for data creation and solving it.

```python
##Using the GD function
x_data=np.random.random(10)
y_data= x_data*20 + 10

w0_init=5
w1_init=10

lr_gd(X=x_data,   y=y_data,   w1=w1_init,   w0=w0_init,   learning_rate=0.01,
epochs=600)
```

The above code gives us the below output.

```
epoch 0 error => 1295.14 w0 =>  6.11 w1 =>  10.74
epoch 1 error => 964.61 w0 =>  7.06 w1 =>  11.38
epoch 2 error => 719.44 w0 =>  7.88 w1 =>  11.94
epoch 3 error => 537.57 w0 =>  8.58 w1 =>  12.42
epoch 4 error => 402.64 w0 =>  9.18 w1 =>  12.84
epoch 5 error => 302.53 w0 =>  9.7 w1 =>  13.2
epoch 6 error => 228.25 w0 =>  10.14 w1 =>  13.52
epoch 7 error => 173.11 w0 =>  10.52 w1 =>  13.79
epoch 8 error => 132.19 w0 =>  10.84 w1 =>  14.03
epoch 9 error => 101.79 w0 =>  11.12 w1 =>  14.24
epoch 10 error => 79.22 w0 =>  11.36 w1 =>  14.43
epoch 11 error => 62.44 w0 =>  11.56 w1 =>  14.59
epoch 12 error => 49.95 w0 =>  11.73 w1 =>  14.73
epoch 13 error => 40.66 w0 =>  11.88 w1 =>  14.86
.......................................................................
epoch 584 error => 0.05 w0 =>  10.17 w1 =>  19.74
epoch 585 error => 0.05 w0 =>  10.17 w1 =>  19.74
epoch 586 error => 0.05 w0 =>  10.16 w1 =>  19.74
epoch 587 error => 0.05 w0 =>  10.16 w1 =>  19.74
epoch 588 error => 0.05 w0 =>  10.16 w1 =>  19.75
epoch 589 error => 0.05 w0 =>  10.16 w1 =>  19.75
epoch 590 error => 0.05 w0 =>  10.16 w1 =>  19.75
epoch 591 error => 0.04 w0 =>  10.16 w1 =>  19.75
```

```
epoch 592 error => 0.04 w0 =>  10.16 w1 =>  19.75
epoch 593 error => 0.04 w0 =>  10.16 w1 =>  19.75
epoch 594 error => 0.04 w0 =>  10.16 w1 =>  19.75
epoch 595 error => 0.04 w0 =>  10.16 w1 =>  19.75
epoch 596 error => 0.04 w0 =>  10.16 w1 =>  19.76
epoch 597 error => 0.04 w0 =>  10.16 w1 =>  19.76
epoch 598 error => 0.04 w0 =>  10.15 w1 =>  19.76
epoch 599 error => 0.04 w0 =>  10.15 w1 =>  19.76
```

We can see from the output that at 600 epochs the error is almost zero and the final weights given by GD are [w_0 = 10.15 and w_1 = 19.76]. This result is nearly perfect if we compare it to the correct value of weights [w_0 = 10 and w_1 = 20].

The error function in the above linear regression example is simple, and optimization is relatively easy. However, in neural networks, the error surface is very complex. The error function in the output layer depends on the weights of the output layer and the hidden layers before it. In the neural network models, we will not solve the above optimization in one shot. We will find the partial derivatives at the output layer, apply the chain rule, and find the error contribution by hidden layers in the backpropagation step. Later we update the weights based on the corresponding negative gradients.

8.7.4   Multiple Solutions for a Problem

The gradient descent algorithm in the neural network does not always give us the global minimum. The algorithm may exit after hitting a local minimum. To avoid the high error associated with local minima, we fine-tune the learning rate. Even after fine-tuning the learning rate, there is no guarantee that the algorithm will finally reach global minima. However, when we use optimal learning rates, we can end up at a local minimum that is very close to global minima. Two people building two neural network models on the same data can have different weights, but they may still have the same accuracy. Those two models would have ended up in two local minima. The diagram in Fig. 8.31 shows the error function with local and global minima on it.

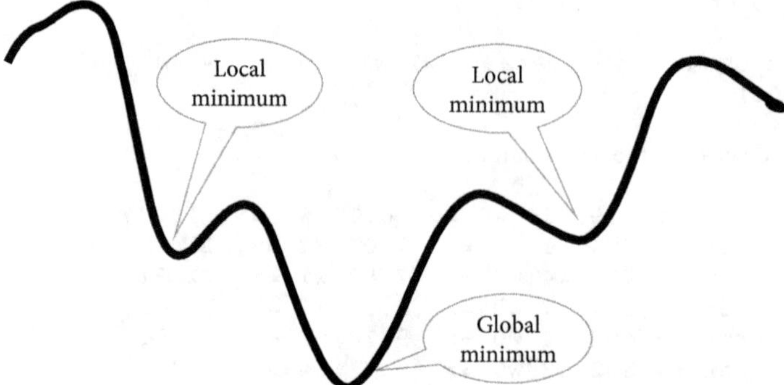

FIGURE 8.31   Local and global minima.

For example, Fig. 8.32 shows a neural network model result for a dataset. We solved it in Sec. 8.6.3.

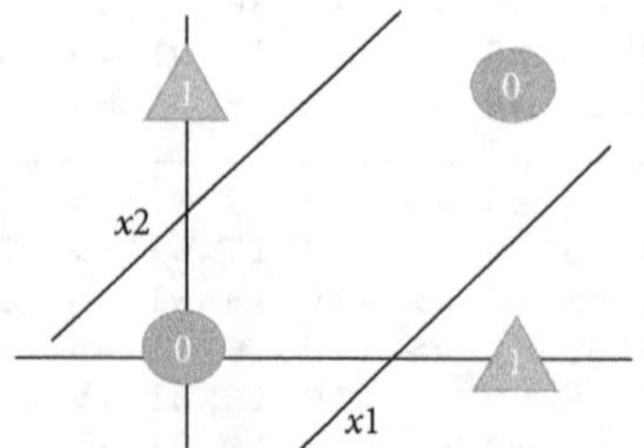

FIGURE 8.32   Neural network model-1 for a same dataset.

We will have a set of weights for the model in Fig. 8.32. This model has 100 percent accuracy. With the same 100 percent accuracy, we can build a model with entirely different weights.

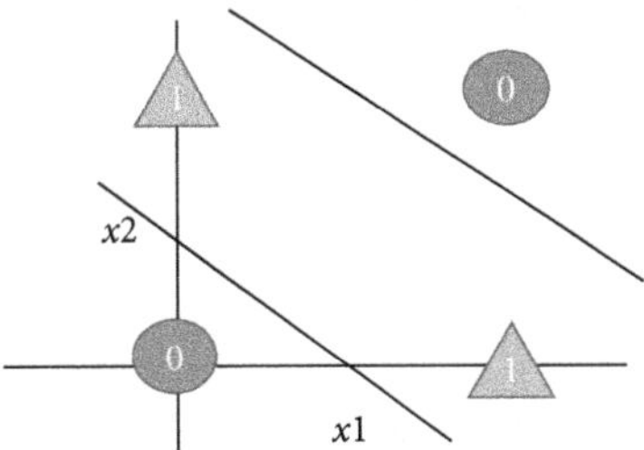

FIGURE 8.33 Neural network model-2 from the same dataset.

The two models in Figs. 8.32 and 8.33 have very different weights, but they have the same accuracy. A neural network may have multiple solutions at a given level of accuracy. Figure 8.34 shows a few more models with the same accuracy.

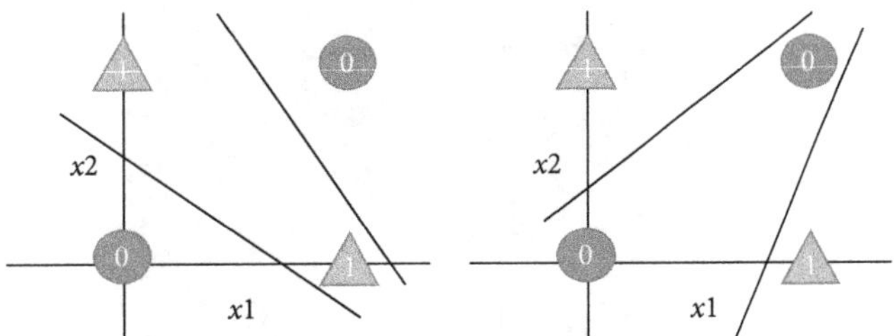

FIGURE 8.34 Neural network solutions from the same dataset.

Until now, we have used simple datasets to understand the concept of neural networks. The real use of the neural network is in finding the complex nonlinear patterns in the data. A neural network with two hidden nodes finds simple patterns in the data, whereas a neural network with 30 hidden nodes learns much more complex patterns in the data. Neural networks are known as universal function approximators. No matter what the pattern is in the data, we can model it using a neural network by configuring it with the right number of hidden nodes and activation functions.

8.8 CASE STUDY: RECOGNIZING HANDWRITTEN DIGITS

In this case study, we will take the images of handwritten digits and build a neural network model that takes the input as these handwritten digits and builds a model to predict the number inside that image. The applications of these models can be implemented, for example, for automatically reading number plates of the vehicles to track a vehicle or to detect traffic violations, reading the account number from a cheque to fasten the banking processes, and automatically reading the contents of a manually filled-out paper form.

8.8.1 Background and Objective

Take an image as input and predict the digit inside that image. An image is nothing but the collection of pixels. If we import and store an image in Python, it will get stored as an array of numbers. If we plot the image, we will see the image if we can print the array of pixel values. A color image is a three-dimensional array: number of pixels in rows (height of the image) and number of pixels in the columns (width of the image) are two dimensions. In each pixel the RGB intensity is the third dimension. Intensity values of red, green, and blue are between 0 and 255. If a pixel value is RGB(0,0,0), it will end up as a pure black pixel. Similarly, RGB(255,255,255) is a pure white pixel, and RGB(255,0,0) is red. Below is the code for importing an image and printing the pixel values (Fig. 8.35).

```
x=plt.imread(r'D:\Chapter8\ANN\5.Datasets\Sample_images\Marketvegetables.
jpg')

plt.rcParams["figure.figsize"] = (12,8)
plt.imshow(x)

print('Shape of the image',x.shape)
print(x)
```

The above code gives us the below output.

FIGURE 8.35   Figure output from the code.

```
print('Shape of the image',x.shape)
print(x)

Shape of the image (2400, 1600, 3)
[[[100   85   82]
  [103   89   86]
  [108   97   91]
  ...
  [111   88   74]
  [ 98   74   64]
  [105   80   73]]

 [[164  122  126]
  [128   87   91]
  [ 80   41   42]
  ...
  [ 98   75   59]
  [104   82   69]
  [105   83   72]]
```

```
[[213 171 173]
 [199 155 156]
 [170 121 124]
 ...
 [105  84  63]
 [109  90  73]
 [ 93  74  59]]

 ...
[[142 123 108]
 [150 131 117]
 [144 124 113]
 ...
 [ 32  22  20]
 [ 47  34  28]
 [ 49  26  20]]

[[140 122 100]
 [143 122 105]
 [142 120 106]
 ...
 [ 41  25  25]
 [ 53  29  29]
 [ 56  18  17]]

[[131 113  89]
 [132 111  90]
 [134 111  95]
 ...
 [ 30   8  10]
 [ 46  11  15]
 [ 59   6  12]]]
```

From the above output, we can see that the image has 2400 rows; those are the number of pixels in the height of the image. One thousand six hundred columns are the number of pixels in the width. The depth is 3, and this number 3 corresponds to RGB intensities. If we print the pixel values of the image, we can see row-by-row values, and in each cell, there are three numbers between 0 and 255.

In this case study, in our dataset, we are considering grayscale images. The grayscale images have height and width, but they do not have three values in depth. There will be one number in the color dimension for grayscale images. The objective in this case study is to take the grayscale image as input and predict the number inside it using the neural network model. Taking images as input indeed means taking pixel intensity numbers as input to predict the numbers.

8.8.2 Data

The data that we are using already has taken the images and converted them into an array of numbers and stored them inside a text file. In the above example, we have seen a 2400×1600 image. That image has 3,840,000 pixels. If we are taking each pixel as an input variable, then for a given image we will have 3.84 million input variables. It will be challenging for our system to build a model with 3.84 million input variables. Our digits dataset contains fewer pixels.

The data we are considering here has 16×16 pixels—overall, 256 pixels in each image. In our model, we will have 256 input variables: $x_1, x_2, x_3 \dots x_{256}$. Our data is a standard dataset known as USPS data; it stands for the United States Postal Services data. This data contains 7291 scanned images of handwritten digits, and these images are converted to CSV files by taking the pixel intensities.

AT&T research labs shared the USPS dataset. The other standard digits dataset is MNIST data with 28×28 pixels and 60,000 records. We will use this data later. We will use USPS data in this case study. We can download these datasets from Dr. Yann Le Cunn's website: http://yann.lecun.com/exdb/mnist/.

The below code helps in importing the data.

```
##USPS Data Importing

digits_data_raw  =  np.loadtxt(r"D:\Chapter8\ANN\5.Datasets\USPS\USPS_train.
txt")

##Input data is in nparry format. we convert it into dataframe for better
handling
digits_data=pd.DataFrame(digits_data_raw)

#Shape of the data
print(digits_data.shape)

##Details of the data
print(digits_data.head())
```

The above code gives us the below output.

```
print(digits_data.shape)
(7291, 257)


print(digits_data.head())
     0    1    2    3     4      5    ...     251     252     253     254     255  256
0  6.0 -1.0 -1.0 -1.0 -1.000 -1.000  ... -0.474 -0.991 -1.000 -1.000 -1.000 -1.0
1  5.0 -1.0 -1.0 -1.0 -0.813 -0.671  ...  0.762  0.126 -0.095 -0.671 -0.828 -1.0
2  4.0 -1.0 -1.0 -1.0 -1.000 -1.000  ...  1.000 -0.179 -1.000 -1.000 -1.000 -1.0
3  7.0 -1.0 -1.0 -1.0 -1.000 -1.000  ... -1.000 -1.000 -1.000 -1.000 -1.000 -1.0
4  3.0 -1.0 -1.0 -1.0 -1.000 -1.000  ...  0.791  0.439 -0.199 -0.883 -1.000 -1.0

[5 rows x 257 columns]
```

As expected, there are 256 pixel values. The extra column is the target column—the label associated with each image. The labels are stored in the first column. For example, the label for the image in the first row is 6. The code below gives us the frequency of each label in the dataset.

```
##Frequency count of target
print(digits_data[0:][0].value_counts())
```

The following is the above code's output.

```
print(digits_data[0:][0].value_counts())
0.0    1194
1.0    1005
2.0     731
6.0     664
3.0     658
4.0     652
7.0     645
9.0     644
5.0     556
8.0     542
Name: 0, dtype: int64
```

From this output, we can see that there are more than 1000 combinations of digit 0. Similarly, there are about 1000 combinations of digit 1. We will now see a few images. While creating this dataset, the 16×16 image is flattened to make it a single row in the data. We need to take a row from this data and build the pixel matrix of size 16×16. We will now portray the images with the help of the following code.

```
#first image
i=0
data_row=digits_data_raw[i][1:]
pixels = np.matrix(data_row)
pixels=pixels.reshape(16,16)
plt.title(["Row number ", i] , fontsize=20)
plt.imshow(pixels, cmap='Greys')

#second image
i=1
data_row=digits_data_raw[i][1:]
pixels = np.matrix(data_row)
pixels=pixels.reshape(16,16)
plt.title(["Row number ", i] , fontsize=20)
plt.imshow(pixels, cmap='Greys')
```

In this code, 'i' is the row number. By changing the value of row number (i) in the code above, we can portray a few more images. Figure 8.36 shows the output from this code.

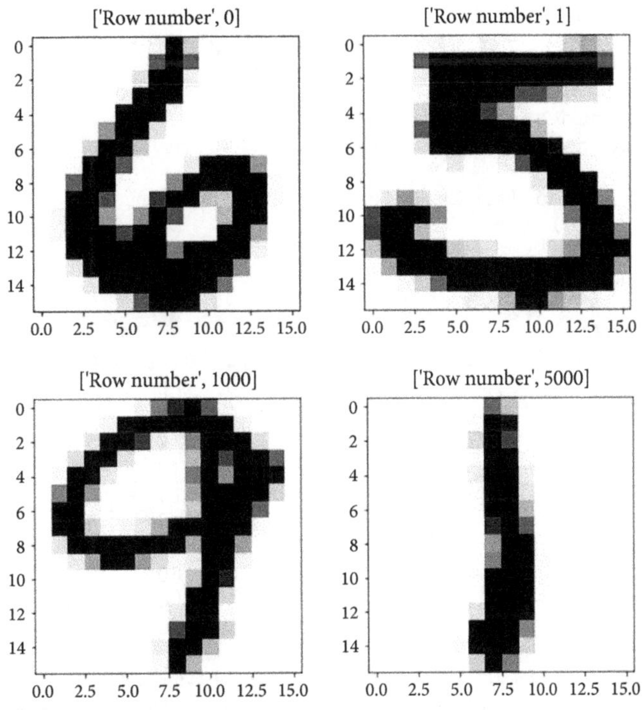

FIGURE 8.36   Output images of the code.

We will now prepare the data for model building.

```
##Train and Test data creation
X=digits_data.drop(digits_data.columns[[0]], axis=1)
y=digits_data[0:][0]

X_train, X_test, y_train, y_test = train_test_split(X, y ,test_size=0.2)

#Shape of the data
print("X_train shape", X_train.shape)
print("y_train shape", y_train.shape)
print("X_test shape", X_test.shape)
print("y_test shape", y_test.shape)
```

The code above gives us the following output.

```
X_train shape (5832, 256)
y_train shape (5832,)
X_test shape (1459, 256)
y_test shape (1459,)
```

8.8.3   Model Building

Before building the model, we need to make an essential data transformation. Until now, we have discussed just one binary output in our target variable. In this case, our output is not binary. There are 10 classes in our output. An image can contain any number between [0,9]. The output is multiclass, and we are going to build a multiclass classification model. We can extend our binary target classification concept to multiclass classification. Imagine that we are building a model that predicts whether the target is zero or not, which is a binary classification model. Imagine the second model that predicts whether the target is one or not. Similarly, we need to build 10 models; each model has a binary target, starting from zero or not zero, until 9 or not 9. It means we will now have 10 nodes in the output layer. There are 10 classes in the target, and we are going to create one binary variable for each class. It is simple one-hot encoding on the target variable. The code below helps us in creating the one-hot encoded variables for the target.

```
##Creating multiple binary columns for multiple outputs
digit_labels=pd.DataFrame()

#Convert target into onehot encoding
digit_labels = pd.get_dummies(y_train)

#see our newly created labels data
digit_labels.head(10)
```

The code above gives the following output.

```
digit_labels.head(10)
```

	0.0	1.0	2.0	3.0	4.0	5.0	6.0	7.0	8.0	9.0
7140	0	0	0	0	0	1	0	0	0	0
2853	0	1	0	0	0	0	0	0	0	0
3941	0	0	0	0	0	1	0	0	0	0
1896	0	1	0	0	0	0	0	0	0	0
3306	0	0	1	0	0	0	0	0	0	0
1548	0	0	0	0	0	0	0	1	0	0
3089	1	0	0	0	0	0	0	0	0	0
511	0	0	0	1	0	0	0	0	0	0
141	0	0	0	0	1	0	0	0	0	0
6188	0	0	0	0	0	0	0	1	0	0

In the above output table, we can see 10 newly created binary variables derived from one target column. Since there are 10 nodes in the output layer, for each new data point, we will get 10 predicted values—in the form of 10 probabilities. Out of these 10 probability values as output, we need to choose only one as our predicted value for the target. We will assign (or choose) the class that has the maximum probability. Now, we have the input layer with 256 nodes and an output layer with 10 nodes. We are all set to build the model. Before building the neural network model, we need to create a list with minimum and maximum values of all the input variables. We will use this list later on in the model. Given below is the code for creating a list with 256 pairs of values—one pair for each of the input variables.

```
#getting minimum and maximum of each column of x_train into a list
min_max_all_cols=[[X_train[i][0:].min(),  X_train[i][0:].max()]  for  i  in
range(1,X_train.shape[1]+1)]
```

```
print(len(min_max_all_cols))
print(min_max_all_cols)
```

The code output is as given below.

```
print(len(min_max_all_cols))
256
```

```
print(min_max_all_cols)
[[-1.0, 0.638], [-1.0, 1.0], [-1.0, 1.0], [-1.0, 1.0], [-1.0, 1.0], [-1.0, 1.0],
[-1.0, 1.0], [-1.0, 1.0], [-1.0, 1.0], [-1.0, 1.0], [-1.0, 1.0], [-1.0, 1.0],
[-1.0, 1.0], [-1.0, 1.0], [-1.0, 1.0], [-1.0, 0.752], [-1.0, 0.776], [-1.0,
1.0], [-1.0, 1.0], [-1.0, 1.0], [-1.0, 1.0], [-1.0, 1.0], [-1.0, 1.0], [-1.0,
1.0], [-1.0, 1.0], [-1.0, 1.0], [-1.0, 1.0], [-1.0, 1.0], [-1.0, 1.0], [-1.0,
1.0], [-1.0, 1.0], [-1.0, 0.997], [-1.0, 0.796], [-1.0, 1.0], [-1.0, 1.0],
...........................................................................................
[-1.0, 1.0], [-1.0, 1.0], [-1.0, 1.0], [-1.0, 1.0], [-1.0, 1.0], [-1.0, 1.0],
[-1.0, 1.0], [-1.0, 1.0], [-1.0, 1.0], [-1.0, 1.0], [-1.0, 0.539]]
```

Given below is the code for building the neural network model.

```
##Configure the network
import neurolab as nl
net = nl.net.newff(minmax=min_max_all_cols, size=[20,10], transf=[nl.trans.
LogSig()]*2)

##Training method is Resilient Backpropagation method
net.trainf = nl.train.train_rprop

##Train the network
net.train(X_train, digit_labels, show=1, epochs=300)
```

Table 8.6 provides the code explanation.

TABLE 8.6 The Code Explanation

`import neurolab`	Package for building neural networks.
`newff()`	Function to configure neural networks.
`minmax`	This parameter takes a list of lists as input. We need to supply the minimum and maximum value of all the input variables. It helps the algorithm in weights initialization.
`size=[20,10]`	Takes a list as input. We need to mention nodes in each layer except the input layer. [Nodes in hidden layer1, Nodes in hidden layer2,…, Nodes in hidden layer k, Nodes in output layer] `[20,10]`: One hidden layer with 20 nodes and an output layer with 10 nodes.
`Transf= nl.trans.LogSig()`	`Transf` is the activation function parameter. `LogSig()` is the sigmoid function's standard syntax. For regression output we can use `PureLin`.
`[nl.trans.LogSig()]*2`	`*2` denotes the two activations from input to hidden; hidden to output in this case. If we have two hidden layers, then we need to mention `*3`.
`net.trainf = nl.train.train_rprop`	Backpropagation method is mentioned in this step.
`net.train`	Function to fit the model. Takes training data as input.
`show=1`	`show=1`: Shows error value in each epoch. `show=0`: Builds the model directly. Epochs do not show errors.
`epochs`	The number of epochs. One epoch is one full run of the data. Mention epochs as a number between 50 and 500.

We now have a neural network with 256 nodes in the input layer, 20 nodes in the hidden layer, and 10 nodes in the output layer. Figure 8.37 is our neural network diagram.

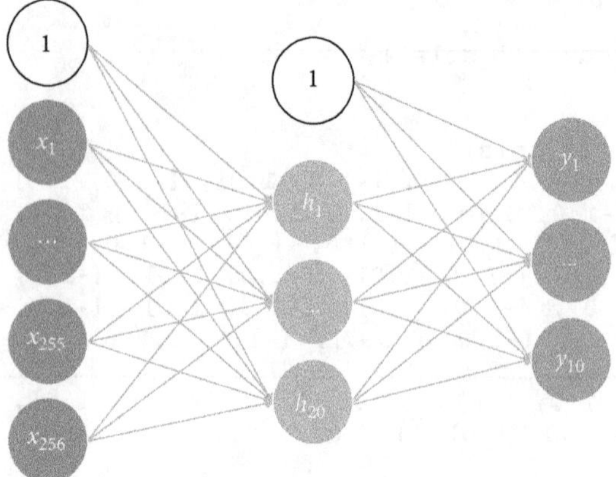

FIGURE 8.37   The resulting neural network diagram with 20 hidden nodes.

This network has several weights. Hence the training process will take much time. Let us calculate the weights to get an idea of the complexity of this optimization problem (Table 8.7).

TABLE 8.7   Count of Weights

Description	Calculation	Details
Input nodes	256	
Hidden nodes	20	
Input to hidden nodes weights	(256 + 1) * 20 = 5140	1 is the bias term
Hidden to output nodes weights	(20 + 1) * 10 = 210	1 is the bias term
Overall weights	5140 + 210 = 5350	

We need to calculate 5350 weights. There are 7291 records in the data. In each epoch, we will perform at least 5350 × 7291 calculations in the feed-forward and backpropagation steps. If there are 300 epochs, then the overall calculations will be at least 300 × 5350 × 7291 = 11,702,055,000. The actual number of calculations is even greater than this. The model in this case study may likely take around 5 to 10 minutes to converge. The model-building code given earlier gives us the following output.

```
Epoch: 1; Error: 9455.013182029094;
Epoch: 2; Error: 5030.587645099068;
Epoch: 3; Error: 3428.0018261229907;
Epoch: 4; Error: 2506.792840890022;
Epoch: 5; Error: 2170.068250358049;
Epoch: 6; Error: 1960.129130528695;
Epoch: 7; Error: 1817.9416148782927;
Epoch: 8; Error: 1681.7147938612607;
.................................................. . .
Epoch: 54; Error: 173.37241914568668;
Epoch: 55; Error: 170.31735671858064;
Epoch: 56; Error: 167.13513606363136;
Epoch: 57; Error: 163.9595221056975;
.................................................. . .
Epoch: 133; Error: 75.19772053990428;
Epoch: 134; Error: 74.88067178682655;
Epoch: 135; Error: 74.58012183106469;
.................................................. . .
```

```
Epoch: 204; Error: 60.81190149673153;
Epoch: 205; Error: 60.701523978641156;
Epoch: 206; Error: 60.576765509324815;
Epoch: 207; Error: 60.42412633897462;
Epoch: 208; Error: 60.26046374620901;
.............................................................. . .
Epoch: 246; Error: 55.64719942209722;
Epoch: 247; Error: 55.58679036996949;
Epoch: 248; Error: 55.52024126099841;
.............................................................. . .
Epoch: 296; Error: 51.62127794531158;
Epoch: 297; Error: 51.55689143321883;
Epoch: 298; Error: 51.47452795411011;
Epoch: 299; Error: 51.357848342655274;
Epoch: 300; Error: 51.246705558021446;
```

We can now print the model. The model is nothing but a set of weights. Given below is the code for fetching the weights from the model.

```python
#Model results
## Input to Hidden layer weights
print(net.layers[0].np['w'])
print(net.layers[0].np['b'])

## Hidden to Output layer weights
print(net.layers[1].np['w'])
print(net.layers[1].np['b'])
```

The following is the code output.

```python
print(net.layers[0].np['w'])
[[ 0.35746658  0.05278264 -0.10881015 ...  0.18619654  0.07496636
   -0.0574272 ]
 [ 0.21625002 -0.0726982  -0.29443584 ... -0.00643208 -0.01758393
   -0.32367234]
 [-0.12848771 -0.22832443  0.03700611 ... -0.04455668 -0.05013741
   -0.29095858]
 ...
 [-0.07392276 -0.34128372 -0.55736005 ... -0.40142089 -0.58627276
   -0.43219117]
 [ 0.16730895  0.00682749 -0.15222545 ... -0.21038047 -0.46275667
   -0.15019614]
 [-0.72170932 -0.86444701 -0.6580617  ... -0.95910974 -0.65948987
   -1.64598035]]

print(net.layers[0].np['b'])
[-2.7919784  -2.4022574   2.40409585  1.97727187 -1.69377248  1.31945481
 -0.84373513  0.93267911 -0.35511918  0.67121341 -0.06856221  0.94108216
 -0.87433817  1.1804362  -1.73962796 -1.48973649  1.70233063  2.6714419
  2.8696523   4.90559859]

print(net.layers[1].np['w'])
[[-7.08558315e+00 -2.21569837e+00 -2.06117061e+00  6.63863589e+00
  -1.24507839e+01  2.27819908e+00  5.23086220e+00 -1.29955523e+01
  -1.67971851e+01 -1.78199979e+00 -3.70834475e+00  1.86135131e+00
  -1.66316034e+00 -6.57511914e-01 -1.70506992e+00 -2.30374410e+00
   5.40778544e+00  1.44604719e+00  7.47212159e-01 -1.44428093e+00]
 ...
```

```
[ 6.81077116e+00   6.60175408e+00   1.71551747e+00  -2.03934684e+01
   4.24100207e+00   8.01552310e-02   1.26657163e+00  -1.67994455e+00
  -7.68184791e-01  -2.20653785e+00  -7.72908416e+00  -8.68998621e+00
  -4.72371621e+00  -3.31745354e+02  -2.44455054e+01  -4.32516689e+00
   1.99581193e+00  -2.44861958e+00   4.13027154e+00  -1.80259303e+00]]

print(net.layers[1].np['b'])
[ 2.90183573   -4.48091727    3.48703738    2.27825865  -16.72443277
  -4.1181299    -1.91023727    5.13861223    9.6709162     8.02037892]
```

The code below gives us an idea of the count of weights.

```
## Shape of the weights
print(net.layers[0].np['w'].shape)
(20, 256)

print(net.layers[0].np['b'].shape)
(20,)

print(net.layers[1].np['w'].shape)
(10, 20)

print(net.layers[1].np['b'].shape)
(10,)
```

The model building is completed. We have to use these weights to get predictions for new data points.

8.8.3.1   Deciding Hidden Nodes   There are several hyperparameters in a neural network model. The most impactful hyperparameter is the number of hidden nodes. If this number is too high, then the model will be overfitted. If it is too low, then the model will be underfitted. We need to look at the train and test data accuracies to fine-tune this parameter. We can use a binary search approach to fine-tune this parameter.

8.8.4   Model Predictions and Validation

Each new data point in the test data gives us 10 probabilities–one probability for each digit. We will take the final result as the digit with maximum probability.

```
## Prediction on test data
predicted_values = net.sim(X_test.as_matrix())
predicted=pd.DataFrame(predicted_values)
print(round(predicted.head(10),1))

## Converting predicted probabilities into numbers
predicted_number=predicted.idxmax(axis=1)
print(predicted_number.head(15))
```

The following is the code output.

```
print(round(predicted.head(10),3))
          0      1      2      3      4      5      6      7      8      9
0    0.000    0.0  0.000    0.0  0.226  0.000  1.0  0.000  0.000    0.0
1    0.000    1.0  0.000    0.0  0.000  0.000  0.0  0.004  0.000    0.0
2    0.008    0.0  0.003    0.0  0.000  0.284  0.0  0.000  0.003    0.0
3    0.000    0.0  0.000    0.0  0.001  0.000  0.0  1.000  0.000    0.0
4    0.000    1.0  0.000    0.0  0.009  0.000  0.0  0.001  0.000    0.0
5    0.000    1.0  0.000    0.0  0.007  0.000  0.0  0.002  0.000    0.0
6    0.000    0.0  0.000    0.0  0.000  0.000  0.0  0.421  0.000    0.0
7    0.000    0.0  0.000    0.0  0.000  0.999  0.0  0.000  0.000    0.0
8    0.000    1.0  0.000    0.0  0.038  0.000  0.0  0.000  0.000    0.0
9    0.000    0.0  0.000    1.0  0.000  0.000  0.0  0.000  0.026    0.0
```

```
print(predicted_number.head(15))
0       6
1       1
2       5
3       7
4       1
5       1
6       7
7       5
8       1
9       3
10      5
11      8
12      0
13      0
14      0
```

In this output, we can see 10 probabilities for each data point. We convert them to single digits based on the class with the highest probability. For example, the first record has a 0.998 probability for Class-6; the last data point has a 1.0 probability for Class-3. We can now create the confusion matrix and calculate the accuracy.

```
ConfusionMatrix = cm(y_test,predicted_number)
print("ConfusionMatrix on test data \n", ConfusionMatrix)
ConfusionMatrix on test data
 [[264   0   2   0   0   0   1   0   0   0]
 [  0 196   0   0   1   0   0   0   0   0]
 [  3   1 126   1   1   1   0   1   4   0]
 [  1   0   4 127   1   1   0   0   3   0]
 [  0   2   3   0 113   0   2   0   1   2]
 [  3   2   5   3   1  91   2   0   4   2]
 [  1   0   0   0   2   0 125   0   0   0]
 [  0   0   1   1   1   0   0 120   0   3]
 [  0   2   3   1   2   1   1   1  96   0]
 [  0   0   0   1   2   0   0   2   0 118]]
```

```
accuracy=np.trace(ConfusionMatrix)/sum(sum(ConfusionMatrix))
print("Test Accuracy", accuracy)
Test Accuracy 0.9431117203564084
```

We can see that the accuracy with the test data is above 90 percent. Let us use this model to get some predictions on a few data points from the test data.

```
#Random number between 0 and 7291
i=500

random_sampel_data=digits_data_raw[[i]]
random_sampel_data1=pd.DataFrame(random_sampel_data)
X_sample=random_sampel_data1.drop(random_sampel_data1.columns[[0]], axis=1)

predicted_values = net.sim(X_sample)
predicted=pd.DataFrame(predicted_values)
predicted_number=predicted.idxmax(axis=1)
predicted_number

data_row=random_sampel_data[0][1:]
pixels = np.matrix(data_row)
pixels=pixels.reshape(16,16)
plt.title(["Row number = ", i, "Predicted Digit ", predicted_number[0]] ,
fontsize=20)
plt.imshow(pixels, cmap='Greys')
```

The code above randomly takes a point from the data and gives us the predicted values. Figure 8.38 shows the output from this code.

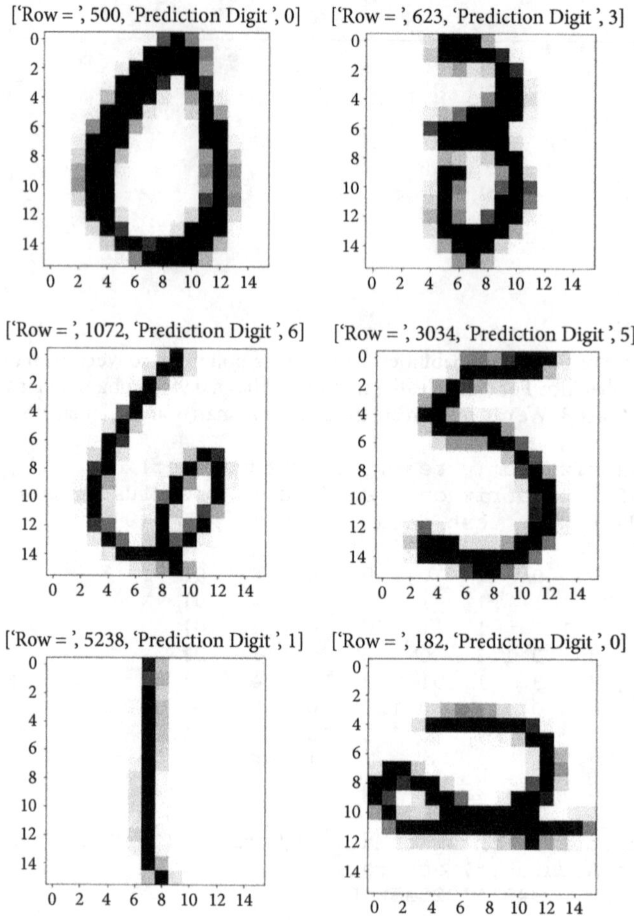

FIGURE 8.38   Image output from the code.

8.9   *DEEP NEURAL NETWORKS*

Until now we have worked on neural networks with a single hidden layer. In the final case study, the input data has 256 pixels; we built a neural network with 20 hidden nodes. Imagine working with some real image processing and computer vision problems where the input datasets have 2 million pixels. We cannot solve such problems with just one hidden layer. We require multiple hidden layers to solve such sophisticated image-processing problems. A neural network with multiple hidden layers is known as a deep neural network.

A deep neural network is much more complicated than a network with a single layer. We have to deal with thousands of weights. We will have to fine-tune many parameters to build the optimal model. Such a deep neural network is a particular case of artificial neural network (ANN), but it has now developed as a separate branch of machine learning called deep learning. When it comes to deep learning, we will be discussing the different types of deep neural networks such as convolutional neural networks (CNNs) and recurrent neural networks (RNNs). Deep neural networks take much time to build the model. Deep neural networks require massive computational power. The standard Python packages that we have discussed until now cannot handle the calculations in deep learning. We need different packages that can perform these calculations efficiently. Fortunately, Google has released an open-source package by the name TensorFlow to work with deep learning algorithms. We will use TensorFlow to work with different types of deep learning algorithms in the upcoming chapters.

8.10   *CONCLUSION*

In this chapter, we have discussed the basic concepts of neural networks. We have also discussed the backpropagation algorithm. This algorithm will be referred to in multiple upcoming chapters. The neural network is the peak point of machine learning and the starting point of deep learning. A good understanding of ANNs is necessary to follow the upcoming chapters on deep learning. Neural network algorithms have gained exponential popularity after 2010, primarily due to the advent of better computation machines. A significant amount of research is going on in this field now. ANNs have a wide variety of exciting applications in the industry and our daily lives.

8.11   *PRACTICE PROBLEMS*

1. Download the Breast Cancer Wisconsin (Diagnostic) dataset
   - Import the data. Complete the necessary exploration and sanitization of the data.
   - Build a machine learning model to classify the breast tumors as malignant or benign.
   - Perform the model validation and measure the accuracy of the model.
   - Check for innovative ways to improve the accuracy of the model.

   Dataset credits—http://archive.ics.uci.edu/ml/datasets/breast+cancer+wisconsin+(diagnostic), 1. Dr. William H. Wolberg, General Surgery Department, University of Wisconsin, Clinical Sciences Center Madison, WI 53792. wolberg@eagle.surgery.wisc.edu 2. W. Nick Street, Computer Sciences Department, University of Wisconsin, 1210 West Dayton St., Madison, WI 53706. street@cs.wisc.edu. 3. Olvi L. Mangasarian, Computer Sciences Department, University of Wisconsin, 1210 West Dayton St., Madison, WI 53706. olvi@cs.wisc.edu.

2. Download Porto Seguro's Safe Driver Prediction dataset
   - Import the data. Complete the necessary exploration and sanitization of the data.
   - Build a machine learning model to predict if an auto insurance policyholder files a claim or not.
   - Perform the model validation and measure the accuracy of the model.
   - Check for innovative ways to improve the accuracy of the model.

   Dataset credits—Log in and download the data from https://www.kaggle.com/c/porto-seguro-safe-driver-prediction/data.

8.12   *REFERENCES*

1. Vegetable images from Wiki Images: https://upload.wikimedia.org/wikipedia/commons/2/24/Marketvegetables.jpg.
2. USPS numbers data: http://yann.lecun.com/exdb/mnist/.

CHAPTER 9
TENSORFLOW AND KERAS

A logistic regression model solves simple problems. It works very well with the data where the target variable classes are linearly separable. We have introduced a neural network as a layered network model to solve the problem of multiple and nonlinear decision boundaries. Neural networks can be configured with multiple hidden nodes and layers. The number of free parameters (weights) in the neural network optimization function can run into the thousands. The standard Python computation libraries based on NumPy may not be sufficient to work with these complicated optimization algorithms. We need dedicated frameworks or tools to work efficiently with large neural networks. In this chapter, we are going to discuss the details of one such popular open-source tool called TensorFlow from Google.

9.1 DEEP NEURAL NETWORKS

A neural network with a single layer is known as a shallow neural network. When the problem is highly nonlinear, we may need to add multiple hidden layers. For example, almost all computer vision problems require multiple hidden layers. A neural network with multiple hidden layers is known as a deep neural network. There are several types of deep neural networks. The subject that deals with deep neural networks is called deep learning. We can take deep learning as a specialized subset of machine learning. The reason behind it is that deep learning is only an advanced version of one machine learning algorithm, i.e., neural networks.

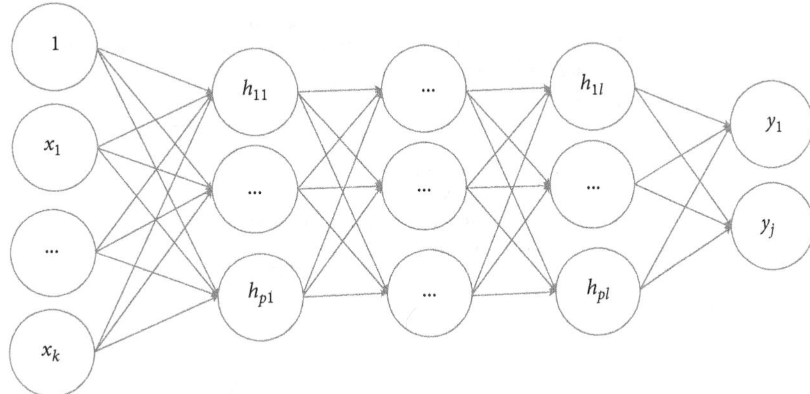

FIGURE 9.1 General illustration of deep neural networks.

The diagram in Fig. 9.1 shows a deep neural network with 'l' hidden layers.

9.1.1 Number of Parameters

The first striking difference between machine learning algorithms and deep learning is the execution time. The deep neural networks take a much higher execution time and massive computational resources compared to machine learning models. The main reason behind an extraordinarily long training time is the number of free parameters in the optimization function. The free parameters are nothing but the weights in the gradient descent algorithm. Let us see a few examples of calculations of the number of weights (Table 9.1).

TABLE 9.1 Calculating the Number of Weights in a Neural Network

Input Layer Nodes	Hidden Layer Nodes	Output Layer Nodes	Number of Weights
100	10	10	1,120
256	[64,64]	10	21,258
576	[128,100,15]	10	88,431
1024	[100,150,100,40]	10	137,200
4096	[256, 256, 256, 256, 256]	100	1,337,700

As we can see in Table 9.1, the number of weight parameters runs into the thousands for deep neural networks. An optimization function has to perform millions of calculations to find the final optimal weights for these deep neural networks. We need efficiently written codes that can perform multiple calculations parallelly and give us the results in the stipulated time. Standard packages such as Scikit-learn can work well for machine learning problems, but they fail on deep neural networks.

9.2   DEEP LEARNING FRAMEWORKS

Almost all tech companies such as Google, Facebook, Microsoft, and Amazon deal with one or the other kinds of deep learning problems. They have created some specialized tools that can effectively solve their specific intricate optimization problems. A few of these companies have made their tools or packages available as an open-source product. These powerful packages are now freely available to the data scientists to crack deep learning problems. There are two major deep learning frameworks: TensorFlow and PyTorch (Table 9.2 and Figs. 9.2 and 9.3).

TABLE 9.2 TensorFlow and PyTorch

TensorFlow

1. Open source
2. Developed by the Google Brain team for internal Google use
3. Released in November 2015
4. Performs computations using data flow graphs
5. Adopted by many companies for their deep learning implementations
6. User languages: Python and C++

PyTorch

7. Open source
8. Developed by the Facebook AI team
9. Released in October 2016
10. Widely used in computer vision and deep learning
11. User language: Python

FIGURE 9.2   TensorFlow logo.

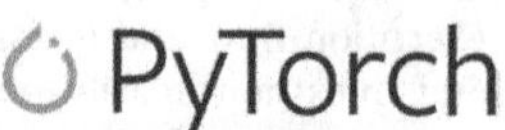

FIGURE 9.3   PyTorch logo.

Apart from TensorFlow and PyTorch, there are a few more deep learning packages such as MXNet, CNTK, Caffe, and Paddle. Many developers use these packages. Nevertheless, TensorFlow and PyTorch have the maximum user base and the support of a rapidly growing user community. Out of these two, TensorFlow has gained enormous popularity soon after its announcement by Google. We are going to use TensorFlow here.

9.2.1 What Is TensorFlow?

TensorFlow is a library that can effectively perform complex mathematical calculations. A tensor is a multidimensional vector. We can roughly imagine a tensor as data and TensorFlow as data flow. As we have discussed earlier, deep learning algorithms involve complex matrix calculations. TensorFlow works best when it comes to matrix computations. TensorFlow is scalable to multi-CPUs and even GPUs (graphics processing units). TensorFlow represents the data and calculations in the form of computation graphs. These computations in the form of graphs and independent nodes can be calculated parallelly. It gives us a chance to run multiple calculations in parallel that undeniably reduces the overall computation time.

In simple words, the TensorFlow package rewrites the code for matrix calculations to allow parallel computation. We can use Python's NumPy to build our models. TensorFlow does the same job as NumPy, but it provides the functions to build models quickly and efficiently. TensorFlow is appropriately documented and enjoys excellent community support.

9.2.2 Computational Graphs

What makes TensorFlow special? Why is TensorFlow much faster than standard machine learning libraries? Computational graphs are the main reason behind TensorFlow's speed. Inside TensorFlow, computations are represented using computational graphs. We can also call them data flow graphs.

A computational graph has nodes and edges. Nodes represent the data or computation. Edges represent connections between computations and data. For example, if $c = a + b$ is the calculation, then a will be one node and b will be another node connected to an operational node $a + b$ (Fig. 9.4).

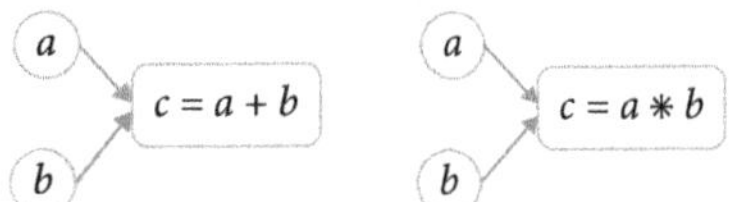

FIGURE 9.4   Nodes and operations.

The nodes are data and operations, and the edges are the direction of the data flow from one node to another. The nodes can contain multidimensional data and matrix calculations. Figure 9.5 is the computational graph representation of the logistic regression line.

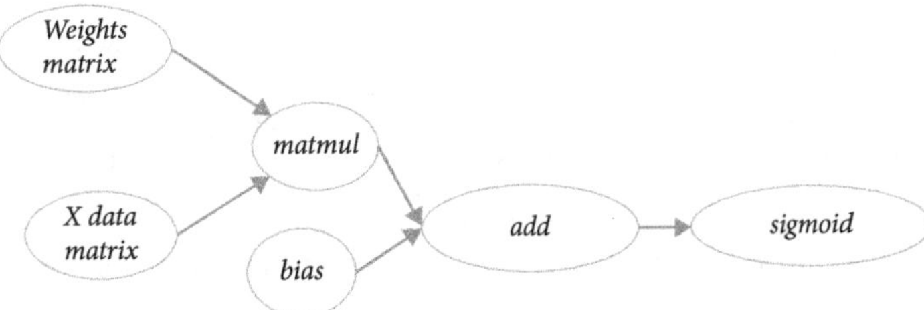

FIGURE 9.5   Computational graph representation of a logistic regression line.

Figure 9.6 is the same graph with the actual data.

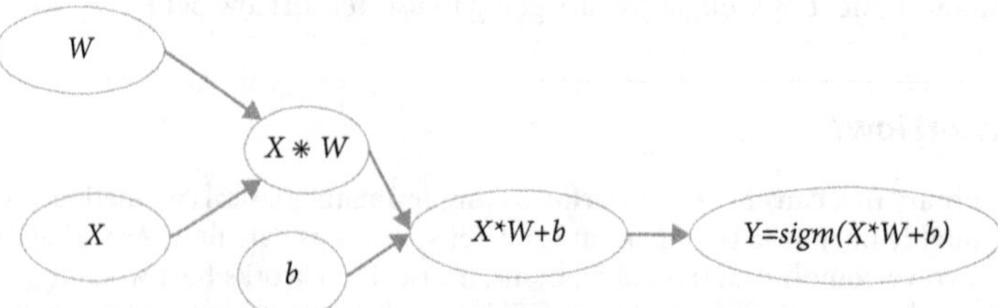

FIGURE 9.6 Computational graph representation of a logistic regression line with data.

If we change the function in the final equation from sigmoid to linear, then the graph in Fig. 9.6 becomes a computational graph for linear regression. Figure 9.7 shows the computational graph for neural networks.

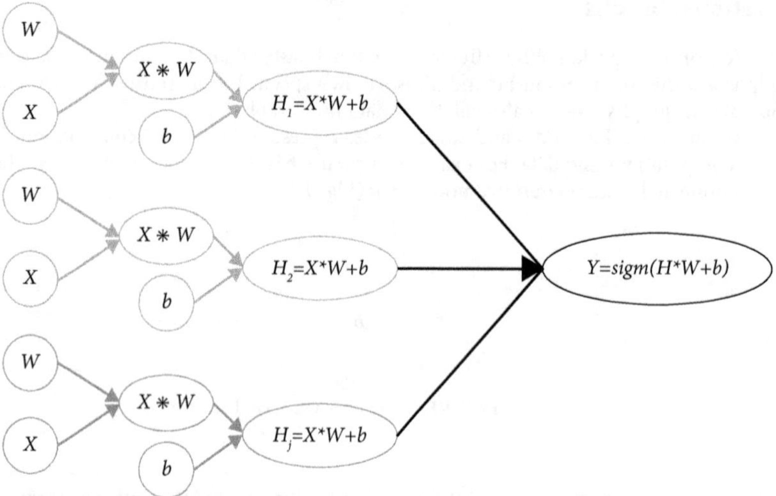

FIGURE 9.7 Computational graph representation for neural networks.

Computational graphs are particularly useful if there are multiple complex operations involved; look at the neural network computational graph in Fig. 9.7. Independent nodes in computation graphs are calculated parallelly. In the graph in Fig. 9.7, when we are calculating H_2, we do not need to wait for H_1. The graph calculates all the values of H_1, H_2, ..., H_j parallelly. If there are 1000 hidden nodes, then all 1000 hidden nodes are getting calculated in parallel. This particular feature of independent nodes getting evaluated parallelly saves much time while performing complex operations. The independent subgraph calculations finally contribute to the overall calculations. This specific feature is beneficial for solving deep learning problems. The partial derivatives and the chain rule applications can be handled efficiently using computational graphs. Also, computational graphs easily facilitate distributed computation, spreading the work across multiple CPUs or GPUs or systems.

9.2.3 Python Notebook

Jupyter Notebook is an alternative integrated development environment (IDE, as you know it) to work with Python. Jupyter is a web browser–based IDE. We can open a Notebook file in a browser and start writing the code in the cells. There are two main reasons for using the Jupyter Notebook. First, personal PCs may not have the resources to build deep learning models. We often build models on a server or cloud. Since Jupyter Notebook is web browser–based, it is easy to code using a server. Second, Google and Microsoft have, fortunately, given free cloud access to data scientists to learn and develop deep learning models. We need to code in notebooks on their cloud. We can use Google Colab notebooks or Azure notebooks to code and learn even if we do not have a good computer (with adequate computing power as required to solve deep learning algorithms). We will learn some basic commands of working with Jupyter Notebook. We need not install the Jupyter Notebook separately now. When we install Anaconda (as we did), it automatically installed both Spyder and Jupyter notebooks. Some basic concepts required while working with Jupyter Notebook are detailed next.

9.2.3.1 Opening a Jupyter Notebook   Open Jupyter Notebook from the Anaconda prompt.

1. Open the Anaconda prompt >> type Jupyter Notebook.
2. The notebook will open in a web browser.
3. Jupyter uses the localhost as the local server.

Figure 9.8 shows the screenshot of Anaconda prompt.

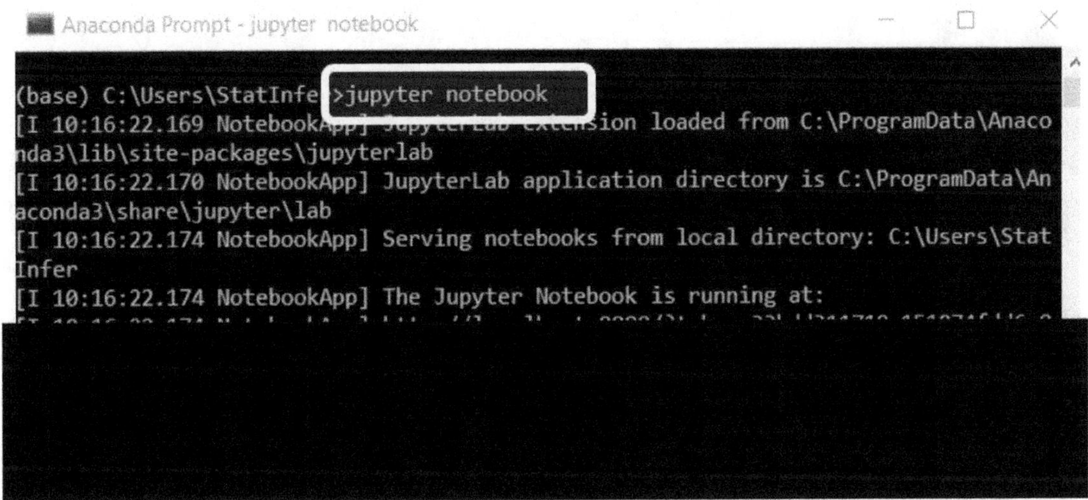

FIGURE 9.8   Anaconda prompt.

The command, as typed in the Anaconda prompt and shown in the screenshot in Fig. 9.8, will open the window shown in Fig. 9.9.

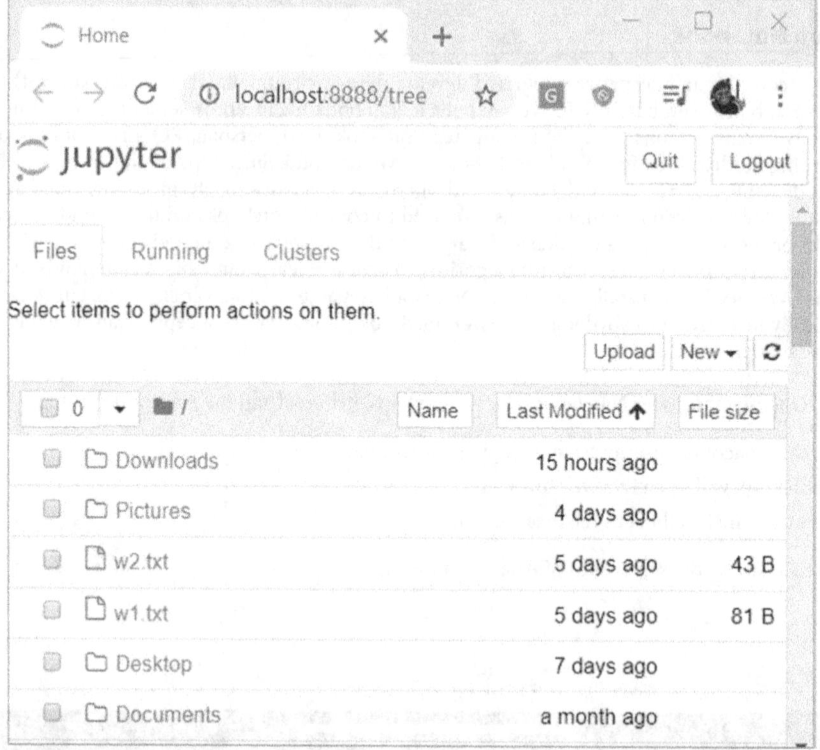

FIGURE 9.9   Jupyter Notebook home page.

This method takes the user home directory as the default home for Jupyter. If you want to open a Jupyter Notebook from a specific location, then you need to follow the steps as given below.

1. Go to the target folder >> Click on path >> type the command "Jupyter Notebook."
2. This command will consider the target folder as home.

Figures 9.10 and 9.11 show the screenshots of how to start the Jupyter Notebook and how to type the command.

FIGURE 9.10   Starting Jupyter Notebook.

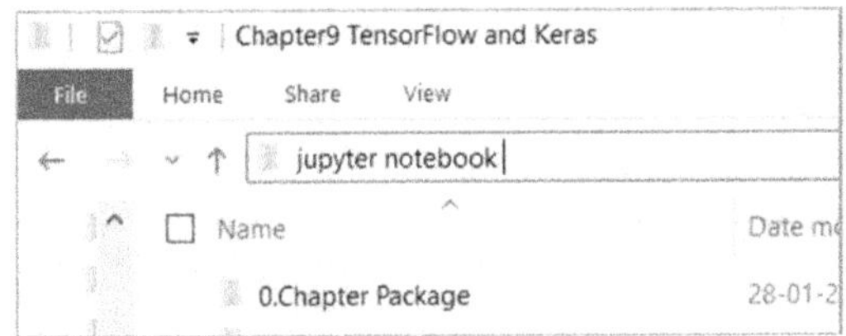

FIGURE 9.11   Starting Jupyter Notebook—command.

The command in Fig. 9.11 opens the Jupyter Notebook home page shown in Fig. 9.12.

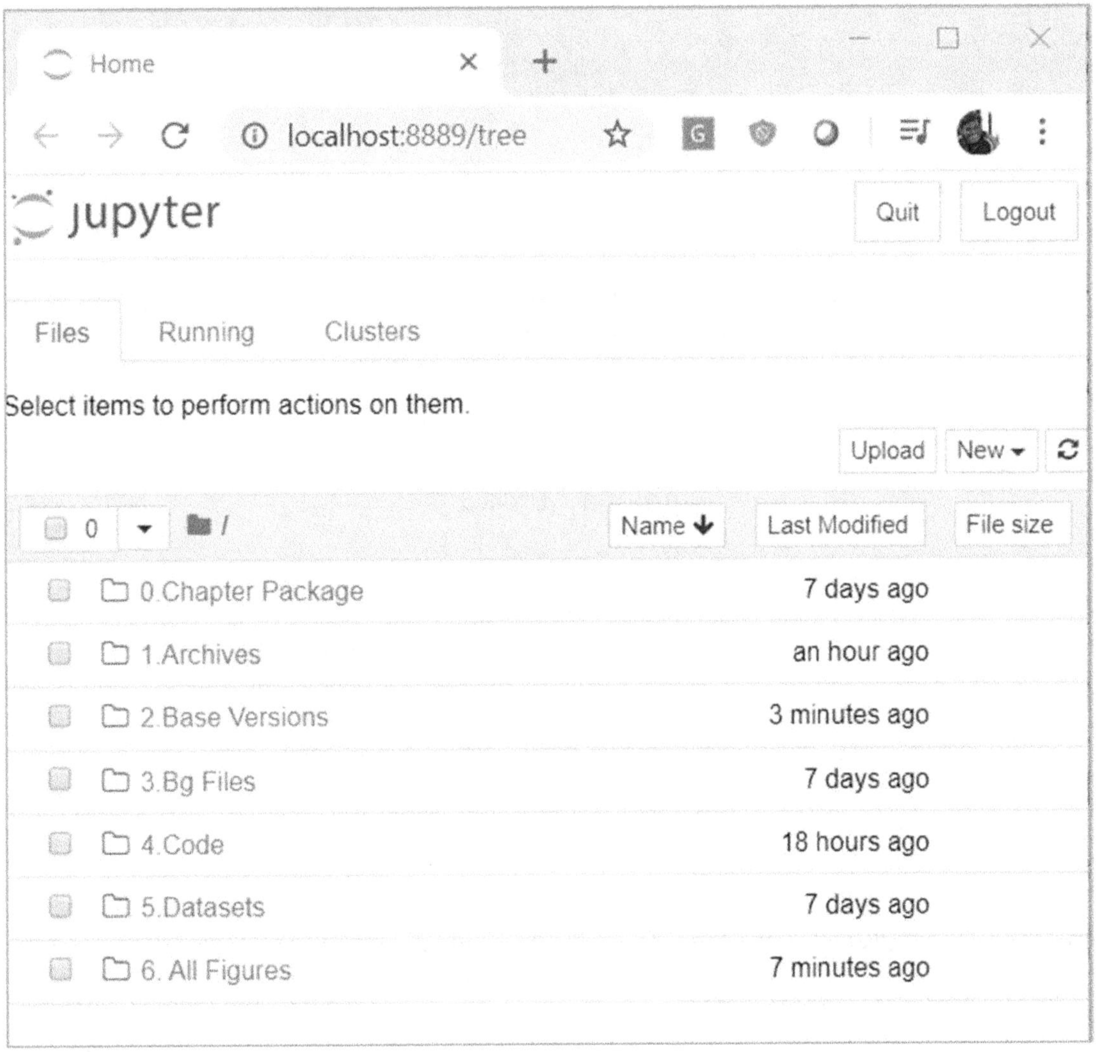

FIGURE 9.12   Jupyter Notebook home.

9.2.3.2 Basic Commands in Jupyter Notebook   Table 9.3 shows the commands that help us to get started with Jupyter Notebook.

TABLE 9.3   Basic Commands in Jupyter Notebook

Creating a new notebook file	Click on "New" to open a new Python 3 notebook file.

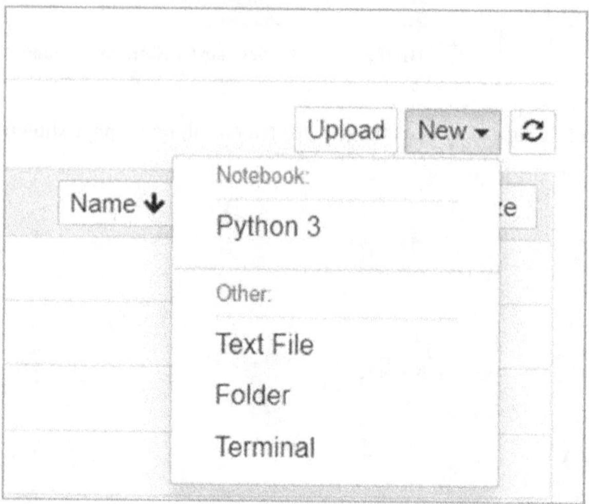

Opening a new Python notebook.

The command above opens the following notebook file:

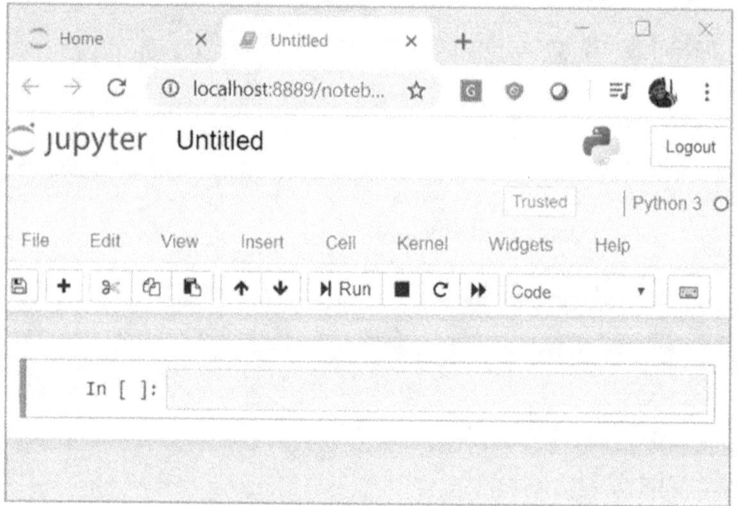

A new Python workbook.

Writing and submitting the code   Once the notebook file is open, try writing these commands inside the cell:

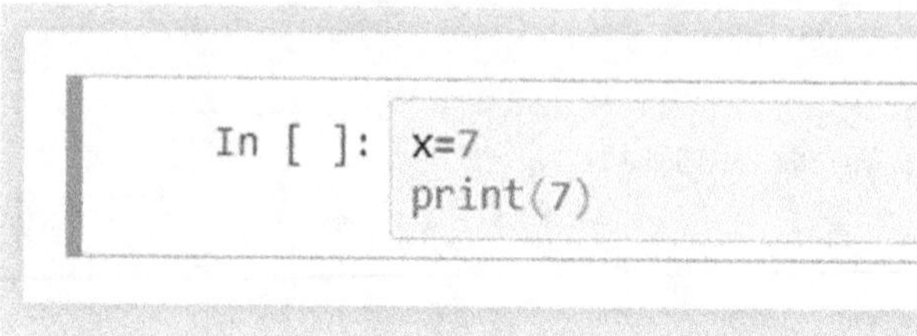

The Python code input.

Execute this code using the Ctrl+Enter command or Shift+Enter command or use the Run button in menu items. This command gives us the output as given below.

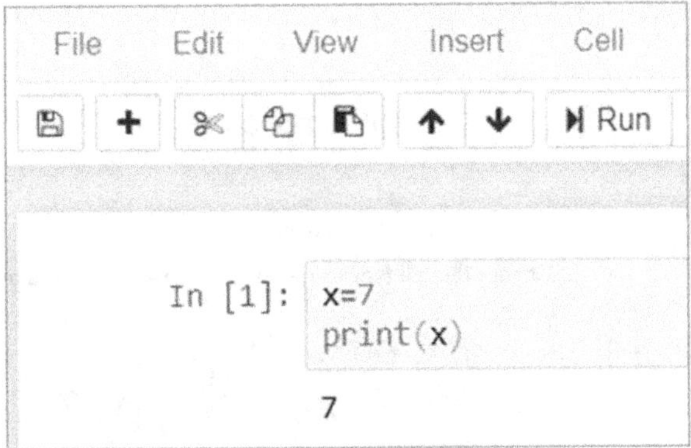

The Python code output.

Add cells

Use the Insert option:

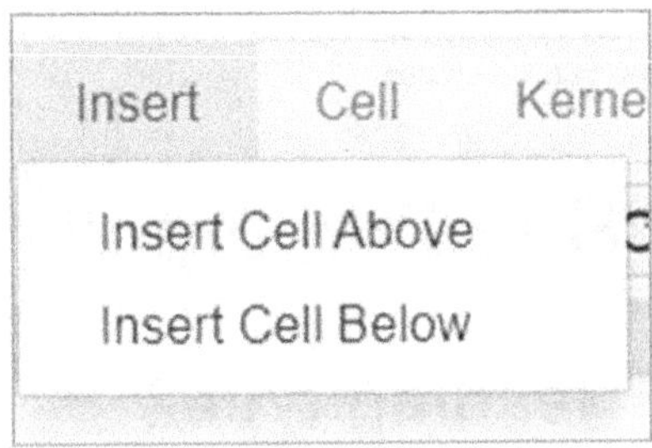

Inserting a new code cell in the notebook.

Or use the '+' sign in the menu options:

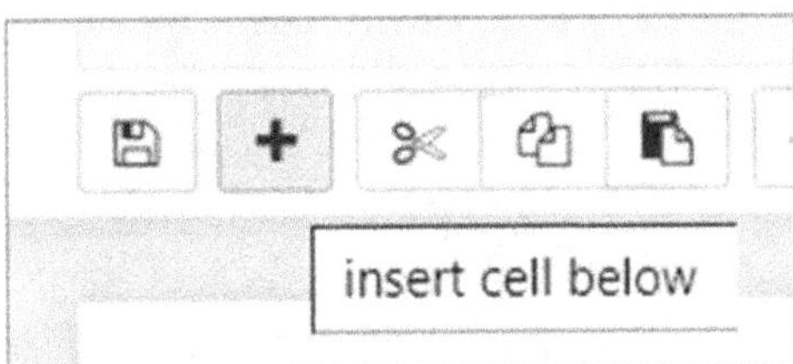

Another way to insert a new code cell.

Add a non-code cell

Change cell type from code to Markdown.

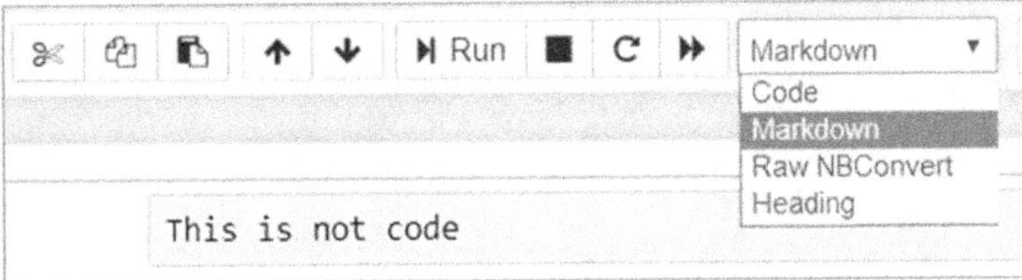

Changing the cell type.

(*Continued*)

TABLE 9.3   Basic Commands in Jupyter Notebook (*Continued*)

The code above gives us the following output:

This is not code

Output of Markdown.

Add a heading   Change the cell type to Heading or use # (# followed by space).

Heading type1
Heading type2
Heading type3

Get headings by changing the cell type.

The following is the output:

Heading type1

Heading type2

Heading type3

Different heading levels by using #, ##, and ###.

Change the notebook file name   Click on the notebook name and edit.

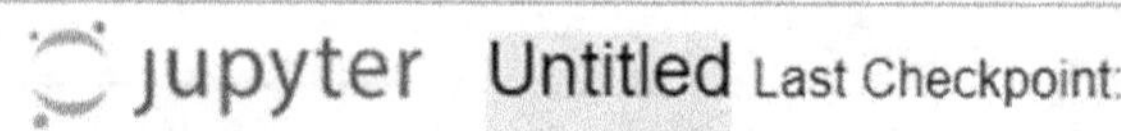

Renaming Jupyter Notebook.

If we click on "Untitled" above, it opens the following window:

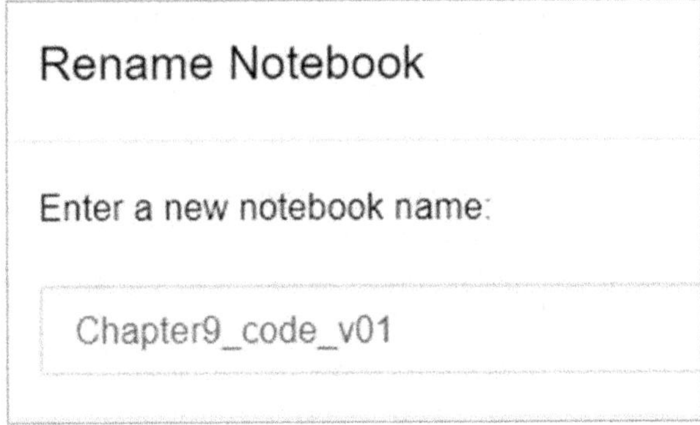

Renaming Jupyter Notebook.

Saving the notebook   Use the save button.

Save and Checkpoint.

Or use the option under "File" in the menu.

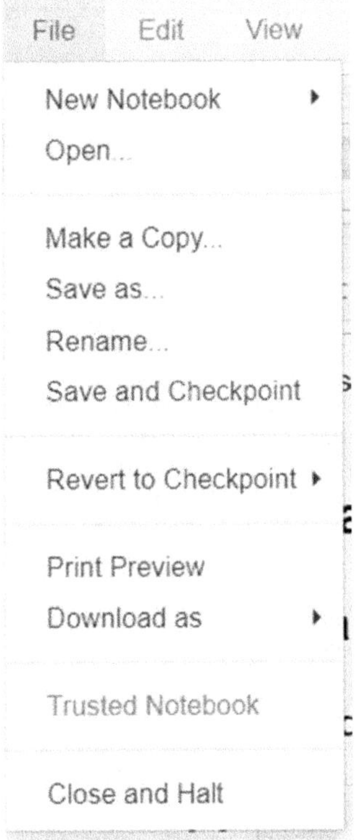

Another way to save.

(*Continued*)

TABLE 9.3   Basic Commands in Jupyter Notebook (*Continued*)

Notebook file extension	The extension is "IPYNB," which stands for interactive Python notebook.

Notebook file extension.

Google "colab" notebook file	https://colab.research.google.com/

1. No Python setup is required. A Google account is sufficient.
2. Free access to GPUs (limited GPUs).
3. Code easily and share easily.
4. The best tool for deep learning education.

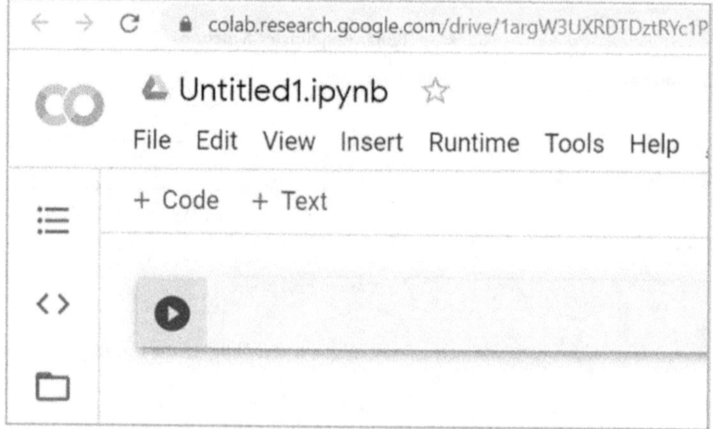

Google Colab notebook.

There are many more commands in Jupyter Notebook. We will learn more of them based on need. The commands described so far are sufficient to get started.

9.2.4   Installing TensorFlow

We use the command "`!pip install`" to install any new package in Python. We can either run `pip install` in the Anaconda prompt or directly execute the installation step in Jupyter Notebook using the "`!pip install`" command.

Given below is the command to install TensorFlow. We can mention the exact version if required.

```
!pip install tensorflow
```

This command requires an Internet connection to download the TensorFlow package from the website. Given below is the code for checking the installed version.

```
!pip show tensorflow
```

Figure 9.13 is the output.

```
!pip show tensorflow
```

```
Name: tensorflow
Version: 2.0.0
Summary: TensorFlow is an open source machine learning
framework for everyone.
Home-page: https://www.tensorflow.org/
Author: Google Inc.
Author-email: packages@tensorflow.org
License: Apache 2.0
Location: c:\users\statinfer\appdata\roaming\python\py
thon37\site-packages
Requires: wheel, protobuf, grpcio, keras-applications,
numpy, tensorboard, astor, gast, google-pasta, wrapt,
absl-py, keras-preprocessing, termcolor, tensorflow-es
timator, opt-einsum, six
Required-by:
```

FIGURE 9.13   The output of `!pip show tensorflow`.

9.3   KEY TERMS IN TENSORFLOW

TensorFlow has a different coding paradigm. We can recognize TensorFlow as a sophisticated version of the NumPy package. Storing the data and handling the calculations may be a bit intricate in TensorFlow. TensorFlow 2.0 has several changes made as compared to its old version. Nevertheless, it is relatively easy to get started with TensorFlow 2.0. In this section, we will examine some critical terms in TensorFlow.

9.3.1   Tensors

Intuitively we can recognize a tensor as a multidimensional array. In TensorFlow, data is represented as tensors. An array or a vector is a collection of scalar values, a matrix is a two-dimensional array, and a tensor is a multidimensional array (Fig. 9.14).

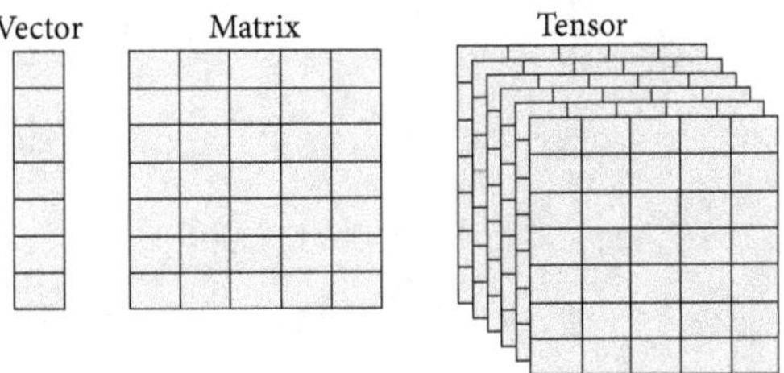

FIGURE 9.14   Vector, matrix, and tensor representations.

Note that the strict definition of vectors, matrices, and tensors in mathematics is slightly different—the explanation given here is based on general coding terminology. Table 9.4 provides some examples of tensors.

TABLE 9.4   Representation of Tensors by Dimensionality

Prices of smartphones from six different companies—a single-dimensional tensor	Prices of smartphones, along with ratings of each phone—a two-dimensional tensor	Prices of smartphones, along with ratings of each phone in three different regions—a three-dimensional tensor	Prices of smartphones, along with ratings of each phone, in three different regions, for the last two years—a four-dimensional tensor

700
450
350
200
580
300

1D tensor

700	4.5
450	5.0
350	3.6
200	4.2
580	4.3
300	2.9

2D tensor

700	4.5
450	5.0
350	3.6
200	4.2
580	4.3
300	2.9

3D tensor

700	4.5
450	5.0
350	3.6
200	4.2
580	4.3
300	2.9

4D tensor

In real-life scenarios, color images are represented as three-dimensional tensors. The first dimension is height, the second dimension is width, and the third dimension is depth. Every pixel has three values—RGB values that constitute depth.

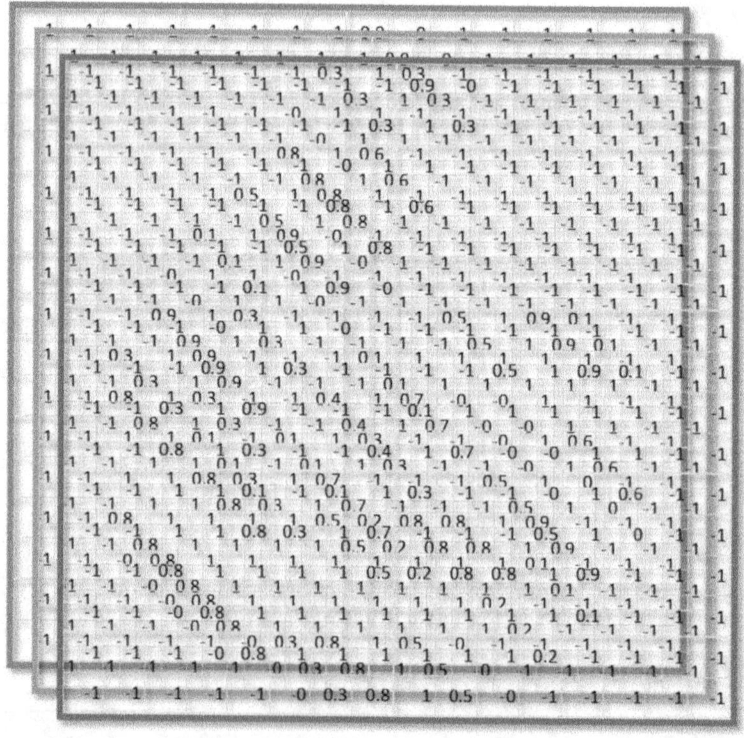

FIGURE 9.15   A 3D-image representation.

Figure 9.15 shows an example of an image with three dimensions. If we import an image and print its shape, we will see [Height, Width, Color].

There are two types of tensors: constant and variable.

9.3.1.1 *Constant Tensors*   Tensors that contain constant values are constant tensors. We define these using the `tf.constant()` function. We can mention their type and initialize them with a value. There is nothing new in these definitions. They are just like constants and variables in any general programming language.

Data values are stored as tensors in TensorFlow. Let us see some examples of tensors. The code given below is used for creating variables and constants.

```
a = tf.constant([50,10])
print(a)

print('a in tensorflow ==>', a)
print('numpy value of a ==>', a.numpy())
print('dtype of a ==>', a.dtype)
print('shape of a ==>', a.shape)
```

The following is the code output.

```
print(a)
tf.Tensor([50 10], shape=(2,), dtype=int32)

a in tensorflow ==> tf.Tensor([50 10], shape=(2,), dtype=int32)
numpy value of a ==> [50 10]
dtype of a ==> <dtype: 'int32'>
shape of a ==> (2,)
```

We can use the inbuilt `tf.XX()` function to create constant tensors, just like numpy.

```
print('Tensor of Ones: \n',tf.ones(shape=(2, 2)))
print('Tensor of Zeros: \n',tf.zeros(shape=(2, 2)))
print('Random Normal Distribution \n', tf.random.normal(shape=(3, 2), mean=5,
stddev=1))
```

The following is the code output.

```
Tensor of Ones:
 tf.Tensor(
[[1. 1.]
 [1. 1.]], shape=(2, 2), dtype=float32)

Tensor of Zeros:
 tf.Tensor(
[[0. 0.]
 [0. 0.]], shape=(2, 2), dtype=float32)

Random normal values
 tf.Tensor(
[[4.7691674 4.7110634]
 [4.4724483 3.3842342]
 [4.229244  4.7824235]], shape=(3, 2), dtype=float32)
```

9.3.1.2 *Variables*　The data types that allow us to change the previously assigned values work as trainable parameters. We can define them using the tf.Variable(), with the type and initial value. We generally create a variable and initialize it. Given below are a few examples of variables.

```
## Simple variable
x = tf.Variable(5)
print(x)
```

```
Output
<tf.Variable 'Variable:0' shape=() dtype=int32, numpy=5>
```

```
##randomly initialized variable, like we do for our weights
w = tf.Variable(tf.random.normal(shape=(2, 2)))
print(w)
```

```
Output
<tf.Variable 'Variable:0' shape=(2, 2) dtype=float32, numpy=
array([[ 1.5475181 , -1.0206841 ],
       [ 0.07939683, -0.6459112 ]], dtype=float32)>
```

```
## Simple variable
m = tf.Variable(5)
print(m)
```

```
Output
<tf.Variable 'Variable:0' shape=() dtype=int32, numpy=5>
```

```
m = tf.Variable(5)
print('New value', m.assign(2))
```

```
Output
New value <tf.Variable 'UnreadVariable' shape=() dtype=int32, numpy=2>
```

```
m = tf.Variable(5)
print('increment by 1', m.assign_add(1))
```

```
Output
increment by 1 <tf.Variable 'UnreadVariable' shape=() dtype=int32, numpy=6>
```

```
m = tf.Variable(5)
print('Decrement by 2', m.assign_sub(2))
```

```
Output
Decrement by 2 <tf.Variable 'UnreadVariable' shape=() dtype=int32, numpy=3>
```

9.4　MODEL BUILDING WITH TENSORFLOW

To work with TensorFlow, we need to follow some necessary steps. First, we define the model equation. Then we proceed to initialize the weights and define the cost function. Finally, we run the parameter training iterations using an optimization algorithm. Figure 9.16 shows the coarse steps in building a model in TensorFlow.

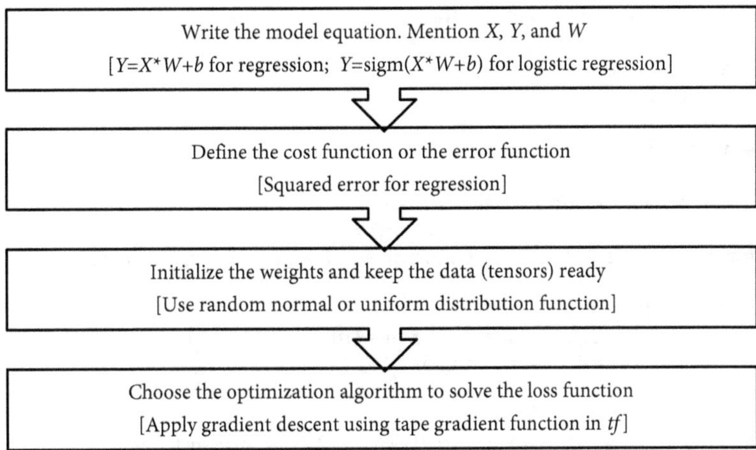

FIGURE 9.16   High-level steps to build a model in TensorFlow.

9.4.1 Building a Regression Model with TensorFlow

Given below is the code for building a regression model in TensorFlow. We will create some sample data and use that for building a model using TensorFlow.

```
#This step is for data creation, x, and y
x_train= np.array(range(5000,5100)).reshape(-1,1)
y_train=[3*i+np.random.normal(500, 10) for i in x_train]

import matplotlib.pyplot as plt
plt.title("x_train vs y_train data")
plt.plot(x_train, y_train, 'b.')
plt.show()
```

We multiply x_train by 3 and add some randomness to create the data. After building the regression model, we expect the final weight of *x* to be "3." Figure 9.17 is the plot generated by this code.

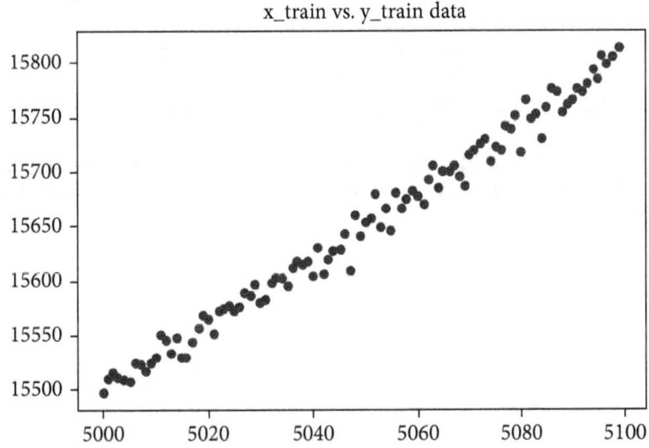

FIGURE 9.17   Scatter plot x_train versus y_train.

The following code is used for building the model.

```python
#Model y=X*W + b
#Model function
def output(x):
    return W*x + b

#Loss function Reduce mean square
def loss_function(y_pred, y_true):
    return tf.reduce_mean(tf.square(y_pred - y_true))

#Initialize Weights
W = tf.Variable(tf.random.uniform(shape=(1, 1)))
b = tf.Variable(tf.ones(shape=(1,)))

#Optimization
## Writing training/learning loop with GradientTape
learning_rate = 0.000000001
steps = 200 #epochs

for i in range(steps):
    with tf.GradientTape() as tape:
        predictions = output(x_train)
        loss = loss_function(predictions,y_train)
        dloss_dw, dloss_db = tape.gradient(loss, [W, b])
    W.assign_sub(learning_rate * dloss_dw)
    b.assign_sub(learning_rate * dloss_db)
    print(f"epoch is: {i}, loss is {loss.numpy()},  W is: {W.numpy()}, b is
{b.numpy()}")
```

The critical function that we need to understand in the above code is `tf.GradientTape()`. This function calculates the gradients. We use these gradients to multiply them with the learning rate and update the weights in each epoch. Even if we do not understand the syntax, it is fine. In the section on Keras later, we will discuss why it is fine not to master TensorFlow syntax. Given below is the output of the code.

```
epoch : 0, loss  212363920.0,  W : [[0.36013526]], b  [1.0000291]
epoch : 1, loss  191256560.0,  W : [[0.49980214]], b  [1.0000567]
epoch : 2, loss  172247104.0,  W : [[0.6323465]], b  [1.000083]
epoch : 3, loss  155127040.0,  W : [[0.75813156]], b  [1.0001079]
epoch : 4, loss  139708608.0,  W : [[0.877502]], b  [1.0001315]
                                        . . . . . . . . .
epoch : 197, loss  124.76454,  W : [[3.0987952]], b  [1.0005703]
epoch : 198, loss  124.74134,  W : [[3.0987997]], b  [1.0005703]
epoch : 199, loss  124.72179,  W : [[3.098804]], b  [1.0005703]
```

As expected, the weight is 3. We do not have any bias in the data. The expected bias is 0, and it has been estimated as 1 by TensorFlow. The overall accuracy of the model is good. We can print and visualize how the overall model has converged (Fig. 9.18).

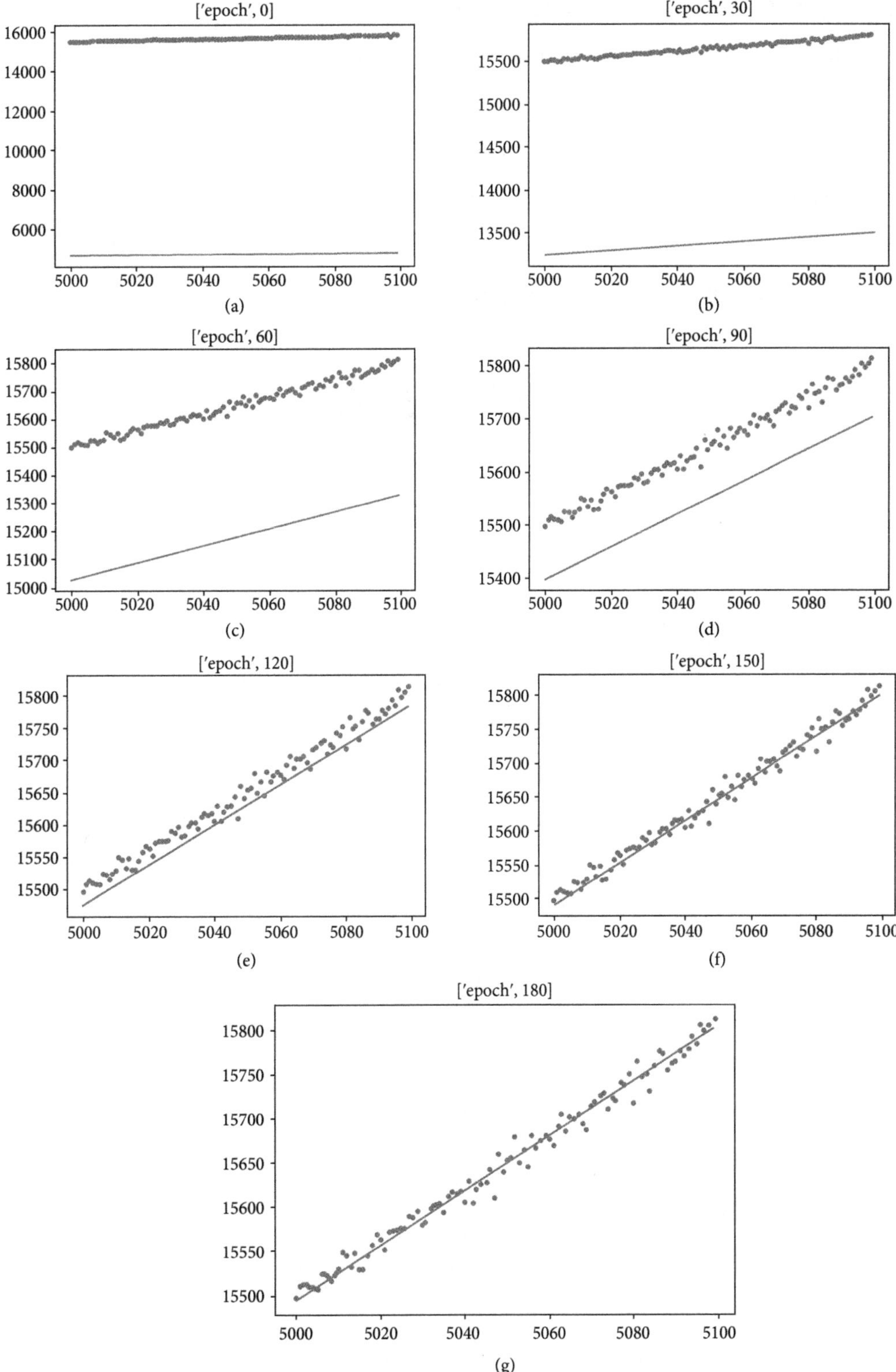

FIGURE 9.18   (*a*) Epoch 0 plot. (*b*) Epoch 30 plot. (*c*) Epoch 60 plot. (*d*) Epoch 90 plot. (*e*) Epoch 120 plot. (*f*) Epoch 150 plot. (*g*) Epoch 180 plot.

9.4.2   Logistic Regression Model Building with TensorFlow

The following code is used for building a logistic regression model in TensorFlow.

```python
# This step is for data creation
x_train= np.random.rand(100,1)
y_train=np.array([0 if i < 0.5 else 1 for i in x_train]).reshape(-1,1)

import matplotlib.pyplot as plt
plt.plot(x_train, y_train, 'b.',)
plt.show()
```

This code gives the output as shown in Fig. 9.19.

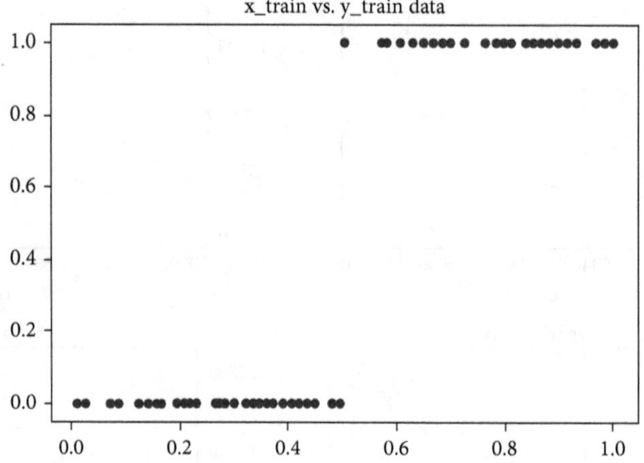

FIGURE 9.19   The code output: x_train versus y_train.

Given below is the code for building the model. The model equation changes, and the rest of the code remains the same as we used for linear regression in the previous section.

```python
#sigmoid wrapped around the linear equation
def output(x):
    return tf.sigmoid(W*x + b)

#Loss function : sum of squares
def loss_function(y_pred, y_true):
    return tf.reduce_sum(tf.square(y_pred - y_true))

#Initialize Weights
W = tf.Variable(tf.random.uniform(shape=(1, 1)))
b = tf.Variable(tf.zeros(shape=(1,)))

## Optimization
learning_rate = 0.1
steps = 300 #epochs

for i in range(steps):
    with tf.GradientTape() as tape:
        predictions = output(x_train)
        loss = loss_function(y_train, predictions)
        dloss_dw, dloss_db = tape.gradient(loss, [W, b])
    W.assign_sub(learning_rate * dloss_dw)
    b.assign_sub(learning_rate * dloss_db)
    print(f"epoch : {i}, loss {loss.numpy()}, W : {W.numpy()}, b {b.numpy()}")
```

The code above gives us the following output.

```
epoch : 0, loss  22.734655,  W : [[1.0051948]], b  [-1.0418472]
epoch : 1, loss  18.946268,  W : [[1.6631205]], b  [-0.69767785]
epoch : 2, loss  16.435150,  W : [[1.8341044]], b  [-1.2462128]
epoch : 3, loss  14.747210,  W : [[2.2962391]], b  [-1.1231506]
epoch : 4, loss  13.4111223, W : [[2.4752746]], b  [-1.4781798]
epoch : 5, loss  12.355062,  W : [[2.8097491]], b  [-1.4635689]
.........................................................................  . .

epoch : 294, loss  2.295719861,  W : [[13.713083]], b  [-7.182775]
epoch : 295, loss  2.2923476696, W : [[13.729434]], b  [-7.191068]
epoch : 296, loss  2.2889902591, W : [[13.745748]], b  [-7.1993423]
epoch : 297, loss  2.2856481075, W : [[13.762024]], b  [-7.2075977]
epoch : 298, loss  2.2823212146, W : [[13.778264]], b  [-7.2158346]
epoch : 299, loss  2.2790091037, W : [[13.794468]], b  [-7.224053]
```

We can print and visualize how the overall model has converged (Fig. 9.20).

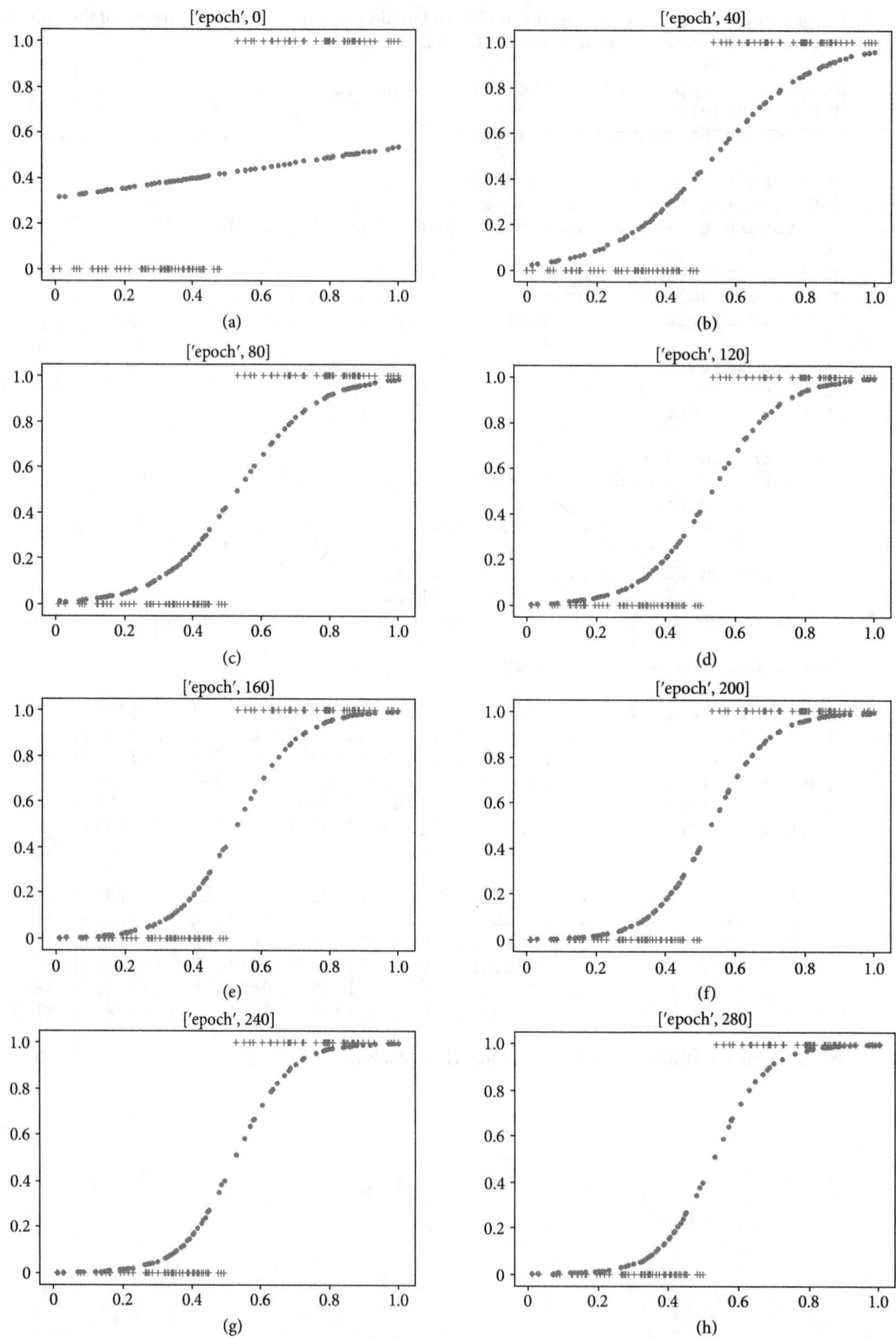

FIGURE 9.20 Steps in model convergence with increasing epoch sets: (*a*) Epoch 0 plot. (*b*) Epoch 40 plot. (*c*) Epoch 80 plot. (*d*) Epoch 120 plot. (*e*) Epoch 160 plot. (*f*) Epoch 200 plot. (*g*) Epoch 240 plot. (*h*) Epoch 280 plot.

9.5   *KERAS*

The programming in TensorFlow is very similar to working with NumPy and other essential functions in Python. The regression algorithm is the same; only the data changes from case to case. While building the machine learning algorithms earlier in this book, we neither created the cost function nor did we write the iterations. There are prebuilt libraries like Scikit-learn. We need to send the data and call the right function from the library. Scikit-learn internally uses NumPy and performs all optimization tasks.

Similarly, is there a package or wrapper on top of TensorFlow that takes only data from the user and does the optimization tasks on its own? Yes, it is; the package name is "Keras." In simple terms, Keras makes TensorFlow coding easy by providing some useful functions.

9.5.1   What Is Keras?

Keras is a high-level API on top of TensorFlow. Keras has several functions and features that will automatically write the TensorFlow code in the background. We need not write the low-level coding in TensorFlow. Keras has a simple syntax and fewer lines of code. Analogously, using TensorFlow is like using Numpy, whereas using Keras is like using the Scikit-learn package. Scikit-learn uses NumPy internally, and Keras internally uses TensorFlow. Most data scientists use Keras to build deep learning models. Keras has fewer lines of code, and learning Keras syntax is easy. Most importantly, Keras gives a lot of useful options while building the model. We need not install Keras separately. With TensorFlow 2.0 onward, Keras automatically gets installed along with TensorFlow. We can use the following command to import Keras:

```
from tensorflow import keras
```

9.5.2   Working with Keras

Keras coding also tries to mimic the network structure of neural networks. In Keras, we configure and build the models using a sequence of layers. The sequential model is a linear stack of layers. The first layer in the stack is the "first hidden Layer." We mention the information on the input shape. The last layer is the "Output Layer." The model gets information on labels from the last layer. We can add all the other "model layers" in between. Based on the input, hidden and output layers, the model will automatically prepare the weight parameters. In TensorFlow, the user has to mention all of it manually. Once we configure the model, we can send the input data and start the training process. Let us see an example of working with Keras.

9.5.3   MNIST on Keras

The dataset MNIST ("Modified National Institute of Standards and Technology") is the most widely used dataset of computer vision. This dataset is available as a sample dataset inside the Keras library.

The goal is to predict the number in the image by taking image pixel values as input. We will now write the code to understand the data and then move to model building.

```
##Import Packages
from tensorflow import keras
from tensorflow.keras import layers
from tensorflow.keras.models import Sequential
from tensorflow.keras.layers import Dense

## The data, shuffled and split between train and test sets
(X_train, Y_train), (X_test, Y_test) = keras.datasets.mnist.load_data()
num_classes=10
x_train = X_train.reshape(60000, 784)
x_test = X_test.reshape(10000, 784)
x_train = x_train.astype('float32')
x_test = x_test.astype('float32')
x_train /= 255
x_test /= 255
print(x_train.shape, 'train input samples')
print(x_test.shape, 'test input samples')
```

```
# convert class vectors to binary class matrices
y_train = keras.utils.to_categorical(Y_train, num_classes)
y_test = keras.utils.to_categorical(Y_test, num_classes)

print(y_train.shape, 'train output samples')
print(y_test.shape, 'test output samples')
```

This code is for importing the data. This data is a part of the Keras sample demo datasets. The following is the code output.

```
(60000, 784) train input samples
(10000, 784) test input samples
(60000, 10) train output samples
(10000, 10) test output samples
```

We have 60,000 images in the training data and 10,000 images in the test data. There are 784 pixels in each image. Let us sketch a few images before starting with the model building. The following code is used for the image sketches.

```
# Plot 4 images
%matplotlib inline
import matplotlib.pyplot as plt
plt.subplot(221)
plt.imshow(X_train[1], cmap=plt.get_cmap('gray'))
plt.subplot(222)
plt.imshow(X_train[6], cmap=plt.get_cmap('gray'))
plt.subplot(223)
plt.imshow(X_train[7], cmap=plt.get_cmap('gray'))
plt.subplot(224)
plt.imshow(X_train[9], cmap=plt.get_cmap('gray'))
plt.show()
```

Figure 9.21 shows the code output.

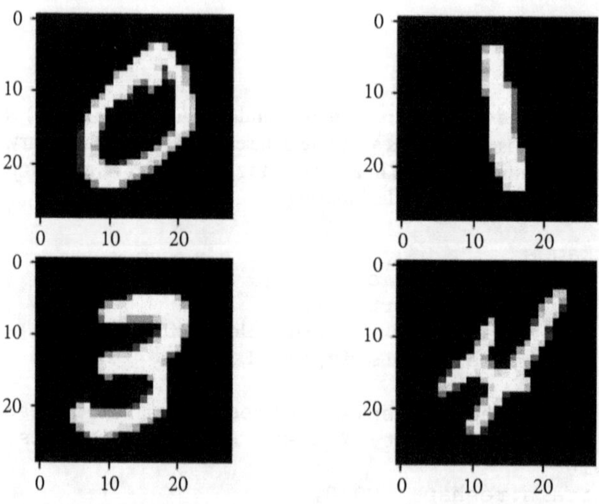

FIGURE 9.21   The code output images.

Now we will go ahead with model building. We need to configure the model. While configuring the model, we need to mention the layers sequentially. Given below is the code for building the model with two hidden layers of 20 nodes each. Our neural network is [784 input nodes; 20 nodes in H_1; 20 nodes in H_2; 10 nodes in output layer]. We expect 15,700 (785 × 20) weight parameters in the first layer, 420 (21 × 20) in the second layer, and 210 (21 × 10) in the final layer; `model.summary()` function gives us a summary of weight parameters in the network.

```
model = keras.Sequential()

model.add(layers.Dense(20, activation='sigmoid', input_shape=(784,)))

model.add(layers.Dense(20, activation='sigmoid'))

model.add(layers.Dense(10, activation='softmax'))

model.summary
```

Note on the code: A dense layer is simply a layer where each node in the previous layer is connected (densely connected) to every node in the next layer.

In the code above, we configure the hidden layer in the first step. The model needs to know what input shape it should expect. For this reason, the first layer in a sequential model needs to receive information about its input shape. Only the first layer needs the shape information because the following layers can make an automatic shape inference. In the final layer, mention the number of output classes as the number of nodes. The code above gives us the output shown in Table 9.5.

TABLE 9.5   The Code Output

```
Model: "sequential"
```

Layer (type)	Output Shape	Param #
dense (Dense)	**(None,** 20)	15700
dense_1 (Dense)	**(None,** 20)	420
dense_2 (Dense)	**(None,** 10)	210

```
Total params: 16,330
Trainable params: 16,330
Non-trainable params: 0
```

Now we are ready to build the model. The code below is used for compiling the model and training the model on the input data.

```
model.compile(loss='categorical_crossentropy', metrics=['accuracy'])

model.fit(x_train, y_train, epochs=10)
```

In the compile function, we need to mention the loss function and the validation metrics. We have discussed "mean squared error" as a loss function. There are some more options in the loss function. We will discuss all possible hyperparameters in the next chapter. For now, we can look at `categorical_crossentropy` as a specific formula for loss. The following is the code output.

```
Train on 60000 samples
Epoch 1/10
60000/60000 [=======] - 4s 61us/sample - loss: 0.9420 - accuracy: 0.7941
Epoch 2/10
60000/60000 [=======] - 3s 48us/sample - loss: 0.3382 - accuracy: 0.9055
Epoch 3/10
60000/60000 [=======] - 3s 45us/sample - loss: 0.2610 - accuracy: 0.9242
Epoch 4/10
60000/60000 [=======] - 3s 47us/sample - loss: 0.2270 - accuracy: 0.9338
Epoch 5/10
60000/60000 [=======] - 3s 48us/sample - loss: 0.2067 - accuracy: 0.9398
Epoch 6/10
60000/60000 [=======] - 4s 64us/sample - loss: 0.1920 - accuracy: 0.9439
Epoch 7/10
60000/60000 [=======] - 3s 47us/sample - loss: 0.1816 - accuracy: 0.9470
Epoch 8/10
60000/60000 [=======] - 3s 51us/sample - loss: 0.1730 - accuracy: 0.9493
Epoch 9/10
60000/60000 [=======] - 3s 49us/sample - loss: 0.1651 - accuracy: 0.9518
Epoch 10/10
60000/60000 [=======] - 3s 47us/sample - loss: 0.1594 - accuracy: 0.9537
```

The model is ready. We can use the model to get the accuracy of the test data using the following code.

```
loss, acc = model.evaluate(x_test,  y_test, verbose=2)
print("Test Accuracy: {:5.2f}%".format(100*acc))
```

The following is the code output for test accuracy.

```
10000/1 - 0s - loss: 0.0838 - accuracy: 0.9594
Test Accuracy: 95.94%
```

The model has achieved nearly 96 percent accuracy. The same model with 16,330 weights takes much more time if we use NumPy-based standard Python packages. Keras and TensorFlow complete the task considerably faster. This section concludes our discussion on model building with Keras.

9.6 CONCLUSION

In this chapter, we discussed a famous deep learning framework called TensorFlow. We learned some of the basic commands in TensorFlow. We then discussed Keras, a wrapper on the top of TensorFlow. This package, Keras, was developed independently; it was not part of TensorFlow. From TensorFlow 2.0 onward, Keras is installed automatically with TensorFlow. Now, TensorFlow documentation uses a lot of Keras code. Keras is probably the best and easiest way to build and implement deep learning models. In the upcoming chapters, we will use Keras in our exercises freely. The alternative option to TensorFlow is PyTorch. Both are equally good. You may choose either one and start building deep learning models.

9.7 REFERENCES

1. TensorFlow logo image: https://en.wikipedia.org/wiki/TensorFlow.
2. MNIST on Keras: http://yann.lecun.com/exdb/mnist/.

CHAPTER 10
DEEP LEARNING HYPERPARAMETERS

Almost all the machine learning algorithms have hyperparameters. In a few algorithms, we can permit them to use default values, while in some other algorithms, we have to fine-tune and find the optimal values (of hyperparameters). For example, in decision trees, we need to fine-tune the depth of the tree or the number of leaf nodes. Most basic machine learning algorithms have a maximum of two or three hyperparameters. Deep learning algorithms have a minimum of half a dozen hyperparameters. Apart from accuracy-related hyperparameters that will impact overfitting and underfitting, many hyperparameters affect the execution time, which is critical in neural networks. In principle, anyone can build a basic deep learning model. However, fine-tuning the hyperparameters and mastering the implications of each hyperparameter requires a significant amount of expertise. In this chapter, we will discuss some hyperparameters that play a critical role in deep neural networks.

10.1 REGULARIZATION

In the previous chapters, we have already discussed the role of the number of hidden layers and hidden nodes as critical hyperparameters in deep neural networks. If the number of hidden nodes is too high, then the model may get overfitted. If the number of hidden nodes is too low, then the model may be underfitted. To achieve the best result from any deep learning model, we need to have an optimal number of hidden nodes. When we are dealing with deep neural network problems, trying different combinations of hidden layers and nodes and building a model for each combination to find the final optimal number of layers and nodes is a tedious and time-consuming process. As an alternative, we can use regularization techniques, as discussed in this section.

Following a regularization technique, we choose a relatively large neural network to start with, which is most likely to lead to overfitting. We then keep some constraints on the weights so that they are always at a lower value. In this way, we have several nodes to explain the complexity (in the data). As each node does not exhibit its full power, the overfitting is also in control. In simple words, instead of dropping the hidden nodes, we keep them but with lower weights.

In a normal model building process, we try to minimize the total error while finding the weights; in regularization models, we minimize errors and weights.

$$\text{Actual error or cost function} = \sum \left(y - g\left(\sum_{k=1}^{m} w_k h_k \right) \right)^2$$

The equation above is the usual squared error. Given below is the new regularized error.

$$\text{New regularized error} = \sum \left(y - g\left(\sum_{k=1}^{m} w_k h_k \right) \right)^2 + \lambda \sum w_i^2$$

Usually, we try to minimize only the error. Here we are minimizing the error and sum of squares of weights. The second term ($\lambda \sum w_i^2$) is imposing some penalty on the weights. In the new regularized error equation, as given above, λ is the regularization parameter. We need to consider two scenarios—when λ is too high and λ is too low. If the value of λ is high, then to keep the overall cost function low, we may have to reduce the weights a lot. Imposing a high penalty on weights will lead to underfitting even if we have several (too many) hidden nodes. On the other hand, if the value of λ is zero (very near to zero), then there will be no regularization term; the weights will achieve their full effect that will ultimately lead to overfitting. We need to fine-tune and find the optimal value of the regularization parameter, λ.

10.1.1   Regularization in Regression

In regression, we try to minimize the sum of squares of errors.

$$\text{Regression cost function} = \sum\left(y_i - \sum_{k=1}^{m}\beta_k x_{ki}\right)^2$$

The regression will lead to overfitting if we have too many polynomial terms in the predictor variables list. If we overdo the feature engineering and create too many derived variables, then also it may lead to overfitting. However, we can still have a way to retain (too) many features by using the regularization method.

$$\text{New cost function} = \sum\left(y_i - \sum_{k=1}^{m}\beta_k x_{ki}\right)^2 + \lambda\sum_{k=1}^{m}\beta_k^2$$

In the above equation,

1. If λ is too low, the second term will be near to zero. Then weights will be calculated as usual. It may lead to overfitting.
2. If λ is too high, it will impose a high penalty on weights to reduce the overall cost function. This may lead to underfitting.
3. This regularization parameter needs to be fine-tuned.

Given below is a worked-out example on a minor dataset to demonstrate how regularization works in regression. The following code is to create the data and give us a plot.

```python
#import the packages
import pandas as pd
import numpy as np

# The data set
x=[-0.99768,-0.69574,-0.40373,-0.10236,0.22024,0.47742,0.82229]
y=[2.0885,1.1646,0.3287,0.46013,0.44808,0.10013,-0.32952]
input_data = pd.DataFrame(list(zip(x, y)), columns =['x', 'y'])
print(input_data)

# Plotting the data
x = np.array(input_data.x)
y = input_data.y
import matplotlib.pyplot as plt
get_ipython().run_line_magic('matplotlib', 'inline')
plt.title("Input data", fontsize=20)
plt.scatter(x,y,s=50,c="g")
plt.xlabel("X")
plt.ylabel("Y")
plt.show()
```

Figure 10.1 shows the code result.

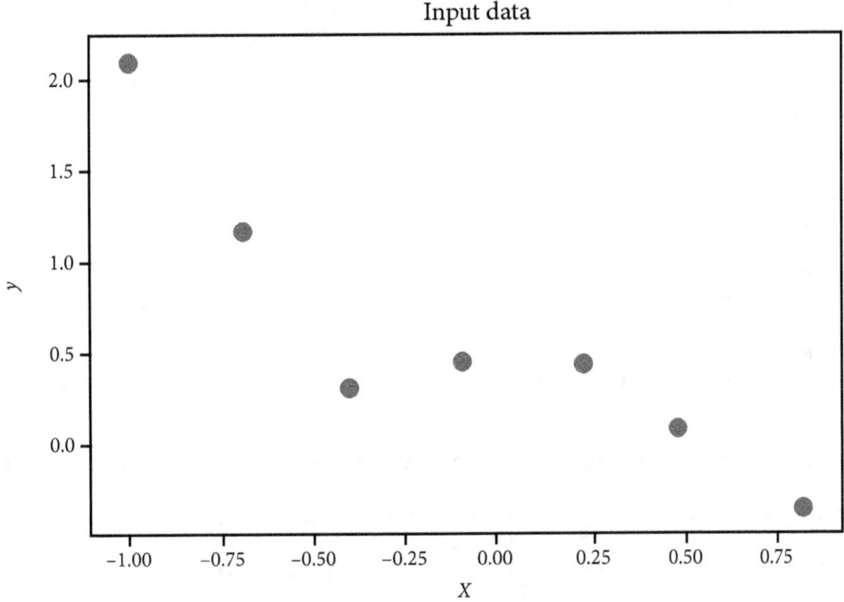

FIGURE 10.1 Input data plot resulting from the code.

We will try to build a simple regression line and a fifth-order polynomial regression. Simple regression (linear) will be an underfitted model for this data; on the other hand, a fifth-order polynomial will be an overfitted model.

```
##### Simple regression
import statsmodels.api as sm
x1 = sm.add_constant(x)
m1 = sm.OLS(y,x1).fit()
print("m1 SSE", m1.ssr)

##### Second Order polynomial regression
x2 = sm.add_constant(np.column_stack([x,np.square(x)]))
m2 = sm.OLS(y,x2).fit()
print("m2 SSE", m2.ssr)

##### Fifth order polynomial
x3 = sm.add_constant(np.column_stack([x,  np.power(x,2),np.power(x,3),np.
power(x,4),np.power(x,5)]))
m3 = sm.OLS(y,x3).fit()
print("m3 SSE", m3.ssr)
```

The following is the code output.

```
print("m1 SSE", m1.ssr)
m1 SSE 0.7107401451797566

print("m2 SSE", m2.ssr)
m2 SSE 0.457231720521299

print("m3 SSE", m3.ssr)
m3 SSE 0.010562888801624625
```

We will take this fifth-order polynomial model, which is already overfitted. The fifth-order polynomial for seven data points is always overfitted.

$$y = w_0 + w_1 x_1 + w_2 x_2 + w_3 x_3 + w_4 x_4 + w_5 x_5$$

We will now use a regularization parameter. The weights will be freshly calculated using a regularized cost function. We can get all the six weights, but as lambda increases, the weights (values) will reduce.

```python
X = x3
y = np.array(y)
n_col = X.shape[1]
d = np.identity(n_col)
d[0,0] = 0
w = []

reg =0
w.append(np.linalg.lstsq(X.T.dot(X) + reg * d, X.T.dot(y))[0])

reg =1
w.append(np.linalg.lstsq(X.T.dot(X) + reg * d, X.T.dot(y))[0])

reg =10
w.append(np.linalg.lstsq(X.T.dot(X) + reg * d, X.T.dot(y))[0])

print("Regularized weights  lambda=0 \n", w[0])
print("Regularized weights  lambda=1 \n", w[1])
print("Regularized weights  lambda=10 \n", w[2])
```

The code above prints the regularized weights.

```
Regularized weights  lambda=0
 [ 0.47252877  0.68135289 -1.38012842
-5.97768747  2.44173268  4.73711433]

Regularized weights  lambda=1
 [ 0.3975953  -0.42066637  0.12959211
-0.3974739   0.17525553 -0.33938772]

Regularized weights  lambda=10
 [ 0.52047074 -0.18250706  0.06064258
-0.14817721  0.07433006 -0.12795737]
```

With lambda = 0 we should see the same weights as the fifth-order polynomial regression. We have built that model previously. Let us compare the two models. Given below is the code for fetching the old and new weights.

```python
print("Regularized Weights With lambda = 0 \n", list(w[0]))
print("Standard Weights With inbuilt package \n",list(m3.params))
```

```
Regularized Weights With lambda = 0
 [0.4725287728743442, 0.6813528948567631, -1.3801284186124971,
-5.9776874674697105, 2.44173268479345 7, 4.737114334831566]

Standard Weights With inbuilt package
 [0.47252877287434003, 0.6813528948567651, -1.3801284186124536,
-5.977687467469684, 2.4417326847934, 4.7371143348315226]
```

As expected, the regularized weights with lambda = 0 are the same as standard weights. Now we will see the plot and observe the results from the three models.

```python
import matplotlib.pyplot as plt
get_ipython().run_line_magic('matplotlib', 'inline')
plt.rcParams["figure.figsize"] = (8,6)
plt.title('Model results for different lambda values', fontsize=20)
plt.scatter(x,y, s = 50, c = "g")
x_new = np.linspace(x.min(), x.max(), 200)
```

```
plt.plot(x_new,  np.poly1d(np.polyfit(x,  X.dot(w[0]),  5))(x_new),label='$\
lambda$ = 0',  c = "b")
plt.plot(x_new,  np.poly1d(np.polyfit(x,  X.dot(w[1]),  5))(x_new),label='$\
lambda$ = 1',  c = "r")
plt.plot(x_new,  np.poly1d(np.polyfit(x,  X.dot(w[2]),  5))(x_new),label='$\
lambda$ = 10',  c = "g")
plt.legend(loc='upper right');
plt.show()
```

Figure 10.2 is the output plot from this code.

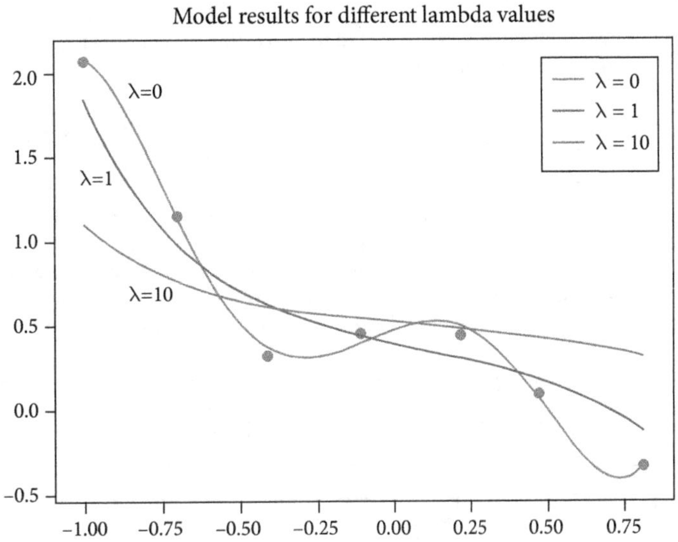

FIGURE 10.2   Model results as the code output.

Given below are the observations from the graph curves shown in Fig. 10.2. The first regression line (line with many curves) is the model with lambda = 0; this model is overfitted. The lambda = 0 model is equivalent of the standard fifth-order polynomial model without regularization. The second regression line (almost flat line) is the model with lambda = 10, and this model is under-fitted. The final regression line (the middle line) is the model with lambda = 1; this model is better than the other two models.

We conclude that in the three models that are built with a fifth-order polynomial, by changing the value of the regularization parameter, we can avoid overfitting. From the output shown in Fig. 10.2, we choose the model with lambda = 1.

```
Final Weights
 [ 0.3975953   -0.42066637   0.12959211
-0.3974739    0.17525553  -0.33938772]
Final SSE  0.24363202160352718
```

10.1.2 L1 and L2 Regularization

In regression, adding too many polynomial terms led to overfitting. In the previous section, we did not reduce the number of polynomial terms, but we used a regularized cost function to regulate the weights. Until now, we minimized the sum of squares of weights (using a regularized cost function). This method is known as L2 norm or L2 regularization. A regression model based on this method is known as ridge regression.

$$\text{L2 regularization} = \sum \left(y_i - \sum_{k=1}^{m} \beta_k x_{ki} \right)^2 + \lambda \sum_{k=1}^{m} \beta_k^2$$

Alternatively, we can use the cost function that minimizes the sum of absolute values of the weights. This method is known as L1 norm or L1 regularization. In regression, it is known as Losso regression.

$$\text{L1 regularization} = \sum\left(y_i - \sum_{k=1}^{m}\beta_k x_{ki}\right)^2 + \lambda\sum_{k=1}^{m}|\beta_k|$$

For a specific type of data, L1 works the best, and for a few other specific data types, L2 works out to be better. Both L1 and L2 regularization reduce overfitting effectively; we can use either one. Experience is the best teacher in this case.

10.1.3   Regularization in Neural Networks

Too many hidden nodes and hidden layers will lead to overfitting in neural networks. We can use the following regularized cost function in the case of neural networks.

$$\text{New L2 regularized error} = \sum\left(y - g\left(\sum_{k=1}^{m}w_k h_k\right)\right)^2 + \lambda\sum w_i^2$$

While building a neural network, if we minimize the above equation, then weights will not be inflated. It will reduce overfitting. To comprehend and visualize the impact on the neural network, we will use simulations available on the TensorFlow playground website: https://playground.tensorflow.org/. This website is a useful resource to try and understand different hyperparameters in neural networks. Once we open the website, we need to configure the parameters on the left-hand side and the top of the page. Figure 10.3 provides some details related to this website.

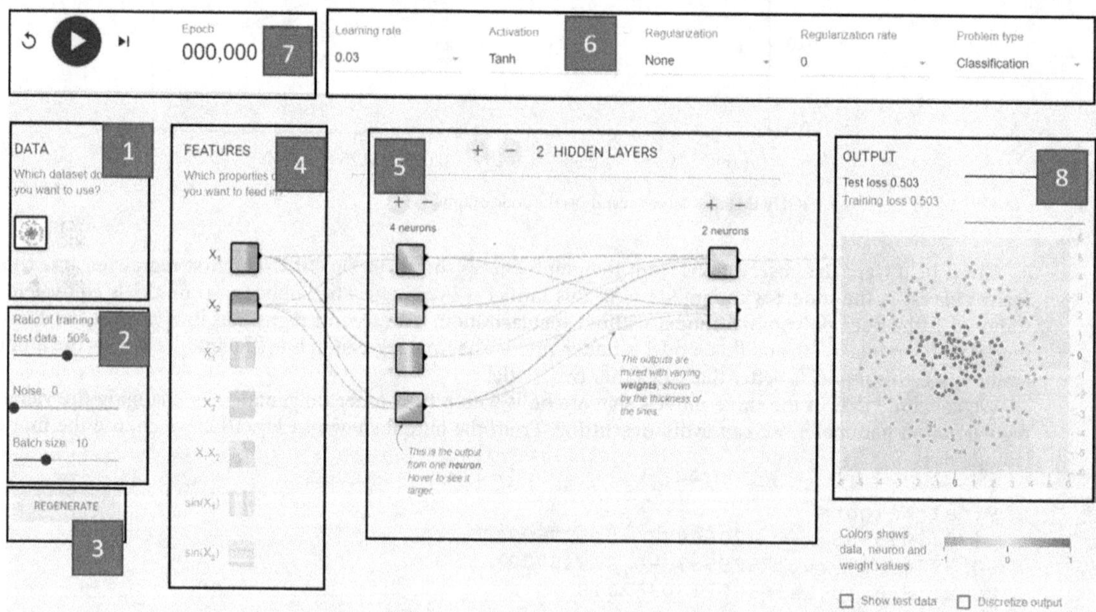

FIGURE 10.3   TensorFlow playground website.

The following is the explanation of each box in the image shown in Fig. 10.3.

1. *Box 1:* Choose the dataset for simulation. There are a total of four datasets: Circle, XOR, Gaussian, and Spiral (Fig. 10.4).

FIGURE 10.4   Datasets on playground website.

2. *Box 2:* Choose the ratio of train and test data. Choose the noise or the irreducible error and then the batch size (Fig. 10.5). We will discuss it later.

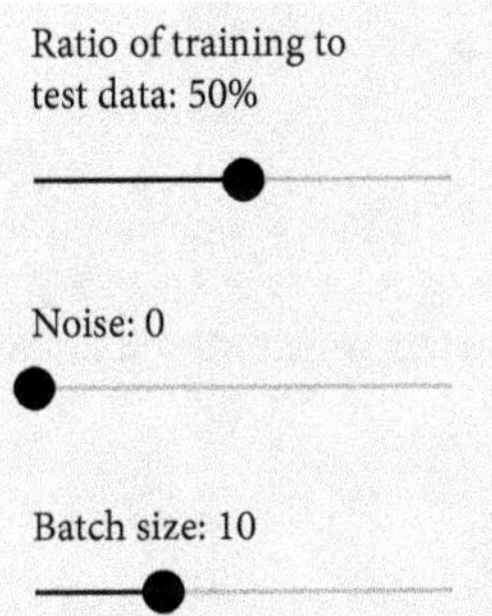

FIGURE 10.5   Choose batch size.

3. *Box 3:* Click this button to regenerate a new random data (Fig. 10.6).

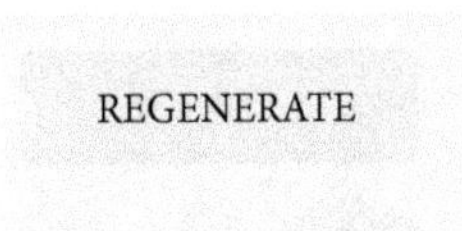

FIGURE 10.6   Regenerate new random data.

4. *Boxes 4 and 5*—Input layer details: The actual variables and their transformations are available. We can choose them as input nodes. In Box 5, we can mention hidden layers and nodes (Fig. 10.7).

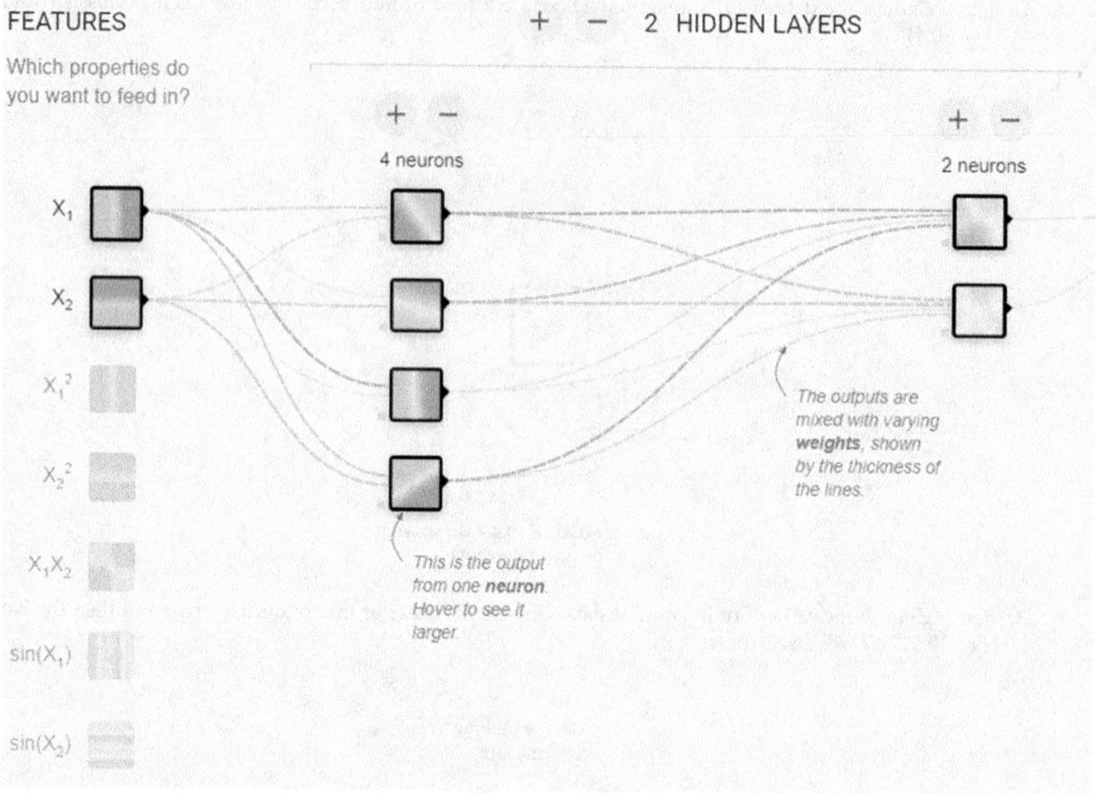

FIGURE 10.7   Network configuration.

5. *Box 6:* It has some more hyperparameters which we will discuss later (Fig. 10.8).

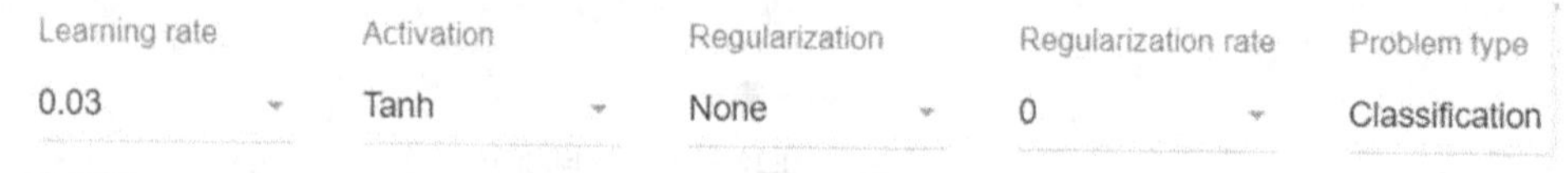

FIGURE 10.8   Hyperparameters configuration.

6. *Box 7:* Click on the play button to start the algorithm execution. Epochs are shown while the algorithm is executing (Fig. 10.9).

FIGURE 10.9   Start the algorithm execution.

7. *Box 8:* This box shows the final output. Train and test errors are shown for each epoch. We can also see the classification graph (Fig. 10.10).

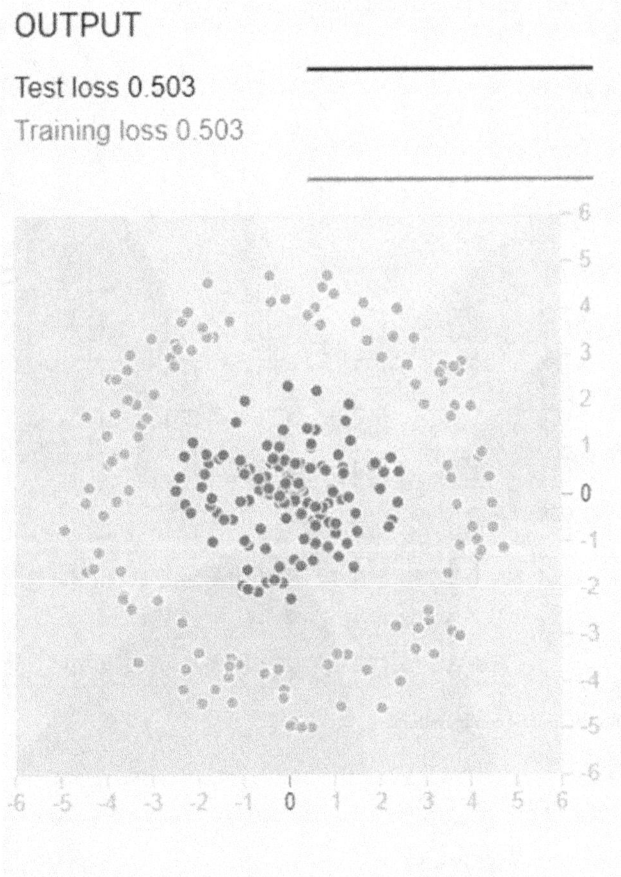

FIGURE 10.10   Train and test loss.

We need to set all the parameters and click on the play button. It will then start the training process. We can stop the training process after certain epochs. We will now build an overfitted neural network model with too many hidden layers and nodes. In the next step, we will apply regularization and avoid overfitting. Follow the configuration as given in Table 10.1, and build a neural network model; stop after executing 100 epochs, and look at the train and test data accuracy.

TABLE 10.1   Configuration Table Showing Different Parameters

Data	Exclusive or (XOR)
Ratio of training to test data	50%
Noise	25
Batch size	10
Features	$x_1, x_2, x_1^2, x_2^2, x_1 * x_2, \sin x_1, \sin x_2, \ldots$
Hidden layers	6
Neurons or hidden nodes	8 in each layer. Total of 48 nodes
Learning rate	0.03
Activation	Tanh
Regularization	None
Regularization rate	0
Problem type	Classification

Choose the configuration listed in Table 10.1 and hit the play button to run it for 100 epochs. Look at the train and test error in the output. Reset and regenerate the sample(Box-3); run it for 100 epochs. Try it multiple times. We can continuously observe the effect of overfitting.

Figure 10.11 shows the result of the configuration given in Table 10.1.

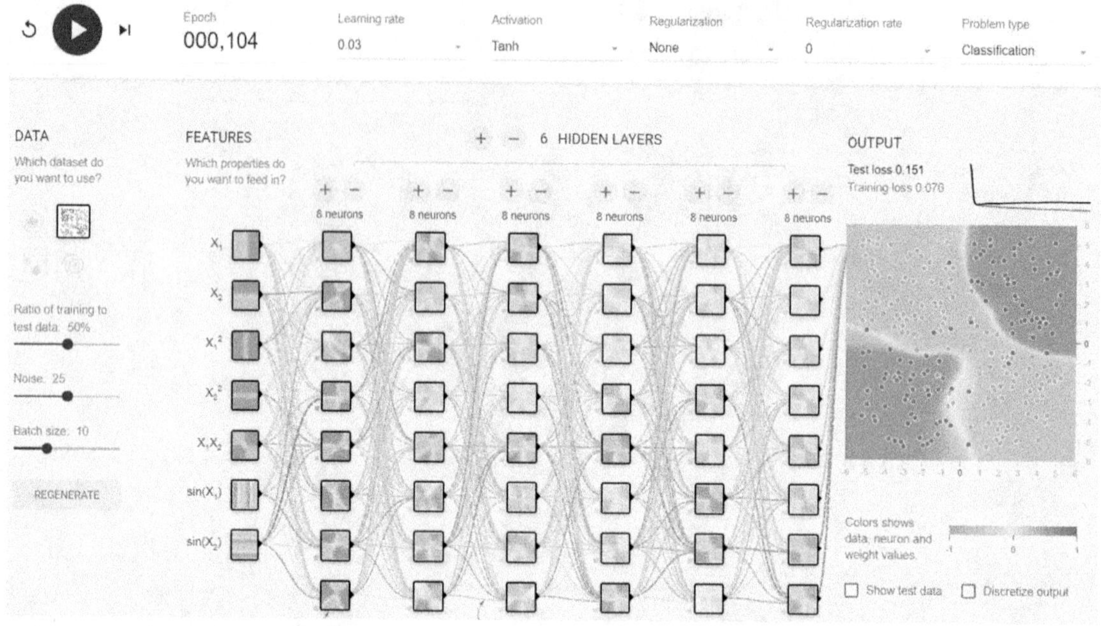

FIGURE 10.11   Results for specific configurations.

Let us focus on the output (Fig. 10.12).

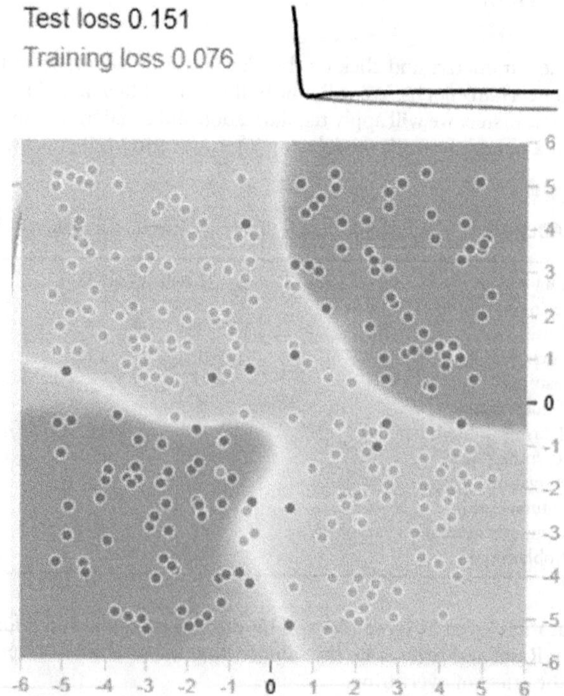

FIGURE 10.12   The output.

As expected, the output shows a high error on test data and low error on train data. We need only a single layer with two nodes for this data, but we have configured it using six layers and eight nodes in each of the layers. If you are not able to observe the same results, then click on the regenerate button and train from epoch 0 again. We will now use regularization to avoid this overfitting in the model by keeping the same number of layers and nodes. Table 10.2 is the configuration table with updated values of regularization. Again, train for 100 epochs with the configuration given in Table 10.2.

TABLE 10.2   Configuration Table with Regularization and Its Rate Activated

Data	Exclusive or (XOR)
Ratio of training to test data	50%
Noise	25
Batch size	10
Features	$x_1, x_2, x_1^2, x_2^2, x_1 * x_2, \sin x_1, \sin x_2,$
Hidden layers	6
Neurons or hidden nodes	8 in each layer. Total of 48 nodes
Learning rate	0.03
Activation	Tanh
Regularization	**L2**
Regularization rate	**0.1**
Problem type	Classification

The configuration in Table 10.2 gives us the output as in Fig. 10.13.

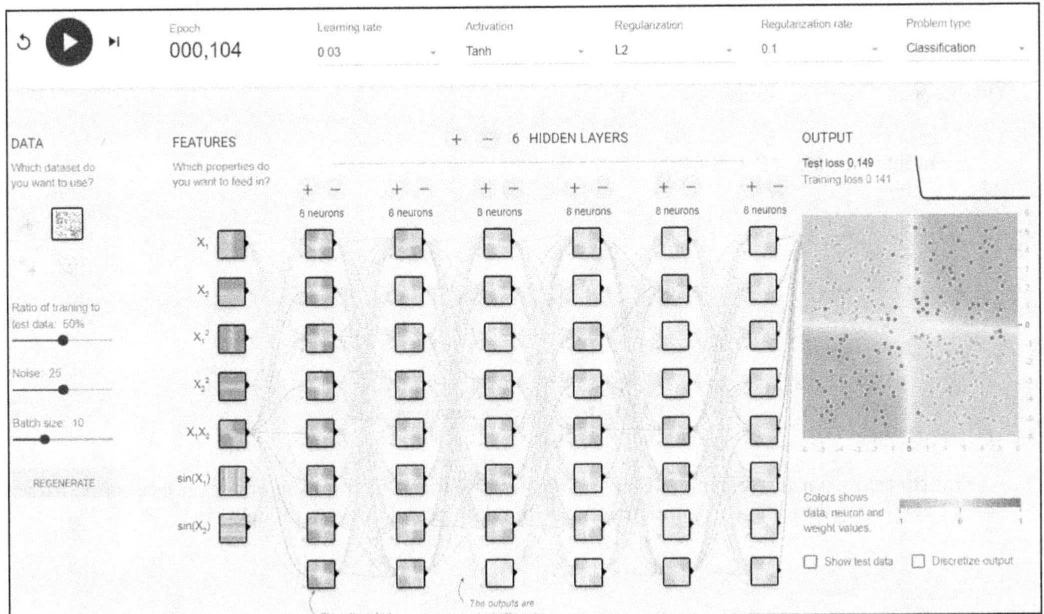

FIGURE 10.13   Model output.

Let us zoom into the output (Fig. 10.14).

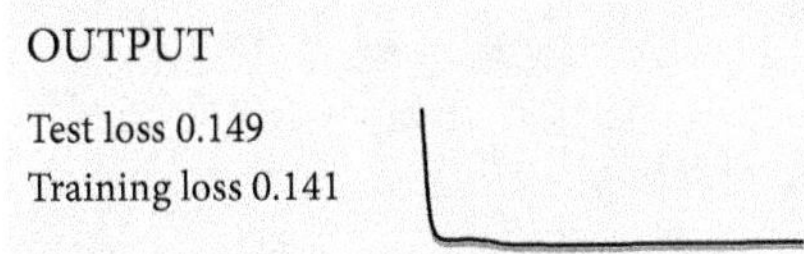

FIGURE 10.14   Train and test loss of model.

In this experiment, we used 48 hidden nodes; still, we avoided overfitting using L2 regularization. We will now rebuild the model with L1 regularization with the lambda value as 0.03. Figure 10.15 shows the result of L1 regularization.

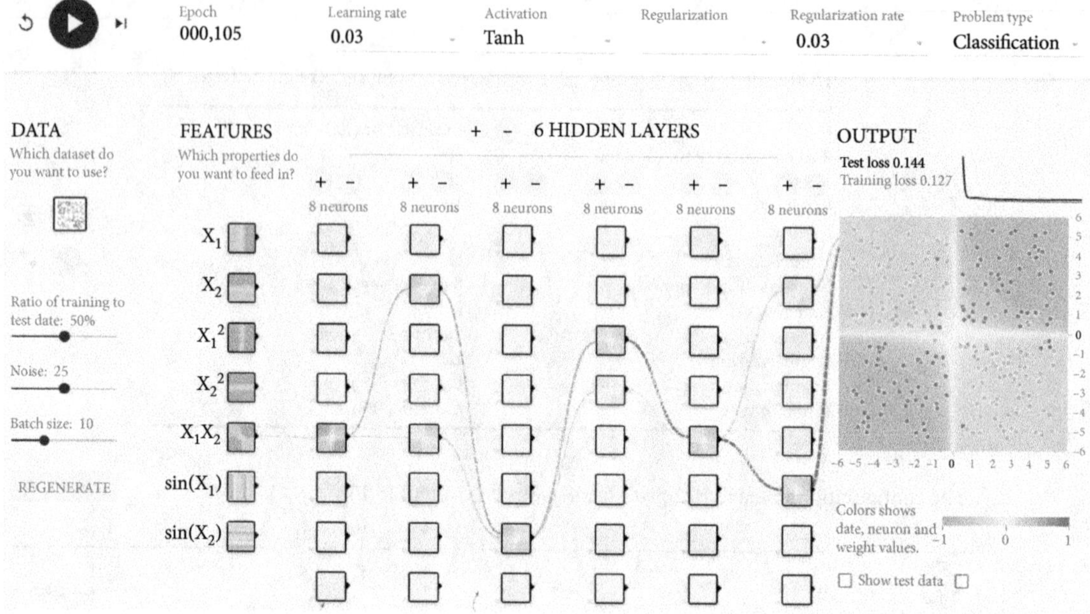

FIGURE 10.15   Results of L1 regularization.

Zoom into the output (Fig. 10.16).

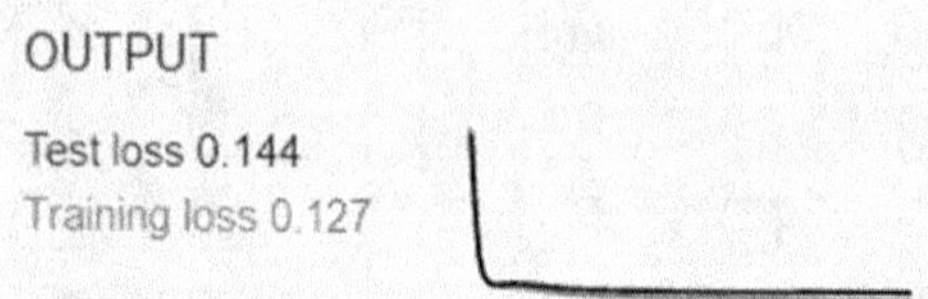

FIGURE 10.16   L1 regularization output focused.

This concludes our discussion of L1 and L2 regularization and their impact on a neural network. Now we will see how L1 and L2 regularization are configured while building neural network models.

10.1.4   L1 and L2 Regularization Code

In Keras (explained in Chap. 9), we can use the `kernel_regularizer` parameter inside the `layers.Dense()` function. Table 10.3 presents the code for building the model without regularization and with regularization.

TABLE 10.3   Code with and without Regularization

Without Regularization

```
model.add(layers.Dense(128, activation='sigmoid'))

model.add(layers.Dense(10, activation='softmax'))
```

With Regularization

```
model_r.add(layers.Dense(256, activation='sigmoid', input_
shape=(784,), kernel_regularizer=regularizers.l2(0.01)))

model_r.add(layers.Dense(128, activation='sigmoid',kernel_
regularizer=regularizers.l2(0.01)))
```

When we build a model with too many hidden nodes, we can see the accuracy of train data is high starting from the first epoch. The same configuration model shows less accuracy in its epochs while training with regularization parameters. Given below are the results from the two models: the first one is without any regularization and the second one is with regularization.

```
model = keras.Sequential()
model.add(layers.Dense(256, activation='sigmoid', input_shape=(784,)))
model.add(layers.Dense(128, activation='sigmoid'))
model.add(layers.Dense(10, activation='softmax'))
model.summary()

model.compile(loss='categorical_crossentropy', metrics=['accuracy'])
model.fit(x_train, y_train,epochs=10)
```

The following is the code output.

```
Train on 60000 samples
Epoch 1/10
60000/60000 [=====]     7s 111us/sample - loss: 0.3734 - accuracy: 0.8954
Epoch 2/10
60000/60000 [=====]     6s 105us/sample - loss: 0.1657 - accuracy: 0.9494
Epoch 3/10
60000/60000 [=====]     7s 112us/sample - loss: 0.1170 - accuracy: 0.9641
Epoch 4/10
60000/60000 [=====]     7s 118us/sample - loss: 0.0903 - accuracy: 0.9724
Epoch 5/10
60000/60000 [=====]     7s 119us/sample - loss: 0.0738 - accuracy: 0.9771
Epoch 6/10
60000/60000 [=====]     7s 116us/sample - loss: 0.0621 - accuracy: 0.9813
Epoch 7/10
60000/60000 [=====]     7s 112us/sample - loss: 0.0532 - accuracy: 0.9847
Epoch 8/10
60000/60000 [=====]     6s 101us/sample - loss: 0.0451 - accuracy: 0.9867
Epoch 9/10
60000/60000 [=====]     7s 117us/sample - loss: 0.0400 - accuracy: 0.9883
Epoch 10/10
60000/60000 [=====]     7s 113us/sample - loss: 0.0347 - accuracy: 0.9902
```

```
#Final Results
loss, acc = model.evaluate(x_train,  y_train, verbose=2)
print("Train Accuracy: {:5.2f}%".format(100*acc))

loss, acc = model.evaluate(x_test,  y_test, verbose=2)
print("Test Accuracy: {:5.2f}%".format(100*acc))

60000/1 - 2s - loss: 0.0142 - accuracy: 0.9920
Train Accuracy: 99.20%
10000/1 - 0s - loss: 0.0424 - accuracy: 0.9774
Test Accuracy: 97.74%
```

We can see slight overfitting in this model. We can see above 90 percent accuracy from the second epoch itself. We will build the same model with regularization now.

```
from tensorflow.keras import regularizers
model_r = keras.Sequential()
model_r.add(layers.Dense(256, activation='sigmoid', input_shape=(784,), kernel_
regularizer=regularizers.l2(0.01)))
model_r.add(layers.Dense(128, activation='sigmoid',kernel_regularizer=
regularizers.l2(0.01)))
model_r.add(layers.Dense(10, activation='softmax'))
model_r.summary()

model_r.compile(loss='categorical_crossentropy', metrics=['accuracy'])
model_r.fit(x_train, y_train,epochs=10)
```

The following is the output.

```
Train on 60000 samples
Epoch 1/10
60000/60000 [=====]   - 7s 119us/sample - loss: 1.7504 - accuracy: 0.6602
Epoch 2/10
60000/60000 [=====]   - 8s 134us/sample - loss: 1.3363 - accuracy: 0.7459
Epoch 3/10
60000/60000 [=====]   - 9s 145us/sample - loss: 1.2472 - accuracy: 0.7570
Epoch 4/10
60000/60000 [=====]   - 8s 139us/sample - loss: 1.1956 - accuracy: 0.7654
Epoch 5/10
60000/60000 [=====]   - 8s 141us/sample - loss: 1.1568 - accuracy: 0.7727
Epoch 6/10
60000/60000 [=====]   - 9s 143us/sample - loss: 1.1159 - accuracy: 0.7875
Epoch 7/10
60000/60000 [=====]   - 7s 116us/sample - loss: 1.0757 - accuracy: 0.8009
Epoch 8/10
60000/60000 [=====]   - 7s 118us/sample - loss: 1.0428 - accuracy: 0.8104
Epoch 9/10
60000/60000 [=====]   - 7s 122us/sample - loss: 1.0114 - accuracy: 0.8158
Epoch 10/10
60000/60000 [=====]   - 8s 127us/sample - loss: 0.9873 - accuracy: 0.8214

#Final Results
loss, acc = model_r.evaluate(x_train,  y_train, verbose=2)
print("Train Accuracy: {:5.2f}%".format(100*acc))
```

```
loss, acc = model_r.evaluate(x_test,  y_test, verbose=2)
print("Test Accuracy: {:5.2f}%".format(100*acc))

60000/1 - 2s - loss: 0.9381 - accuracy: 0.7984
Train Accuracy: 79.84%
10000/1 - 0s - loss: 0.8485 - accuracy: 0.8025
Test Accuracy: 80.25%
```

We can now see the impact of regularization on the weights. Epoch by epoch comparison on the train data reveals the impact of the penalty on weights. The final model also shows no signs of overfitting. We can apply L1 regularization also in a similar manner using the parameter `kernel_regularizer=regularizers.l1(0.01)`.

10.1.5   Data Standardization in L1 and L2 Regularization

There is one crucial point that we need to be extremely careful while working with L1 and L2 regularization—since we are directly imposing penalties on weights and we are using a single regularization parameter lambda on all the weights together. We need to bring all the features on to the same scale. We need to standardize the data before applying regularization in deep neural networks. If the data is not standardized, then the regularization parameter may not have any impact on some of the input and hidden nodes. In the example given in Sec. 10.1.4, we used the following code for data standardization. We divided all the input data by 255, which is the maximum of the training data column. This code brings all the values between 0 and 1.

```
x_train /= 255
x_test /= 255
```

Here we did the data standardization by dividing all the data values by 255. However, for more complex sets of data, we may need to perform a different type of operation(s) (to standardize), which will depend upon the type of data.

10.2   *DROPOUT REGULARIZATION*

There is one more regularization that helps us to reduce the dominance of individual hidden nodes. The dropout method is another effective way of avoiding overfitting. In the dropout method, we ignore a few hidden nodes while training the model. In the backpropagation, we randomly drop a few nodes in each of the forward passes. This random dropping of nodes will make sure that no single hidden node is dominating in the network. We have to mention the probability (p) of dropping each node. If p is near to 0, then there will be too many hidden nodes, and the model may come out as overfitted. If p is near to 1, then we can avoid overfitting even if there are too many hidden nodes. Table 10.4 shows an example of a dropout method with $p = 0.5$. As discussed, here p is the probability of dropping a node.

TABLE 10.4 Dropout Regularization

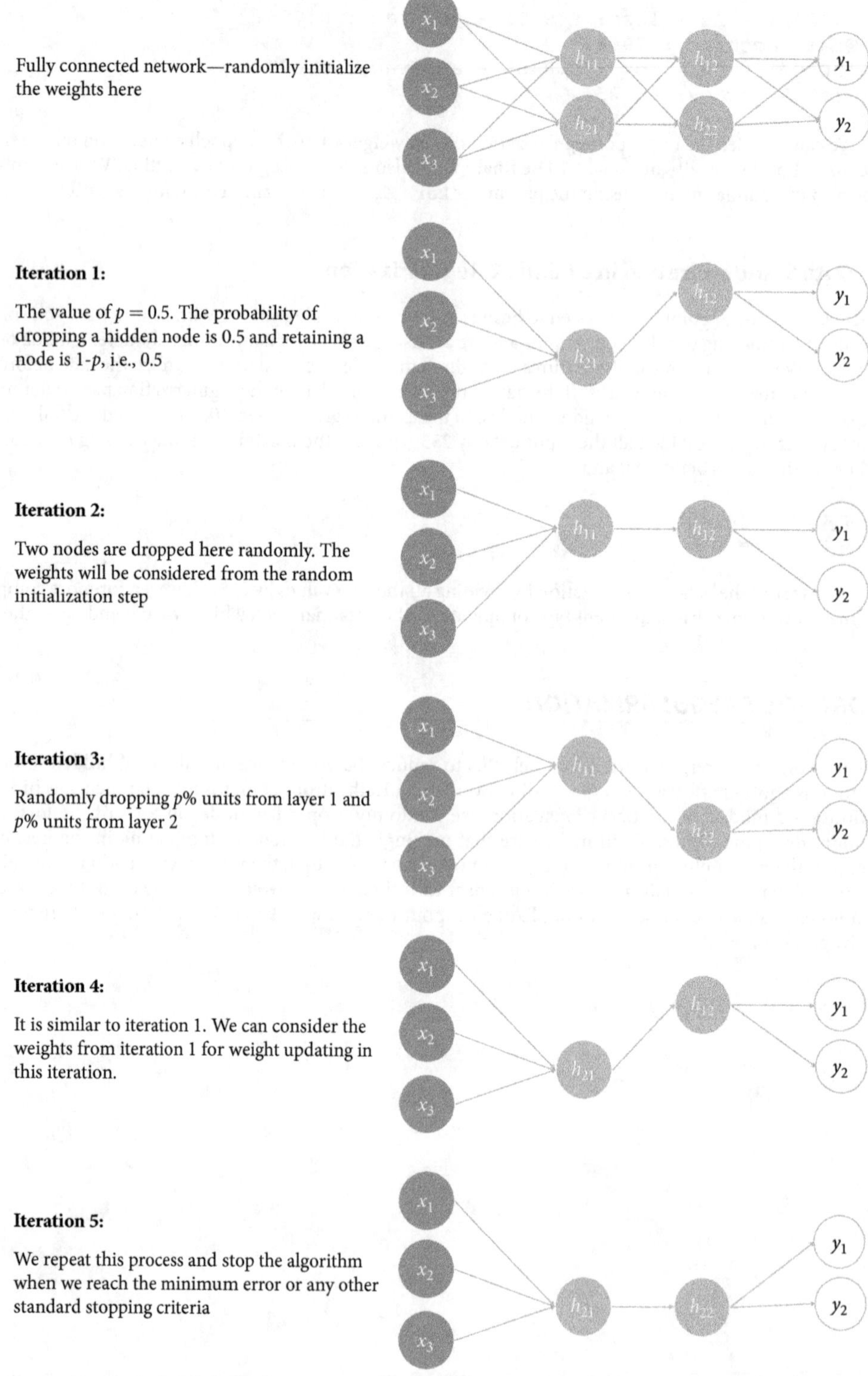

Fully connected network—randomly initialize the weights here

Iteration 1:

The value of $p = 0.5$. The probability of dropping a hidden node is 0.5 and retaining a node is 1-p, i.e., 0.5

Iteration 2:

Two nodes are dropped here randomly. The weights will be considered from the random initialization step

Iteration 3:

Randomly dropping $p\%$ units from layer 1 and $p\%$ units from layer 2

Iteration 4:

It is similar to iteration 1. We can consider the weights from iteration 1 for weight updating in this iteration.

Iteration 5:

We repeat this process and stop the algorithm when we reach the minimum error or any other standard stopping criteria

While training the model, we have not considered all the four hidden nodes. We always dropped $p\%$ of the nodes in each hidden layer. Once we reach the stopping criteria, how do we use this model for prediction? For example, the original model has 14 $(6 + 4 + 4)$ weights. In each iteration, we have trained only 6 $(3 + 1 + 2)$ weights. We will take the final values of all the 14 weights from the trained model. The available weights will be considered from the last iteration, and the remaining weights can be picked from its previous iterations when they last appeared. While calculating the predicted values, we multiply each weight by q $(q = 1-p)$, where q is the probability of retaining a node. Since we have not trained the network with all nodes, we should not use all of them in testing. If we are using all the weights, then we have to multiply the weights by q to incorporate the impact of dropout. Consider the following expressions:

$$\text{While training } y = \text{sigmoid}\left(w_{22}h_{22}\right) \text{ or } y = \text{sigmoid}\left(w_{12}h_{12}\right)$$

$$\text{While using the model for prediction } y = \text{sigmoid}\left(q * w_{22}h_{22} + q * w_{12}h_{12}\right)$$

Similarly, we apply the same multiplication factor q on the hidden layer level as well. In our network, we have 14 weights, and we can use all the weights in the prediction, but we need to multiply each weight by q.

The following three crucial points are to be noted in the dropout method:

1. The dropout is not applied at the overall network level; it is applied to each hidden layer level. We can even apply the dropout on a few layers and keep all nodes in the rest of the layers.

2. The dropout happens at each iteration. We do not drop the nodes once and train the network. We drop the weights randomly in each iteration. We cannot guess what is the network architecture in a given iteration.

3. The third point is about weights. All the weights are considered at the time of prediction. Each weight is multiplied by q, where $q = 1 - p$.

10.2.1   Dropout Method's Code

We need to mention dropout as a layer. It is an imaginary layer that is applied to the hidden layers. The dropout rate can vary from one layer to another.

```python
from tensorflow.keras.layers import Dropout

model_rd = keras.Sequential()

model_rd.add(layers.Dense(256, activation='sigmoid', input_shape=(784,)))
model_rd.add(Dropout(0.7))

model_rd.add(layers.Dense(128, activation='sigmoid'))
model_rd.add(Dropout(0.6))

model_rd.add(layers.Dense(10, activation='softmax'))
model_rd.summary()
```

From this code, we can figure out the dropout layer after the hidden layer. We choose $p = 0.7$, which means 70 percent of the nodes will be dropped from the first hidden layer, and 60 percent of the nodes will be dropped from the second hidden layer. At any given iteration, we will see only 77 nodes in the first hidden layer and 51 nodes in the second layer. The code given above gives us the results in Table 10.5.

TABLE 10.5   The Code Output

Layer (type)	Output Shape	Param #
dense_6 (Dense)	**(None,** 256)	200960
dropout (Dropout)	**(None,** 256)	0
dense_7 (Dense)	**(None,** 128)	32896
dropout_1 (Dropout)	**(None,** 128)	0
dense_8 (Dense)	**(None,** 10)	1290

```
Total params: 235,146
Trainable params: 235,146
Non-trainable params: 0
```

Dropout is an abstract layer with zero nodes. We are now all set to train the model. For an individual epoch, we will see less accuracy, as we are using only fewer nodes. Finally, when we calculate the accuracy of train and test data, we will see a higher value.

```
model_rd.compile(loss='categorical_crossentropy', metrics=['accuracy'])
model_rd.fit(x_train, y_train,epochs=10)
```

The following is the code output.

```
60000/60000 [=====] - 7s 120us/sample - loss: 0.8283 - accuracy: 0.7303
Epoch 2/10
60000/60000 [=====] - 8s 126us/sample - loss: 0.4358 - accuracy: 0.8732
Epoch 3/10
60000/60000 [=====] - 7s 115us/sample - loss: 0.3768 - accuracy: 0.8950
Epoch 4/10
60000/60000 [=====] - 7s 122us/sample - loss: 0.3405 - accuracy: 0.9054
Epoch 5/10
60000/60000 [=====] - 7s 119us/sample - loss: 0.3188 - accuracy: 0.9146
Epoch 6/10
60000/60000 [=====] - 7s 120us/sample - loss: 0.3047 - accuracy: 0.9173
Epoch 7/10
60000/60000 [=====] - 8s 127us/sample - loss: 0.2980 - accuracy: 0.9218
Epoch 8/10
60000/60000 [=====] - 7s 115us/sample - loss: 0.2903 - accuracy: 0.9244
Epoch 9/10
60000/60000 [=====] - 7s 109us/sample - loss: 0.2783 - accuracy: 0.9290
Epoch 10/10
60000/60000 [=====] - 7s 109us/sample - loss: 0.2734 - accuracy: 0.9315

#Final Results
loss, acc = model_rd.evaluate(x_train, y_train, verbose=2)
print("Train Accuracy: {:5.2f}%".format(100*acc))

loss, acc = model_rd.evaluate(x_test, y_test, verbose=2)
print("Test Accuracy: {:5.2f}%".format(100*acc))

60000/1 - 2s - loss: 0.0781 - accuracy: 0.9603
Train Accuracy: 96.03%
10000/1 - 1s - loss: 0.0869 - accuracy: 0.9575
Test Accuracy: 95.75%
```

We can see from the above output that the model is not showing any signs of overfitting.

10.3   *EARLY STOPPING METHOD*

The early stopping method follows a simple approach to avoid overfitting. If there are too many hidden nodes and layers, then the accuracy of train data increases as the number of epochs increases. With sufficient hidden nodes and adequate epochs, the accuracy might even reach 100 percent. The story is different in test data. The accuracy of test data might increase for the first few epochs; after that it might decrease, when we are about to enter the zone of overfitting. We can stop the model building (iterations) at the epoch, where we do not see any significant growth in the accuracy of the test data. This process is known as early stopping. While building the model itself, we can save the model for each epoch inside a file and finally take the model that has the highest accuracy on train data and the matching accuracy on test data.

Early stopping is very similar to what we manually do to avoid overfitting. We manually observe the point where the test data shows a dip in the accuracy level. If we incorporate the same method inside the code, then it turns out to be early stopping.

First, we need to learn how to store the model and its weights in each epoch. For storing the model weights, we use the h5py package. Using this package, we can store the model weights in a file. The model weight file will have the hdf5 extension.

The code given below is used for saving the model (in each epoch).

```
model_re = keras.Sequential()
model_re.add(layers.Dense(256, activation='sigmoid', input_shape=(784,)))
model_re.add(layers.Dense(128, activation='sigmoid'))
model_re.add(layers.Dense(10, activation='softmax'))
model_re.summary()

model_re.compile(loss='categorical_crossentropy', metrics=['accuracy'])

from tensorflow.keras.callbacks import ModelCheckpoint
import h5py

checkpoint = ModelCheckpoint(r"D:\Chapter10 Deep Learning Hyperparameters\4.
Code\epoch-{epoch:02d}.hdf5")

model_re.fit(x_train, y_train,epochs=10,validation_data=(x_test, y_test),
callbacks=[checkpoint])
```

The code above saves all the model weight files inside the directory that we have mentioned in the code. This code gives us the following output. The vital step in the above code is saving model checkpoints. We first mention the model checkpoints directory and then mention it in the callbacks parameter in the `fit()` function.

```
Train on 60000 samples, validate on 10000 samples
Epoch 1/10
60000/60000 [=====] - 8s 128us/sample - loss: 0.3773 - accuracy: 0.8932 -
val_loss: 0.1967 - val_accuracy: 0.9388
Epoch 2/10
60000/60000 [=====] - 7s 116us/sample - loss: 0.1664 - accuracy: 0.9490 -
val_loss: 0.1353 - val_accuracy: 0.9598
Epoch 3/10
60000/60000 [=====] - 7s 117us/sample - loss: 0.1160 - accuracy: 0.9650 -
val_loss: 0.1047 - val_accuracy: 0.9677
Epoch 4/10
60000/60000 [=====] - 7s 116us/sample - loss: 0.0881 - accuracy: 0.9729 -
val_loss: 0.0987 - val_accuracy: 0.9718
Epoch 5/10
60000/60000 [=====] - 7s 117us/sample - loss: 0.0729 - accuracy: 0.9782 -
val_loss: 0.0840 - val_accuracy: 0.9756
Epoch 6/10
60000/60000 [=====] - 7s 118us/sample - loss: 0.0606 - accuracy: 0.9817 -
val_loss: 0.0792 - val_accuracy: 0.9771
Epoch 7/10
```

```
60000/60000 [=====] - 7s 117us/sample - loss: 0.0522 - accuracy: 0.9840 -
val_loss: 0.0793 - val_accuracy: 0.9764
Epoch 8/10
60000/60000 [=====] - 7s 117us/sample - loss: 0.0448 - accuracy: 0.9864 -
val_loss: 0.0763 - val_accuracy: 0.9786
Epoch 9/10
60000/60000 [=====] - 7s 118us/sample - loss: 0.0391 - accuracy: 0.9881 -
val_loss: 0.0900 - val_accuracy: 0.9743
Epoch 10/10
60000/60000 [=====] - 8s 127us/sample - loss: 0.0341 - accuracy: 0.9900 -
val_loss: 0.0786 - val_accuracy: 0.9787
```

In the above output, loss and accuracy are calculated on train data. `val_loss` and `val_accuracy` are results from test data. Figure 10.17 is the list of hdf5 files created by the above code.

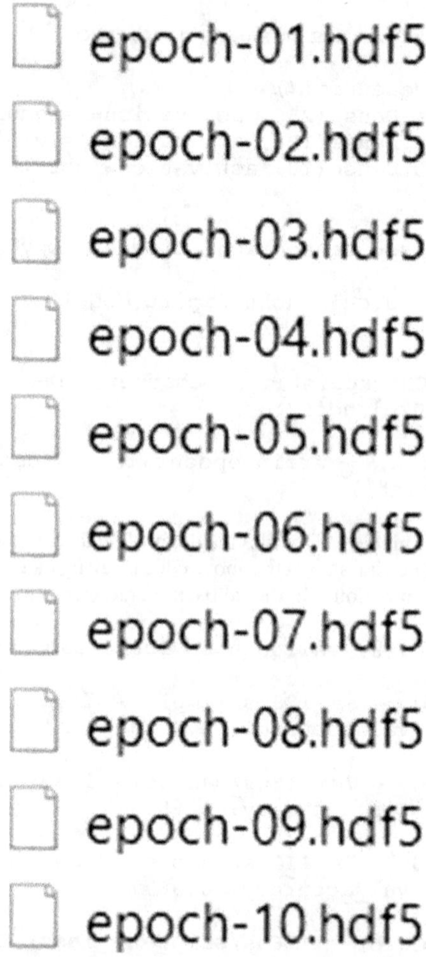

FIGURE 10.17   Saving weights in each epoch.

We can load the model weights from a particular epoch using the code as given below. Imagine that after epoch 7, the model is getting into overfitting; then we can load the weights from epoch 7 using the `load_weights()` function.

```
model_re.load_weights(r"D:\Chapter10 Deep Learning Hyperparameters\4.Code\
epoch-07.hdf5")
```

Either we can manually perform early stopping regularization by using the above weights storing (and `load_weights`) approach, or we can directly use the `keras.callbacks.EarlyStopping()` function. In this function, we have to mention the validation measure and minimum improvement in the test data that we want to see in each iteration.

```
es = keras.callbacks.EarlyStopping(monitor='val_accuracy',
                                   min_delta=0.01,
                                   patience=2)
```

In the code above, note the following:

1. `monitor`: Monitor accuracy or loss from either train data or test data.

2. `min_delta`: It is the minimum improvement required in each step.

3. `patience`: It refers to the number of epochs to wait for, once the monitor termination condition is reached. Sometimes test accuracy seems to be decreased in an epoch, but it increases upward after that. To avoid this strict rule, we can use the `patience` parameter; it will make a model wait for a few more epochs before exiting.

4. In general the code tells the following: terminate the model building when the accuracy improvement is less than 0.01 in the validation data for two consecutive epochs.

```
model_re = keras.Sequential()
model_re.add(layers.Dense(256, activation='sigmoid', input_shape=(784,)))
model_re.add(layers.Dense(128, activation='sigmoid'))
model_re.add(layers.Dense(10, activation='softmax'))
model_re.summary()

model_re.compile(loss='categorical_crossentropy', metrics=['accuracy'])

es = keras.callbacks.EarlyStopping(monitor='val_accuracy',
                                   min_delta=0.01,
                                   patience=2)

#train the model with call back method
model_re.fit(x_train, y_train, epochs=30,validation_data=(x_test, y_test),
callbacks=[es])
```

Though we have mentioned 30 epochs, the model will exit when it reaches the `min_delta` termination criteria. The following is the code output.

```
Train on 60000 samples, validate on 10000 samples
Epoch 1/30
60000/60000 [=====] - 7s 125us/sample - loss: 0.3745 - accuracy: 0.8940 -
val_loss: 0.1940 - val_accuracy: 0.9401
Epoch 2/30
60000/60000 [=====] - 7s 112us/sample - loss: 0.1665 - accuracy: 0.9495 -
val_loss: 0.1350 - val_accuracy: 0.9585
Epoch 3/30
60000/60000 [=====] - 6s 105us/sample - loss: 0.1176 - accuracy: 0.9646 -
val_loss: 0.1062 - val_accuracy: 0.9657
Epoch 4/30
60000/60000 [=====] - 6s 104us/sample - loss: 0.0908 - accuracy: 0.9726 -
val_loss: 0.0941 - val_accuracy: 0.9714
Epoch 5/30
60000/60000 [=====] - 7s 113us/sample - loss: 0.0740 - accuracy: 0.9776 -
val_loss: 0.0841 - val_accuracy: 0.9739
Epoch 6/30
60000/60000 [=====] - 7s 119us/sample - loss: 0.0616 - accuracy: 0.9814 -
val_loss: 0.0799 - val_accuracy: 0.9770
```

The model has exited at the sixth epoch because the accuracy of validation data has not improved more than 1 percent for two consecutive epochs. Hence the training got stopped.

10.4 *LOSS FUNCTIONS*

Until now, in all our discussions so far, we have used the squared error or average squared error as the error function. The same error function is known as the cost function or loss function. While training the model, we consider the distance between the actual value and predicted value as a loss. We try to find the weights that minimize this loss. Given below is the sum of squares of error in a neural network.

$$\text{Squared loss } (E) = \frac{1}{2}\sum(y - \hat{y})^2, \text{ where } \hat{y} = g\left(\sum w_j\left(g\left(\sum w_{ij}x_i\right)\right)\right)$$

We search for the optimal values of weights that will minimize the final error function. We use a gradient descent (GD) method to find the optimal weights. Given below are a few more examples of loss functions.

$$\text{Mean squared error (MSE) } \frac{1}{n}\sum(y - \hat{y})^2$$

$$\text{Mean absolute error (MAE) } \frac{1}{n}\sum|y - \hat{y}|$$

$$\text{Mean absolute percentage error (MAPE) } \frac{1}{n}\sum\frac{|y - \hat{y}|}{y}$$

Most functions, as given above, are aggregating the difference between actual and predicted values. Not all loss functions are mathematically convenient to optimize or to find partial derivatives. Hence, we use some standard loss functions. The above functions work on regression problems, where the output value of y is continuous.

There are dedicated loss functions for classification problems. The most popular loss function for classification problems is cross-entropy.

$$\text{Binary cross-entropy} = -\sum(y log(\hat{y}) + (1 - y)log(1 - \hat{y}))$$

We will check whether this cost function captures the error or not. Let us take an example of a few data points to get an idea of the loss function. In Table 10.6, we calculate loss at individual points.

TABLE 10.6 Binary Cross-Entropy Formula

Actual Value (y)	Predicted Value ($\hat{y}$)	Cross-Entropy Formula	Loss (Log base-10)	Loss (Log base-e)
0	0.01	$-((0)log(0.01) + (1)log(0.99))$	0.004	0.01
0	0.99	$-((0)log(0.99) + (1)log(0.01))$	2	4.605
1	0.01	$-((0)log(0.01) + (0)log(0.99))$	2	4.605
1	0.99	$-((0)log(0.99) + (0)log(0.01))$	0.004	0.01

From Table 10.6, we can observe that the cross-entropy is maximum when actual and predicted values are far away from each other. That makes it a perfect loss function. Table 10.7 is an extended version of Table 10.6.

TABLE 10.7   Binary Cross-Entropy Calculation

Actual Value (y)	Predicted Value ($\hat{y}$)	Cross-Entropy (Log base-10)	Cross-Entropy (Log base-e)
0	0.01	0.00	0.01
0	0.11	0.05	0.12
0	0.21	0.10	0.24
0	0.31	0.16	0.37
0	0.41	0.23	0.53
0	0.51	0.31	0.71
0	0.61	0.41	0.94
0	0.71	0.54	1.24
0	0.81	0.72	1.66
0	0.91	1.05	2.41
0	0.99	2.00	4.61
1	0.01	2.00	4.61
1	0.11	0.96	2.21
1	0.21	0.68	1.56
1	0.31	0.51	1.17
1	0.41	0.39	0.89
1	0.51	0.29	0.67
1	0.61	0.21	0.49
1	0.71	0.15	0.34
1	0.81	0.09	0.21
1	0.91	0.04	0.09
1	0.99	0.00	0.01

Table 10.7 shows the values of cross-entropy for different values of actual and predicted values. Cross-entropy is minimum when the actual value is near to the predicted value. We can observe the minimum values of cross-entropy in the first few and last few rows. Rows from the middle of the table show that the predicted values are far away from the actual values; hence, they have a high loss value.

Until now, we have discussed a binary class as the target. If we consider the output classes as y_1 and y_2 then the same binary cross-entropy can be rewritten as

$$\text{Binary cross-entropy} = -\sum (y_1 log(\hat{y}_1) + y_2 log(\hat{y}_2))$$

If there are three classes in the output, then the cross-entropy formula will be

$$\text{Cross-entropy for three classes} = -\sum (y_1 log(\hat{y}_1) + y_2 log(\hat{y}_2) + y_3 log(\hat{y}_3))$$

For multiple classes, we need to use a generalized version of the above equation that is known as categorical cross-entropy.

$$\text{Categorical cross-entropy} = -\sum_{j=1}^{m} y_j log(\hat{y}_j)$$

where $j = 1$ to m classes.

The below equation is the formula for categorical cross-entropy on a full dataset with n records.

$$\text{Categorical cross-entropy} = -\sum_{i=1}^{n} \left(\sum_{j=1}^{m} y_j log(\hat{y}_j) \right)$$

where $i = 1$ to n records and $j = 1$ to m classes.

For classification problems, cross-entropy-based loss functions work effectively. For regression problems, the deviation-based loss functions work effectively. We can mention the loss function in the `model.compile()` function. Given below is an example of that:

```
model.compile(loss='categorical_crossentropy', metrics=['accuracy'])
```

10.5   *ACTIVATION FUNCTIONS*

Let us quickly revisit the neural network model equation.

$$y = g\left(\sum w_i h_i\right), \text{ where } h_i = g\left(\sum w_{ij} x_j\right)$$

$$y = g\left(\sum w_i \left(g\left(\sum w_{ij} x_j\right)\right)\right)$$

The function $g(x)$ in the above equation is called the activation function. Until now, we have discussed only the sigmoid activation function in this and the previous chapters. There are a few alternative activation functions that we are taking up in this section.

10.5.1   Sigmoid

We use sigmoid activation for classification problems (Fig. 10.18). It quashes the input values between 0 and 1.

$$\text{sigmoid}\left(x\right) = \frac{e^x}{1 + e^x}$$

$$\text{sigmoid}\left(x\right) = \frac{1}{1 + e^{-x}}$$

Sigmoid function

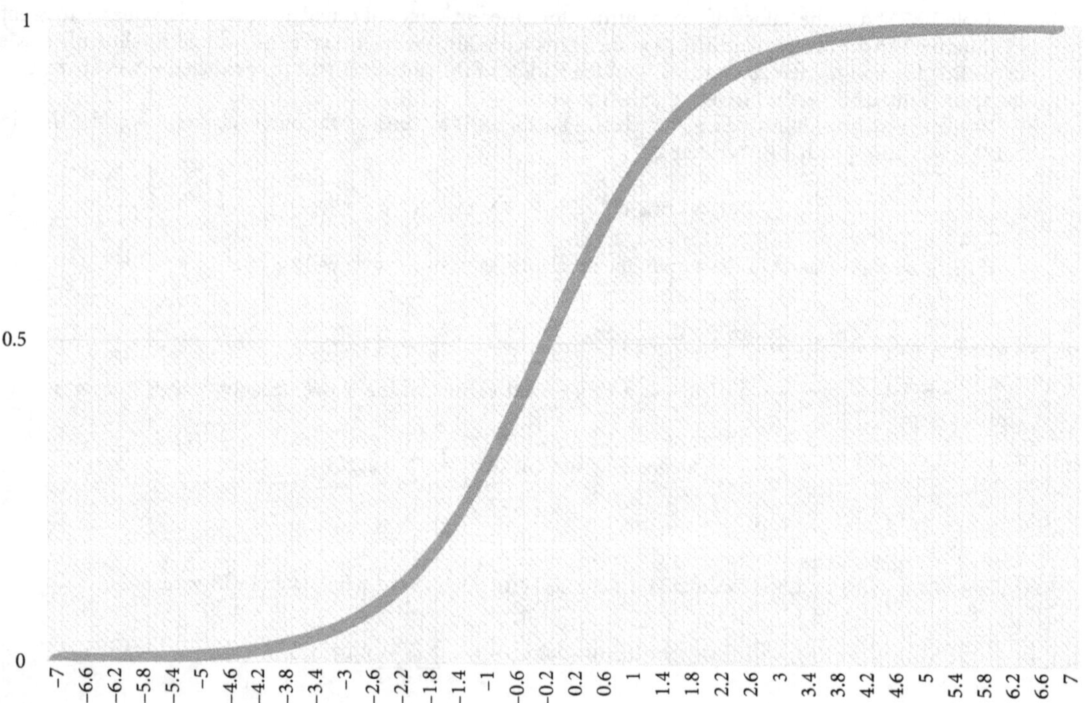

FIGURE 10.18   Sigmoid function.

10.5.2 Tanh

The sigmoid function looks like an "S"-shaped curve, and it quashes the values between 0 and 1. There is one more function that also looks like an "S"-shaped curve; it is the tan hyperbolic function formally known as the tanh function (Fig. 10.19). It is written as "tanh" or "TanH" or "Tanh." We can comprehend the tanh as a rescaled version of the sigmoid function. Tanh squashes the values between –1 and +1.

$$\text{Tanh}(x) = \frac{e^x - e^{-x}}{e^x + e^{-x}}$$

$$\text{Tanh}(x) = \frac{e^{2x} - 1}{e^{2x} + 1}$$

Tanh function

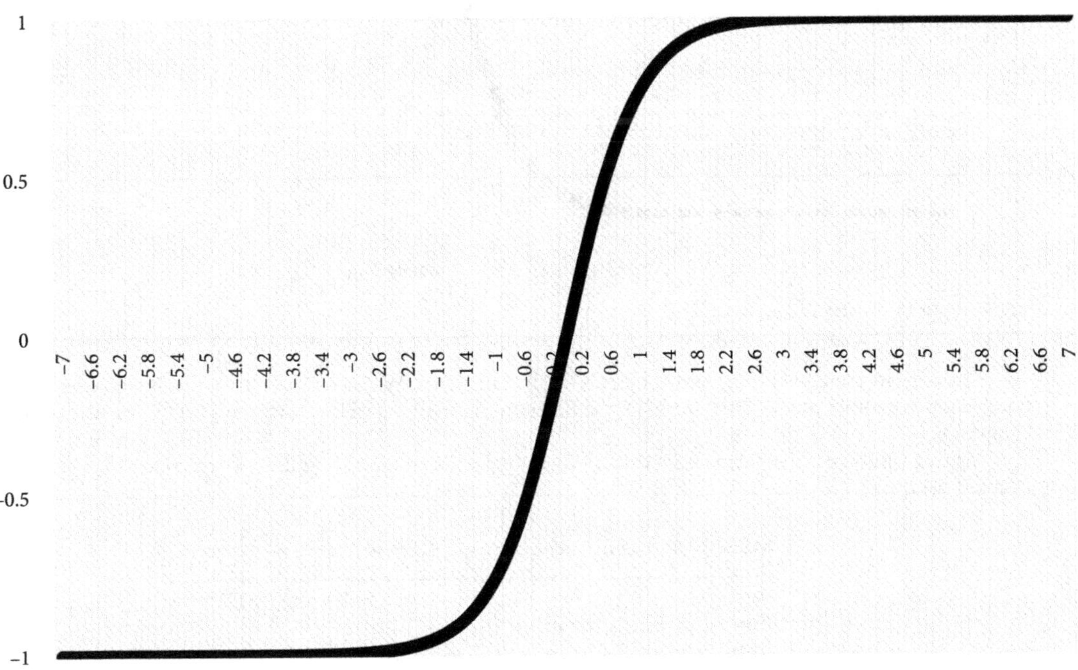

FIGURE 10.19 Tanh function.

10.5.2.1 Sigmoid vs. Tanh   Plots of both sigmoid and tanh look similar, but there is a difference in their output range. Sigmoid is a nonzero-centered curve. Its center is at 0.5, whereas hyperbolic tan is a zero-centered curve. The derivatives of the sigmoid and hyperbolic tan functions are different. When tested on several classification problems, the hyperbolic tan function seems to be converging faster (Fig. 10.20).

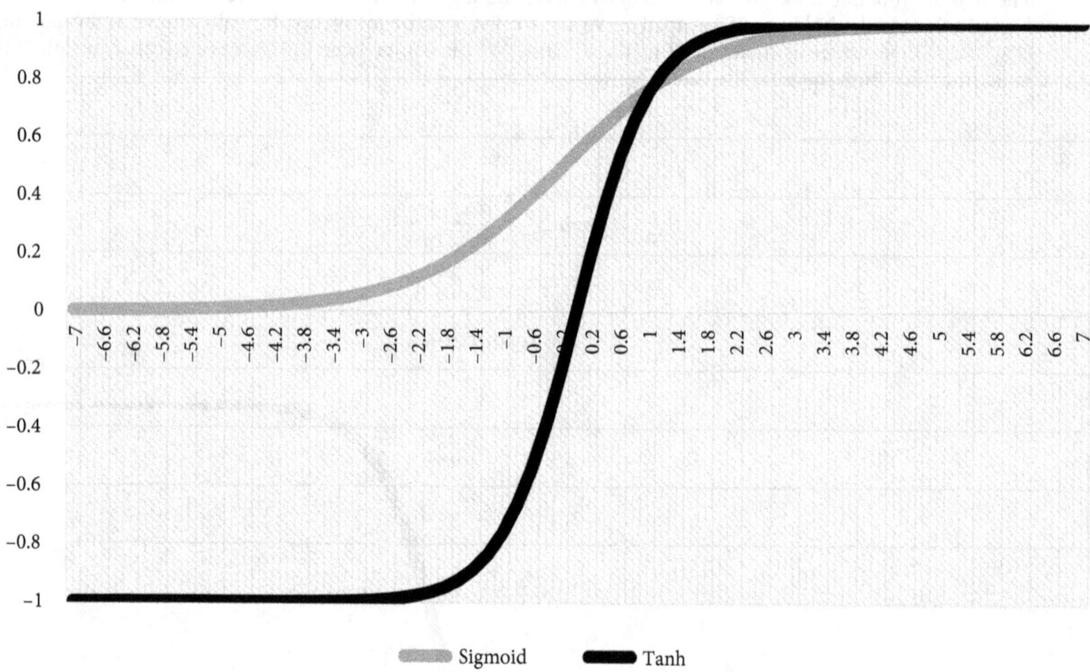

FIGURE 10.20   Functions compared.

Given below is a comparison of two models on the same data. We used sigmoid activation in one model and tanh activation in another model. First, we will try simulations on the TensorFlow playground at https://playground.tensorflow.org/.

Model 1 with Sigmoid Activation: Set this configuration and run the epochs until we reach less than 10 percent error on test data.

TABLE 10.8   Configuration Table with Sigmoid Activation Function

Data	Exclusive or (XOR)
The ratio of training to test data	50%
Noise	5
Batch size	10
Features	x_1, x_2
Hidden layers	1
Neurons or hidden nodes	3
Learning rate	0.03
Activation	**Sigmoid**
Regularization	None
Regularization rate	0
Problem type	Classification

Figure 10.21 shows results from the configuration in Table 10.8.

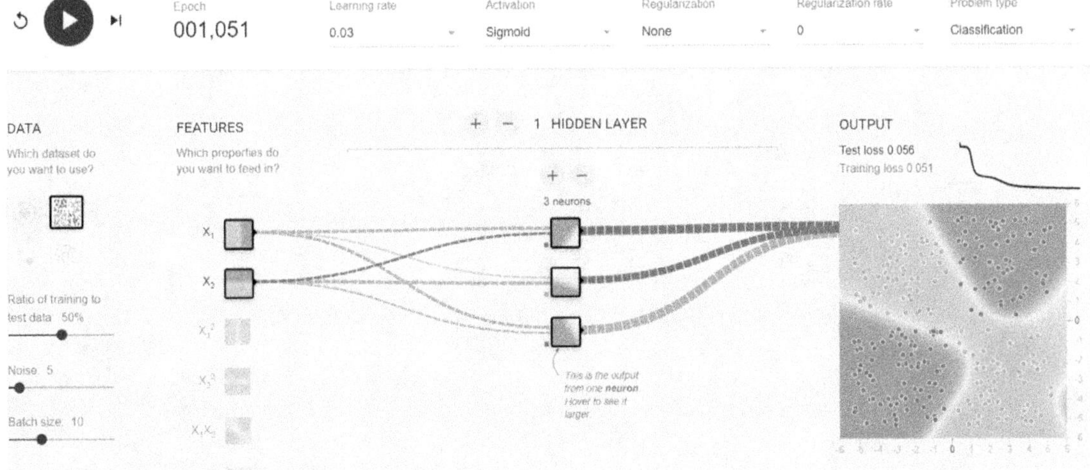

FIGURE 10.21   Results for Model 1 with sigmoid activation.

The configuration in Table 10.8 takes about 1000 epochs to reach less than 0.05 test loss. We can regenerate the data and rerun the same configuration multiple times. Table 10.9 shows a few more attempts.

TABLE 10.9   Sigmoid Activation Results

Epoch	Output
1039	Test loss 0.076
	Training loss 0.045
Epoch	Output
1008	Test loss 0.065
	Training loss 0.050

Now for the same data with the same network configuration, we use tanh activation and check the number of epochs required to reach 0.05 test loss (Table 10.10).

TABLE 10.10   Configuration Table with Tanh Activation Function

Data	Exclusive or (XOR)
The ratio of training to test data	50%
Noise	5
Batch size	10
Features	x_1, x_2
Hidden layers	1
Neurons or hidden nodes	3
Learning rate	0.03
Activation	**Tanh**
Regularization	None
Regularization rate	0
Problem type	Classification

Figure 10.22 shows the results of tanh activation.

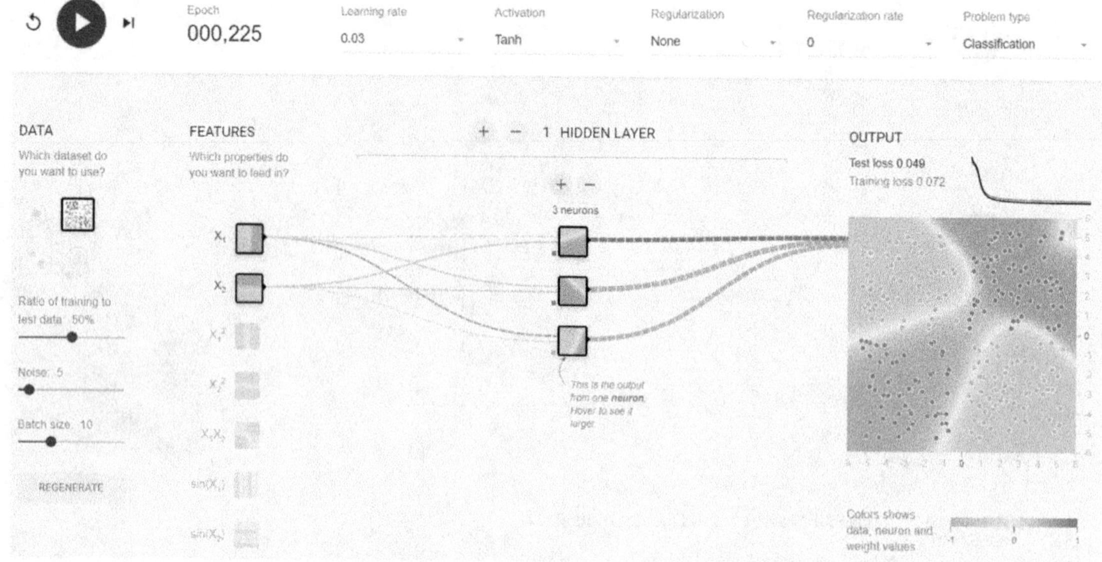

FIGURE 10.22 Results of tanh activation.

With tanh activation, we can reach 0.05 loss within 300 epochs, which is less compared to sigmoid. We can regenerate the data and rerun the same configuration multiple times. Table 10.11 shows a few more attempts.

TABLE 10.11 Tanh Activation Results

Epoch	Output
128	Test loss 0.068
	Training loss 0.056
Epoch	Output
227	Test loss 0.077
	Training loss 0.068

This difference in execution time has a significant impact on the overall neural network training with large datasets.

10.5.3 ReLU Activation

ReLU is another activation function. It is often preferred in deep neural networks with multiple hidden layers. ReLU activation solves the problem of vanishing gradients, as discussed in the following section.

10.5.3.1 Problem of Vanishing Gradients   Deep neural networks with too many hidden layers often suffer from a problem called vanishing gradients. The weights are updated based on the gradient values. If the gradient values are already less at the output layer, then they will further shrink by the time we reach the hidden layers near the input layer. Computationally the derivative of the sigmoid function is near to zero on both ends (Fig. 10.23).

Sigmoid Function and Derivative

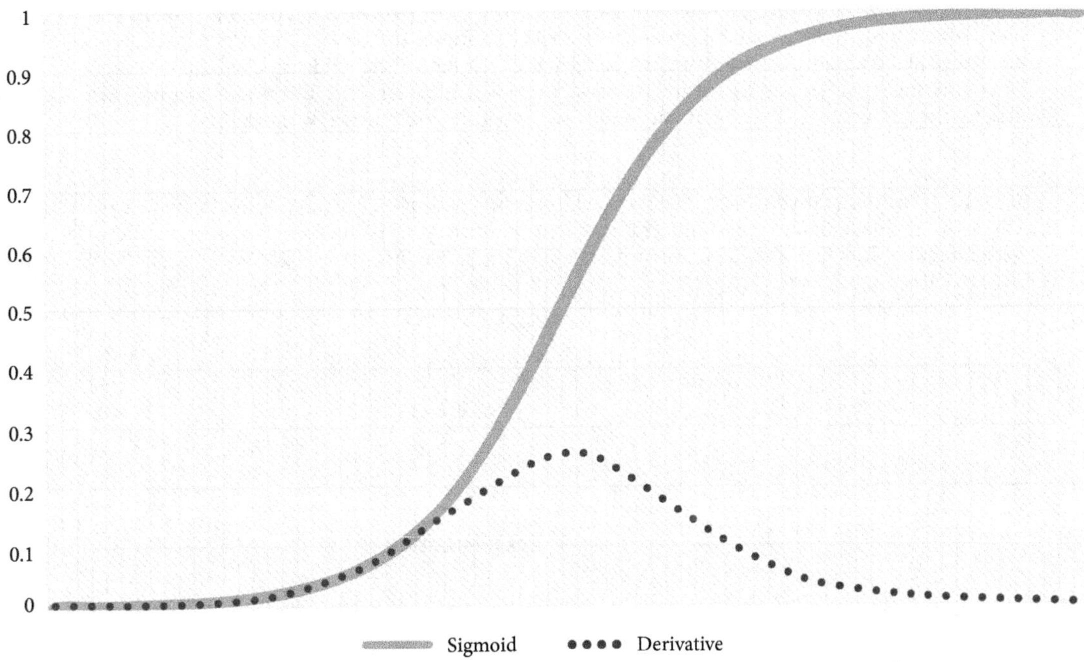

FIGURE 10.23   The dotted curve as the derivative of the sigmoid function.

When the sigmoid value is close to zero or one, near the output layer, then its gradient is near to zero. It means the gradients will further reduce and become insignificant near the input layer. This leads to negligible weight change near the input layer. The weights near the input layer will never change. This issue is known as the problem of vanishing gradients.

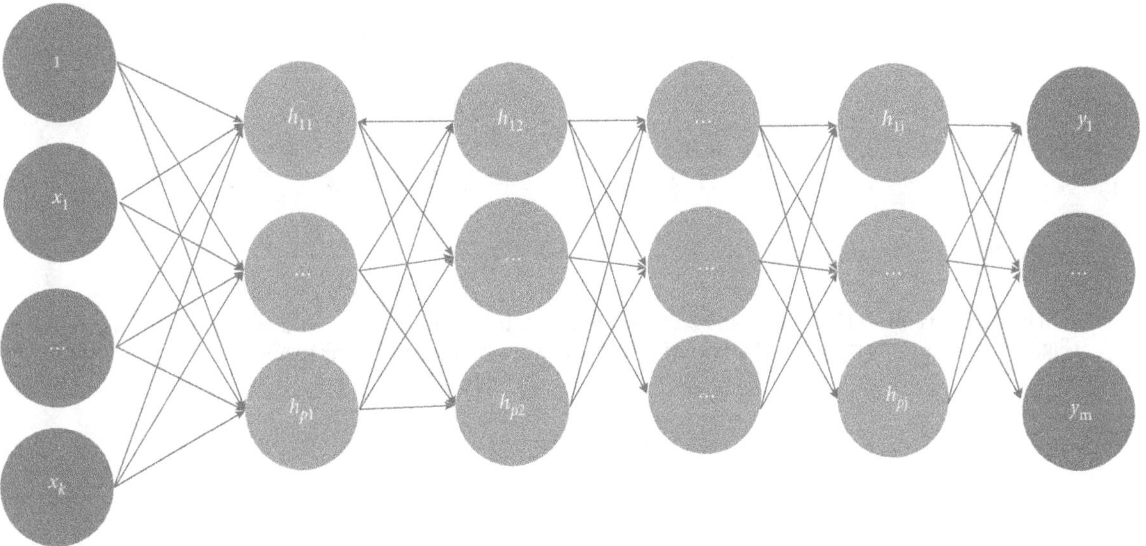

FIGURE 10.24   Deep neural network.

For example, the network shown in Fig. 10.24 has p hidden nodes in each hidden layer, and there are a total of j hidden layers. If the gradients are already low at the last hidden layer, we will be further multiplying them with fractional values to calculate the gradients for the former layers. By the time we perform all the matrix calculations and reach layer 1 from layer j, this final gradient value will be almost zero for the first hidden layer. It means there will be no change in the weights. If weights are unchanged, then the model will not learn anything in that epoch. It is a problem of vanishing gradients. We need an activation function that has larger gradients and does not fall into this problem of vanishing gradients for deep neural networks. Let us take it up in the following section.

10.5.3.2   *Rectified Linear Unit*   The rectified linear unit activation function, formally known as ReLU activation, solves the problem of the vanishing gradient that we discussed in the previous section. ReLU activation is a straightforward function; the ReLU value is zero for all negative input values and identity for the rest of the values (Fig. 10.25).

$$ReLU(x) = \max(0, x)$$

$$ReLU(x) = \begin{cases} 0, & x < 0 \\ x, & x \geq 0 \end{cases}$$

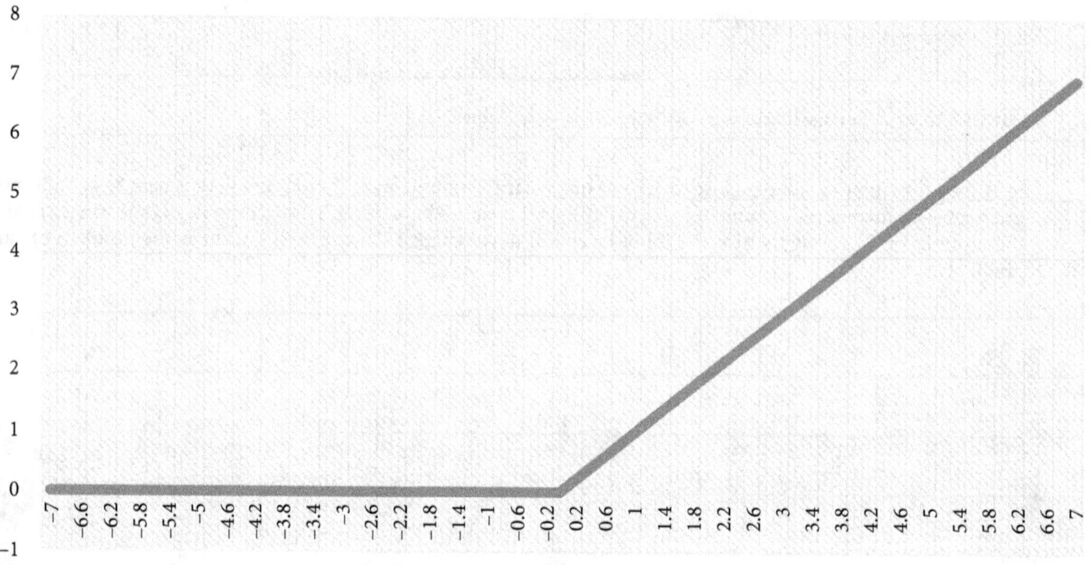

FIGURE 10.25   ReLU function.

The derivative of ReLU functions is zero for the values less than 0, and when x is positive, the derivative is 1 (Fig. 10.26). This property of ReLU helps us in resolving the problem of vanishing gradients.

ReLU Function and Derivative

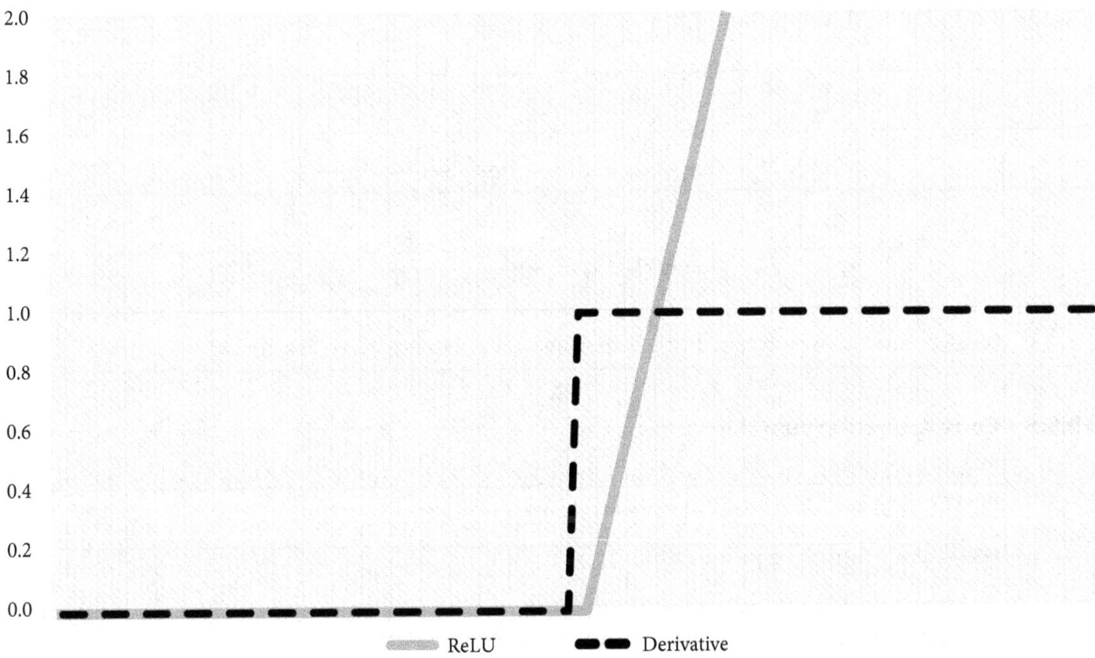

FIGURE 10.26   ReLU function and derivative.

The ReLU function is the most preferred activation function when building computer vision models. In most cases, these image processing models are very deep networks. In the next section, we will take up the activation function at the output layer.

10.5.4   Softmax

Until now, we have talked about the activation function for the hidden layers. What about the activation function at the final output layer? If the output is a simple binary classification, then we can use sigmoid or tanh at the output layer. If the output is a multiclass classification, then we need to use a generalized version of the sigmoid function known as the softmax activation function. If there are 10 classes in the output, then there will be 10 nodes in the output layer and we will receive 10 probabilities from the neural networks result. We collate them using softmax. Each class is one-hot encoded, and the softmax value is calculated by dividing the output value of class-j with the sum of outputs from all classes.

$$\text{Softmax}\left(y_j\right) = \frac{e^{y_j}}{\sum_{k=1}^{K} e^{y_k}}$$

If there are four classes in the output, then the softmax values of each class can be defined as

$$y_1 = \begin{bmatrix} 1 \\ 0 \\ 0 \\ 0 \end{bmatrix}, y_2 = \begin{bmatrix} 0 \\ 1 \\ 0 \\ 0 \end{bmatrix}, y_3 = \begin{bmatrix} 0 \\ 0 \\ 1 \\ 0 \end{bmatrix}, y_4 = \begin{bmatrix} 0 \\ 0 \\ 0 \\ 1 \end{bmatrix}$$

$$\text{Output class } y_1 = \frac{e^{w_1 x}}{e^{w_1 x} + e^{w_2 x} + e^{w_3 x} + e^{w_4 x}}$$

$$\text{Output class } y_2 = \frac{e^{w_2 x}}{e^{w_1 x} + e^{w_2 x} + e^{w_3 x} + e^{w_4 x}}$$

$$\text{Output class } y_3 = \frac{e^{w_3 x}}{e^{w_1 x} + e^{w_2 x} + e^{w_3 x} + e^{w_4 x}}$$

$$\text{Output class } y_4 = \frac{e^{w_4 x}}{e^{w_1 x} + e^{w_2 x} + e^{w_3 x} + e^{w_4 x}}$$

For all the multiclass problems, the activation function at the output layer is softmax.

10.5.5   Code Activation Functions

Given below is the example code that shows how to configure a network with different activation functions. We can mention different activation functions in different layers.

```
model2 = keras.Sequential()

model2.add(layers.Dense(15, activation='sigmoid', input_shape=(784,)))
model2.add(layers.Dense(15, activation='relu'))
model2.add(layers.Dense(15, activation='tanh'))
model2.add(layers.Dense(15, activation='relu'))
model2.add(layers.Dense(10, activation='softmax'))
model2.summary()

model2.compile(loss='categorical_crossentropy', metrics=['accuracy'])
model2.fit(x_train, y_train,epochs=10)
```

Table 10.12 and Fig. 10.27 show the code output.

TABLE 10.12   The Code Output

Layer (type)	Output Shape	Param #
dense_21 (Dense)	(None, 15)	11775
dense_22 (Dense)	(None, 15)	240
dense_23 (Dense)	(None, 15)	240
dense_24 (Dense)	(None, 15)	240
dense_25 (Dense)	(None, 10)	160

```
Total params: 12,655
Trainable params: 12,655
Non-trainable params: 0
```

```
model2.compile(loss='categorical_crossentropy', metrics=['accuracy'])
model2.fit(x_train, y_train,epochs=10)

Train on 60000 samples
Epoch 1/10
60000/60000 [==============================] - 4s 59us/sample - loss: 0.1521 - accuracy: 0.9554
Epoch 2/10
60000/60000 [==============================] - 3s 51us/sample - loss: 0.1477 - accuracy: 0.9574
Epoch 3/10
60000/60000 [==============================] - 3s 50us/sample - loss: 0.1465 - accuracy: 0.9575
Epoch 4/10
60000/60000 [==============================] - 3s 48us/sample - loss: 0.1430 - accuracy: 0.9584
Epoch 5/10
60000/60000 [==============================] - 3s 48us/sample - loss: 0.1415 - accuracy: 0.9589
Epoch 6/10
60000/60000 [==============================] - 3s 47us/sample - loss: 0.1392 - accuracy: 0.9595
Epoch 7/10
60000/60000 [==============================] - 3s 48us/sample - loss: 0.1379 - accuracy: 0.9601
Epoch 8/10
60000/60000 [==============================] - 3s 47us/sample - loss: 0.1356 - accuracy: 0.9609
Epoch 9/10
60000/60000 [==============================] - 3s 48us/sample - loss: 0.1341 - accuracy: 0.9611
Epoch 10/10
60000/60000 [==============================] - 3s 47us/sample - loss: 0.1329 - accuracy: 0.9616
```

FIGURE 10.27   Model training results.

Linear is another activation function that can be used for solving problems with continuous output. We can use a simple linear activation function to solve regression problems.

10.6   LEARNING RATE

In neural networks, while solving the optimization problem using backpropagation, we use the below formula for updating the weights.

$$W := W + \Delta W$$

$$\Delta W = \eta * \left(-\frac{\partial E}{\partial W} \right)$$

$$W := W + \eta * \left(-\frac{\partial E}{\partial W} \right)$$

To reach the minimum value of the error, we move the weights in that direction where the overall error reduces. A negative gradient of error gives us direction. We are multiplying the actual gradient ΔW by η (Eta). The η is known as the learning rate. By increasing or decreasing the values of the learning rate, we can dictate how much the weights should move in one iteration.

The surface of the error function is extremely complex. There is a high risk of ending up in a local minimum. If this local minimum is near to the global minimum, then it is still acceptable. However, the real issue is falling into a local minimum with still a lot of error left in the model. To avoid falling into a local minimum, we can use a slightly higher learning rate to jump over the local minimum. If the learning rate is high, then we will be taking more significant leaps while moving the weights.

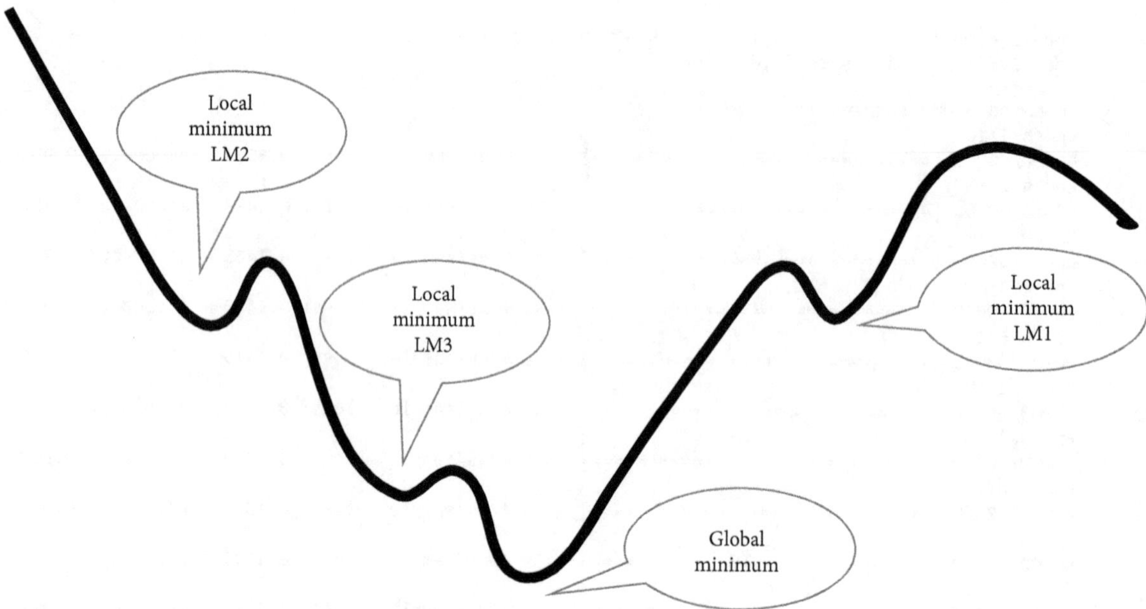

FIGURE 10.28 Local and global minima in the final cost function.

The diagram in Fig. 10.28 shows the final cost function. There are three local minimum points. If we end up in LM1 or LM2, then the error will be too high in the model. If the learning rate is low, then we may fall into one of these local minimums. By having an optimal learning rate, we can jump over the local minimum. The diagram (Fig. 10.29) illustrates the effect of a small learning rate. A small learning rate will make the whole training process slow (Fig. 10.29).

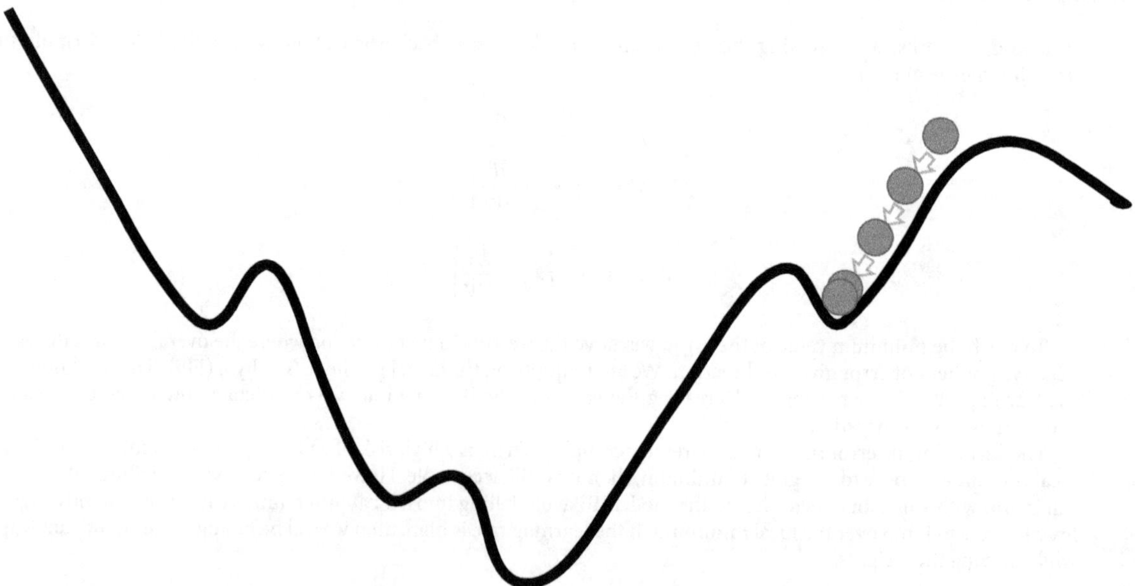

FIGURE 10.29 Cost function with a small learning rate.

Having a more significant learning rate is also a problem. If the learning rate is too high, then we may never get a chance to enter the global minimum. The weights are changing so significantly that the optimization function keeps oscillating between a set of weights; it will never converge to a minimum value. Figure 10.30 shows an example where the learning rate is too high.

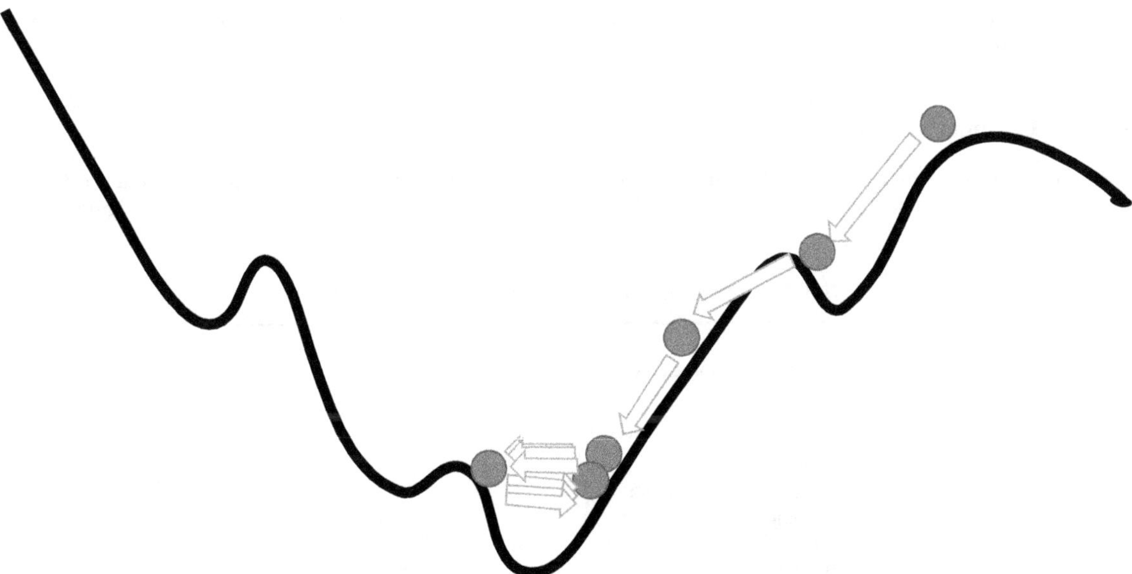

FIGURE 10.30   Cost function with very high learning rate.

An optimal learning rate will make sure that the system is avoiding the local minimum and finally converging to a global minimum or anywhere near to the global minimum. Figure 10.31 shows an example of an optimal learning rate convergence.

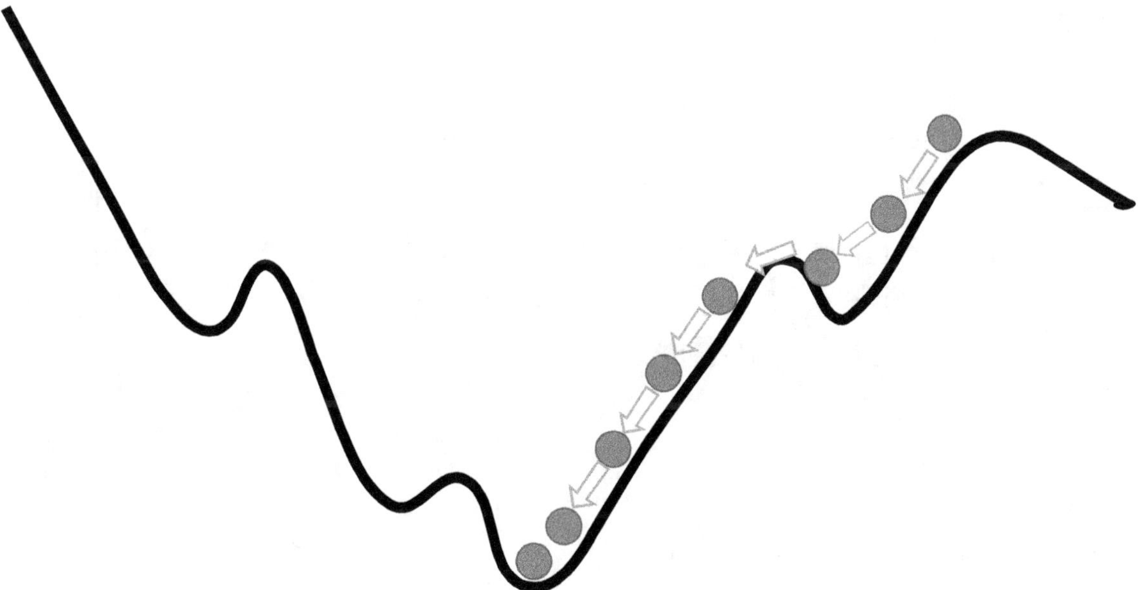

FIGURE 10.31   Cost function with an optimal learning rate.

We usually try learning rates between 0.0001 and 0.1. There is no shortcut to finding the optimal learning rate. While building a model for the first time on a dataset, it may take much time to find the optimal learning rate. We can build the model with a learning rate (usually less than 1) and then observe how the model is converging. Based on the observation, we can increase or decrease the learning rate. If the model is underfitted and taking much time in convergence, then we need to increase our learning rate. If the model shows a reduction in error for the first few epochs and shows no improvement after a few epochs, then we may have to decrease the learning rate. We will be able to develop this intuition on the learning rate only by practical experience—best gained on real-time projects that solve some business problems.

10.6.1   Learning Rate Demo

We will experiment with different learning rates on the TensorFlow playground at https://playground.tensorflow.org/. In this experiment, we will try building a model with a high learning rate. Table 10.13 presents the configuration of the model.

TABLE 10.13   Configuration Table with Very High Learning Rate

Data	Circle
The ratio of training to test data	50%
Noise	10
Batch size	10
Features	x_1, x_2
Hidden layers	2
Neurons or hidden nodes	[3,3]
Learning rate	**3 (Very high)**
Activation	Tanh
Regularization	None
Regularization rate	0
Problem type	Classification

What follows is the result of the configuration presented in Table 10.13 (Fig. 10.32).

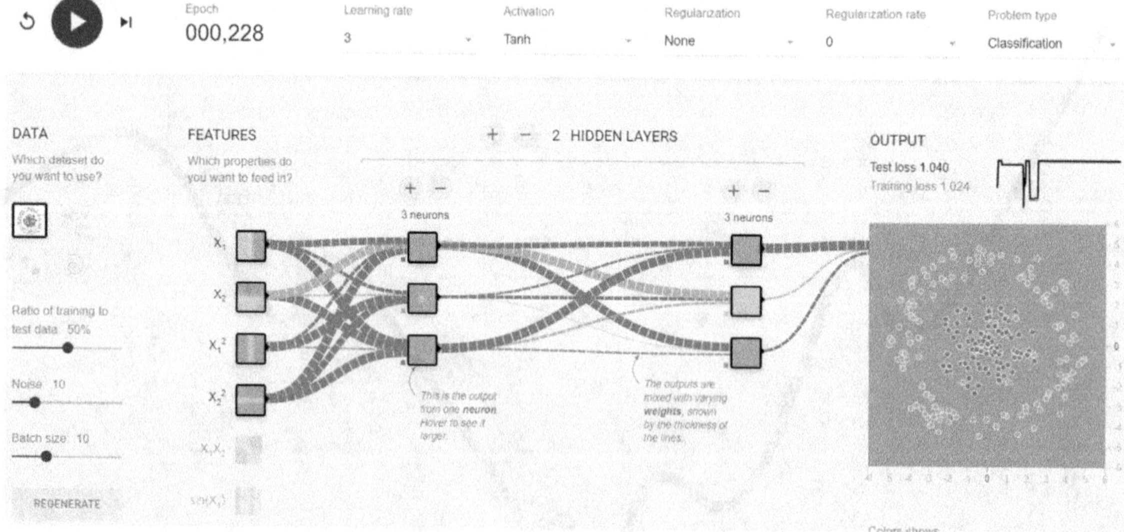

FIGURE 10.32   Results with a high learning rate set to 3.

As expected, the optimization did not converge to global minima. The loss stopped reducing after a few epochs. Even if we leave this for many more epochs, the system will not improve. We can observe the same by regenerating the data multiple times (Fig. 10.33).

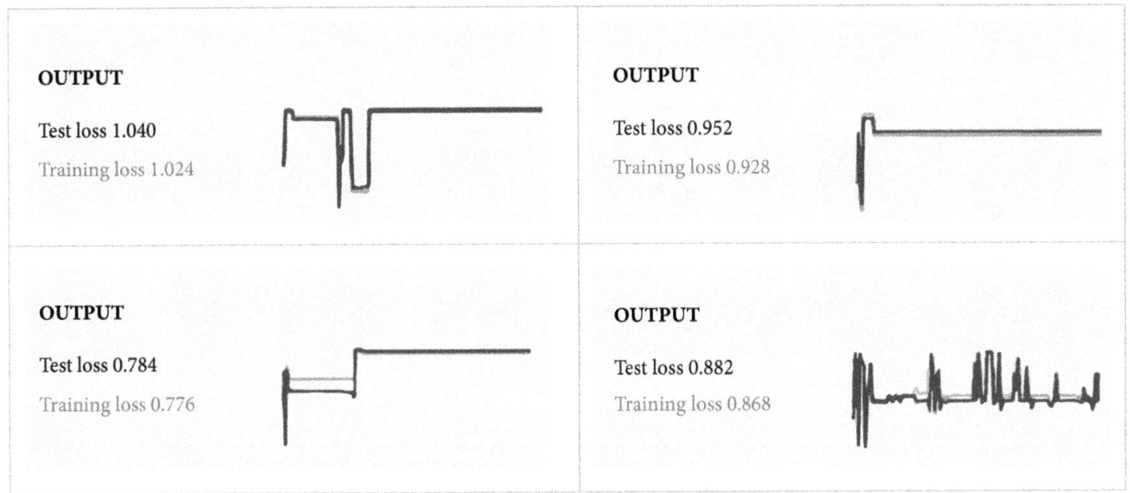

FIGURE 10.33   Results with a high learning rate.

Now we will retry the same dataset with the same configuration with an insignificant learning rate. The first impact on the model will be on the training time. The improvement in the loss will be negligible even after thousands of epochs (Table 10.14).

TABLE 10.14   Configuration Table with Learning Rate Set to a Very Low Value

Data	Circle
The ratio of training to test data	50%
Noise	10
Batch size	10
Features	x_1, x_2
Hidden layers	2
Neurons or hidden nodes	[3,3]
Learning rate	**0.00001 (Very Low)**
Activation	Tanh
Regularization	None
Regularization rate	0
Problem type	Classification

Figure 10.34 shows the result of the configuration presented in Table 10.14.

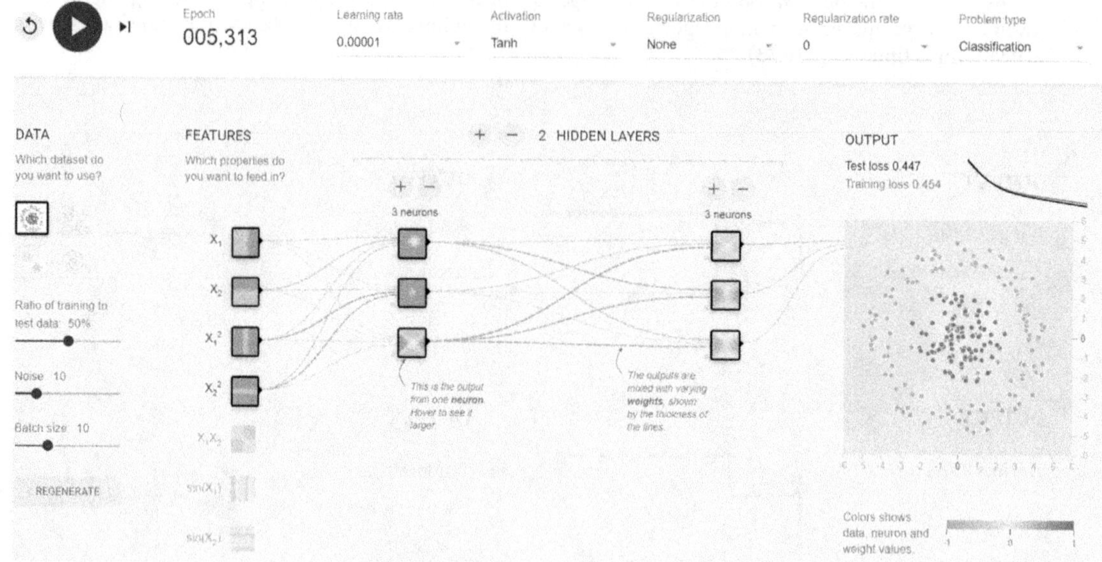

FIGURE 10.34   Results with a low learning rate set to 0.00001.

As expected, the learning process is prolonged. We ran for 50,000 epochs, but still the loss is 0.45 (Fig. 10.34). There is a high chance for us to get stuck inside a local minimum. We can observe the same in multiple experiments (Fig. 10.35).

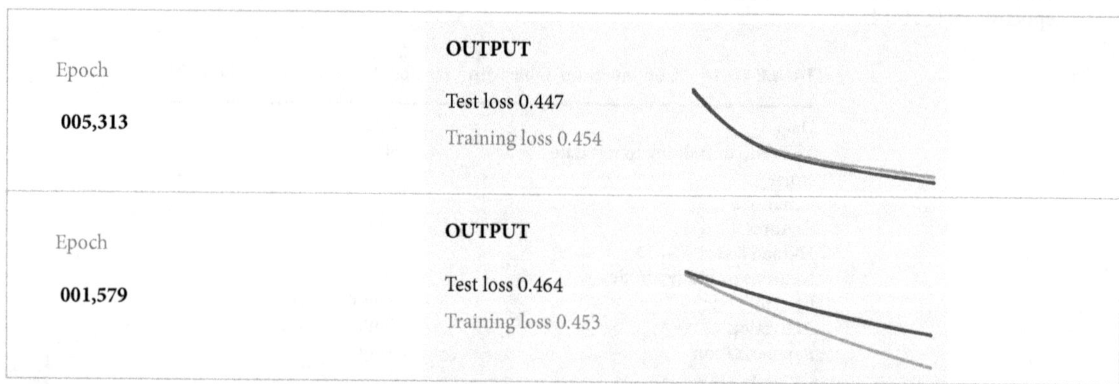

FIGURE 10.35   Results with a low learning rate.

Now we will try to search for the optimal learning rate between 0.00001 and 3. Let us build the model with 0.1 as the learning rate (Table 10.15).

TABLE 10.15   Configuration Table with Learning Rate Set to an Optimal Value

Data	Circle
The ratio of training to test data	50%
Noise	10
Batch size	10
Features	x_1, x_2
Hidden layers	2
Neurons or hidden nodes	[3,3]
Learning rate	**0.1 (optimal)**
Activation	Tanh
Regularization	None
Regularization rate	0
Problem type	Classification

Figure 10.36 shows the result of the configuration presented in Table 10.15.

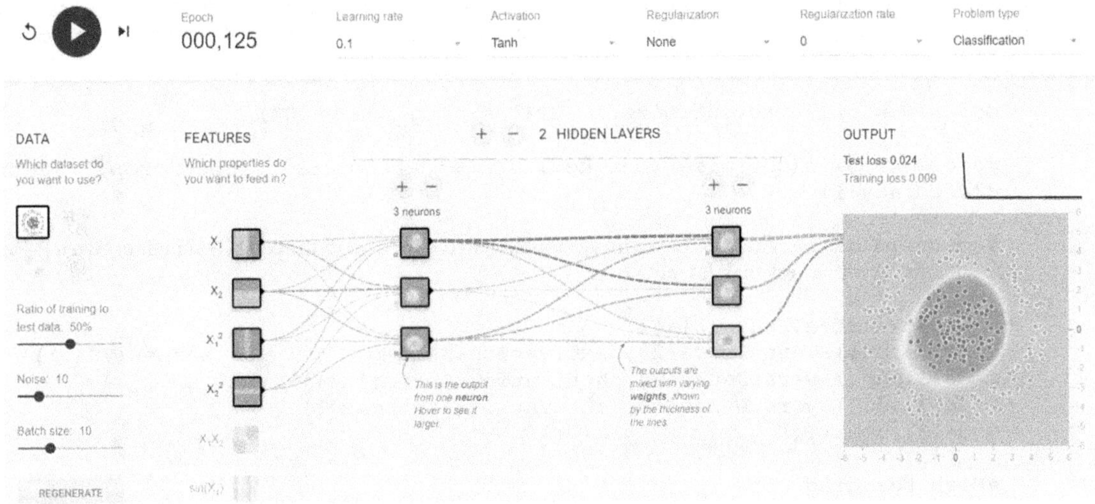

FIGURE 10.36   Results with a learning rate of 0.1.

There are two points to observe in the result presented in Fig. 10.36. The first point is about the loss. The loss is 0.02; this means that the model was not stuck at a local minimum. The second point is about epochs; we achieved 0.02 loss within 150 epochs, so the learning process is not slow. We can repeat the same process multiple times with different samples using the regenerate option. We will observe the same results.

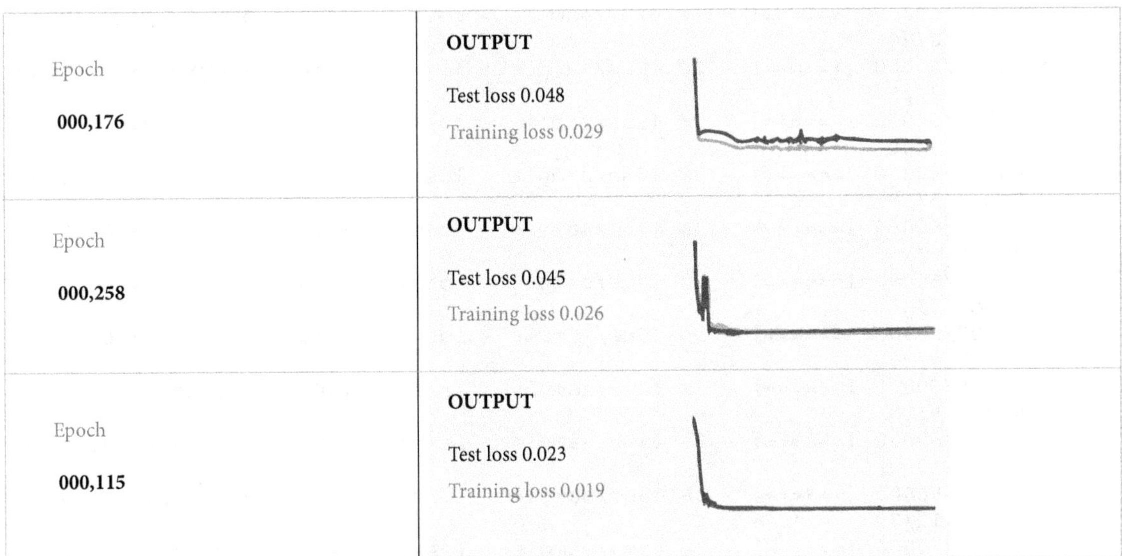

FIGURE 10.37   Results with optimal learning rate.

From the results discussed in Fig. 10.37, the learning rate of 0.1 seems to be working correctly for this data. We can repeat the same experiment with a learning rate of 0.3 and 0.01. They give us similar results, as discussed above. Only extreme learning rates are causing issues. The optimal learning rate is not a single point; it is a small range of values. We can choose any learning rate in that range. In our case, we can choose any number between 0.01 and 0.3.

10.6.2   Learning Rate Code

The learning rate is a part of the gradient descent function. We need to mention the learning rate in the optimizer function. For example, `tf.keras.optimizers.SGD()` is an optimizer function. SGD is a stochastic gradient descent, and it is a specific type of gradient descent function. Given below is the code to mention the learning rate in the optimizer function.

```
opt_new = tf.keras.optimizers.SGD(learning_rate=0.01)

model3.compile(optimizer=opt_new, loss='categorical_crossentropy', metrics=
['accuracy'])
```

We mention the learning rate in the optimizer function and pass it on to the compile function. Given below is the code for a model with a very high learning rate.

```
model3 = keras.Sequential()
model3.add(layers.Dense(20, activation='sigmoid', input_shape=(784,)))
model3.add(layers.Dense(20, activation='sigmoid'))
model3.add(layers.Dense(10, activation='softmax'))
model3.summary()

#High Learning Rate
opt_new = tf.keras.optimizers.SGD(learning_rate=10)
model3.compile(optimizer=opt_new, loss='categorical_crossentropy', metrics=
['accuracy'])
model3.fit(x_train, y_train,epochs=20)
```

The following is the output from the above code.

```
Train on 60000 samples
Epoch 1/20
60000/60000 [======] - 3s 51us/sample - loss: 3.9900 - accuracy: 0.1026
Epoch 2/20
60000/60000 [======] - 3s 45us/sample - loss: 3.5914 - accuracy: 0.1680
Epoch 3/20
60000/60000 [======] - 3s 45us/sample - loss: 3.4296 - accuracy: 0.1986
Epoch 4/20
60000/60000 [======] - 3s 46us/sample - loss: 3.3977 - accuracy: 0.2019
Epoch 5/20
60000/60000 [======] - 3s 44us/sample - loss: 2.8417 - accuracy: 0.1979
Epoch 6/20
60000/60000 [======] - 3s 47us/sample - loss: 2.1760 - accuracy: 0.2013
Epoch 7/20
60000/60000 [======] - 3s 46us/sample - loss: 2.1548 - accuracy: 0.2046
Epoch 8/20
60000/60000 [======] - 3s 44us/sample - loss: 2.1463 - accuracy: 0.2056
Epoch 9/20
60000/60000 [======] - 3s 44us/sample - loss: 2.1469 - accuracy: 0.2064
Epoch 10/20
60000/60000 [======] - 3s 43us/sample - loss: 2.1370 - accuracy: 0.2084
Epoch 11/20
60000/60000 [======] - 3s 45us/sample - loss: 2.1403 - accuracy: 0.2059
Epoch 12/20
60000/60000 [======] - 3s 44us/sample - loss: 2.1521 - accuracy: 0.2031
Epoch 13/20
60000/60000 [======] - 3s 43us/sample - loss: 2.1633 - accuracy: 0.2060
Epoch 14/20
60000/60000 [======] - 3s 44us/sample - loss: 2.1453 - accuracy: 0.2051
Epoch 15/20
60000/60000 [======] - 3s 44us/sample - loss: 2.1397 - accuracy: 0.2078
```

```
Epoch 16/20
60000/60000 [======] - 3s 43us/sample - loss: 2.1454 - accuracy: 0.2063
Epoch 17/20
60000/60000 [======] - 3s 44us/sample - loss: 2.1363 - accuracy: 0.2066
Epoch 18/20
60000/60000 [======] - 3s 43us/sample - loss: 2.1498 - accuracy: 0.2088
Epoch 19/20
60000/60000 [======] - 3s 44us/sample - loss: 2.1559 - accuracy: 0.2072
Epoch 20/20
60000/60000 [======] - 3s 43us/sample - loss: 2.1491 - accuracy: 0.2046
```

From the output given above we can see that the model got stuck at an accuracy of 0.2. The weights must have been oscillating between two points and should not be able to penetrate further inside a minimum. Now we will try with a very low learning rate.

```python
model3 = keras.Sequential()
model3.add(layers.Dense(20, activation='sigmoid', input_shape=(784,)))
model3.add(layers.Dense(20, activation='sigmoid'))
model3.add(layers.Dense(10, activation='softmax'))
model3.summary()

#Low Learning Rate
opt_new = tf.keras.optimizers.SGD(learning_rate=0.00001)
model3.compile(optimizer=opt_new, loss='categorical_crossentropy', metrics=
['accuracy'])
model3.fit(x_train, y_train,epochs=20)
```

The following are the results of the above code.

```
Train on 60000 samples
Epoch 1/20
60000/60000 [======] - 3s 51us/sample - loss: 2.5737 - accuracy: 0.0992
Epoch 2/20
60000/60000 [======] - 3s 46us/sample - loss: 2.5655 - accuracy: 0.0992
Epoch 3/20
60000/60000 [======] - 3s 47us/sample - loss: 2.5577 - accuracy: 0.0992
Epoch 4/20
60000/60000 [======] - 3s 48us/sample - loss: 2.5501 - accuracy: 0.0992
Epoch 5/20
60000/60000 [======] - 3s 48us/sample - loss: 2.5429 - accuracy: 0.0992
Epoch 6/20
60000/60000 [======] - 3s 51us/sample - loss: 2.5358 - accuracy: 0.0992
Epoch 7/20
60000/60000 [======] - 3s 44us/sample - loss: 2.5291 - accuracy: 0.0992
Epoch 8/20
60000/60000 [======] - 3s 46us/sample - loss: 2.5226 - accuracy: 0.0992
Epoch 9/20
60000/60000 [======] - 3s 46us/sample - loss: 2.5163 - accuracy: 0.0992
Epoch 10/20
60000/60000 [======] - 3s 58us/sample - loss: 2.5102 - accuracy: 0.0992
Epoch 11/20
60000/60000 [======] - 3s 50us/sample - loss: 2.5044 - accuracy: 0.0992
Epoch 12/20
60000/60000 [======] - 3s 49us/sample - loss: 2.4987 - accuracy: 0.0992
Epoch 13/20
60000/60000 [======] - 3s 50us/sample - loss: 2.4932 - accuracy: 0.0992
Epoch 14/20
60000/60000 [======] - 3s 48us/sample - loss: 2.4880 - accuracy: 0.0992
```

```
Epoch 15/20
60000/60000 [======] - 3s 48us/sample - loss: 2.4829 - accuracy: 0.0992
Epoch 16/20
60000/60000 [======] - 3s 49us/sample - loss: 2.4779 - accuracy: 0.0992
Epoch 17/20
60000/60000 [======] - 3s 48us/sample - loss: 2.4732 - accuracy: 0.0992
Epoch 18/20
60000/60000 [======] - 3s 49us/sample - loss: 2.4685 - accuracy: 0.0992
Epoch 19/20
60000/60000 [======] - 3s 48us/sample - loss: 2.4641 - accuracy: 0.0992
Epoch 20/20
60000/60000 [======] - 3s 49us/sample - loss: 2.4597 - accuracy: 0.0992
```

We can see from the above output that the model is either stuck at a local minimum or the model is extremely slow in learning. Now we will try building the model with a medium learning rate.

```
model3 = keras.Sequential()
model3.add(layers.Dense(20, activation='sigmoid', input_shape=(784,)))
model3.add(layers.Dense(20, activation='sigmoid'))
model3.add(layers.Dense(10, activation='softmax'))
model3.summary()

#Optimal Learning Rate
opt_new = tf.keras.optimizers.SGD(learning_rate=0.01)
model3.compile(optimizer=opt_new, loss='categorical_crossentropy', metrics=
['accuracy'])
model3.fit(x_train, y_train,epochs=20)
```

The following is the result of the above code.

```
Train on 60000 samples
Epoch 1/20
60000/60000 [======] - 3s 50us/sample - loss: 2.2487 - accuracy: 0.2494
Epoch 2/20
60000/60000 [======] - 3s 47us/sample - loss: 2.0170 - accuracy: 0.4581
Epoch 3/20
60000/60000 [======] - 3s 44us/sample - loss: 1.6577 - accuracy: 0.5755
Epoch 4/20
60000/60000 [======] - 3s 45us/sample - loss: 1.3164 - accuracy: 0.6801
Epoch 5/20
60000/60000 [======] - 4s 64us/sample - loss: 1.0575 - accuracy: 0.7509
Epoch 6/20
60000/60000 [======] - 3s 53us/sample - loss: 0.8729 - accuracy: 0.8001
Epoch 7/20
60000/60000 [======] - 3s 45us/sample - loss: 0.7379 - accuracy: 0.8337
Epoch 8/20
60000/60000 [======] - 2s 41us/sample - loss: 0.6395 - accuracy: 0.8535
Epoch 9/20
60000/60000 [======] - 3s 42us/sample - loss: 0.5673 - accuracy: 0.8673
Epoch 10/20
60000/60000 [======] - 3s 44us/sample - loss: 0.5138 - accuracy: 0.8769
Epoch 11/20
60000/60000 [======] - 3s 48us/sample - loss: 0.4735 - accuracy: 0.8843
Epoch 12/20
60000/60000 [======] - 3s 47us/sample - loss: 0.4423 - accuracy: 0.8889
Epoch 13/20
60000/60000 [======] - 3s 43us/sample - loss: 0.4178 - accuracy: 0.8922
Epoch 14/20
60000/60000 [======] - 3s 45us/sample - loss: 0.3979 - accuracy: 0.8962
```

```
Epoch 15/20
60000/60000 [======] - 3s 42us/sample - loss: 0.3814 - accuracy: 0.8993
Epoch 16/20
60000/60000 [======] - 4s 59us/sample - loss: 0.3674 - accuracy: 0.9021
Epoch 17/20
60000/60000 [======] - 3s 51us/sample - loss: 0.3554 - accuracy: 0.9042
Epoch 18/20
60000/60000 [======] - 2s 41us/sample - loss: 0.3447 - accuracy: 0.9065
Epoch 19/20
60000/60000 [======] - 3s 47us/sample - loss: 0.3353 - accuracy: 0.9084
Epoch 20/20
60000/60000 [======] - 3s 42us/sample - loss: 0.3267 - accuracy: 0.9107s
```

There is no shortcut for reaching the optimal learning rate. Since the optimal learning rate is a range of values, it will not be an immense challenge to find the optimal learning rate. It is entirely achievable with reasonable efforts.

10.6.3 Momentum

The learning rate can be improved by adding one more additional factor called momentum. In the initial epochs, the weights will change by larger values. As the error function is reaching the minimum value, the delta weights will be smaller. If we are reaching a local minimum, then the momentum can help us to push out of local minima and help the algorithm to converge much faster.

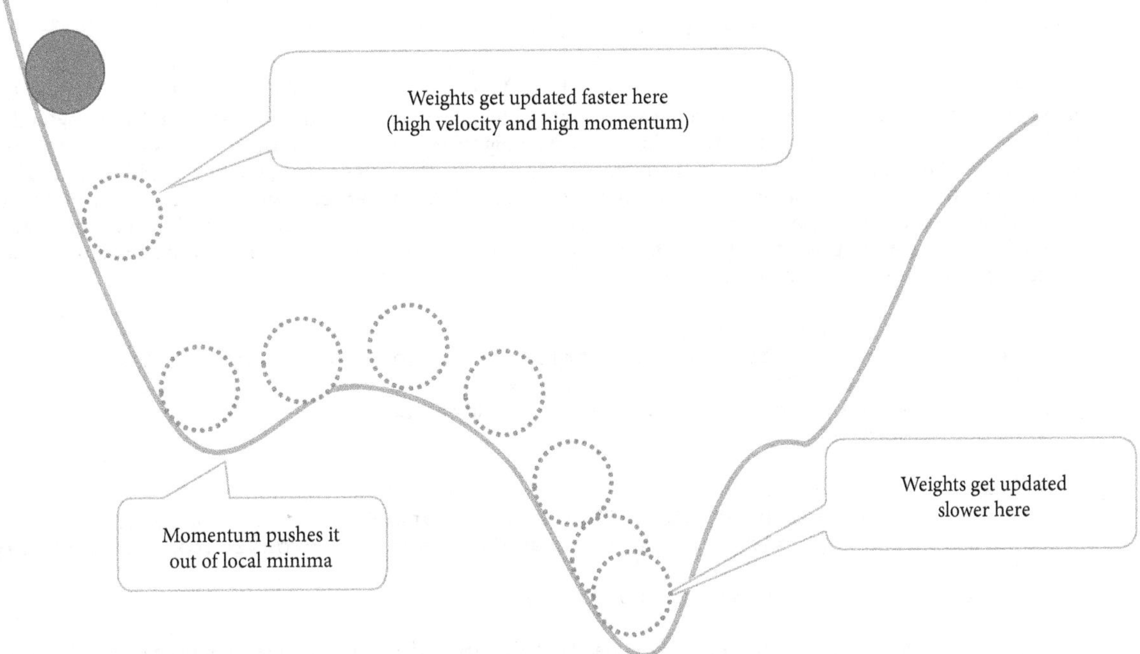

FIGURE 10.38   The error function with different rate of velocity and momentum.

In the diagram in Fig. 10.38, we can see the error function and movement of weights and corresponding movement of error value in each epoch. Gradient descent optimization looks like a ball rolling on a surface. The ball will gain some momentum while rolling. We can capture that momentum and make use of it to pull out of local minima. Given below is the mathematical formula for including momentum in learning rate and further impact the overall weights training.

The original weight-updating formula is

$$W := W + \Delta W$$

Since we have to keep track of the current epoch's and previous epoch's rate of change, we will introduce the time dimension in the equation.

$$W(t) := W(t-1) + \Delta W(t)$$

$$\Delta W(t) = \eta * \left(-\frac{\partial E}{\partial W}\right)$$

Until now, the weights are updating by a factor $\eta * \left(-\dfrac{\partial E}{\partial W}\right)$. Given below is the new updating factor.

$$\eta * \left(-\frac{\partial E}{\partial W}\right) + \alpha * \Delta W(t-1)$$

The new formula for updating the weights is

$$W(t) := W(t-1) + \eta * \left(-\frac{\partial E}{\partial W}\right) + \alpha * \Delta W(t-1)$$

In the simplified form

$$W(t) := W(t-1) + \Delta W(t) + \alpha * \Delta W(t-1)$$

The additional term in the equation $\alpha * \Delta W(t-1)$ is the momentum term. This term will carry forward the momentum to the next step. If there was a massive change in the previous epoch (time), a factor of that is now added to the new weight-updating formula. It will help us in converging faster, and most importantly, it may help us in getting out of the local minimum. This term α is the momentum parameter that can be configured by us while building the model. If we include momentum in the model, we can observe a slightly faster convergence and a better exclusion of the local minimum. Stand-alone momentum does not exist; it always comes along with the learning rate. Momentum should be seen as a helping aid in the learning rate; it is not an alternative to the learning rate. Given below is the code to include momentum along with the learning rate.

```
model3 = keras.Sequential()
model3.add(layers.Dense(20, activation='sigmoid', input_shape=(784,)))
model3.add(layers.Dense(20, activation='sigmoid'))
model3.add(layers.Dense(10, activation='softmax'))
model3.summary()

#Optimal learning rate
opt_new = tf.keras.optimizers.SGD(learning_rate=0.01, momentum=0.5)
model3.compile(optimizer=opt_new, loss='categorical_crossentropy', metrics=
['accuracy'])
model3.fit(x_train, y_train,epochs=20)
```

The code above is the same as the previous model. We have included a momentum term in the above code. Given below is the result of that code.

```
Train on 60000 samples
Epoch 1/20
60000/60000 [======] - 3s 48us/sample - loss: 2.1933 - accuracy: 0.3189
Epoch 2/20
60000/60000 [======] - 3s 46us/sample - loss: 1.5562 - accuracy: 0.6173
Epoch 3/20
60000/60000 [======] - 3s 44us/sample - loss: 0.9885 - accuracy: 0.7617
```

```
Epoch 4/20
60000/60000 [======] - 3s 46us/sample - loss: 0.7133 - accuracy: 0.8293
Epoch 5/20
60000/60000 [======] - 3s 46us/sample - loss: 0.5652 - accuracy: 0.8588
Epoch 6/20
60000/60000 [======] - 3s 48us/sample - loss: 0.4825 - accuracy: 0.8761
Epoch 7/20
60000/60000 [======] - 3s 48us/sample - loss: 0.4303 - accuracy: 0.8874
Epoch 8/20
60000/60000 [======] - 3s 48us/sample - loss: 0.3934 - accuracy: 0.8954
Epoch 9/20
60000/60000 [======] - 3s 47us/sample - loss: 0.3654 - accuracy: 0.9014
Epoch 10/20
60000/60000 [======] - 3s 49us/sample - loss: 0.3431 - accuracy: 0.9069
Epoch 11/20
60000/60000 [======] - 3s 49us/sample - loss: 0.3245 - accuracy: 0.9116
Epoch 12/20
60000/60000 [======] - 3s 47us/sample - loss: 0.3085 - accuracy: 0.9154
Epoch 13/20
60000/60000 [======] - 3s 46us/sample - loss: 0.2948 - accuracy: 0.9186
Epoch 14/20
60000/60000 [======] - 3s 46us/sample - loss: 0.2826 - accuracy: 0.9218
Epoch 15/20
60000/60000 [======] - 3s 47us/sample - loss: 0.2717 - accuracy: 0.9243
Epoch 16/20
60000/60000 [======] - 3s 48us/sample - loss: 0.2619 - accuracy: 0.9268
Epoch 17/20
60000/60000 [======] - 3s 47us/sample - loss: 0.2529 - accuracy: 0.9294
Epoch 18/20
60000/60000 [======] - 3s 48us/sample - loss: 0.2447 - accuracy: 0.9319
Epoch 19/20
60000/60000 [======] - 3s 50us/sample - loss: 0.2372 - accuracy: 0.9337
Epoch 20/20
60000/60000 [======] - 3s 50us/sample - loss: 0.2301 - accuracy: 0.9357
```

We can observe a faster convergence after including the momentum term. This parameter is very handy for broad and deep neural networks.

10.7 OPTIMIZERS

Until now, we have discussed the gradient descent algorithm for updating the weights. The original theory of gradient descent uses all the data to calculate the gradients to update the weights.

$$W := W + \eta * \left(-\frac{\partial E}{\partial W} \right)$$

If there are millions of points, then calculating the gradient for all those points in one go takes much time. Generally speaking, this formula works, but for massive datasets it has some practical implementation challenges.

10.7.1 SGD—Stochastic Gradient Descent

Stochastic gradient descent, formally known as SGD, uses an estimate of gradient instead of an actual gradient. SGD approximates the overall gradient using a single data point. It will make sure that the individual gradients are calculated faster and weights are getting updated rapidly. Earlier, in gradient descent, the weight update will happen after going through complete training data; now, weights are getting updated for every record. By the time we reach the end of the training data, we already have reasonable estimates of the weights.

One epoch is one complete run over train data. If there are N records in the data, then in each epoch, there will be N iterations if we are using SGD. If we are running 10 epochs and there are 1 million records, then there will be a total of 10 million iterations for the weight updates to happen in the SGD algorithm. SGD is suitable for smaller datasets, but it is prolonged for bigger or massive datasets. Moreover, we are not confident about updating weights for every single record. GD uses N records for gradient calculation, and SGD uses a single record at a time. Neither of them is perfect for massive datasets. We need a method that is in between these two.

10.7.2   Mini-Batch Gradient Descent

We will take SGD and make a small modification to it. Instead of calculating gradients for every single record, we will make a batch of records, a small subset of the data, and calculate the gradient to update the weights. For example, there are 20,000 records, and the batch size is 100, then we will take the first 100 records to calculate the gradients and update the weights. We move to the next 100 records and repeat the same. So within one epoch, there will be 200 iterations (20,000/100). This subset of data is known as a batch. This method is known as mini-batch gradient descent. In all practical problems, we use mini-batch gradient descent. This concept introduces a new hyperparameter called batch size. In GD, the batch size is set based on the total number of records in the data. Typically, professionals try to keep the batch size at about 1 to 5 percent of the total records. For the best results, batch size should be kept in powers of 2, such as 32, 64, 128, 256, 512, and so on. Internally, the memory allocation is the same for a batch size of 65, 100 as for a batch size of 128. So it is better to have a batch size of 128 instead of 100. Similarly, a batch size 33 to 64 gets the same internal memory allocation, so in place of keeping a batch size of 50 it is better to keep it as 64.

All three methods, GD, SGD, and mini-batch GD, are the same. The only differentiating factor is the batch size. If batch size $= 1$, then it is SGD; if batch size $= N$, then it is GD; if the batch size is between 1 and N, it is mini-batch GD. Given below is the code for mentioning the optimizer in Keras. We will build our first model with the batch size equal to the number of rows in the data.

```
model4 = keras.Sequential()
model4.add(layers.Dense(20, activation='sigmoid', input_shape=(784,)))
model4.add(layers.Dense(20, activation='sigmoid'))
model4.add(layers.Dense(10, activation='softmax'))
model4.summary()

opt_new = tf.keras.optimizers.SGD(learning_rate=0.01, momentum=0.5)
model4.compile(loss='categorical_crossentropy', metrics=['accuracy'])

#Batch size=full data(GD)
model4.fit(x_train, y_train,batch_size=x_train.shape[0], epochs=10)
```

We need to pass the `batch-size` parameter in the `model.fit()` function. The following is the result of the above code.

```
Train on 60000 samples
Epoch 1/10
60000/60000 [=====] - 1s 15us/sample - loss: 2.4585 - accuracy: 0.0987
Epoch 2/10
60000/60000 [=====] - 0s 3us/sample - loss: 2.4167 - accuracy: 0.0987
Epoch 3/10
60000/60000 [=====] - 0s 3us/sample - loss: 2.3907 - accuracy: 0.0987
Epoch 4/10
60000/60000 [=====] - 0s 3us/sample - loss: 2.3712 - accuracy: 0.0989
Epoch 5/10
60000/60000 [=====] - 0s 3us/sample - loss: 2.3553 - accuracy: 0.0999
Epoch 6/10
60000/60000 [=====] - 0s 3us/sample - loss: 2.3417 - accuracy: 0.1026
Epoch 7/10
60000/60000 [=====] - 0s 3us/sample - loss: 2.3298 - accuracy: 0.1073
Epoch 8/10
60000/60000 [=====] - 0s 3us/sample - loss: 2.3190 - accuracy: 0.1149
Epoch 9/10
60000/60000 [=====] - 0s 3us/sample - loss: 2.3091 - accuracy: 0.1249
Epoch 10/10
60000/60000 [=====] - 0s 3us/sample - loss: 2.2998 - accuracy: 0.1375
```

From the above output, we can observe that the loss has not reduced drastically even after 10 epochs. Now we will build the model with SGD where the batch size is 1.

```
model4 = keras.Sequential()
model4.add(layers.Dense(20, activation='sigmoid', input_shape=(784,)))
model4.add(layers.Dense(20, activation='sigmoid'))
model4.add(layers.Dense(10, activation='softmax'))
model4.summary()

opt_new = tf.keras.optimizers.SGD(learning_rate=0.01, momentum=0.5)
model4.compile(loss='categorical_crossentropy', metrics=['accuracy'])

#Batch size=1 (SGD)
model4.fit(x_train, y_train,batch_size=1, epochs=2)
```

The following is the output of the above code.

```
Train on 60000 samples
Epoch 1/2
60000/60000 [=====] - 80s 1ms/sample - loss: 0.5104 - accuracy: 0.8616
Epoch 2/2
60000/60000 [=====] - 80s 1ms/sample - loss: 0.3648 - accuracy: 0.9124
```

From the above output, we can observe that within two epochs, we achieved a high accuracy level. However, the problem is in execution time. Each epoch takes a significantly higher amount of time. Now we will build the third model with batch size between 1 and N.

```
model4 = keras.Sequential()
model4.add(layers.Dense(20, activation='sigmoid', input_shape=(784,)))
model4.add(layers.Dense(20, activation='sigmoid'))
model4.add(layers.Dense(10, activation='softmax'))
model4.summary()

opt_new = tf.keras.optimizers.SGD(learning_rate=0.01, momentum=0.5)
model4.compile(loss='categorical_crossentropy', metrics=['accuracy'])

#Batch size = 512
model4.fit(x_train, y_train,batch_size=512, epochs=10)
```

The following is the output of the above code.

```
Train on 60000 samples
Epoch 1/10
60000/60000 [=====] - 1s 14us/sample - loss: 2.0902 - accuracy: 0.4012
Epoch 2/10
60000/60000 [=====] - 0s 8us/sample - loss: 1.6120 - accuracy: 0.6944
Epoch 3/10
60000/60000 [=====] - 0s 6us/sample - loss: 1.2406 - accuracy: 0.7579
Epoch 4/10
60000/60000 [=====] - 0s 7us/sample - loss: 0.9523 - accuracy: 0.8217
Epoch 5/10
60000/60000 [=====] - 0s 6us/sample - loss: 0.7359 - accuracy: 0.8597
Epoch 6/10
60000/60000 [=====] - 0s 6us/sample - loss: 0.5813 - accuracy: 0.8793
Epoch 7/10
60000/60000 [=====] - 0s 8us/sample - loss: 0.4759 - accuracy: 0.8921
Epoch 8/10
60000/60000 [=====] - 0s 7us/sample - loss: 0.4037 - accuracy: 0.9022
```

```
Epoch 9/10
60000/60000 [=====] - 0s 7us/sample - loss: 0.3535 - accuracy: 0.9102
Epoch 10/10
60000/60000 [=====] - 0s 8us/sample - loss: 0.3181 - accuracy: 0.9160
```

This output shows better results than GD and SGD. Mini-batch GD is used in solving real business problems since it is better than the other two in terms of execution time and accuracy. There are other optimizer functions too; we will learn them in later chapters.

10.8   CONCLUSION

In this chapter, we discussed some of the essential hyperparameters in deep neural networks. We discussed regularization, activation functions, learning rate, and batch size. Although there are many more hyperparameters, we have limited our scope to the most widely used hyperparameters. There are two significant challenges while working with deep neural networks. The first one is building an optimal model that is neither underfitted nor overfitted. The second challenge is building a model within the stipulated time. By setting these hyperparameters to their default values, we may not be building an optimal model. Also, the model configuration with default values may take much time to converge. There are no shortcuts to find the optimal values of these hyperparameters. The values of hyperparameters entirely depend on the data, its information content, and the complexity involved (in the data). What may work best for computer vision problems may not work well for banking or marketing-related data. We need to develop a sense of intuition by solving many different problems. Anybody can build a deep learning model. What distinguishes an expert from a beginner is the expertise level in fine-tuning hyperparameters and being fully conversant with their implications.

CHAPTER 11
CONVOLUTIONAL NEURAL NETWORKS

Image processing or computer vision is one primary application where deep learning is used very frequently in a big way. Computer vision applications have made deep learning very famous in recent times. Some popular deep learning applications are face detection, object recognition, digit recognition, and extracting text from images. Deep learning finds its applications in almost every industry and many parts of our day-to-day lives.

The health care industry uses deep learning for x-ray image analysis. Deep learning is used by autonomous cars in sensing the road by analyzing input video images. Deep learning is used in agriculture for segregating good versus rotten vegetables. Many other industry applications include counting the number of faces in an input image, detecting intruders using a security cam video footage, age, and gender prediction. The utilization of deep learning–based computer vision is endless. Figure 11.1 shows an image from a deep learning application that uses object recognition to solve a business case.

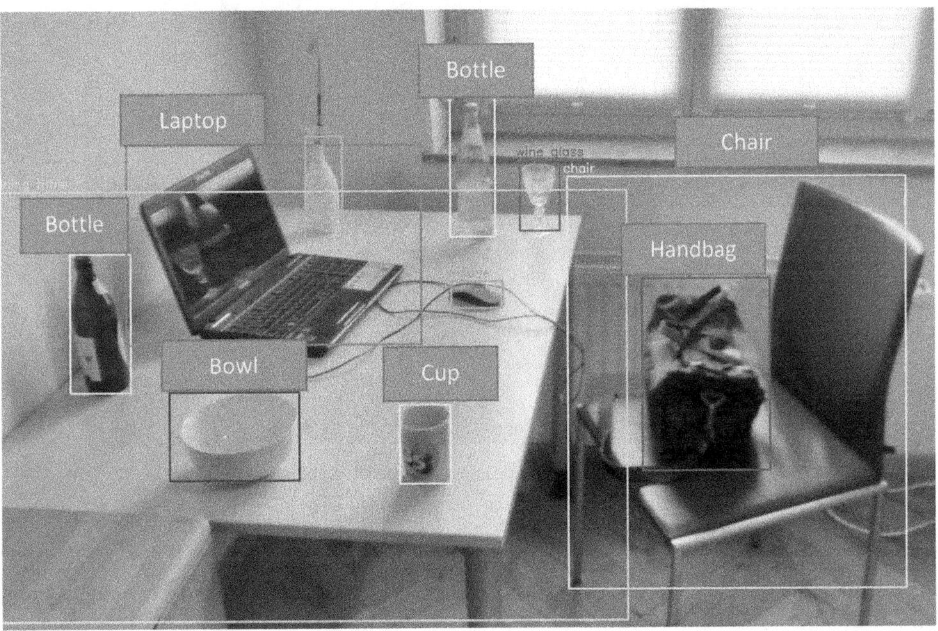

FIGURE 11.1   Image recognition to solve a business problem.

The standard artificial neural networks (ANNs) do an excellent job of detecting the intricate patterns in the data. We need to make some modifications to standard ANNs to make them work effectively on image data. There is a dedicated computer vision algorithm in deep learning called the convolution neural network (CNN), which is the topic of discussion in this chapter.

11.1 *ANNs FOR IMAGES*

The standard ANNs can learn almost any pattern in the data. In ANNs, we take each pixel as an input while building the model. Considering each pixel as an input is not practically viable. In ANNs, as part of the data pre-processing step, we flatten the images. That makes images lose a fundamental property called spatial dependency. Image pixels also have a local correlation. All these are the topics of discussion in this section.

11.1.1 Spatial Dependence

If we take a picture (image) of a dog, then we find pixels around the dog's eye very comparable. Similarly, pixels around the ear are nearly identical to each other. Pixels around the nose are different from the ones in other areas. We need to consider this spatial dependence while building the model. This local correlation of pixels helps in classifying image features accurately (Fig. 11.2).

FIGURE 11.2   Classifying images needs local correlation of pixels.

In ANNs, the first step we perform is flattening the image to a single row. For example, look at the image in Fig. 11.3, which depicts the numeric character 6.

-1	-1	-1	-1	-1	-1	-1	-1	0.9	-0	-1	-1	-1	-1	-1	-1
-1	-1	-1	-1	-1	-1	-1	0.3		0.3	-1	-1	-1	-1	-1	-1
-1	-1	-1	-1	-1	-1	-0			-1	-1	-1	-1	-1	-1	-1
-1	-1	-1	-1	-1	-1	0.8		0.6	-1	-1	-1	-1	-1	-1	-1
-1	-1	-1	-1	-1	0.5		0.8	-1	-1	-1	-1	-1	-1	-1	-1
-1	-1	-1	-1	0.1			-0	-1	-1	-1	-1	-1	-1	-1	-1
-1	-1	-1	-0		-0	-1	-1	-1	-1	-1	-1	-1	-1	-1	-1
-1	-1	-1			0.3	-1	-1	-1	-1	0.5		0.9	0.1	-1	-1
-1	-1	0.3		0.9	-1	-1	-1	0.1						-1	-1
-1	-1	0.8		0.3	-1	-1	0.4		0.7	-0	-0			-1	-1
-1	-1		0.1	-1	0.1		0.3	-1	-1	-0		0.6	-1	-1	-1
-1	-1		0.8	0.3		0.7	-1	-1	-1	0.5		0	-1	-1	-1
-1	-1	0.8				0.5	0.2	0.8	0.8		0.9	-1	-1	-1	-1
-1	-1	-0	0.8						0.1	-1	-1	-1	-1	-1	-1
-1	-1	-1	-0	0.8				0.2	-1	-1	-1	-1	-1	-1	-1
-1	-1	-1	-1	-1	-0	0.3	0.8		0.5	-0	-1	-1	-1	-1	-1

FIGURE 11.3   First step in the digital representation of an image.

While building the ANN model, we do not use the image in Fig. 11.3 as it is. We convert it into a row and pass it on to the model as input. The image in Fig. 11.3 is flattened as one row (Fig. 11.4).

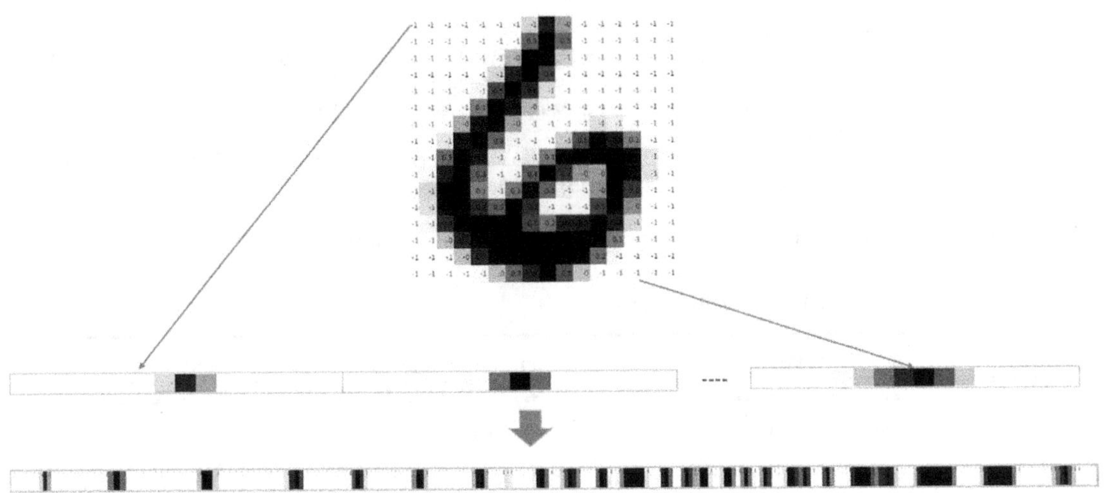

FIGURE 11.4   Flattened image.

Looking at the matrix-shaped image in Fig. 11.4, we can easily say it is the image of a 6. However, when we look at the flattened array, it is impossible to guess the number it represents. In MNIST data, the images are of size 28 × 28. MNIST train data has 60,000 images. The first step is to flatten the 28 × 28 matrix–shaped images to a 1 × 784 array. We generally use the reshape function before starting ANN model building.

```python
(X_train, Y_train), (X_test, Y_test) = keras.datasets.mnist.load_data()
print("X_train shape", X_train.shape)
print("X_test shape", X_test.shape)

x_train = X_train.reshape(60000, 784)
x_test = X_test.reshape(10000, 784)

print("X_train new shape", x_train.shape)
print("X_test new shape", x_test.shape)
```

The above code gives us the below output.

```
X_train shape (60000, 28, 28)
X_test shape (10000, 28, 28)
X_train new shape (60000, 784)
X_test new shape (10000, 784)
```

The original data has 60,000 images, and each image has a size of 28 × 28. The original data is a three-dimensional tensor. We used the reshape function to flatten the 28 × 28 image into a single row of 784 pixels. The reshaped tensor has only one dimension now. We can use this reshaped data to build the model as given in the following code snippet.

```python
num_classes=10
x_train = x_train.astype('float32')
x_test = x_test.astype('float32')
x_train /= 255
x_test /= 255

## Convert class vectors to binary class matrices
y_train = keras.utils.to_categorical(Y_train, num_classes)
y_test = keras.utils.to_categorical(Y_test, num_classes)

model = keras.Sequential()
model.add(layers.Dense(20, activation='sigmoid', input_shape=(784,)))
model.add(layers.Dense(20, activation='sigmoid'))
model.add(layers.Dense(10, activation='softmax'))
model.summary()
model.compile(loss='categorical_crossentropy', metrics=['accuracy'])
model.fit(x_train, y_train,epochs=10)
```

Table 11.1 presents the model summary resulting from the above code.

TABLE 11.1   The Model Summary

Layer (type)	Output Shape	Param #
dense (Dense)	(**None,** 20)	15700
dense_1 (Dense)	(**None,** 20)	420
dense_2 (Dense)	(**None,** 10)	210

Total params: 16,330

Trainable params: 16,330

Non-trainable params: 0

Converting images to single rows—also known as flatting the images—removes the spatial correlation between the surrounding pixels. Somehow, we need to preserve the local correlation. Loss of spatial dependency is the first issue that we observed with standard ANNs.

11.1.2 Number of Free Parameters in ANNs

The second issue with standard ANNs is the number of weight parameters. Until now, we have worked with MNIST data, which has $28 \times 28 = 784$ pixels. The real-life images have millions of pixels per image. Consider a smartphone camera with a 2-megapixel resolution. It has images that have 2000 pixels in width and 1000 pixels in height, overall 2,000,000 pixels. Furthermore, the color images have depth as the third dimension. The RGB value is the depth factor for each pixel. In a 2-megapixel image, there will be 6 million numeric values. If we take each pixel as an input node, then we will have 6 million nodes as input. We are going to add hidden layers and nodes to this input. The network will have free parameters (weights) running into the billions. Practically, it is difficult to build such models in a specified timeframe. Even if we build a model, then implementing and using it in live applications may not be easy. Standard ANNs are fully connected dense layers resulting in an exponential number of weights. The number of free parameters and dense connectivity (fully connected layer) are the second issue that we need to address in image processing.

We need to find a solution to the above stated two issues. The first is preserving the local correlation. The second is fully connected dense layers. Out of these two, preserving local correlation is of utmost importance. Somehow, we can handle the calculations, but managing spatial dependence requires an innovative solution. Let us move on to the next sections for more details.

11.2 FILTERS

We can think of filters as subregions on the image. Until now, we have considered pixels inside the image, which is the most primitive subregion. Instead of taking the input information from the pixels, we try to take the information from the larger subregions of the image. We use filters to get the information on image subregions. A filter is a numerical matrix. We can run a filter on top of the image and extract the regional information. This way, we make sure that the spatial dependence is intact and local correlations between the pixels are not lost.

11.2.1 How a Filter Works

A black and white image is a matrix of numbers—with rows and columns. A filter is also a matrix but has a slightly smaller size. Usually, a filter is a 3×3 or a 5×5 matrix. We run a filter over the image and calculate the dot product of values from the image pixels with filter values. For example, Fig. 11.5 shows an image of size 7×8 with some values in it. These values inside the image matrix are pixel values; they depend on the input image. We are considering a filter of a 3×3 matrix. We roll this filter matrix over the rows and columns of the image that creates a new image by taking the dot product.

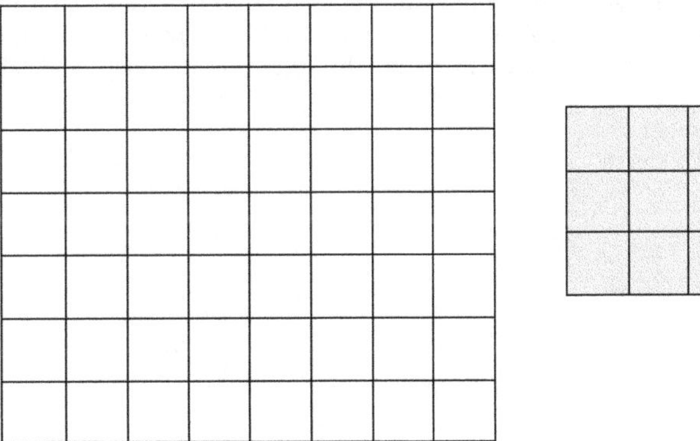

FIGURE 11.5 Image and the filter.

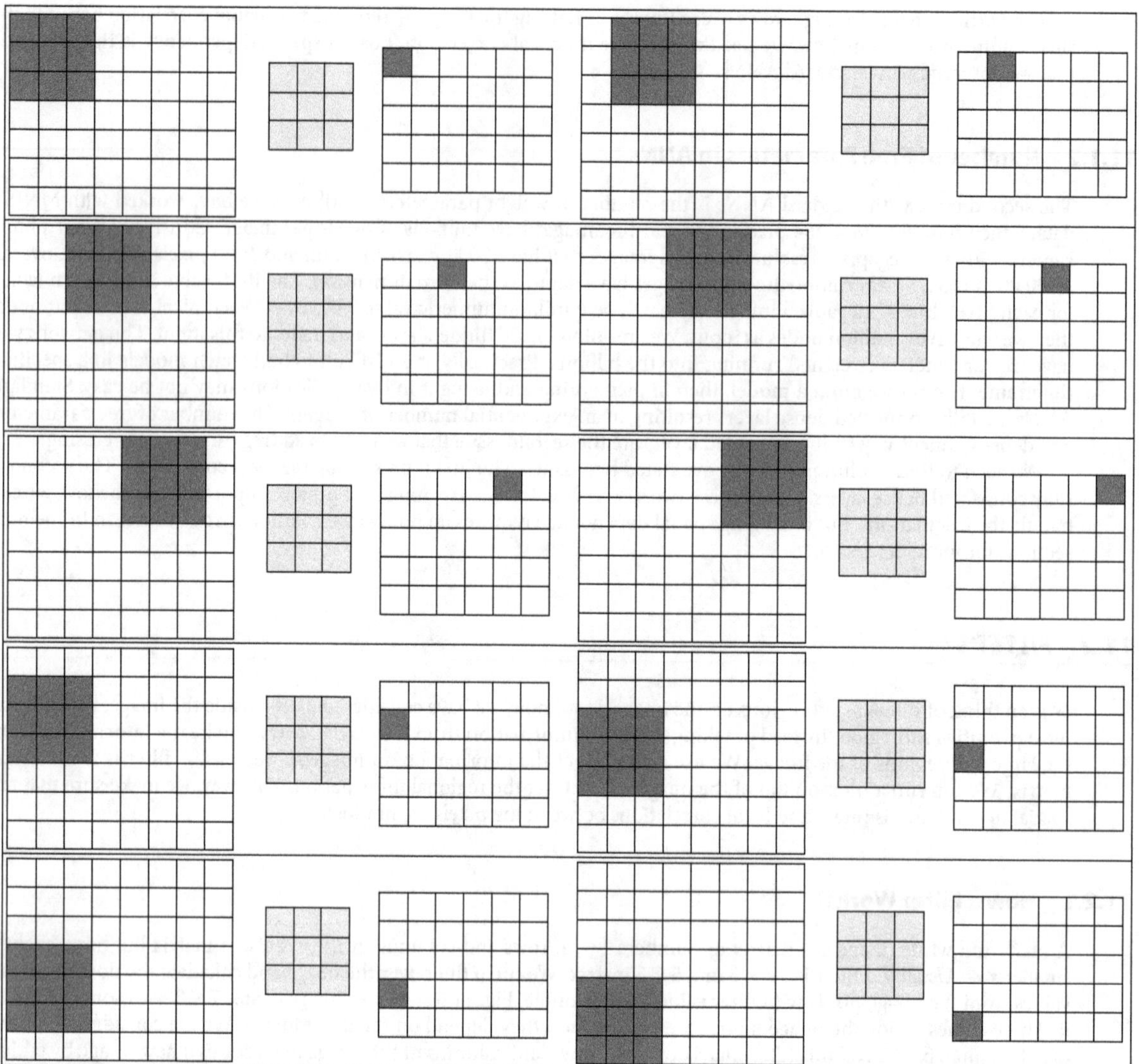

FIGURE 11.6 Applying filter on an image.

We will repeat the same demonstration of rolling a filter over an image to create new images (Fig. 11.6). Let the image and filter matrices areas be as shown in Table 11.2; in this example, we have filled all the cells of the filter matrix with 1. We will now create the new image by rolling this filter over the image and calculating the dot product.

TABLE 11.2   Creating a New Image by Rolling Over the Filter

Image, filter, and resultant matrices

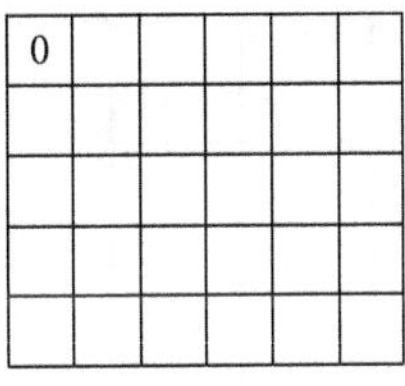

Filter and image pixels dot product

$0*1+0*1+0*1+0*1+0*1+0*1+0*1+0*1+0*1 \rightarrow 0$

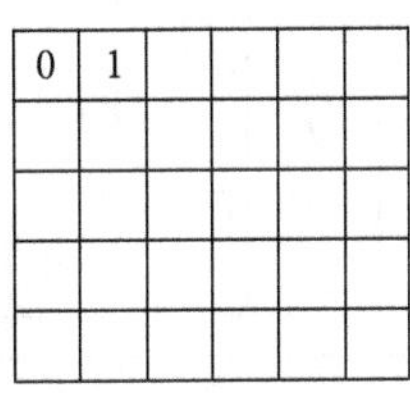

Filter and image pixels dot product

$0*1+0*1+0*1+0*1+0*1+1*1+0*1+0*1+0*1 \rightarrow 1$

Filter and image pixels dot product

$0*1+0*1+1*1+0*1+1*1+1*1+0*1+0*1+1*1 \rightarrow 4$

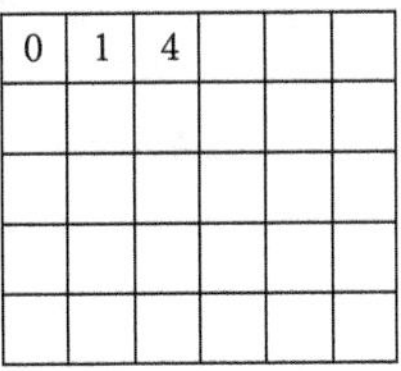

The final result

A 7 × 8 image has now been re-created as a 5 × 6 image. This new resultant matrix is known as a convoluted image or convoluted feature. This filter is known as the kernel matrix. The new convoluted image now preserves the local correlation. Depending on the values in the kernel matrix, we will be able to extract a particular feature from the image.

11.2.2   Kernel Matrix for Detecting Features

The values of a kernel matrix decide the resultant convoluted features. In the example given above, we considered a kernel matrix that had all 1's in it. Depending on the values in the matrix, each kernel captures a specific type of feature. Some kernels capture straight lines in the image. Some other kernels capture circles. A few kernels capture the sharp edges, while others capture the curves (Fig. 11.7).

Kernel for smoothing the images

1	1	1
1	1	1
1	1	1

Kernel for sharpening the images

0	−1	0
−1	10	−1
0	−1	0

Kernel for detecting horizontal lines in the images

0	0	0
1	1	1
0	0	0

Kernel for detecting vertical lines in the images

0	1	0
0	1	0
0	1	0

FIGURE 11.7   Kernel matrix for detecting features.

We already use filters in many of our smartphone camera apps or Instagram app. A few examples are vintage filters, sharpen filters, and filters for black and white images. Have a close look at the example images shown in Fig. 11.8 with different filters.

FIGURE 11.8   Image re-created with the application of a variety of filters.

However, these Instagram filters are not useful for our classification model. We need filters that can capture some specific features from the data. Every image is made up of only small numbers of features. Most images have straight lines, curves, and circles. If we can somehow capture these features using kernels, then we can easily classify the images based on the unique features present in them. Let us consider a basic example to understand the image features captured by kernels. We will consider number data for an explanation because it is intuitive and easy to understand. In the later sections of this chapter, we will consider more practical CNN case studies. Have a look at the images of numbers shown in Fig. 11.9.

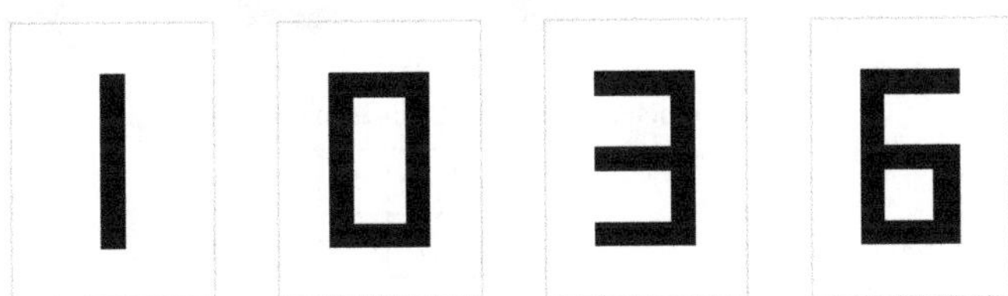

FIGURE 11.9   Example number data to discuss how features are captured by kernels.

Each of the four images in Fig. 11.9 has some basic features. All the images in this dataset are formed from horizontal and vertical lines. Some of them have "L"-shaped corners, and the inverse "L"-shaped edges only. Every digit is a combination of these basic features (or shapes). If we somehow extract the basic features out of the data, then we can easily classify them by utilizing these features. We will apply filters to extract specific features (Fig. 11.10).

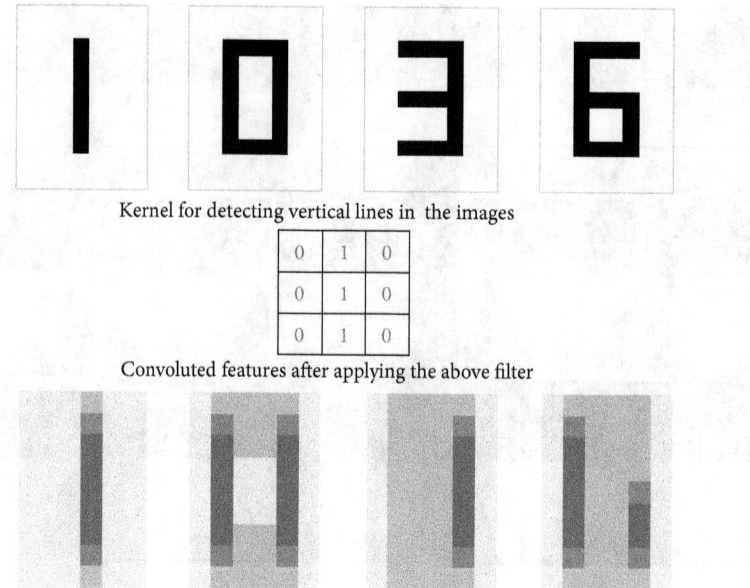

FIGURE 11.10   Application of filters to extract specific features.

We can observe in the diagram in Fig. 11.10 that the kernel matrix that we have used has captured the vertical lines. However, this kernel alone will not suffice for the classification of numbers. There are a few more features. The other important feature is horizontal lines. Figure 11.11 shows the kernel to detect horizontal lines.

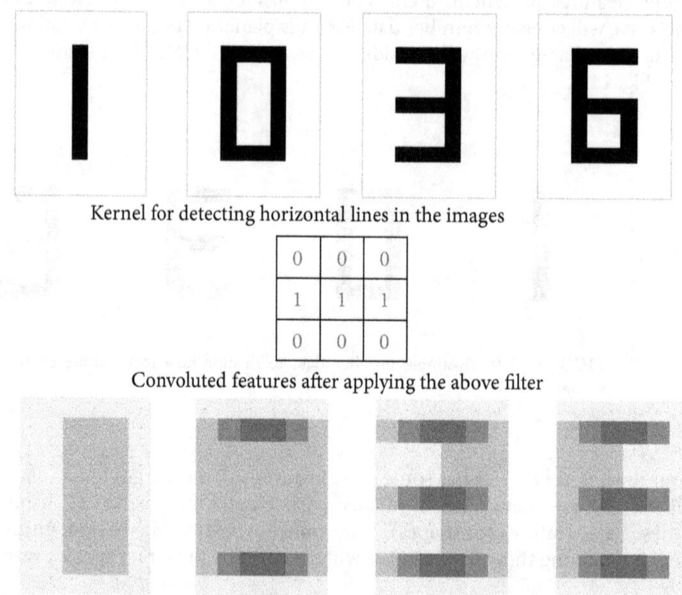

FIGURE 11.11   Kernel to detect horizontal lines.

From the diagrams in Fig. 11.11, we can see that the digit 1 has no horizontal lines. Images of 3 and 6 have three horizontal lines. These are a few examples of capturing features at a low level. A combination of these basic features completes the image. We can further apply filters on these low-level features (or convoluted images) to extract the mid-level features. We tend to add as many filters as possible so that we can capture all the low-level, mid-level, and high-level features. The screenshots shown in Fig. 11.12 are examples of various features learned by different filters.

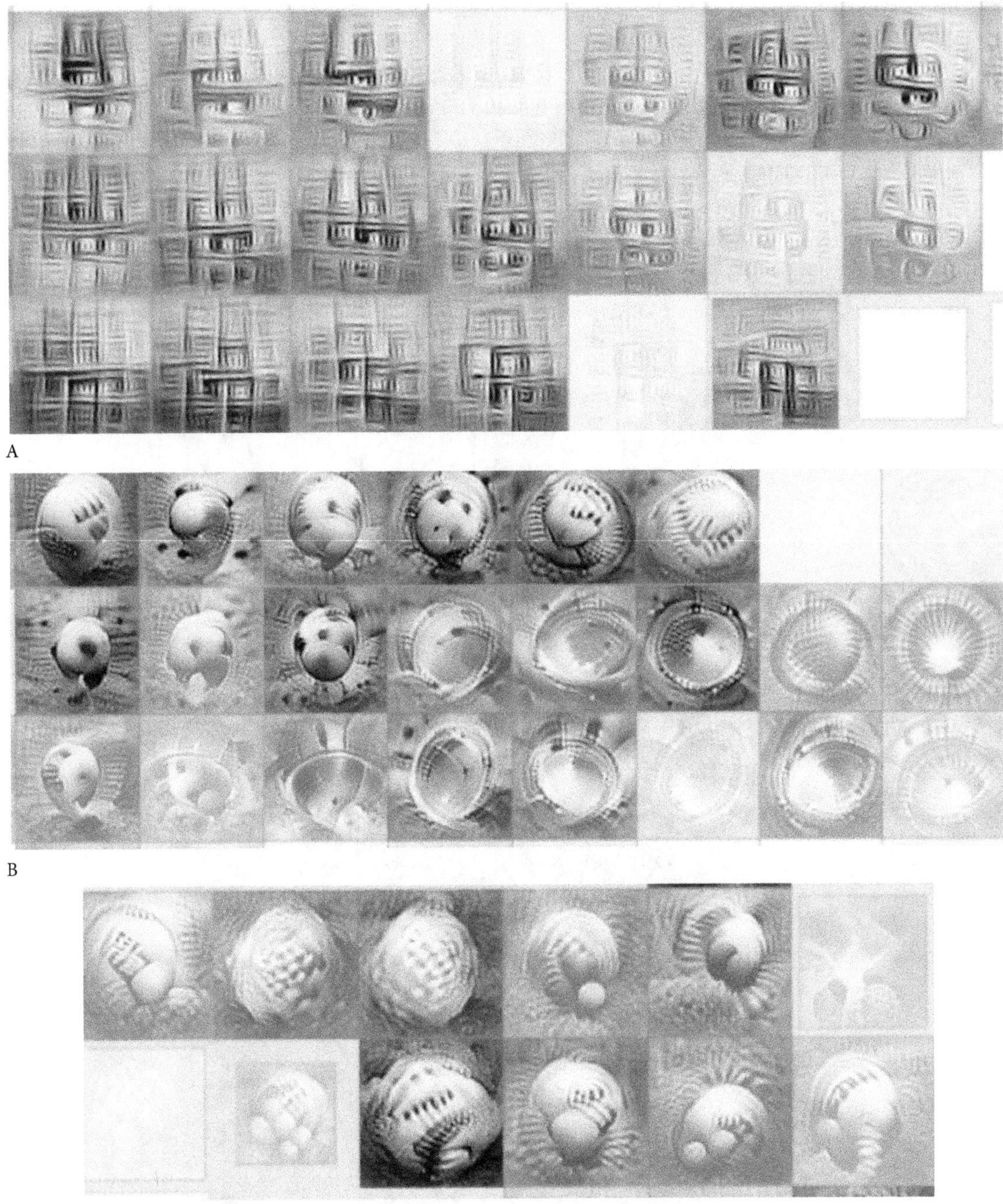

FIGURE 11.12 (*a*) Low-level features. (*b*) Mid-level features. (*c*) High-level features. Final prediction: volleyball, 90%; golf ball, 30%; balloons, 25%; parachute, 20%.

Figure 11.13 shows another example of feature detection. Now let's get more details on the kernel matrix.

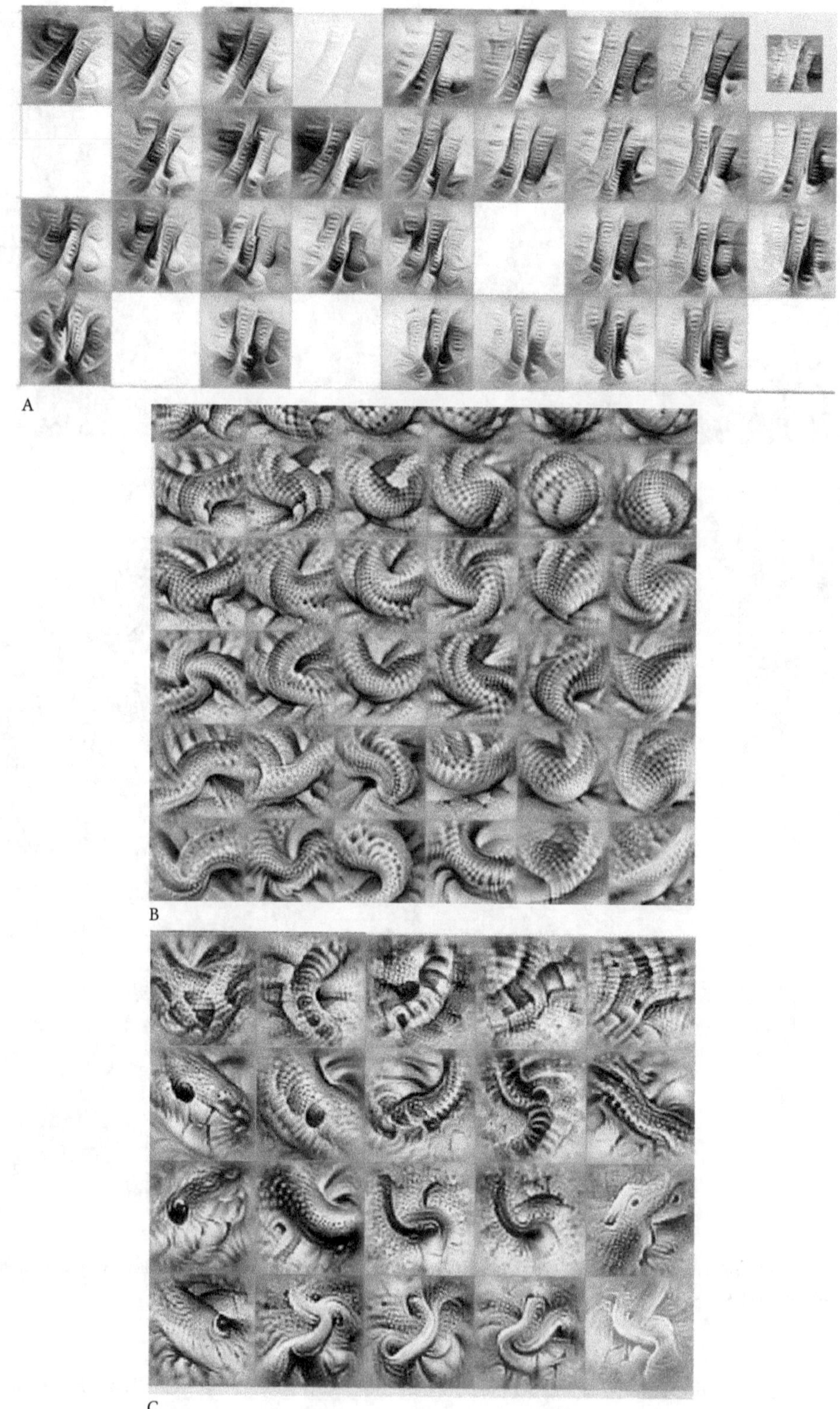

FIGURE 11.13 (*a*) Low-level features. (*b*) Mid-level features. (*c*) High-level features. Final prediction: Indian cobra, 67%; green mamba, 60%; alligator lizard, 34%; mud turtle, 10%.

11.2.3   Weights in the Kernel Matrix

The discussion until now is about saving the spatial dependency in the images. We can use the concept of the filter and capture all the essential features of an image. One question that comes to our mind is, how do we know which kernel to use? How can we fill the values in a kernel matrix? Real images have hundreds of features. How do we detect all the features in any given dataset? We do not need to detect the features manually. We need not fill the values in the kernel matrix. We will let the model detect the features in its training process. We just need to initialize the values of a kernel randomly. We can call the values in a kernel as weights. A 3×3 kernel will have nine weights and a bias term, a total of 10 weights (Fig. 11.14).

w1	w2	w3
w4	w5	w6
w7	w8	w9

FIGURE 11.14   Weights to initialize kernels.

The kernel matrix is a random weight matrix at the start of the model building. At the end of the model training process, each kernel matrix will transpire as a feature detection filter. If there are straight lines in our data, then a kernel will end up in a line detection filter. If there are circles in our data, then a kernel will end up as a circle detection filter. There are many more abstract features and patterns in the data that are tough to imagine. We configure the model by assigning enough kernel matrices so that we can capture all the features from the images. These features will help us in the final classification. So, the first step here is to apply the random weights matrix on the input layer and create new convoluted features. We would want to conclude this discussion with one final point. The feature detection happens during the training process. We do not manually supply features.

11.3   THE CONVOLUTION LAYER

In ANN, we have an input layer, and the layer after the input layer is a fully connected hidden layer. However, in the case of CNN, we apply a filter on the input layer and create convoluted features. The filter is nothing but a matrix of weights. These weights are determined while training. The first layer in a CNN is the convolution layer. The convolution layer keeps the local correlation intact. It captures all the features from the input. Unlike ANN, where the hidden layer is fully connected to the input layer, the convolution layer is sparsely connected. For a 3×3 kernel matrix, we have only nine weights and one bias term. These 10 weights are shared across all the input space. However, in ANN, each pixel will have one weight. The free parameters are also less in CNN as compared to a fully connected deep ANN.

- Take a black and white image with 1000 pixels; if we build an ANN with 10 hidden nodes in the first layer, then we will have $(1000 + 1) * 10 = 10{,}010$ weights in the first layer.
- Suppose we take the same data and build a CNN model by applying 10 filters, with each filter of the size 5×5, then the number of weights will be $(5 \times 5 + 1) * 10 = 260$. With these 260 weights, we could gain much more insights and features than the 10,000 weights of an ANN model.

In the convolution layer, we are tweaking the input data. Are we introducing any error in this step? Are we losing any data at this layer? Yes, we lose some data in this process. Since each filter captures the spatial dependence, by keeping a sufficient number of filters we can minimize the loss of information.

The resultant matrix after applying the filter is also known as an activation map. We take multiple filters that will result in multiple convoluted features or activation maps. All these convoluted features make the convolution layer (Fig. 11.15).

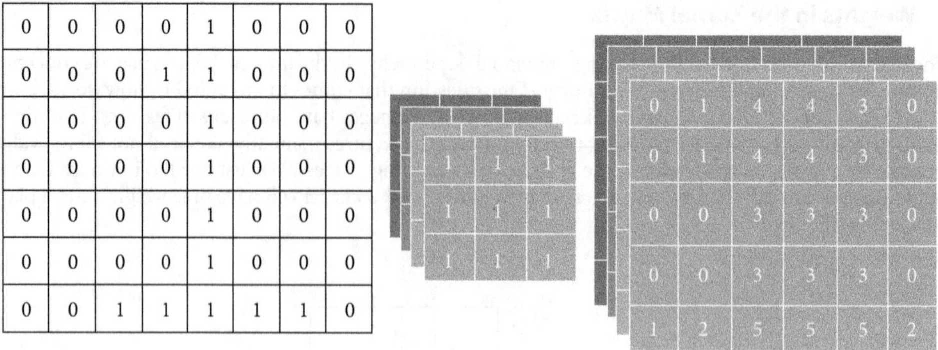

FIGURE 11.15   Activation maps and the convolution layer.

In the diagram in Fig. 11.15, we are showing four filters of size 3 × 3, which will give us four activation maps—one each for the four filters. All these four activation maps together are known as the convolution layer. Let us count the number of weights in the convolution layer in Fig. 11.15. Each filter has nine weights and one bias term, making it a total of 10 weights per filter multiplied by four filters. So overall, there are 40 weights in this layer.

11.3.1 The Convolution Layer in Keras

The following is an example of how to add a convolution layer. We use the function `Conv2D()`. This function moves the kernel matrix along the rows and columns of the image, hence it is called `Conv2D`. We usually initialize random weights in the kernel. Given below is the code used for adding a convolution layer to the model.

```
from tensorflow.keras.layers import Conv2D
model=Sequential()
model.add(Conv2D(filters=1,
              kernel_size=7,
              input_shape=(28,28,1),
              kernel_initializer='random_uniform'))
```

Given below are the parameters from the above code.

- `Conv2D()`: We are moving the kernel matrix in two dimensions, along rows and columns.
- `filters`: Number of filters or kernel matrices. We need to declare sufficient filters to capture all the features. Here we have mentioned it as 1. Usually, in practical problems, it is 8 or 16 or 32.
- `kernel_size`: Size of the kernel matrix. Here we have mentioned 7, which gives us a 7 × 7 matrix. Usually, it is 3 × 3 or 5 × 5 or 7 × 7.
- `input_shape`: Required parameter for the first layer only. From the second convolution layer onward, the shape will be derived automatically. In our example, the input image shape is (28,28,1).
- `kernel_initializer`: Initial values of the kernel matrix. Usually, it is randomly initialized.

The code above takes the image as input and delivers convoluted features as output. While solving real-life problems we need not visualize the result of all the convoluted features. Most kernels show some abstract hidden features, and it is difficult to interpret with visual inspection. However, in this case, we are printing it to get an intuition about how the filter works. Given below is the code for printing the convolution layer images.

```
img_reshape=np.expand_dims(x, axis=0)
img_reshape=np.expand_dims(img_reshape, axis=3)
img_reshape=model.predict(img_reshape)
pixels = np.matrix(img_reshape[:][:][:][0])
plt.imshow(pixels,cmap=plt.get_cmap('gray'))
plt.show()
```

The above code gives us the output shown in Fig. 11.16.

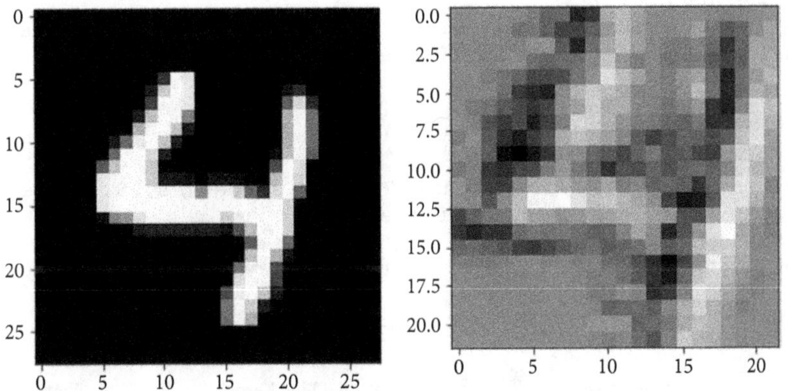

FIGURE 11.16   Actual image and code output for a randomly initialized kernel.

In the output shown in Fig. 11.16, we see the result of a randomly initialized kernel. During the training process these weights in the matrix will be adjusted. Finally, each kernel matrix will identify a feature. Now we will create customized kernels for detecting horizontal and vertical lines. These kernels are known as constant initializers in Keras. The following is the code to create the constant kernel matrices.

```python
import numpy as np
filter1=np.array([[1,1,1,1,1,1,1],
          [1,1,1,1,1,1,1],
          [100,100,100,100,100,100,100],
          [100,100,100,100,100,100,100],
          [100,100,100,100,100,100,100],
          [1,1,1,1,1,1,1],
          [1,1,1,1,1,1,1]])
print("filter1 \n", filter1)

filter2=np.transpose(filter1)
print("filter2 \n",filter2)
```

This code gives us the output as given below.

```
filter1
 [[  1   1   1   1   1   1   1]
 [  1   1   1   1   1   1   1]
 [100 100 100 100 100 100 100]
 [100 100 100 100 100 100 100]
 [100 100 100 100 100 100 100]
 [  1   1   1   1   1   1   1]
 [  1   1   1   1   1   1   1]]
```

```
filter2
 [[   1    1 100 100 100    1    1]
  [   1    1 100 100 100    1    1]
  [   1    1 100 100 100    1    1]
  [   1    1 100 100 100    1    1]
  [   1    1 100 100 100    1    1]
  [   1    1 100 100 100    1    1]
  [   1    1 100 100 100    1    1]]
```

We will now apply these constant kernel functions using the following code:

```
model=Sequential()
model.add(Conv2D(1, kernel_size=7,input_shape=(28,28,1), kernel_initializer=
keras.initializers.Constant(filter1)))
```

We can use similar code for `filter2`.

```
model=Sequential()
model.add(Conv2D(1, kernel_size=7,input_shape=(28,28,1), kernel_initializer=
keras.initializers.Constant(filter2)))
```

We can print the resultant activation maps from the kernels, as shown in Fig. 11.17.

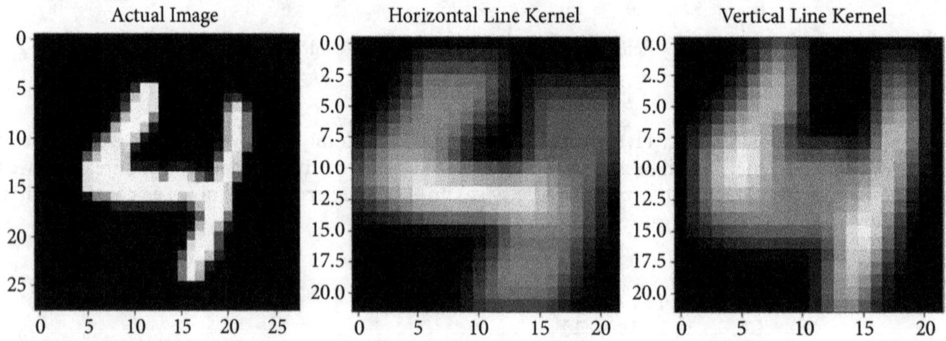

FIGURE 11.17   Actual image and filter results.

From the result shown in Fig. 11.17, we can see the highlighted portions of horizontal lines and vertical lines. Usually, it is not easy to observe such patterns in the images just by looking at the plots. Figure 11.18 shows a few more examples of horizontal and vertical line feature detection kernels.

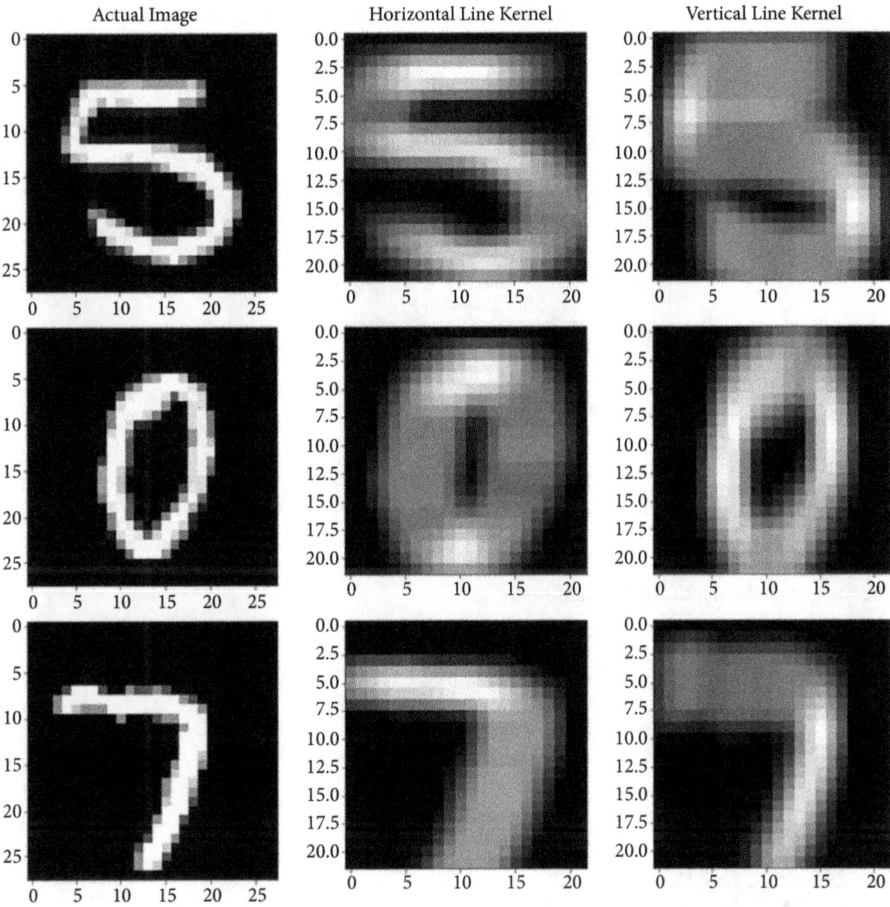

FIGURE 11.18   Examples of horizontal and vertical line feature detection kernels.

11.3.2   Filters for Color Images

The color images have the depth, so the filter will also need to have a depth parameter. The filter is applied just like the black and white images, but the dot product will be considering the depth also (Fig. 11.19).

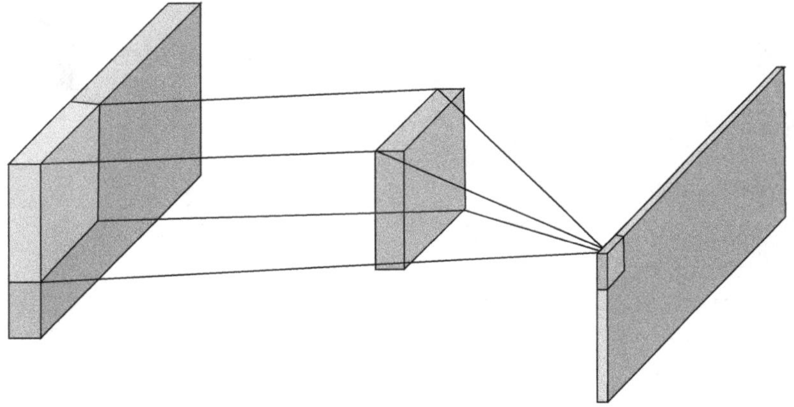

FIGURE 11.19   Kernel filter with depth.

The diagram in Fig. 11.19 shows the filter in depth. The entire calculations will end up as a single number in the convoluted image. Figure 11.20 shows the dot product calculation.

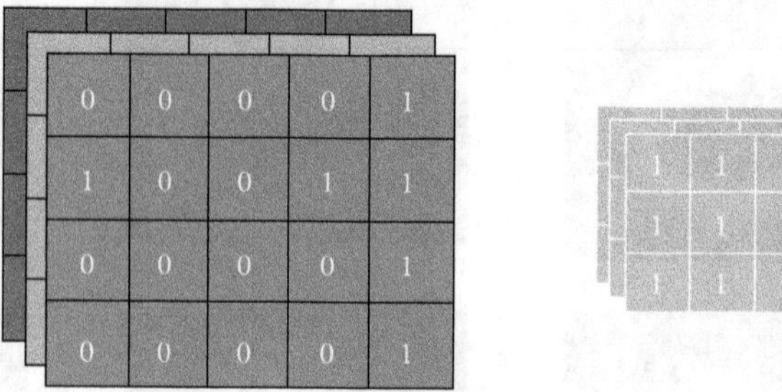

FIGURE 11.20   Kernel filter with depth.

A few important facts to note before we proceed any further:

1. The depth has three-channels, RGB—the channel is a synonym of depth in this context.
2. The image has the depth, so has the filter.
3. The depth of a filter is the same as the depth of the image.
4. The weight values or the kernel matrix values in each dimension need NOT be the same.

Figure 11.21 provides the expanded view of the image and filter.

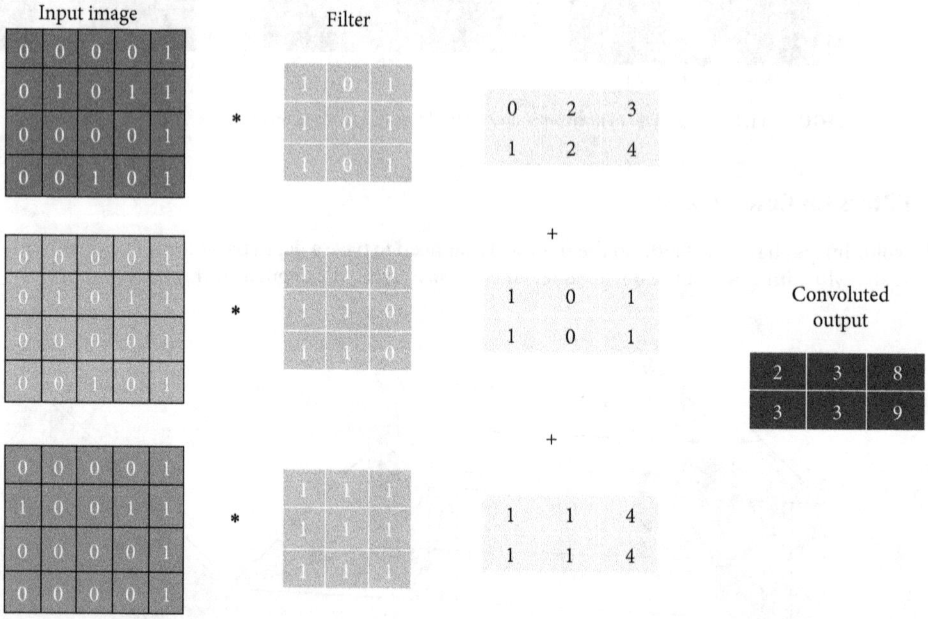

FIGURE 11.21   Detailed view of filters with depth.

We apply dot products between each channel from the input and corresponding channels from the filter. The complete dot product summation is our final result. We can visualize these calculations as a dot product between two cubes. The final output depth in the above example is just a single number. We applied one filter of shape $3 \times 3 \times 3$ on an input image of shape $4 \times 5 \times 3$, which finally gives an output of shape $2 \times 3 \times 1$. For the ease of visualization, we are not showing the bias term here. In real-world problems, there will be an additional bias term. Figure 11.22 shows the final result of the calculation.

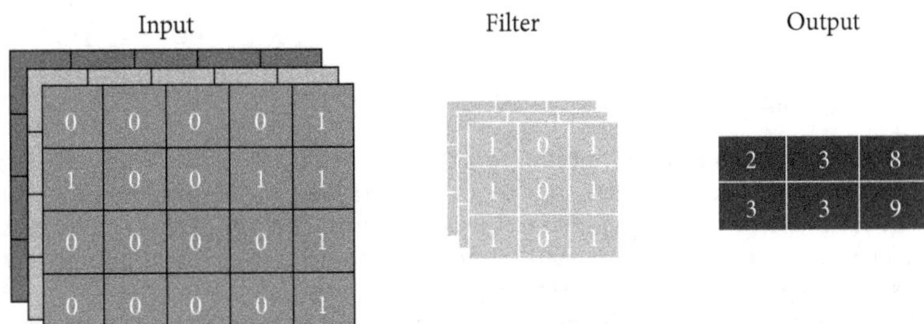

FIGURE 11.22   The input image, applied filter, and the output.

If we apply multiple filters, then it will result in depth in the output as well. Usually, we have multiple filters in each of the convolution layers. Figure 11.23 is a diagrammatic representation of the application of multiple filters.

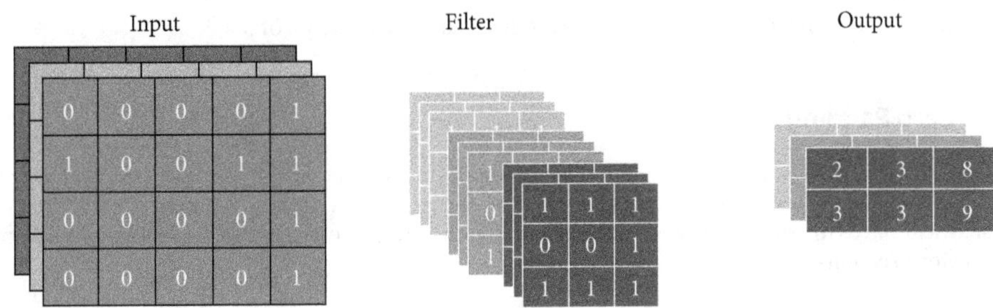

FIGURE 11.23   Application of multiple filters.

Since there is depth involved in the input layer and the convolution layer, we cannot sketch the usual two-dimensional network diagrams as we do in standard ANNs. In this case, we will need 3D network diagrams to represent the channels or depth. The diagram in Fig. 11.23 with input shape $4 \times 5 \times 3$, filter shape $3 \times 3 \times 3 \times 3$, and output shape $2 \times 3 \times 3$ is usually represented as in Fig. 11.24.

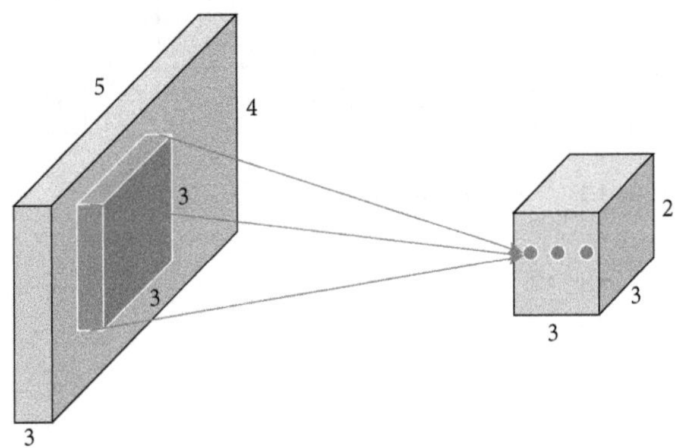

FIGURE 11.24   A 3D network diagram is necessary to represent the depth.

Usually, the number of filters is not shown in the diagram; this is inferred by looking at the output shape, particularly the depth of output. With this convolution layer, we are now ready for the model building.

11.3.2.1 *Code Creating the Convolution Layer*   Given below is the code for creating convolution layers.

```
model = models.Sequential()
model.add(layers.Conv2D(32, (5, 5), activation='relu', input_shape=(32, 32, 3)))

model.add(layers.Conv2D(64, (3, 3), activation='relu'))
```

The following is an explanation of this code.

- `Conv2D()` is the function used for performing the convolution operation (dot product) on input images. Input data has three dimensions: height, width, and depth. Nevertheless, the convolution filter moves in 2D only—height and width. In each kernel, we compute the dot product across the full depth. If the input data is a single series, then `Conv1D()` is used. In an array, we need to move the kernel in one direction only. However, for two-dimensional matrices and images, we use `Conv2D()`.

  Videos are made up of four dimensions. Each video is made of frames (or images). For example, if there are 24 frames in one-second of footage, then the four dimensions would be height, width, depth (RGB), and the number of images. In this case, we move the kernel matrix on each image and do it for all images. Then we use `Conv3D` in that case.

- There are a total of 32 filters or features or kernel matrices. It will end up as depth in the next layer.
- Kernel matrix size is (5,5); it is filled with random weights by default.
- The input shape is the shape of the input image.
- The second convolution layer has 64 filters. Each filter has a shape of 3×3.

11.3.3 Zero Padding

Until now, we have seen the effect of a filter on input images. Each filter tries to extract a feature from the input data. After applying a filter, we observed that the output image shrinks slightly. If we use a 3×3 filter, then the output generated will have two rows and two columns less. Figure 11.25 shows the results from some examples discussed in the previous sections.

0	0	0	0	1	0	0	0
0	0	0	1	1	0	0	0
0	0	0	0	1	0	0	0
0	0	0	0	1	0	0	0
0	0	0	0	1	0	0	0
0	0	0	0	1	0	0	0
0	0	1	1	1	1	1	0

1	1	1
1	1	1
1	1	1

0	1	4	4	3	0
0	1	4	4	3	0
0	0	3	3	3	0
0	0	3	3	3	0
1	2	5	5	5	2

FIGURE 11.25   Kernel filter on input image.

The 3×3 filter on 7×8 input gives us 5×6 in the output. Figure 11.26 is an example from MNIST data. The input shape is 28×28, and the filter size is 7×7, which gives us a 22×22 matrix in the output. Similarly, a 5×5 filter on 28×28 input gave us 24×24 output.

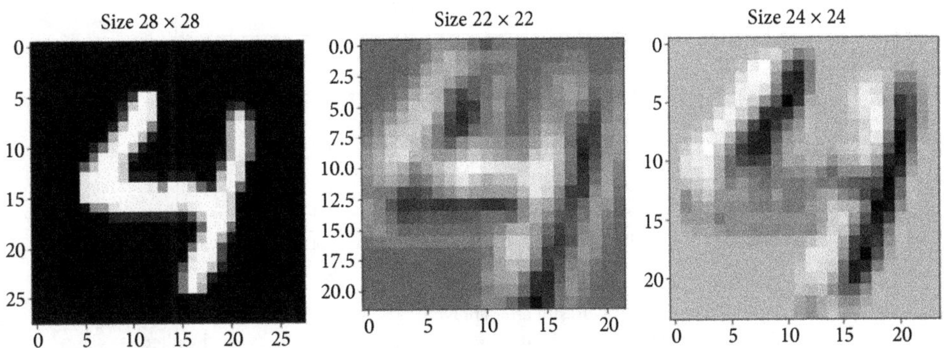

FIGURE 11.26   Actual image and 7×7 and 5×5 filter and kernel outputs.

Table 11.3 gives us a general formula for the reduction in dimensions of the output.

TABLE 11.3   Filter and Output Shapes Calculated

Input Shape	Filter Shape	Output Shape
7×8	3×3	5×6
28×28	5×5	24×24
28×28	7×7	22×22
Any shape	$F \times F$	Reduced by $(F - 1)$
7×8	3×3	$(7 - [3 - 1]) \times (8 - [3 - 1])$ $(7 - [2]) \times (8 - [2])$ 5×6
28×28	5×5	$(28 - [5 - 1]) \times (28 - [5 - 1])$ $(28 - [4]) \times (28 - [4])$ 24×24
28×28	7×7	$(28 - 6) \times (28 - 6)$ 22×22
$n \times n$	$F \times F$	$(n - [F - 1]) \times (n - [F - 1])$

0	0	0	0	1	0	0	0
0	0	0	1	1	0	0	0
0	0	0	0	1	0	0	0
0	0	0	0	1	0	0	0
0	0	0	0	1	0	0	0
0	0	0	0	1	0	0	0
0	0	1	1	1	1	1	0

1	1	1
1	1	1
1	1	1

0	1	4	4	3	0
0	1	4	4	3	0
0	0	3	3	3	0
0	0	3	3	3	0
1	2	5	5	5	2

FIGURE 11.27   Image size reduced after applying filter.

As we see in the diagram in Fig. 11.27, the reduction in the matrix size can be a big problem if there is some information on the edges of the input image. Look at the following example; we are applying a filter to detect the horizontal lines. If the images are in the middle and with no significant information on the edges, then there is no real issue. Figure 11.28 presents the filter for detecting horizontal lines.

Filter-1		
0	0	0
1	1	1
0	0	0

FIGURE 11.28   A filter for detecting horizontal lines.

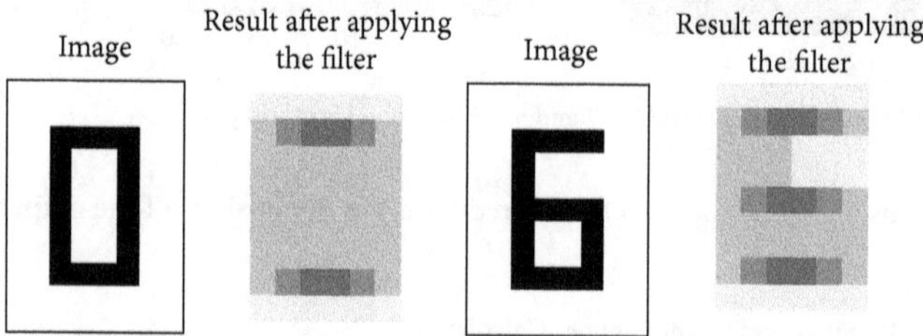

FIGURE 11.29   Images and results after applying filter-1 (the horizontal line detection filter).

Let us see what happens if these images are not in the middle and there is some information at the edges. We will apply the same horizontal line detection kernel and see the results (Fig. 11.30).

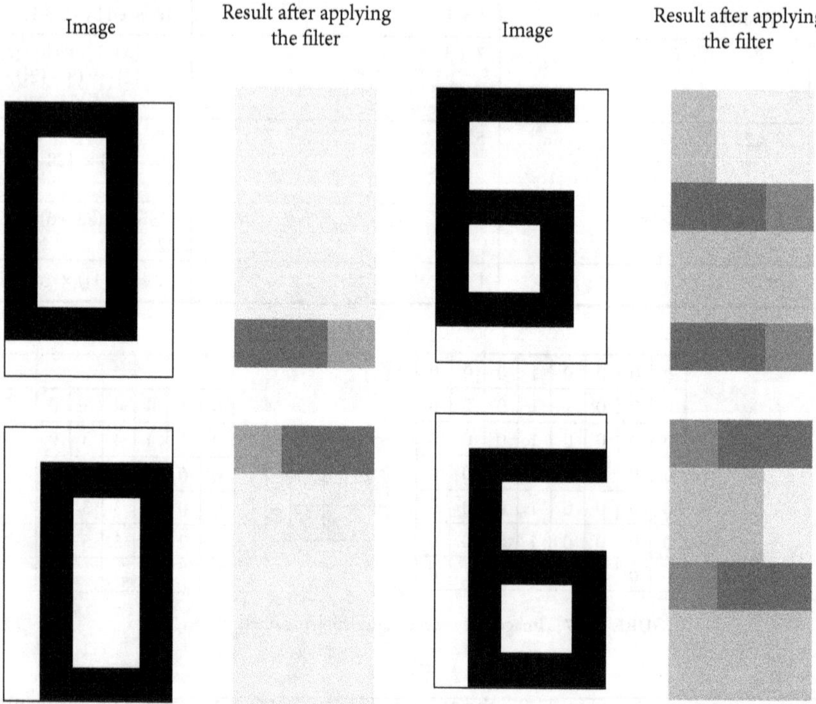

FIGURE 11.30   Information lost at the edges.

In the original image of the zero, there are two horizontal lines, which were captured by the kernel when the image of zero was in the middle with no information on the edges. Consider the image table given in Fig. 11.30. Here, the image of zero is at either the top edge or the bottom edge. In both cases, we lose the information on the edge in the output. We can see from the output that the kernel has identified only one horizontal line. The same is the case with the image of 6. We can see only two horizontal lines in the output when the image of 6 is either at the top edge or at the bottom edge. The same problem is repeated when we apply a filter to detect the vertical lines. Figure 11.31 shows the kernel for the detection of vertical lines.

Filter-2		
0	1	0
0	1	0
0	1	0

FIGURE 11.31   Filter to detect vertical lines.

FIGURE 11.32   Images and results after applying filter-2 (the vertical line detection filter).

We can see from the images in Fig. 11.32 that the vital information on the edges will be lost if we apply filters. How to preserve this information and carry it forward? There is a simple solution called zero padding. Simply push the image to the middle by adding zeros to the edges.

11.3.3.1 How Zero Padding Works Zero padding is nothing but adding a transparent border to the input image. We add a row of zeros and a column of zeros at the edges of the image. It will not create any error or bias in the input image. This zero padding will make sure that the size of the image will not shrink after applying the filter. Most importantly, zero padding will make sure that the information on the edges is intact. Have a look at the pictorial presentation in Fig. 11.33.

Input image

0	0	0	0	1	0	0	0
0	0	0	1	1	0	0	0
0	0	0	0	1	0	0	0
0	0	0	0	1	0	0	0
0	0	0	0	1	0	0	0
0	0	0	0	1	0	0	0
0	0	1	1	1	1	1	0

Image after zero padding (P = 1)

0	0	0	0	0	0	0	0	0	0
0	0	0	0	0	1	0	0	0	0
0	0	0	0	1	1	0	0	0	0
0	0	0	0	0	1	0	0	0	0
0	0	0	0	0	1	0	0	0	0
0	0	0	0	0	1	0	0	0	0
0	0	0	0	0	1	0	0	0	0
0	0	0	1	1	1	1	1	0	0
0	0	0	0	0	0	0	0	0	0

Image after zero padding (P = 2)

0	0	0	0	0	0	0	0	0	0	0	0
0	0	0	0	0	0	0	0	0	0	0	0
0	0	0	0	0	0	1	0	0	0	0	0
0	0	0	0	0	1	1	0	0	0	0	0
0	0	0	0	0	0	1	0	0	0	0	0
0	0	0	0	0	0	1	0	0	0	0	0
0	0	0	0	0	0	1	0	0	0	0	0
0	0	0	0	0	0	1	0	0	0	0	0
0	0	0	0	1	1	1	1	1	0	0	0
0	0	0	0	0	0	0	0	0	0	0	0
0	0	0	0	0	0	0	0	0	0	0	0

FIGURE 11.33 Input image and images after zero padding ($P = 1$) and after zero padding ($P = 2$).

Usually, we apply a padding of size-1 for a 3×3 filter to preserve the shape of the original image. The padding of size-2 will be good for the 5×5 filter. We need to carefully fine-tune the size of padding to build an efficient model. Figures 11.34 and 11.35 show the comparison of results without and with zero padding.

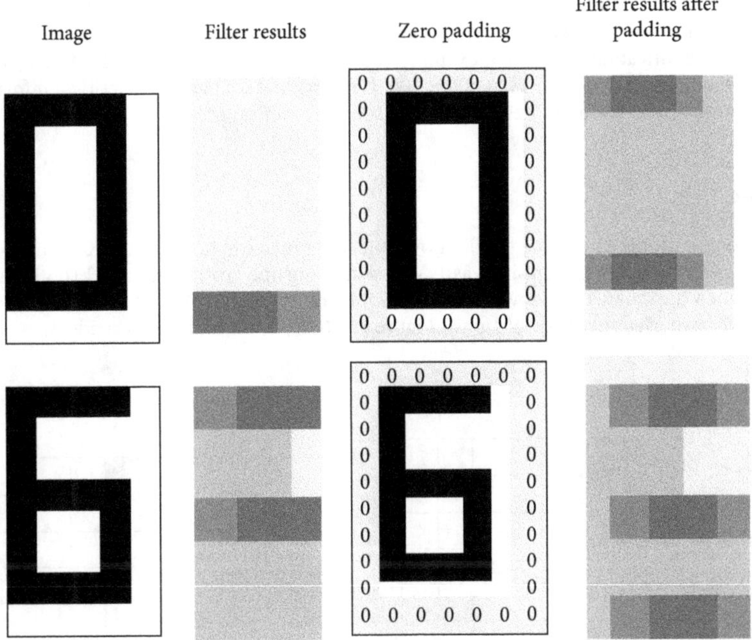

FIGURE 11.34 Results compared without and with zero padding (Filter-1).

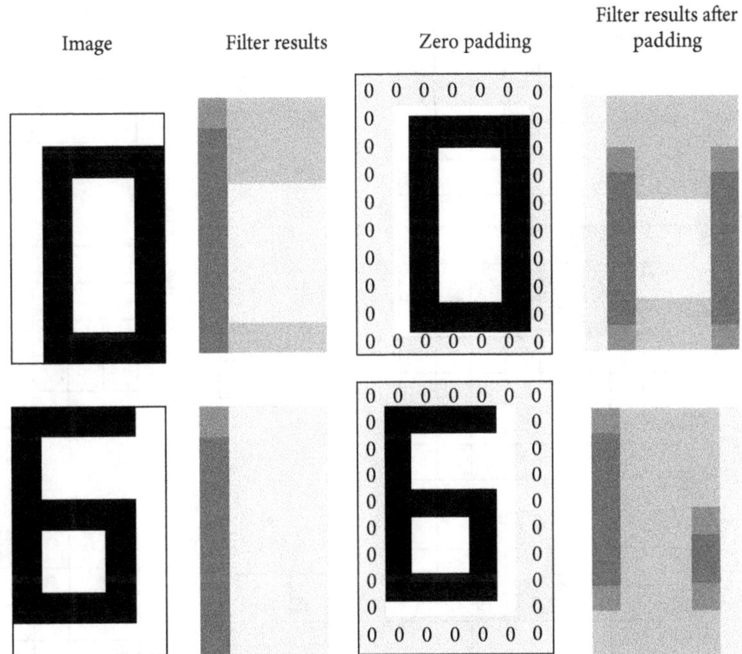

FIGURE 11.35 Results compared without and with zero padding (Filter-2).

We can see the effect of padding on the output. In each image, we are losing some vital information on the edges if we do not apply the padding. While solving real-world problems, we tend to reduce the size of the original image and then build the classification model. For example, if the original image is 1000 × 2000 size, then we will possibly resize and take a sample of 250 × 500 or even 100 × 200. It means that we have some critical information on the edges. This indeed means that zero padding is essential while solving practical problems.

11.3.4 Strides

In the previous sections, we applied the filter over the input image in horizontal and vertical directions. While running these filters on the image, our step size was 1. We are moving one pixel at a time. This is known as stride-1. If we move by stride-1, then the data will be down-sampled slightly on the edges. We can avoid data loss by adding zero padding. Until now, we have used the stride of 1 only. Figure 11.36 shows an example of stride-1.

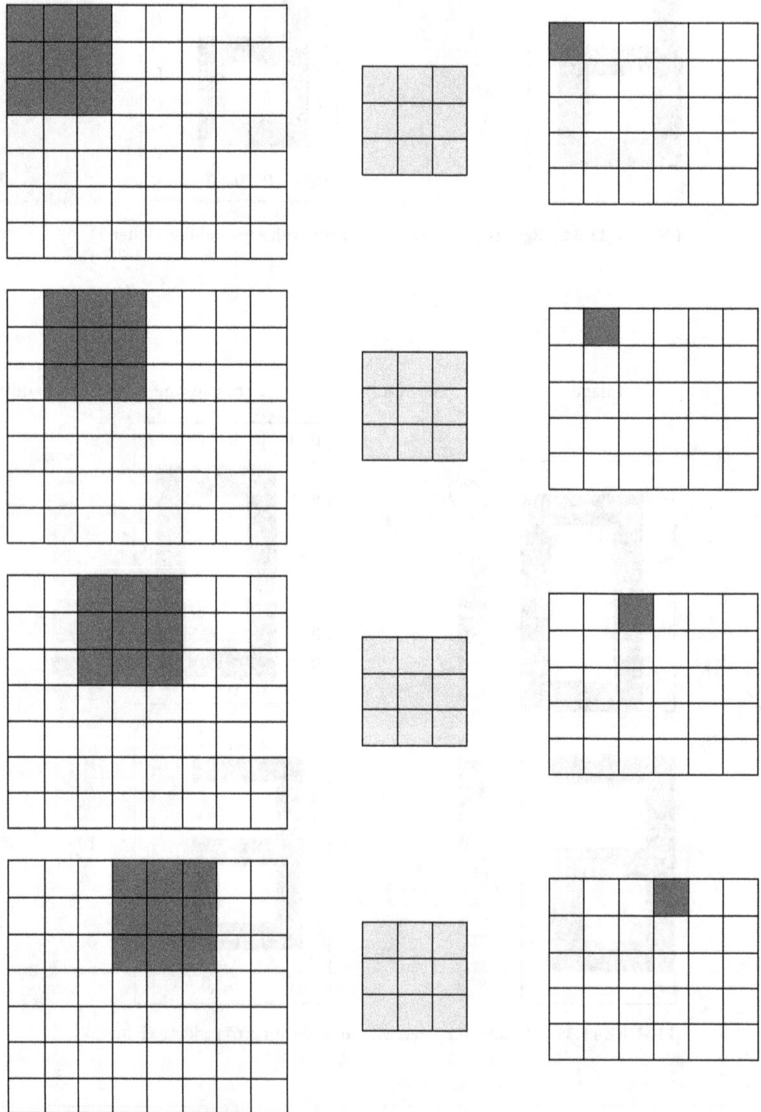

FIGURE 11.36 An example of stride-1.

If we move the filter by stride-2, then the resultant image will be almost halved. At each movement, we are skipping two pixels. Figure 11.37 shows an example of stride-2.

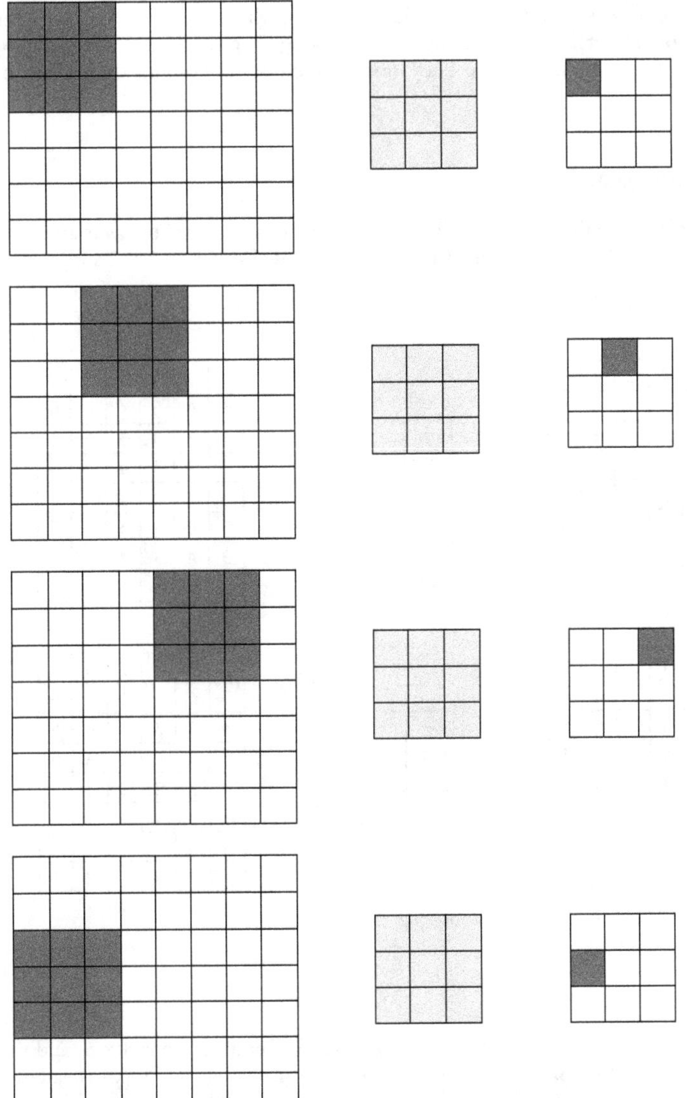

FIGURE 11.37   An example of stride-2.

A 7×8 image has given us an output of size 3×3 with a 3×3 filter of stride-2. Large strides will shrink the original image significantly. Sometimes we use strides to save the computational time. We may lose some information in terms of feature maps. We can make a trade-off between the time saved and the accuracy lost to pick the optimal number of strides. In real-world problems, we tend to leave strides as the default value that is stride-1. If the input data is extensive and the convolution layer is too heavy with large images, then we can down-sample in a different step called the pooling layer. Let us see what the pooling layer is.

11.4 *POOLING LAYER*

The first layer of CNN is the convolution layer. Furthermore, we add as many filters as possible to detect all the features inside an image. First, we will detect the basic features such as straight lines, circles, and curves; later, we will add some more convolution layers to detect the mid-level features that are a combination of the low-level features. We add some more convolution layers to detect the high-level features, which are a combination of mid-level features. In between these convolutional layers, we add pooling layers. Pooling is nothing but down-sampling of the data. We add pooling layers after adding convolution layers. There are two methods of pooling. One is max pooling and another is average pooling.

11.4.1 How Pooling Works

Usually, we perform pooling with a 2×2 matrix using stride-2. Given below is an example of max pooling. We move the 2×2 matrix over the input and get the maximum value from each 2×2 subregion.

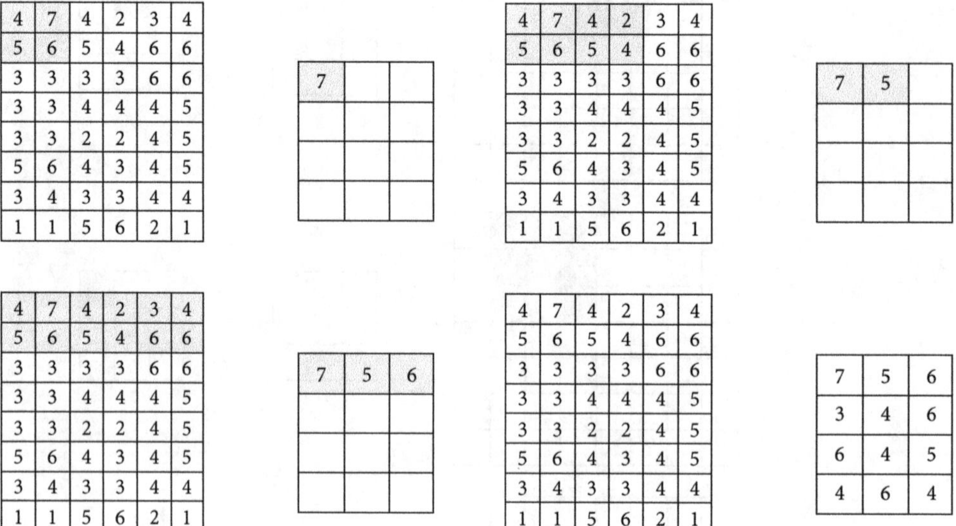

FIGURE 11.38   An example of max pooling.

From the depictions in Fig. 11.38, we can see that the input shape is 8×6, and the output after max pooling is precisely half of the input dimensions. We can also perform average pooling. Instead of taking the maximum number, we consider the average of the values in the matrix. Figure 11.39 shows an example of average pooling.

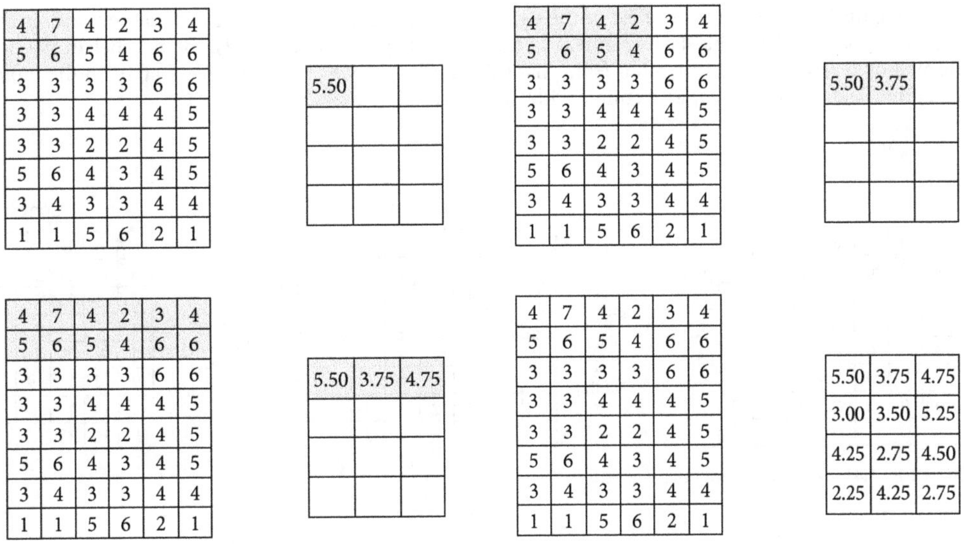

FIGURE 11.39 An example of average pooling.

11.4.2 Why Pooling Is Done

We add pooling layers mainly for two reasons. First, there are too many parameters in the network. The second reason is related to the receptive field.

Facts to be noted before we proceed with the details of the receptive field in the next section:

1. Since each convolution layer gains sufficient information from its previous layer, the down-sampling of the convolution layer does not hurt the overall accuracy.

2. Down-sampling is a useful trick for more massive datasets or deep networks with numerous parameters.

11.4.2.1 *Receptive Field*   In simple terms, a receptive field is a subregion of the input image. A specific feature map from a convolution layer has access to the receptive field. This feature can be from the first convolution layer or the second layer or any other layer. Figure 11.40 has examples of the receptive field. If the filter size is 3 × 3, then the receptive field of the first layer is always the size of the filter.

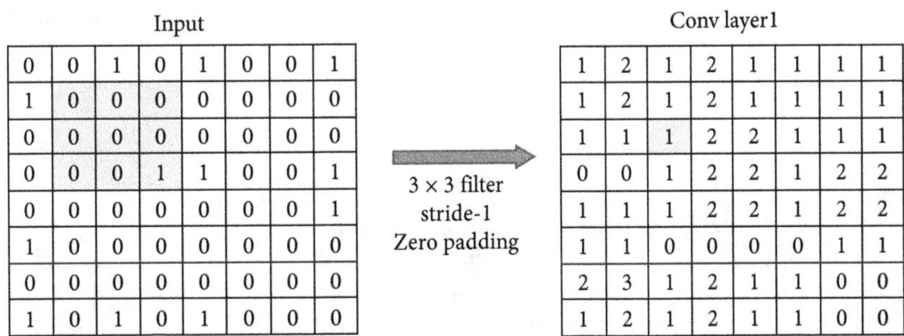

FIGURE 11.40 Examples of the receptive field.

In Figure 11.40, the receptive field of the element at the position (3,3) in the convolution layer is the matrix [2:4, 2:4] (or [1:3, 1:3] if 0 is the starting index) in the input image. In other words, the number in the position (3,3) is formed by receiving the information from that subregion of pixels [2:4, 2:4]. This matrix on the input is called the receptive field. Now let us see an example with two convolution layers (Fig. 11.41).

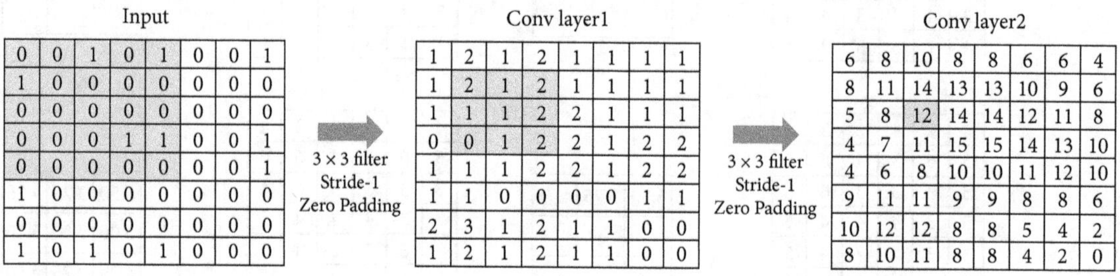

Input

0	0	1	0	1	0	0	1
1	0	0	0	0	0	0	0
0	0	0	0	0	0	0	0
0	0	0	1	1	0	0	1
0	0	0	0	0	0	0	1
1	0	0	0	0	0	0	0
0	0	0	0	0	0	0	0
1	0	1	0	1	0	0	0

3×3 filter, Stride-1, Zero Padding

Conv layer1

1	2	1	2	1	1	1	1
1	2	1	2	1	1	1	1
1	1	1	2	2	1	1	1
0	0	1	2	2	1	2	2
1	1	1	2	2	1	2	2
1	1	0	0	0	0	1	1
2	3	1	2	1	1	0	0
1	2	1	2	1	1	0	0

3×3 filter, Stride-1, Zero Padding

Conv layer2

6	8	10	8	8	6	6	4
8	11	14	13	13	10	9	6
5	8	12	14	14	12	11	8
4	7	11	15	15	14	13	10
4	6	8	10	10	11	12	10
9	11	11	9	9	8	8	6
10	12	12	8	8	5	4	2
8	10	11	8	8	4	2	0

FIGURE 11.41 An example with two convolution layers.

The receptive field of (3,3) element in convolution layer 2 is a matrix of [2:4, 2:4] from convolution layer 1, which indeed has a receptive field of size 5×5 matrix [1:5,1:5] on the input image. So the receptive field of an element in layer 2 is a 5×5 matrix. If we extend this to one more layer, then an element in the convolution layer 3 will have a receptive field of 7×7 from the input image.

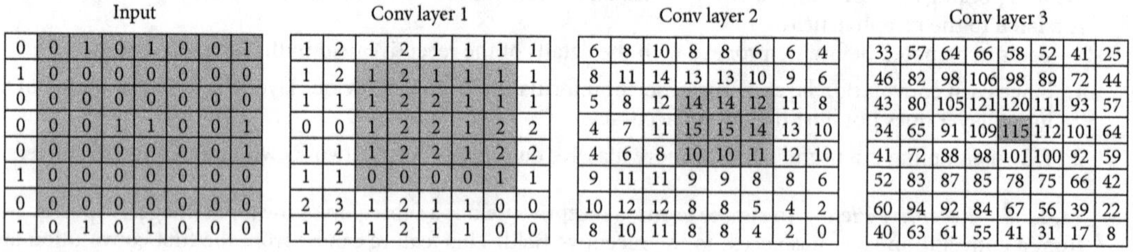

Input

0	0	1	0	1	0	0	1
1	0	0	0	0	0	0	0
0	0	0	0	0	0	0	0
0	0	0	1	1	0	0	1
0	0	0	0	0	0	0	1
1	0	0	0	0	0	0	0
0	0	0	0	0	0	0	0
1	0	1	0	1	0	0	0

Conv layer 1

1	2	1	2	1	1	1	1
1	2	1	2	1	1	1	1
1	1	1	2	2	1	1	1
0	0	1	2	2	1	2	2
1	1	1	2	2	1	2	2
1	1	0	0	0	0	1	1
2	3	1	2	1	1	0	0
1	2	1	2	1	1	0	0

Conv layer 2

6	8	10	8	8	6	6	4
8	11	14	13	13	10	9	6
5	8	12	14	14	12	11	8
4	7	11	15	15	14	13	10
4	6	8	10	10	11	12	10
9	11	11	9	9	8	8	6
10	12	12	8	8	5	4	2
8	10	11	8	8	4	2	0

Conv layer 3

33	57	64	66	58	52	41	25
46	82	98	106	98	89	72	44
43	80	105	121	120	111	93	57
34	65	91	109	115	112	101	64
41	72	88	98	101	100	92	59
52	83	87	85	78	75	66	42
60	94	92	84	67	56	39	22
40	63	61	55	41	31	17	8

FIGURE 11.42 An example of a receptive field with three convolution layers.

In Fig. 11.42, we have applied 3×3 filters in each layer and used zero padding. An element of the third layer has a receptive field of 7×7. Now let us apply a convolution layer followed by pooling and calculate the receptive field (Fig. 11.43).

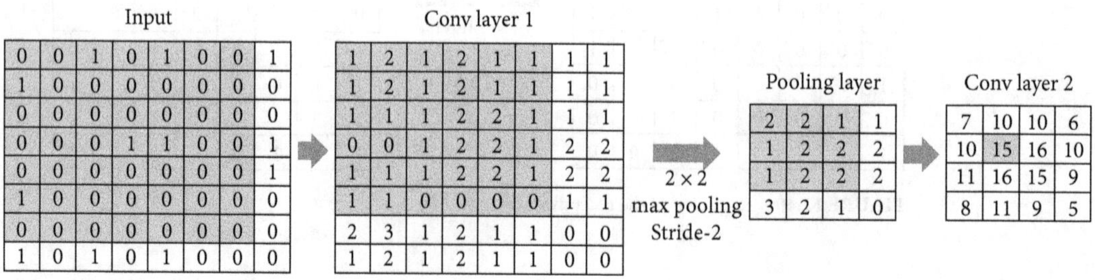

Input

0	0	1	0	1	0	0	1
1	0	0	0	0	0	0	0
0	0	0	0	0	0	0	0
0	0	0	1	1	0	0	1
0	0	0	0	0	0	0	1
1	0	0	0	0	0	0	0
0	0	0	0	0	0	0	0
1	0	1	0	1	0	0	0

Conv layer 1

1	2	1	2	1	1	1	1
1	2	1	2	1	1	1	1
1	1	1	2	2	1	1	1
0	0	1	2	2	1	2	2
1	1	1	2	2	1	2	2
1	1	0	0	0	0	1	1
2	3	1	2	1	1	0	0
1	2	1	2	1	1	0	0

2×2 max pooling, Stride-2

Pooling layer

2	2	1	1
1	2	2	2
1	2	2	2
3	2	1	0

Conv layer 2

7	10	10	6
10	15	16	10
11	16	15	9
8	11	9	5

FIGURE 11.43 Receptive field for the convolution layer followed by the pooling layer.

Now, the receptive field of an element in convolution layer 2 is 7 × 7 on the input image. We needed three layers without pooling, and with pooling, we achieved a 7 × 7 receptive field in two layers. That means we can have less depth in the overall network if we add pooling layers in between. With small receptive fields, we can learn low-level features. If we have broad receptive fields with fewer convolution layers, then the network can learn mid-level and high-level features very early. If we do not have a pooling layer, then we need multiple convolution layers for learning low-, mid-, and high-level features. That will make the whole network very deep, which will hurt the computation time. One more important point is that we do not have any parameters in the pooling layer. It is a simple down-sampling process. Data scientists often include a pooling layer after a convolution layer in CNN models.

11.5   CNN ARCHITECTURE

Until now, we have discussed the convolution layer and pooling layer. The convolution layer is the main difference between standard ANNs and CNNs. ANN has an input layer followed by hidden layers and, finally, the output layer. In CNN, the input layer will be followed by a convolution layer to capture features. The convolution layer is followed by the pooling layer, followed by many more pairs of convolution and pooling layers. Finally, once all the features are captured (learned), we flatten everything and add a couple of fully connected dense hidden layers. We flatten the data into one vector before sending it to a dense layer later. Flattening is nothing but converting a 3D array into a 1D array.

After learning all the essential features, the problem is almost similar to standard ANN. We can flatten the data to a one-dimensional vector and connect it to output using softmax activation. Even in CNNs, we have dense hidden layers, but they are added at the end, just before the output layers (Fig. 11.44).

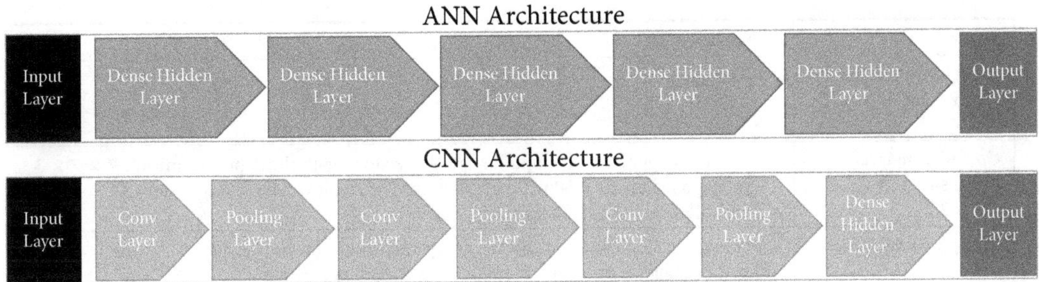

FIGURE 11.44   ANN and CNN architectures.

Usually, the pooling layer is not shown in the final architecture diagram. Edges typically represent weights in the neural network diagram. There are no weights for the pooling layer and flatten layer. That is probably the reason why they are not shown in the final CNN architecture diagram. However, if we want to show an explicit and descriptive architecture diagram, then we can include them as well. Figure 11.45 shows an example of a simple CNN model.

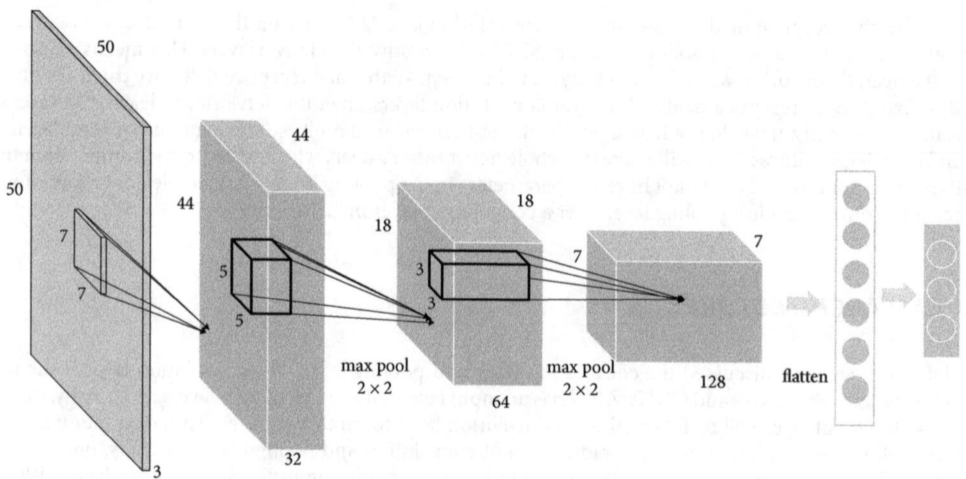

FIGURE 11.45   An exploded view of a simple CNN model.

Table 11.4 explains the CNN architecture diagram shown in Fig. 11.45.

TABLE 11.4   CNN Architecture Description

Layer	Explanation
Input layer: Image with shape 50 × 50 × 3	• 50 height • 50 width • 3 RGB
Convolution layer 1 with shape 44 × 44 × 32	• 7 × 7 filter applied to the input image. 7 × 7 along with the depth technically 7 × 7 × 3 • 7 × 7 applied on 50 × 50 without padding. The result will be of dimensions 44 × 44 • Overall, 32 such 7 × 7 filters are applied
Max pool layer 2 × 2	• Pooling is applied but not shown as a layer in the architecture diagram. • 2 × 2 pooling on 44 × 44 will result in 22 × 22 shape
Convolution layer 2 with shape 18 × 18 × 64	• After the max pool, the shape of the matrix is 22 × 22. If we apply 5 × 5 filter on it, then it will give us 18 × 18 output (without padding) • There are 64 such 5 × 5 filters applied on the 22 × 22 matrix. Which finally gives us the shape of 18 × 18 × 64
Max pool layer 2 × 2	• Again, pooling is applied but not shown explicitly as a layer in the architecture diagram. 2 × 2 pooling on the 18 × 18 matrix will result in a 9 × 9 matrix.
Convolution layer 3 with shape 18 × 18 × 64	• After the max pool, the shape of the image is 9 × 9. If we apply 3 × 3 filter on it, then it will give us 7 × 7 output (without padding) • There are 128 such 3 × 3 filters applied
Flatten layer	• Flattening is done just before adding the dense later • 18 × 18 × 64 3D array will be fitted to make a single vector or 1D array • The shape will be 20,736 × 1
Dense layer	• 20,736 nodes will be connected to the dense layer • The exact number of nodes in the dense hidden layer is not shown in the diagram
Output layer	• The hidden layer is connected to the output layer. We use the standard softmax activation at the output

11.5.1   Weights in a CNN Model

We will now discuss how to calculate the number of weights on CNN. Figure 11.46 shows an example of the CNN model.

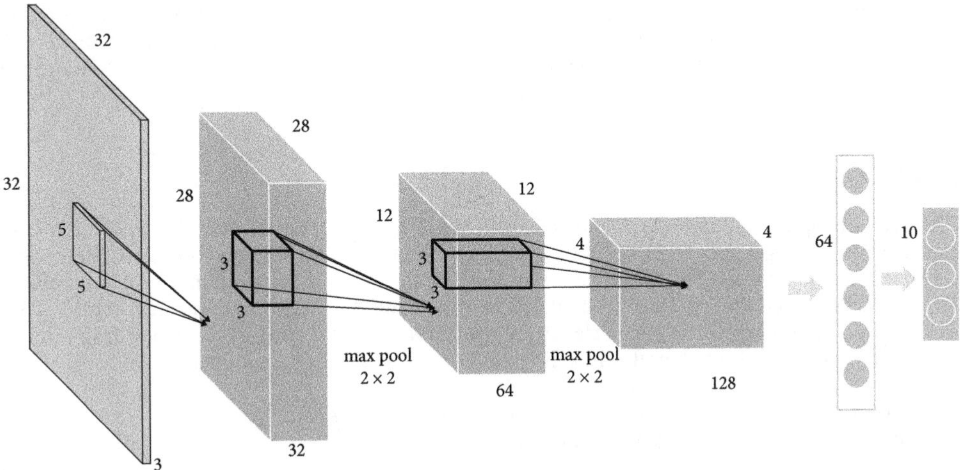

FIGURE 11.46   An example of a CNN model for calculating the number of weights.

TABLE 11.5   Shapes and Weights Calculated for Different Layers

Layer	Shape Calculation	Weights Calculation
Input layer	$32 \times 32 \times 3$	
Convolution layer 1 from 32 filters of size 5×5	$28 \times 28 \times 32$	• $5 \times 5 \times 3 = 75$ weights in the filter • 1 bias per filter = 76 weights • 32 filters => 76×32 weights • 2432 weights in conv layer1
Max pool layer 2×2	$14 \times 14 \times 32$	• Zero weights
Convolution layer 2 from 64 filters of size 3×3	$12 \times 12 \times 64$	• $3 \times 3 \times 32 = 288$ weights in the filter • 1 bias per filter = 289 weights • 64 filters => 289×64 weights • 18,496 weights in conv layer2
Max pool layer 2×2	$6 \times 6 \times 64$	• Zero weights
Convolution layer-3 from 128 filters of size 3×3	$4 \times 4 \times 128$	• $3 \times 3 \times 64 = 576$ weights in the filter • 1 bias per filter = 577 weights • 128 filters => 577×128 weights • 73,856 weights in conv layer3
Flatten layer	2048×1	• Zero weights
Dense layer with 64 nodes	64×1	• 2048 nodes after flattening and each node is connected to every node in a dense layer • 2048×64 weights • $(2048 + 1)*64$ including bias • 131,136 from a dense layer
Output layer	10×1	• Each node from the dense layer is connected to output nodes • $64*10$ weights • $(64 + 1)*10$ including bias • 650 weights from the output layer
Overall weights		2432 weights in conv layer1 + 18,496 weights in conv layer2 + 73,856 weights in conv layer3 + 131,136 from a dense layer + 650 weights from the output layer = 226,570

The total number of weights in the CNN architecture in Fig. 11.46 is 226,570 (Table 11.5). This number is smaller compared to the number of parameters required in ANNs. Imagine a standard ANN model with $32 \times 32 \times 3$ input with just one hidden layer with 128 nodes. Then we will need to train 393,344 (((32*32*3)+1)*128) weights in the first layer itself. We may need many more layers after that.

11.5.2   CNN Code

The CNN example used in Sec. 11.5.1 for calculating weights is not a random CNN model. It is a model used for the classification of images in CIFAR10 data. CIFAR10 is a subset of CIFAR100 data. The CIFAR100 consists of 80 million tiny image datasets collected by Alex Krizhevsky, Vinod Nair, and Geoffrey Hinton. CIFAR10 is available as part of the Keras sample datasets library. The CIFAR10 data consists of 60,000 images; each image size is $32 \times 32 \times 3$, and there are 10 classes in the output. The output classes are airplane, automobile, bird, cat, deer, dog, frog, horse, ship, and truck.

The following code is used for downloading the data and visualizing a few images.

```python
from tensorflow.keras import datasets, layers, models
import matplotlib.pyplot as plt

(X_train, y_train), (X_test, y_test) = datasets.cifar10.load_data()

# Normalize the input data
X_train=X_train/255
X_test=X_test/255

print("X_train.shape", X_train.shape)
print("y_train.shape", y_train.shape)
print("X_test.shape", X_test.shape)
print("y_test.shape", y_test.shape)

#Drawing Few images
class_names = ['airplane', 'automobile', 'bird', 'cat', 'deer', 'dog', 'frog',
'horse', 'ship', 'truck']
plt.figure(figsize=(10,10))
for i in range(16):
    plt.subplot(4,4,i+1)
    plt.imshow(X_train[i], cmap=plt.cm.binary)
    plt.xlabel(class_names[y_train[i][0]])
    plt.xticks([])
    plt.yticks([])
plt.show()
```

This code results are given below.

```
X_train.shape (50000, 32, 32, 3)
y_train.shape (50000, 1)
X_test.shape (10000, 32, 32, 3)
y_test.shape (10000, 1)
```

FIGURE 11.47 Sample images from the data.

The images in Fig. 11.47 are neither pixelated nor low-quality screenshots. The dataset itself has such low-quality pixelated images of size 32×32. We can now go ahead and build the model. We will use the same architecture that we have used in the example of the calculation of weights. Figure 11.48 shows a copy of the architecture.

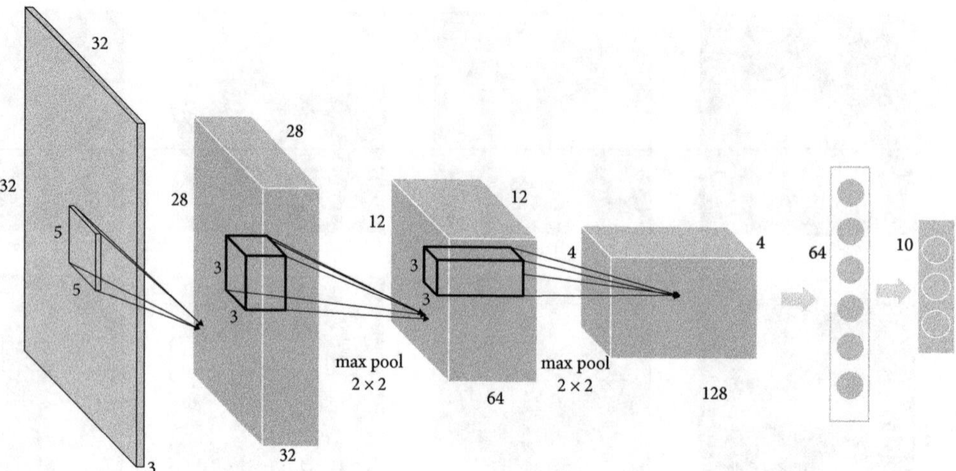

FIGURE 11.48   CNN architecture.

The following is the code for creating the CNN model.

```
model = models.Sequential()

model.add(layers.Conv2D(32, (5, 5), activation='relu', input_shape=(32, 32, 3)))

model.add(layers.MaxPooling2D((2, 2)))

model.add(layers.Conv2D(64, (3, 3), activation='relu'))

model.add(layers.MaxPooling2D((2, 2)))

model.add(layers.Conv2D(128, (3, 3), activation='relu'))

model.add(layers.Flatten())

model.add(layers.Dense(64, activation='relu'))

model.add(layers.Dense(10))

model.summary()
```

Typical image processing networks are very deep. Data scientists often use ReLU activation in image processing examples. ReLU works better than sigmoid and tanh activation.

We have already discussed each layer in depth, and we have also calculated the number of weights in this network. Table 11.6 shows the output.

TABLE 11.6 Model Summary for CIFAR10

```
Model: "sequential"

Layer (type)                 Output Shape              Param #
=================================================================
conv2d (Conv2D)              (None, 28, 28, 32)        2432

max_pooling2d (MaxPooling2D) (None, 14, 14, 32)        0

conv2d_1 (Conv2D)            (None, 12, 12, 64)        18496

max_pooling2d_1 (MaxPooling2 (None, 6, 6, 64)          0

conv2d_2 (Conv2D)            (None, 4, 4, 128)         73856

flatten (Flatten)            (None, 2048)              0

dense (Dense)                (None, 64)                131136

dense_1 (Dense)              (None, 10)                650
=================================================================
Total params: 226,570
Trainable params: 226,570
Non-trainable params: 0
```

We can cross-validate this weight summary with our previous manual calculations. Now we will go ahead and train the model. Given below is the code for training the model.

```
model.compile(
optimizer=tf.keras.optimizers.SGD(),  loss=tf.keras.losses.SparseCategorical
Crossentropy(from_logits=True),

metrics=['accuracy'])

model.fit(X_train, y_train,
        batch_size=32,
        epochs=20,
        validation_data=(X_test, y_test))
```

In this code, we have used the SGD optimizer. Loss is usually categorical cross-entropy. In this data, we have not done one-hot encoding for the target variable. The target variable has integers in it. We have to use `SparseCategori-calCrossentropy()` for the cases where one-hot encoding is not performed; `from_logits=True` will consider the logits for the predicted values instead of probabilities. This option works better for faster execution. The following is the result of this code.

```
Train on 50000 samples, validate on 10000 samples
Epoch 1/12
50000/50000 [==========] - 53s 1ms/sample - loss: 1.8569 - accuracy: 0.3249
- val_loss: 1.5443 - val_accuracy: 0.4411
Epoch 2/12
50000/50000 [==========] - 49s 982us/sample - loss: 1.4183 - accuracy: 0.4879
- val_loss: 1.3052 - val_accuracy: 0.5274
Epoch 3/12
```

```
50000/50000 [==========] - 51s 1ms/sample - loss: 1.2444 - accuracy: 0.5590 -
val_loss: 1.2131 - val_accuracy: 0.5616
Epoch 4/12
50000/50000 [==========] - 50s 994us/sample - loss: 1.1180 - accuracy: 0.6056 -
val_loss: 1.1202 - val_accuracy: 0.5992
Epoch 5/12
50000/50000 [==========] - 49s 985us/sample - loss: 1.0173 - accuracy: 0.6425 -
val_loss: 1.0124 - val_accuracy: 0.6427
Epoch 6/12
50000/50000 [==========] - 50s 998us/sample - loss: 0.9323 - accuracy: 0.6755 -
val_loss: 1.0272 - val_accuracy: 0.6407
Epoch 7/12
50000/50000 [==========] - 49s 984us/sample - loss: 0.8568 - accuracy: 0.7032 -
val_loss: 0.9878 - val_accuracy: 0.6605
Epoch 8/12
50000/50000 [==========] - 51s 1ms/sample - loss: 0.7909 - accuracy: 0.7236 -
val_loss: 0.9460 - val_accuracy: 0.6766
Epoch 9/12
50000/50000 [==========] - 49s 984us/sample - loss: 0.7292 - accuracy: 0.7473 -
val_loss: 0.9363 - val_accuracy: 0.6803
Epoch 10/12
50000/50000 [==========] - 49s 984us/sample - loss: 0.6762 - accuracy: 0.7636 -
val_loss: 0.8751 - val_accuracy: 0.7026
Epoch 11/12
50000/50000 [==========] - 49s 983us/sample - loss: 0.6189 - accuracy: 0.7844 -
val_loss: 0.9910 - val_accuracy: 0.6807
Epoch 12/12
50000/50000 [==========] - 49s 983us/sample - loss: 0.5689 - accuracy: 0.8021 -
val_loss: 0.9346 - val_accuracy: 0.6972
Execution time is 603 seconds
```

From the output above, we can see that the accuracy of the test data is around 70 percent after 10 epochs. If we run a few more epochs, the model will be exceedingly overfitted. We can fine-tune the parameters and add regularization to increase accuracy. We will see the details of how to construct an optimal CNN model in upcoming sections.

11.6   CASE STUDY: SIGN LANGUAGE READING FROM IMAGES

This case study is about predicting the number based on the sign shown with fingers. We are going to use the sign-language dataset in this case study. It is publicly available under the CC BY-SA 4.0 License. This dataset was initially prepared by Turkey Ankara Ayrancı Anadolu High School students. We are thankful to project executives Zeynep Dikle and Arda Mavi for sharing this dataset. The dataset GitHub page is available at https://github.com/ardamavi/Sign-Language-Digits-Dataset.

11.6.1   Background and Objective

The samples are collected from handmade gestures of the digits from 218 participants. Each image includes a one-handed display of numbers 0 to 9. These are color images with size 100×100 pixels. The objective is to build a model that will predict the number based on the hand gesture. The advanced application of this model can be operating devices based on gestures or hand signals. The scope of this model is to predict the number based on the symbol. Figure 11.49 shows a few examples of the images. We should be careful with the sign of the number 6. It looks like three, but it is six in sign language.

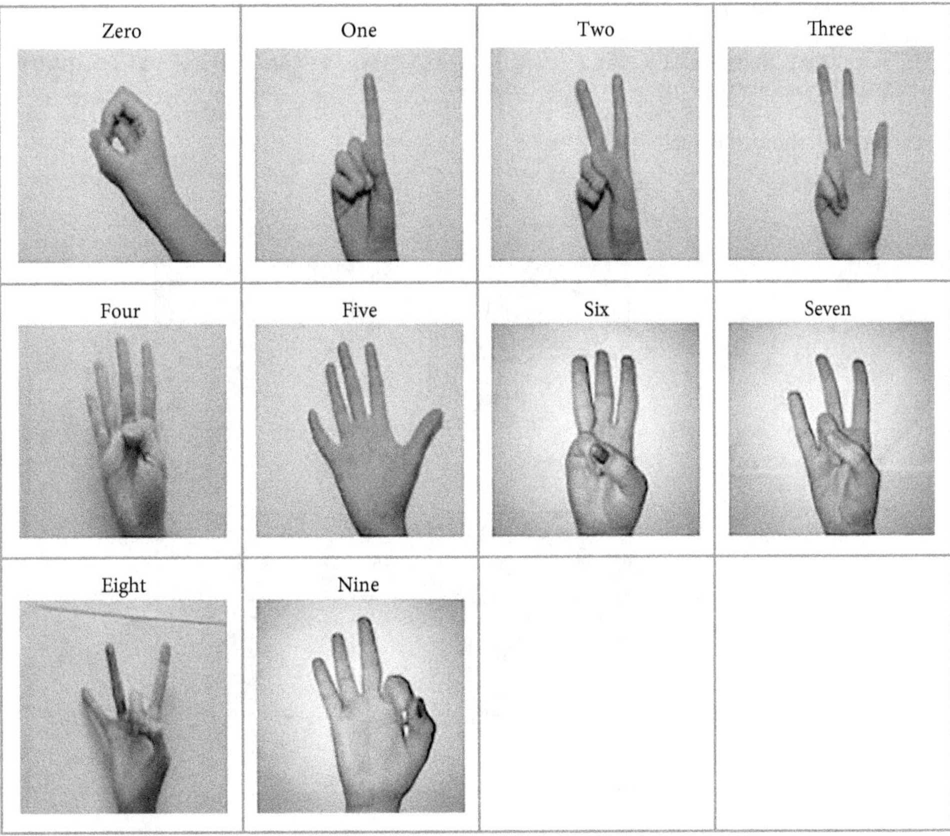

FIGURE 11.49   Samples of handmade gestures representing different digits.

11.6.2 Data

Before going ahead with importing the data, we will verify some random images from the data using the following code.

```
fig, ax = plt.subplots(2,2)
location='D:\\Chapter11 CNN\\5.Datasets\\Sign_Language_Digits\\Sign-
Language-Digits-Dataset-master\\Dataset\\'

i=random.randint(0, 9)
img_id=18+i
img=imageio.imread(location+str(i)+"\\IMG_11"+str(img_id)+".JPG")
ax[0,0].imshow(img)

i=random.randint(0, 9)
img_id=18+i
img=imageio.imread(location+str(i)+"\\IMG_11"+str(img_id)+".JPG")
ax[0,1].imshow(img)

i=random.randint(0, 9)
img_id=18+i
img=imageio.imread(location+str(i)+"\\IMG_11"+str(img_id)+".JPG")
ax[1,0].imshow(img)
```

```
i=random.randint(0, 9)
img_id=18+i
img=imageio.imread(location+str(i)+"\\IMG_11"+str(img_id)+".JPG")
ax[1,1].imshow(img)
```

Figure 11.50 shows the results.

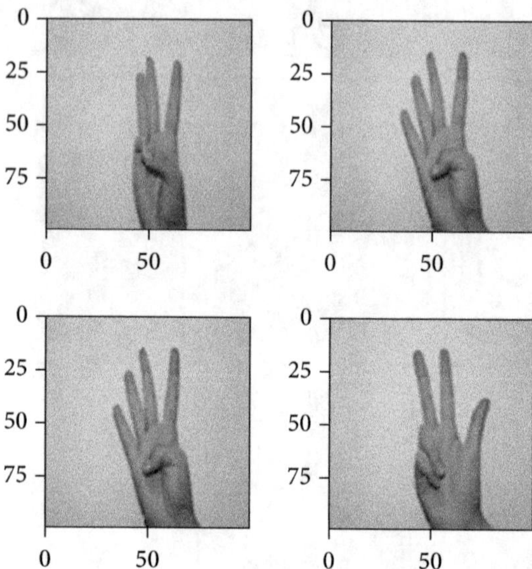

FIGURE 11.50   Sample images from the input data.

The output in Fig. 11.50 shows a few sample images from the data. We will now start importing the data.

11.6.2.1 Data Importing   The importing of image data is different from importing standard numerical tabular data. Usually, the collection of images is stored in their respective folders. In our input dataset, there are 10 folders, and each folder has around 200 sample images. Here the folder names represent the labels. The function `flow_from_directory()` iterates through the data directory and creates random batches for us. There are three major steps.

- **`tf.keras.preprocessing.image.ImageDataGenerator`**: For generating batches of tensor image data. It has image scaling, reshaping, and various other commands.
- **`flow_from_directory`**: Used for creating train, validation, and test data.
- **`fit_generator`**: This function takes input as data and fits the model.

Let us see how to write the code for importing the data. The syntax is not new—just a new way of importing.

```
### Generators
from tensorflow.keras.preprocessing.image import ImageDataGenerator

batch_size = 256
target_size = (100,100)

### Data Directory
data_dir = location

### Data generator
datagen = ImageDataGenerator(rescale = 1./255,
                             validation_split=0.2)
```

```
### Train generator
train_generator = datagen.flow_from_directory(
    data_dir,
    target_size=target_size,
    batch_size=batch_size,
    color_mode = 'grayscale',
    class_mode='categorical',
    subset="training")

#### Validation generator
validation_generator = datagen.flow_from_directory(
    data_dir,
    target_size=target_size,
    batch_size=batch_size,
    color_mode = 'grayscale',
    class_mode='categorical',
    subset="validation")
```

The commands used above are explained in Table 11.7.

TABLE 11.7   Code Explanation

`data_dir = location`	This is the image dataset directory
`Data generator`	Any preprocessing options/steps can be defined here.
	`rescale`—scaling the pixel values in the image matrix. This is a standard pre-processing step
	`validation_split`—set the validation split
`train_generator()`	`target_size`—resizing the input images to a specific size
	`color_mode`—keeping the channel to grayscale for easy calculations
	`batch_size`—it is the batch size. The iterator will generate a random batch with this size
	`subset`—set as train or test data

The following is the code output.

```
Found 1653 images belonging to 10 classes.

Found 409 images belonging to 10 classes.
```

The output shows the number of images from train and test data.

11.6.3   Model Building and Validation

Given below is the code for building the CNN model.

```
model1 = Sequential()

# Convolution layer
model1.add(Conv2D(64,  (3,  3),  input_shape = (100, 100, 1), activation =
'relu'))

# Pooling layer
model1.add(MaxPooling2D(pool_size = (2, 2)))

# Adding second convolutional layer
model1.add(Conv2D(64,  (3,  3),  activation = 'relu'))
```

```
# Pooling layer
model1.add(MaxPooling2D(pool_size = (2, 2)))

# Adding third convolutional layer
model1.add(Conv2D(64, (3, 3), activation = 'relu'))

# Pooling layer
model1.add(MaxPooling2D(pool_size = (2, 2)))

# Flattening
model1.add(Flatten())

# Step 4 - Fully connected dense layers
model1.add(Dense(units = 256, activation = 'relu'))
model1.add(Dense(units = 10, activation = 'softmax'))

model1.summary()
```

In the code above, we have added three convolution layers. Each layer has 64 filters, and each filter size is 3×3. We have also added a few pooling layers. The final dense layer has 256 hidden nodes. Table 11.8 presents the summary of this model, and we need to take a look at the number of weights before going ahead with the model building.

TABLE 11.8 Model Summary for Hand Gestures Data

```
Model: "sequential_2"
```

Layer (type)	Output Shape	Param #
conv2d_6 (Conv2D)	(None, 98, 98, 64)	640
max_pooling2d_6 (MaxPooling2	(None, 49, 49, 64)	0
conv2d_7 (Conv2D)	(None, 47, 47, 64)	36928
max_pooling2d_7 (MaxPooling2	(None, 23, 23, 64)	0
conv2d_8 (Conv2D)	(None, 21, 21, 64)	36928
max_pooling2d_8 (MaxPooling2	(None, 10, 10, 64)	0
flatten_2 (Flatten)	(None, 6400)	0
dense_4 (Dense)	(None, 256)	1638656
dense_5 (Dense)	(None, 10)	2570

```
Total params: 1,715,722
Trainable params: 1,715,722
Non-trainable params: 0
```

There are nearly 1.7 million parameters. We are now all set to compile and fit the model. Training 1.7 million parameters will take much time. We will measure the time, and we will also save the final model.

```python
# model1 compilation
model1.compile(optimizer =SGD(lr=0.01, momentum = 0.9), loss = 'categorical_
crossentropy', metrics = ['accuracy'])


##########################
# fit model and train
##########################

import time
start = time.time()

model1.fit_generator(
        train_generator,
        steps_per_epoch = len(train_generator),
        epochs=20,
        validation_data = validation_generator,
        validation_steps = len(validation_generator),
        verbose=1)

model1.save_weights('m1_Sign_Language_20epochs.h5')

end = time.time()
print("Execution time is", int(end - start), "seconds")
```

We supply the train data, validation data, and epochs. Steps per epoch are the total number of batches in one train epoch. It is (training records ÷ batch size), also called iterations per epoch. The following is the code output. `model1.save_weights()` is used for saving the model weights. It is important to save the weights while building models with prolonged execution time.

```
Epoch 1/20
7/7 [====] - 34s 5s/step - loss: 2.3045 - accuracy: 0.0992 - val_loss: 2.3017 -
val_accuracy: 0.0831
Epoch 2/20
7/7 [====] - 34s 5s/step - loss: 2.2991 - accuracy: 0.1446 - val_loss: 2.2996 -
val_accuracy: 0.1760
Epoch 3/20
7/7 [====] - 35s 5s/step - loss: 2.2957 - accuracy: 0.1337 - val_loss: 2.2973 -
val_accuracy: 0.1125
Epoch 4/20
7/7 [====] - 34s 5s/step - loss: 2.2915 - accuracy: 0.1404 - val_loss: 2.2944 -
val_accuracy: 0.1785
Epoch 5/20
7/7 [====] - 33s 5s/step - loss: 2.2868 - accuracy: 0.2269 - val_loss: 2.2899 -
val_accuracy: 0.2274
Epoch 6/20
7/7 [====] - 35s 5s/step - loss: 2.2794 - accuracy: 0.3388 - val_loss: 2.2854 -
val_accuracy: 0.2274
Epoch 7/20
7/7 [====] - 38s 5s/step - loss: 2.2711 - accuracy: 0.3454 - val_loss: 2.2775 -
val_accuracy: 0.2225
Epoch 8/20
7/7 [====] - 33s 5s/step - loss: 2.2545 - accuracy: 0.2541 - val_loss: 2.2630 -
val_accuracy: 0.2861
Epoch 9/20
7/7 [====] - 29s 4s/step - loss: 2.2289 - accuracy: 0.3430 - val_loss: 2.2373 -
val_accuracy: 0.3374
Epoch 10/20
7/7 [====] - 31s 4s/step - loss: 2.1789 - accuracy: 0.4670 - val_loss: 2.1915 -
val_accuracy: 0.3692
```

```
Epoch 11/20
7/7 [====] - 33s 5s/step - loss: 2.0885 - accuracy: 0.4204 - val_loss: 2.1035 -
val_accuracy: 0.3472
Epoch 12/20
7/7 [====] - 35s 5s/step - loss: 1.8657 - accuracy: 0.4955 - val_loss: 1.9547 -
val_accuracy: 0.3178
Epoch 13/20
7/7 [====] - 34s 5s/step - loss: 1.4969 - accuracy: 0.5396 - val_loss: 1.6527 -
val_accuracy: 0.3936
Epoch 14/20
7/7 [====] - 32s 5s/step - loss: 1.6009 - accuracy: 0.5009 - val_loss: 1.5533 -
val_accuracy: 0.4450
Epoch 15/20
7/7 [====] - 32s 5s/step - loss: 0.9421 - accuracy: 0.7048 - val_loss: 1.2894 -
val_accuracy: 0.5550
Epoch 16/20
7/7 [====] - 33s 5s/step - loss: 0.6899 - accuracy: 0.7725 - val_loss: 1.4266 -
val_accuracy: 0.5330
Epoch 17/20
7/7 [====] - 32s 5s/step - loss: 0.5730 - accuracy: 0.8076 - val_loss: 1.8965 -
val_accuracy: 0.5281
Epoch 18/20
7/7 [====] - 33s 5s/step - loss: 0.5097 - accuracy: 0.8312 - val_loss: 1.8776 -
val_accuracy: 0.5623
Epoch 19/20
7/7 [====] - 30s 4s/step - loss: 0.4508 - accuracy: 0.8560 - val_loss: 1.7863 -
val_accuracy: 0.5721
Epoch 20/20
7/7 [====] - 29s 4s/step - loss: 0.3907 - accuracy: 0.8754 - val_loss: 1.8954 -
val_accuracy: 0.5721
Execution time is 658 seconds
```

We can see from the output that the model is overfitted. The model takes around 10 minutes to execute 20 epochs. We can build one more model with 50 epochs. Given below is the result of the model after 50 epochs.

```
Epoch 1/50
7/7 [====] - 3s 485ms/step - loss: 2.3026 - accuracy: 0.0992 - val_loss:
2.3013 - val_accuracy: 0.0905
Epoch 2/50
7/7 [====] - 3s 429ms/step - loss: 2.2984 - accuracy: 0.1361 - val_loss:
2.2988 - val_accuracy: 0.1271
Epoch 3/50
7/7 [====] - 3s 439ms/step - loss: 2.2956 - accuracy: 0.1216 - val_loss:
2.2972 - val_accuracy: 0.1125
Epoch 4/50
7/7 [====] - 3s 439ms/step - loss: 2.2914 - accuracy: 0.1603 - val_loss:
2.2937 - val_accuracy: 0.1540
    =================================================================
Epoch 41/50
7/7 [====] - 3s 409ms/step - loss: 0.0231 - accuracy: 0.9958 - val_loss:
2.0978 - val_accuracy: 0.6406
Epoch 42/50
7/7 [====] - 3s 409ms/step - loss: 0.0127 - accuracy: 0.9994 - val_loss:
2.1736 - val_accuracy: 0.6430
Epoch 43/50
7/7 [====] - 3s 409ms/step - loss: 0.0104 - accuracy: 0.9994 - val_loss:
2.2443 - val_accuracy: 0.6381
Epoch 44/50
7/7 [====] - 3s 405ms/step - loss: 0.0082 - accuracy: 1.0000 - val_loss:
2.2888 - val_accuracy: 0.6406
```

```
Epoch 45/50
7/7 [====] - 3s 412ms/step - loss: 0.0064 - accuracy: 1.0000 - val_loss:
2.1680 - val_accuracy: 0.6553
Epoch 46/50
7/7 [====] - 3s 415ms/step - loss: 0.0059 - accuracy: 1.0000 - val_loss:
2.2846 - val_accuracy: 0.6455
Epoch 47/50
7/7 [====] - 3s 440ms/step - loss: 0.0046 - accuracy: 1.0000 - val_loss:
2.3375 - val_accuracy: 0.6406
Epoch 48/50
7/7 [====] - 3s 419ms/step - loss: 0.0041 - accuracy: 1.0000 - val_loss:
2.2884 - val_accuracy: 0.6479
Epoch 49/50
7/7 [====] - 3s 408ms/step - loss: 0.0038 - accuracy: 1.0000 - val_loss:
2.3382 - val_accuracy: 0.6430
Epoch 50/50
7/7 [====] - 3s 408ms/step - loss: 0.0033 - accuracy: 1.0000 - val_loss:
2.3458 - val_accuracy: 0.6504
```

Using the following code, we can load the above 50-epochs model and execute 2 epochs on top of it.

```
model1.load_weights(r"D:\Chapter11 CNN\Datasets\Pre_trained_models\ m1_Sign_
Language_50epochs.h5")

model1.fit_generator(
        train_generator,
        steps_per_epoch = len(train_generator),
        epochs=2,
        validation_data = validation_generator,
        validation_steps = len(validation_generator),
        verbose=1)
```

The following is the code output.

```
Epoch 1/2
7/7 [===============================] - 30s 4s/step - loss: 0.0054 - accuracy:
1.0000 - val_loss: 2.0783 - val_accuracy: 0.6381
Epoch 2/2
7/7 [===============================] - 32s 5s/step - loss: 0.0076 - accuracy:
1.0000 - val_loss: 2.0829 - val_accuracy: 0.6504
```

From the results above, it is evident that the model is overfitted. We can follow specific rules and build an optimal CNN model. In the next section, we will see how to configure a CNN model.

11.7 SCHEMING THE IDEAL CNN ARCHITECTURE

While building the CNN model in the previous example, we randomly took some filters and added three convolution layers. We had no logic or rationale for choosing the number of convolution layers and the number of filters. We can use the following tricks while constructing the CNN model.

11.7.1 Number of Convolution and Pooling Layers

The number of convolution layers should NOT be decided before building the model. We need to choose the number of layers based on the size of the image. The number of layers depends on the receptive field of the last layer. The general rule of thumb is that the last convolution layer's receptive field size should be the same as the size of the input data. In the previous section, we discussed the receptive field. Figure 11.51 shows an example of the receptive field.

Input

0	0	1	0	1	0	0	1
1	0	0	0	0	0	0	0
0	0	0	0	0	0	0	0
0	0	0	1	1	0	0	1
0	0	0	0	0	0	0	1
1	0	0	0	0	0	0	0
0	0	0	0	0	0	0	0
1	0	1	0	1	0	0	0

Conv layer1

1	2	1	2	1	1	1	1
1	2	1	2	1	1	1	1
1	1	1	2	2	1	1	1
0	0	1	2	2	1	2	2
1	1	1	2	2	1	2	2
1	1	0	0	0	0	1	1
2	3	1	2	1	1	0	0
1	2	1	2	1	1	0	0

Conv layer2

6	8	10	8	8	6	6	4
8	11	14	13	13	10	9	6
5	8	12	14	14	12	11	8
4	7	11	15	15	14	13	10
4	6	8	10	10	11	12	10
9	11	11	9	9	8	8	6
10	12	12	8	8	5	4	2
8	10	11	8	8	4	2	0

Conv layer3

33	57	64	66	58	52	41	25
46	82	98	106	98	89	72	44
43	80	105	121	120	111	93	57
34	65	91	109	115	112	101	64
41	72	88	98	101	100	92	59
52	83	87	85	78	75	66	42
60	94	92	84	67	56	39	22
40	63	61	55	41	31	17	8

FIGURE 11.51   An example of the receptive field.

An element of the third layer has a receptive field of 7×7. The rule is that before flattening the convolution layers, we should make sure that every element in the last layer has access to the whole image. The pooling layer increases the area of the receptive field significantly. In fact, after every pooling layer with a 2×2 matrix, the receptive field doubles. Also, add a pooling layer after every two convolution layers; it gives the best results in practical scenarios. Let us calculate the receptive field of our previous network (Table 11.9).

TABLE 11.9   Receptive Field Calculations

Layer	Output Shape	Receptive Field	Weights
Input layer	$100 \times 100 \times 1$		
Convolution layer 1 from 64 filters of size 3×3	$98 \times 98 \times 64$	3×3	• $3 \times 3 = 9 + 1$ bias • 64 filters • $10 \times 64 = 640$ weights
Max pool layer 2×2	$98 \times 98 \times 64$	6×6	
Convolution layer 2 from 64 filters of size 3×3	$47 \times 47 \times 64$	8×8	• $3 \times 3 \times 64 = 576 + 1$ bias • 64 filters • 577×64 • 36,928 weights
Max pool layer 2×2	$23 \times 23 \times 64$	16×16	
Convolution layer 3 from 64 filters of size 3×3	$21 \times 21 \times 64$	18×18	• $3 \times 3 \times 64 = 576 + 1$ bias • 64 filters • 577×64 • 36,928 weights
Max pool layer 2×2	$10 \times 10 \times 64$	36×36	
Flatten	6400×1		
Dense layer 256 nodes	256×1		• 6400 input nodes + 1 bias • 256 nodes • 1,638,656 weights
Dense layer 10 nodes	10×1		• $256 + 1$ bias • 10 nodes • 2570 weights
Overall weights			• **1,715,722**

From Table 11.9, we can see the receptive field of the last layer is only 36×36. The image size is 100×100. This model needs to be modified such that the receptive field is near to 100. The network needs to be much deeper than what it is now. A deeper network would mean many more parameters. Many more parameters will increase the execution time. From the sample images, we can see that the images have an original shape of 100×100; we can reshape them to 64×64. It will help us in reducing the network depth and improving the execution time. It is a general practice in CNN to resize the images before building a model.

11.7.2   Number of Filters in the Convolution Layer

In our model, we have three convolution layers. Each layer has 64 filters. Each filter in the convolution layer finally transpires to a feature. The initial convolution layer learns low-level features; middle convolution layers learn mid-level feature, and the layers toward the output learn high-level features. In most datasets, the low-level features are a bunch of lines, circles, curves, and simple shapes. Generally, 16 or 32 filters are sufficient to capture the basic shapes at a low level. In our data, we can consider 16 filters in the first layer. The second layer takes the combinations of the basic shapes to detect mid-level features. In our data, the mid-level features can be the finger shapes such as the shape of a thumb, index finger, and shape of fingernails. Since the next-level (mid-level) features have much more permutations and combinations, we need to add almost double the filters when compared to previous layers. Usually, add two to three layers, and the number of filters should be 64 or 128. A similar logic follows for detecting high-level features. In our example, the high-level features are the real symbols of the digits. These signs are formed from multiple combinations of fingers. The suggested number of filters should be 256 to 512. After learning these features thoroughly, we do not need a lot of dense layers. In the CNN models, a significant portion of the overall number of weights is added from the dense layers. In our example, out of an overall 1.7 million weights, 1.6 million weights are contributed by dense layers. If possible, it is a good idea to get rid of the dense layer. Flattening the network and connecting directly to the output also work perfectly fine. Given below are the final suggestions for our network layers:

- Reshape the images from 100×100 to 64×64.
- Low-level features:
  - Add convolution layer 1 with 16 filters. Kernel size 3×3.
  - Add convolution layer 2 with 32 filters. Kernel size 3×3.
  - Max pooling layer 2×2
- Mid-level features
  - Add convolution layer 3 with 64 filters. Kernel size 3×3.
  - Add convolution layer 4 with 64 filters. Kernel size 3×3.
  - Max pooling layer 2×2
- High-level features:
  - Add convolution layer 5 with 128 filters. Kernel size 3×3.
  - Add convolution layer 6 with 128 filters. Kernel size 3×3.
  - Max pooling layer 2×2
- Flatten and add dense layer with fewer nodes
  - Dense layer with 32 nodes

Table 11.10 provides the calculation of the receptive field.

TABLE 11.10   Receptive Field Calculations

Layer	Receptive Field
Input layer	
Convolution layer 1 from 16 filters of size 3×3	3×3
Convolution layer 2 from 32 filters of size 3×3	5×5
Max pool layer 2×2	10×10
Convolution layer 3 from 64 filters of size 3×3	12×12
Convolution layer 4 from 64 filters of size 3×3	14×14
Max pool layer 2×2	28×28
Convolution layer 5 from 128 filters of size 3×3	30×30
Convolution layer 4 from 128 filters of size 3×3	32×32
Max pool layer 2×2	64×64

The following is the code for building the CNN model.

```
model2 = Sequential()

# Convolution and Pooling layers
model2.add(Conv2D(16, (3, 3), input_shape = (64, 64, 1), activation = 'relu'))
model2.add(Conv2D(32, (3, 3), activation = 'relu'))
model2.add(MaxPooling2D(pool_size = (2, 2)))
```

```
model2.add(Conv2D(64, (3, 3), activation = 'relu'))
model2.add(Conv2D(64, (3, 3), activation = 'relu'))
model2.add(MaxPooling2D(pool_size = (2, 2)))

model2.add(Conv2D(128, (3, 3), activation = 'relu'))
model2.add(Conv2D(128, (3, 3), activation = 'relu'))
model2.add(MaxPooling2D(pool_size = (2, 2)))


# Flattening and  Fully connected dense layers
model2.add(Flatten())
model2.add(Dense(units = 32, activation = 'relu'))
model2.add(Dense(units = 10, activation = 'softmax'))

model2.summary()
```

The above code gives us the output shown in Table 11.11.

TABLE 11.11   The Code Output

```
Model: "sequential_7"
```

Layer (type)	Output Shape	Param #
conv2d_30 (Conv2D)	(None, 62, 62, 16)	160
conv2d_31 (Conv2D)	(None, 60, 60, 32)	4640
max_pooling2d_21 (MaxPooling	(None, 30, 30, 32)	0
conv2d_32 (Conv2D)	(None, 28, 28, 64)	18496
conv2d_33 (Conv2D)	(None, 26, 26, 64)	36928
max_pooling2d_22 (MaxPooling	(None, 13, 13, 64)	0
conv2d_34 (Conv2D)	(None, 11, 11, 128)	73856
conv2d_35 (Conv2D)	(None, 9, 9, 128)	147584
max_pooling2d_23 (MaxPooling	(None, 4, 4, 128)	0
flatten_7 (Flatten)	(None, 2048)	0
dense_14 (Dense)	(None, 32)	65568
dense_15 (Dense)	(None, 10)	330

```
Total params: 347,562
Trainable params: 347,562
Non-trainable params: 0
```

There are 347,562 parameters in this model. We have now reduced the parameters from 1.7 to 0.34 million. We now compile and build the model using the following code.

```
# Model compilation
model2.compile(optimizer =SGD(lr=0.01, momentum = 0.9), loss = 'categorical_
crossentropy', metrics = ['accuracy'])
```

```
########################
# fit model and train
########################

import time
start = time.time()

model2.fit_generator(
        train_generator,
        steps_per_epoch = len(train_generator),
        epochs=50,
        validation_data = validation_generator,
        validation_steps = len(validation_generator),
        verbose=1)

model2.save_weights('m2_ Receptive_field_50epochs.h5')

end = time.time()
print("Execution time is", int(end - start), "seconds")
```

The code above gives us the following output.

```
Epoch 1/50
7/7 [====]- 17s 2s/step - loss: 2.1598 - accuracy: 0.2456 - val_loss: 2.1324 -
val_accuracy: 0.2665
Epoch 2/50
7/7 [====]- 16s 2s/step - loss: 1.9181 - accuracy: 0.3775 - val_loss: 1.8004 -
val_accuracy: 0.4254
Epoch 3/50
7/7 [====]- 15s 2s/step - loss: 1.9567 - accuracy: 0.4035 - val_loss: 2.0482 -
val_accuracy: 0.2714
Epoch 4/50
7/7 [====]- 15s 2s/step - loss: 1.8628 - accuracy: 0.4580 - val_loss: 1.7952 -
val_accuracy: 0.3765
Epoch 5/50
7/7 [====]- 16s 2s/step - loss: 1.3513 - accuracy: 0.5233 - val_loss: 1.7946 -
val_accuracy: 0.4499
Epoch 6/50
7/7 [====]- 15s 2s/step - loss: 1.1147 - accuracy: 0.6134 - val_loss: 1.5928 -
val_accuracy: 0.4963
Epoch 7/50
7/7 [====]- 15s 2s/step - loss: 0.8437 - accuracy: 0.7181 - val_loss: 1.3423 -
val_accuracy: 0.5648
Epoch 8/50
7/7 [====]- 15s 2s/step - loss: 0.7133 - accuracy: 0.7828 - val_loss: 1.3853 -
val_accuracy: 0.5721
Epoch 9/50
7/7 [====]- 14s 2s/step - loss: 0.5564 - accuracy: 0.8227 - val_loss: 1.4136 -
val_accuracy: 0.5892
Epoch 10/50
7/7 [====]- 15s 2s/step - loss: 0.4774 - accuracy: 0.8475 - val_loss: 1.3425 -
val_accuracy: 0.5990
                    ===============================================

Epoch 44/50
7/7 [====]- 15s 2s/step - loss: 0.0095 - accuracy: 0.9994 - val_loss: 1.8748 -
val_accuracy: 0.7408
Epoch 45/50
7/7 [====]- 15s 2s/step - loss: 0.0094 - accuracy: 0.9994 - val_loss: 1.8046 -
val_accuracy: 0.7359
```

```
Epoch 46/50
7/7 [====]- 14s 2s/step - loss: 0.0094 - accuracy: 0.9994 - val_loss: 1.7886 -
val_accuracy: 0.7359
Epoch 47/50
7/7 [====]- 14s 2s/step - loss: 0.0094 - accuracy: 0.9994 - val_loss: 1.8710 -
val_accuracy: 0.7335
Epoch 48/50
7/7 [====]- 15s 2s/step - loss: 0.0094 - accuracy: 0.9994 - val_loss: 1.7576 -
val_accuracy: 0.7335
Epoch 49/50
7/7 [====]- 14s 2s/step - loss: 0.0094 - accuracy: 0.9994 - val_loss: 1.7355 -
val_accuracy: 0.7359
Epoch 50/50
7/7 [====]- 14s 2s/step - loss: 0.0094 - accuracy: 0.9994 - val_loss: 1.9889 -
val_accuracy: 0.7408
Execution time is 734 seconds
```

From the output, we can see that the above model is also overfitted (100 percent accuracy on train data and 74 percent on test data). We have two options now. We can reduce the number of convolution layers and nodes; alternatively, we can keep the same architecture and introduce regularization. Regularization is often the preferred option. The code below introduces dropout regularization layers in the network diagram. We can also decrease the batch size to 128 or 64 to increase the iterations per epoch. It will help us in reducing the total number of epochs. Given below is the code for the model with regularization.

```python
model2 = Sequential()

# Convolution and Pooling layers
model2.add(Conv2D(16, (3, 3), input_shape = (64, 64, 1), activation = 'relu'))
model2.add(Conv2D(32, (3, 3), activation = 'relu'))
model2.add(MaxPooling2D(pool_size = (2, 2)))
model2.add(Dropout(0.5))

model2.add(Conv2D(64, (3, 3), activation = 'relu'))
model2.add(Conv2D(64, (3, 3), activation = 'relu'))
model2.add(MaxPooling2D(pool_size = (2, 2)))
model2.add(Dropout(0.5))

model2.add(Conv2D(128, (3, 3), activation = 'relu'))
model2.add(Conv2D(128, (3, 3), activation = 'relu'))
model2.add(MaxPooling2D(pool_size = (2, 2)))
model2.add(Dropout(0.5))


# Flattening and  Fully connected dense layers
model2.add(Flatten())
model2.add(Dense(units = 32, activation = 'relu'))
model2.add(Dropout(0.5))

model2.add(Dense(units = 10, activation = 'softmax'))

model2.summary()
```

We can see the dropout layers in the above model code. The same can be observed in its output (Table 11.12).

TABLE 11.12   The Code Output

```
Model: "sequential"
```

Layer (type)	Output Shape	Param #
conv2d (Conv2D)	(None, 62, 62, 16)	160
conv2d_1 (Conv2D)	(None, 60, 60, 32)	4640
max_pooling2d (MaxPooling2D)	(None, 30, 30, 32)	0
dropout (Dropout)	(None, 30, 30, 32)	0
conv2d_2 (Conv2D)	(None, 28, 28, 64)	18496
conv2d_3 (Conv2D)	(None, 26, 26, 64)	36928
max_pooling2d_1 (MaxPooling2	(None, 13, 13, 64)	0
dropout_1 (Dropout)	(None, 13, 13, 64)	0
conv2d_4 (Conv2D)	(None, 11, 11, 128)	73856
conv2d_5 (Conv2D)	(None, 9, 9, 128)	147584
max_pooling2d_2 (MaxPooling2	(None, 4, 4, 128)	0
dropout_2 (Dropout)	(None, 4, 4, 128)	0
flatten (Flatten)	(None, 2048)	0
dense (Dense)	(None, 32)	65568
dropout_3 (Dropout)	(None, 32)	0
dense_1 (Dense)	(None, 10)	330

```
Total params: 347,562
Trainable params: 347,562
Non-trainable params: 0
```

We can now compile the model and execute it. We can also track the accuracy change in each epoch by saving the model epochs in the history object.

```
# model compilation
model2.compile(optimizer =SGD(lr=0.01, momentum = 0.9), loss = 'categori-
cal_crossentropy', metrics = ['accuracy'])


########################
# fit model and train
########################

import time
start = time.time()

history=model2.fit_generator(
        train_generator,
        steps_per_epoch = len(train_generator),
        epochs=50,
        validation_data = validation_generator,
        validation_steps = len(validation_generator),
        verbose=1)
```

```
model2.save_weights('m2_Dropout_Rec_fld_50epochs.h5')

end = time.time()
print("Execution time is", int(end - start), "seconds")
```

The code above gives us the following output.

```
Epoch 1/50
26/26 [====] - 18s 690ms/step - loss: 2.3038 - accuracy: 0.0865 - val_loss:
2.3027 - val_accuracy: 0.0978
Epoch 2/50
26/26 [====] - 17s 670ms/step - loss: 2.3031 - accuracy: 0.0926 - val_loss:
2.3024 - val_accuracy: 0.1002
Epoch 3/50
26/26 [====] - 18s 705ms/step - loss: 2.3045 - accuracy: 0.0847 - val_loss:
2.3025 - val_accuracy: 0.1002
Epoch 4/50
26/26 [====] - 18s 703ms/step - loss: 2.3028 - accuracy: 0.0895 - val_loss:
2.3029 - val_accuracy: 0.0978
Epoch 5/50
26/26 [====] - 18s 707ms/step - loss: 2.3023 - accuracy: 0.1053 - val_loss:
2.3030 - val_accuracy: 0.1002
Epoch 6/50
26/26 [====] - 20s 760ms/step - loss: 2.3032 - accuracy: 0.0938 - val_loss:
2.3024 - val_accuracy: 0.0978
Epoch 7/50
26/26 [====] - 20s 757ms/step - loss: 2.3035 - accuracy: 0.0938 - val_loss:
2.3022 - val_accuracy: 0.1002
Epoch 8/50
26/26 [====] - 19s 737ms/step - loss: 2.3036 - accuracy: 0.0932 - val_loss:
2.3024 - val_accuracy: 0.0978
Epoch 9/50
26/26 [====] - 20s 768ms/step - loss: 2.3034 - accuracy: 0.0986 - val_loss:
2.3027 - val_accuracy: 0.1369
Epoch 10/50
26/26 [====] - 19s 749ms/step - loss: 2.3035 - accuracy: 0.1047 - val_loss:
2.3024 - val_accuracy: 0.1002
              ==========================================
Epoch 44/50
26/26 [====] - 17s 660ms/step - loss: 0.6490 - accuracy: 0.7750 - val_loss:
0.7262 - val_accuracy: 0.7800
Epoch 45/50
26/26 [====] - 17s 666ms/step - loss: 0.5881 - accuracy: 0.7973 - val_loss:
0.8850 - val_accuracy: 0.7237
Epoch 46/50
26/26 [====] - 18s 695ms/step - loss: 0.5349 - accuracy: 0.8191 - val_loss:
1.0071 - val_accuracy: 0.7164
Epoch 47/50
26/26 [====] - 17s 656ms/step - loss: 0.5593 - accuracy: 0.8125 - val_loss:
0.6546 - val_accuracy: 0.7873
Epoch 48/50
26/26 [====] - 17s 650ms/step - loss: 0.5278 - accuracy: 0.8246 - val_loss:
0.7862 - val_accuracy: 0.7482
Epoch 49/50
26/26 [====] - 17s 646ms/step - loss: 0.5205 - accuracy: 0.8342 - val_loss:
0.7355 - val_accuracy: 0.7555
Epoch 50/50
26/26 [====] - 17s 650ms/step - loss: 0.4767 - accuracy: 0.8367 - val_loss:
0.6292 - val_accuracy: 0.8142
Execution time is 880 seconds
```

From the output, we can see that each epoch has 26 iterations now. Accuracy on train data is 83 percent, and validation data accuracy is hovering above 80 percent. We can plot and visualize the history of train and validation accuracy using the code below.

```
plt.plot(history.history['accuracy'], label='accuracy')
plt.plot(history.history['val_accuracy'], label = 'val_accuracy')
plt.title("Train and Valid Accuracy by Epochs")
plt.xlabel('Epoch')
plt.ylabel('Accuracy')
plt.ylim([0,1])
plt.legend(loc='lower right')
```

Figure 11.52 shows the output from this code.

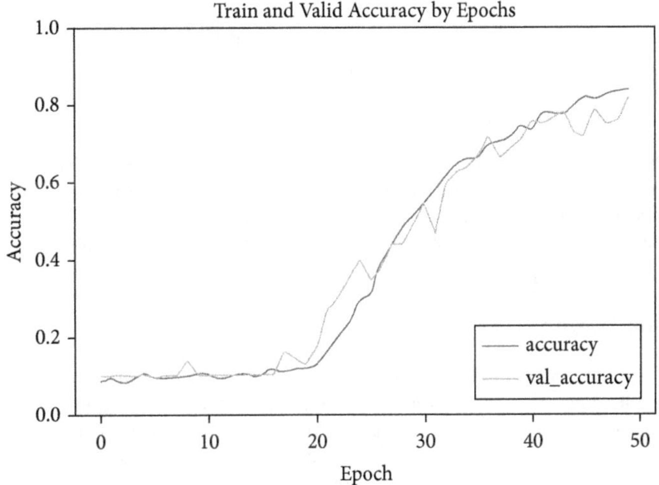

FIGURE 11.52   Train and validation accuracy plot.

The graph in Fig. 11.52 shows the train and test accuracy in each epoch. There is no sign of overfitting. After 50 epochs, both train and validation data show an accuracy of 80 percent. After 100 epochs, this model gives us an accuracy of above 85 percent on train and test data. Since batch size is small, the number of iterations is more for each epoch. A single epoch takes much more time than previous occasions. Given below is the output of 100 epochs. See also Fig. 11.53.

```
Epoch 96/100
26/26 [====] - 19s 742ms/step - loss: 0.2034 - accuracy: 0.9286 - val_loss:
0.4783 - val_accuracy: 0.8484
Epoch 97/100
26/26 [====] - 18s 710ms/step - loss: 0.2024 - accuracy: 0.9298 - val_loss:
0.6207 - val_accuracy: 0.8240
Epoch 98/100
26/26 [====] - 20s 767ms/step - loss: 0.1846 - accuracy: 0.9401 - val_loss:
0.4215 - val_accuracy: 0.8704
Epoch 99/100
26/26 [====] - 20s 760ms/step - loss: 0.1428 - accuracy: 0.9486 - val_loss:
0.5452 - val_accuracy: 0.8484
Epoch 100/100
26/26 [====] - 25s 971ms/step - loss: 0.1866 - accuracy: 0.9383 - val_loss:
0.3812 - val_accuracy: 0.8509
```

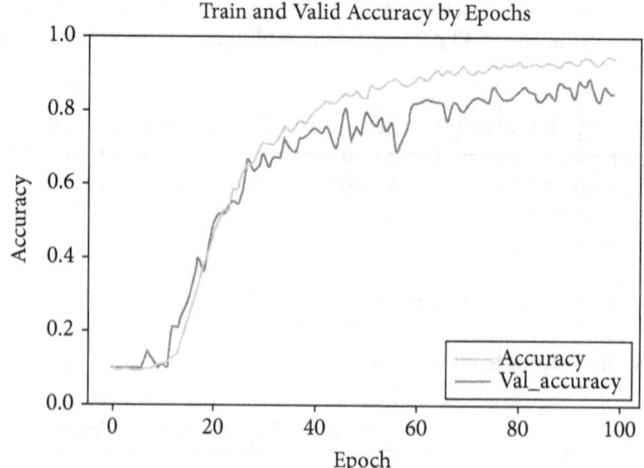

FIGURE 11.53   The code output.

There are a few more tips and tricks to improve model performance. Improving the model is not always about increasing accuracy. Reducing the number of parameters, reducing the execution time, and reducing the overfitting are also considered as an improvement in the model. We can try different optimization functions. A few functions work better for particular types of datasets. Batch normalization is one more method to improve the model.

11.7.3   Batch Normalization

Before building the model, we usually normalize or scale all the input values. For example, if some of the inputs are on the scale of thousands and others are in decimals, then we normalize the input. Different distributions for the inputs are generally considered a problem; we normalize the inputs to avoid it. The formula for normalization is simple: $(x - \text{mean}(x))/\text{sd}(x)$. This normalization of input helps us in faster calculations, and it also avoids the dominance of a few inputs. Input normalization is applied for almost all the models. In some cases, the input normalization is done by default in the training data. What if there is a similar problem in the intermediate layers?

While training the model, intermediate layers change their values (or distribution) to learn and adapt to the new batch of inputs. If the network is deep enough, then this change of distributions makes it very hard for the network to learn, and it slows down the entire training process. This issue is known as the internal covariate shift. In simple words, the change in the distribution of intermediate activations due to the change in the prior layers' parameters is the internal covariate shift. If normalizing the values works on input nodes, why can't we apply the same normalization on the intermediate layers? The batch normalization method tries to reduce the fluctuations in the intermediate values by applying normalization on them.

In batch normalization, we normalize the outputs of previously hidden layers using the formula $(h_i - \text{mean}(h_i))/\text{sd}(h_i)$. We do normalization for every batch of inputs. There is one more step called scaling in batch normalization. Given below are the equations for batch normalization.

$$\hat{h}_i = \frac{h_i - mean(h_i)}{s.d(h_i)}$$

$$y_i = \gamma(\hat{h}_i) + \beta$$

where h_i is incoming activations in a minibatch B.

The normalized and scaled output y_i is used in place of h_i by the next layer. The parameters γ and β are two learnable parameters. These two parameters are randomly initialized, and they are also learned during the training process, along with the original weights.

Batch normalization is a relatively new concept introduced in 2015. On the surface, it is a simple trick of normalizing the activations to reduce the vast fluctuations of hidden layers. Batch normalization works very well for a specific type of problem. We may not see a considerable improvement in some other types of datasets. Batch normalization is added as a new layer while building the neural network models. Thus, this layer also has parameters γ and β. There is no guarantee that batch normalization will always give better accuracy or better speed. Batch normalization, along with other optimal configuration of the network, gives the best results.

Before we go ahead and apply batch normalization, we try to understand the number of parameters calculated in the batch normalization layer. Every batch normalization layer adds four new parameters. Out of these four, two are trainable and two are non-trainable. Trainable parameters are γ and β, and non-trainable parameters are the mean and standard deviations of each batch. γ and β are randomly initialized and learned during the training process. Nontrainable parameters calculated based on data are not derived from backpropagation. Given below is sample code to understand the number of parameter calculations.

```
model = Sequential()
model.add(Conv2D(1, (3, 3), input_shape = (32, 32, 1)))
model.add(BatchNormalization())

model.add(MaxPooling2D(pool_size = (2, 2)))
model.add(BatchNormalization())

model.add(Conv2D(2, (3, 3)))
model.add(BatchNormalization())

model.add(MaxPooling2D(pool_size = (2, 2)))
model.add(BatchNormalization())

model.add(Conv2D(3, (3, 3)))
model.add(BatchNormalization())

model.summary()
```

Table 11.13 provides the code output.

TABLE 11.13   The Code Output

```
Model: "sequential_11"
```

Layer (type)	Output Shape	Param #
conv2d_15 (Conv2D)	(**None**, 30, 30, 1)	10
batch_normalization_29 (Batc	(**None**, 30, 30, 1)	4
max_pooling2d_14 (MaxPooling	(**None**, 15, 15, 1)	0
batch_normalization_30 (Batc	(**None**, 15, 15, 1)	4
conv2d_16 (Conv2D)	(**None**, 13, 13, 2)	20
batch_normalization_31 (Batc	(**None**, 13, 13, 2)	8
max_pooling2d_15 (MaxPooling	(**None**, 6, 6, 2)	0
batch_normalization_32 (Batc	(**None**, 6, 6, 2)	8
conv2d_17 (Conv2D)	(**None**, 4, 4, 3)	57
batch_normalization_33 (Batc	(**None**, 4, 4, 3)	12

```
Total params: 123
Trainable params: 105
Non-trainable params: 18
```

Table 11.14 presents the explanation of this output.

TABLE 11.14   The Output Parameters Explained

Layer	Output Shape	Weights
Input layer	$32 \times 32 \times 1$	
Convolution layer 1 from one filter of size 3×3	$30 \times 30 \times 1$	• $3 \times 3 = 9 + 1$ bias • 10 weights • All parameters are trainable
Batch normalization	$30 \times 30 \times 1$	• 4 weights • 2 trainable • 2 non-trainable
Max pool layer 2×2	$15 \times 15 \times 1$	
Batch normalization	$15 \times 15 \times 1$	• 4 weights • 2 trainable • 2 non-trainable
Convolution layer 2 from two filters of size 3×3	$13 \times 13 \times 2$	• $3 \times 3 = 9 + 1$ bias • 2 filters • 20 weights • All parameters are trainable
Batch normalization	$13 \times 13 \times 2$	• 4 weights $\times$ 2 channels • 4 trainable • 4 non-trainable
Max pool layer 2×2	$6 \times 6 \times 2$	
Batch normalization	$6 \times 6 \times 2$	• 4 weights $\times$ 2 channels • 4 trainable • 4 non-trainable
Convolution layer 3 from three filters of size 3×3	$4 \times 4 \times 3$	• $3 \times 3 \times 2 = 18 + 1$ bias • 3 filters • $19 \times 3 = 57$ weights • All parameters are trainable
Batch normalization	$4 \times 4 \times 3$	• 4 weights $\times$ 3 channels • 6 trainable • 6 non-trainable
Overall weights		• 123 parameters • 105 trainable • 18 non-trainable

Now that we understand batch normalization, we now get back to our case study and include batch normalization layers. After adding batch normalization, we can afford to reduce the depth of the network. The following is the updated code with batch normalization layers.

```
model3 = Sequential()

model3.add(Conv2D(16, (3, 3), input_shape = (64, 64, 1), activation = 'relu'))
model3.add(BatchNormalization())
model3.add(Dropout(0.5))

model3.add(Conv2D(16, (3, 3), activation = 'relu'))
model3.add(MaxPooling2D(pool_size = (2, 2)))
model3.add(BatchNormalization())
model3.add(Dropout(0.5))

model3.add(Conv2D(32, (3, 3), activation = 'relu'))
model3.add(MaxPooling2D(pool_size = (2, 2)))
model3.add(BatchNormalization())
model3.add(Dropout(0.5))

model3.add(Conv2D(32, (3, 3), activation = 'relu'))
model3.add(MaxPooling2D(pool_size = (2, 2)))
model3.add(BatchNormalization())
model3.add(Dropout(0.5))
```

```python
model3.add(Conv2D(64, (3, 3), activation = 'relu'))
model3.add(BatchNormalization())
model3.add(Dropout(0.5))

model3.add(Flatten())
model2.add(Dense(units = 16, activation = 'relu'))
model2.add(Dropout(0.5))
model2.add(Dense(units = 10, activation = 'softmax'))

model3.summary()
```

We can observe the significant reduction in the number of convolution layers in this code. The last model had 347,562 weights. The new model has a smaller number of weights since we have removed two convolution layers with 128 nodes each. The code above gives us the output shown in Table 11.15.

TABLE 11.15   The Code Output of CNN Model

```
Model: "sequential_15"
```

Layer (type)	Output Shape	Param #
conv2d_29 (Conv2D)	(None, 62, 62, 16)	160
batch_normalization_49 (Batc	(None, 62, 62, 16)	64
dropout_5 (Dropout)	(None, 62, 62, 16)	0
conv2d_30 (Conv2D)	(None, 60, 60, 16)	2320
max_pooling2d_23 (MaxPooling	(None, 30, 30, 16)	0
batch_normalization_50 (Batc	(None, 30, 30, 16)	64
dropout_6 (Dropout)	(None, 30, 30, 16)	0
conv2d_31 (Conv2D)	(None, 28, 28, 32)	4640
max_pooling2d_24 (MaxPooling	(None, 14, 14, 32)	0
batch_normalization_51 (Batc	(None, 14, 14, 32)	128
dropout_7 (Dropout)	(None, 14, 14, 32)	0
conv2d_32 (Conv2D)	(None, 12, 12, 32)	9248
max_pooling2d_25 (MaxPooling	(None, 6, 6, 32)	0
batch_normalization_52 (Batc	(None, 6, 6, 32)	128
dropout_8 (Dropout)	(None, 6, 6, 32)	0
conv2d_33 (Conv2D)	(None, 4, 4, 64)	18496
batch_normalization_53 (Batc	(None, 4, 4, 64)	256
dropout_9 (Dropout)	(None, 4, 4, 64)	0
flatten_1 (Flatten)	(None, 1024)	0
dense (Dense)	(None, 16)	16400
dropout_10 (Dropout)	(None, 16)	0
dense_1 (Dense)	(None, 10)	170

```
Total params: 52,074
Trainable params: 51,754
Non-trainable params: 320
```

From the model summary, we can see a considerable reduction in the number of parameters. Here we are attempting to build a simpler model that can give the same accuracy as the previous model. Given below is the code for training the model and plotting the results.

```
model3.compile(optimizer =SGD(lr=0.03, momentum = 0.9), loss = 'categorical_
crossentropy', metrics = ['accuracy'])


########################
# fit model and train
########################

import time
start = time.time()

history=model3.fit_generator(
        train_generator,
        steps_per_epoch = len(train_generator),
        epochs=200,
        validation_data = validation_generator,
        validation_steps = len(validation_generator),
        verbose=1)

model3.save_weights('m3_BatchNorm_200epochs.h5')

end = time.time()
print("Execution time is", int(end - start), "seconds")

## Plotting the results
plt.plot(history.history['accuracy'], label='accuracy')
plt.plot(history.history['val_accuracy'], label = 'val_accuracy')
plt.title("Train and Valid Accuracy by Epochs")
plt.xlabel('Epoch')
plt.ylabel('Accuracy')
plt.ylim([0,1])
plt.legend(loc='lower right')
```

The following is the output from this code.

```
Epoch 1/200
26/26 [====]- 27s 1s/step - loss: 2.4620 - accuracy: 0.1307 - val_loss:
12.3212 - val_accuracy: 0.1027
Epoch 2/200
26/26 [====]- 25s 951ms/step - loss: 2.1956 - accuracy: 0.1664 - val_loss:
2.8450 - val_accuracy: 0.1345
                        ================================

Epoch 42/200
26/26 [====]- 16s 609ms/step - loss: 0.8787 - accuracy: 0.6824 - val_loss:
0.8359 - val_accuracy: 0.7531
Epoch 43/200
26/26 [====]- 17s 660ms/step - loss: 0.8932 - accuracy: 0.6776 - val_loss:
1.0753 - val_accuracy: 0.6088
Epoch 44/200
                        ================================
```

```
Epoch 71/200
26/26 [====]- 14s 548ms/step - loss: 0.7183 - accuracy: 0.7508 - val_loss:
1.0862 - val_accuracy: 0.6919
                    =================================

Epoch 103/200
26/26 [====]- 14s 546ms/step - loss: 0.6082 - accuracy: 0.7925 - val_loss:
0.8440 - val_accuracy: 0.7775
Epoch 104/200
26/26 [====]- 14s 542ms/step - loss: 0.5951 - accuracy: 0.7895 - val_loss:
0.8129 - val_accuracy: 0.7677
Epoch 105/200
26/26 [====]- 14s 547ms/step - loss: 0.5392 - accuracy: 0.8119 - val_loss:
0.8771 - val_accuracy: 0.7775
                    =================================

Epoch 139/200
26/26 [====]- 14s 544ms/step - loss: 0.5376 - accuracy: 0.8113 - val_loss:
0.8353 - val_accuracy: 0.7628
Epoch 140/200
26/26 [====]- 14s 554ms/step - loss: 0.5791 - accuracy: 0.8022 - val_loss:
0.8834 - val_accuracy: 0.7579
                    =================================

Epoch 162/200
26/26 [====]- 19s 747ms/step - loss: 0.5458 - accuracy: 0.8185 - val_loss:
0.6304 - val_accuracy: 0.7971
Epoch 163/200
26/26 [====]- 22s 847ms/step - loss: 0.4964 - accuracy: 0.8282 - val_loss:
0.6141 - val_accuracy: 0.7873
Epoch 164/200
                    =================================

Epoch 198/200
26/26 [====]- 17s 654ms/step - loss: 0.4645 - accuracy: 0.8445 - val_loss:
0.6623 - val_accuracy: 0.8117
Epoch 199/200
26/26 [====]- 17s 650ms/step - loss: 0.5333 - accuracy: 0.8330 - val_loss:
0.7620 - val_accuracy: 0.7702
Epoch 200/200
26/26 [====]- 17s 655ms/step - loss: 0.4684 - accuracy: 0.8475 - val_loss:
0.6996 - val_accuracy: 0.7897
Execution time is 3476 seconds
```

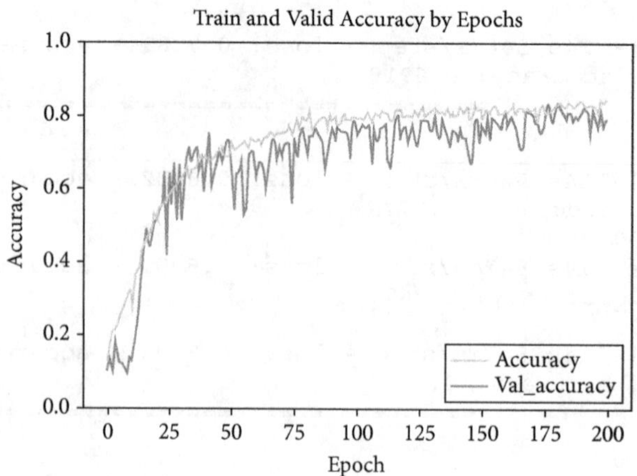

FIGURE 11.54   The code output.

From the output shown in Fig. 11.54, we can see that accuracy has not improved significantly; we got the same accuracy of 80 percent on the train and test data. The point to note here is the simplicity of the model. We achieved the same accuracy with almost seven times fewer parameters. As discussed earlier, batch normalization alone may not give us perfect results. Batch normalization, along with other optimal parameters, gives us the best results. It brings us to the next topic of choosing the right optimizer.

11.7.4   Choosing the Optimizers

Until now, we have seen a couple of variants of gradient descent algorithm as optimizers SGD and mini-batch gradient descent. We also frequently used SGD with momentum. SGD has worked well until now. A few other optimizer functions are also available. They may give us better results for certain types of problems. There are minor variations between these optimizers. Each optimizer handles the momentum and learning rate differently to arrive at the minimum value quickly and also avoid the local minimum.

Given below is the equation for the weight update in SGD with momentum.

$$W(t) := W(t - 1) + \Delta W(t)$$

$$\Delta W(t) = \eta * \left(-\frac{\partial E}{\partial W} \right)$$

$$W(t) := W(t - 1) + \eta * \left(-\frac{\partial E}{\partial W} \right) + \alpha * \Delta W(t - 1)$$

In simple form,

$$W(t) := W(t - 1) + \Delta W(t) + \alpha * \Delta W(t - 1)$$

where η is the learning rate and α is the momentum.

Until now, we have the same learning rate for all the parameters. How about different learning rates for different parameters? Table 11.16 discusses a few adaptive learning rate methods.

TABLE 11.16   A Few Adaptive Learning Rate Methods

Adagrad	The adaptive gradient means adopting the learning rate to parameters individually. The learning rate should be low for frequently occurring (updating) features, and the learning rate should be high for sparse (rarely updating) features in the data. This trick helps in faster convergence. This adaptive learning rate is calculated by dividing the actual learning rate by the sum of the previous gradients. $$W(t) := W(t-1) + \frac{\eta}{\sqrt{G_{t-1}}} * \left(-\frac{\partial E}{\partial W}\right)$$ $$W(t) := W(t-1) - \frac{\eta}{\sqrt{G_{t-1}}} * g_t$$ where $g_t = -\frac{\partial E}{\partial W}$ and $G_{t-1} = \sum_{i=0}^{t-1} g_i^2$ $\frac{\eta}{\sqrt{G_{t-1}}}$ represents the adaptive learning rate If the updates are more for a parameter, then $\sqrt{G_{t-1}}$ will be high and $\frac{\eta}{\sqrt{G_{t-1}}}$ will be low for that parameter
Adadelta / RMSprop	In Adagrad, the learning rate is divided by gradients, and this causes a problem sometimes. With the learning rate being a small fraction and further divided by the sum of gradients, to be used in the weight update step may fall into the problem of vanishing gradients. We slightly modify the adaptive learning rate, and that gives us RMSprop. In the Adadelta method, instead of changing the learning rate based on the previous sum of gradients, we use an exponential weighted average. Instead of using all the gradients until the current iteration, we can restrict it to use only the past few gradients. $$E\left[g^2\right]_t = \gamma E\left[g^2\right]_{t-1} + (1-\gamma)g_t^2$$ $$W(t) := W\left(t-1\right) - \frac{\eta}{\sqrt{E\left[g^2\right]_t}} * g_t$$ $$W(t) := W(t-1) - \frac{\eta}{RMS(g_t)} * g_t$$ Here the term γ is similar to the momentum term. This method is also known as root mean square propagation, formally known as RMSProp.
Adam	Adaptive moment estimation: Adam keeps the advantages of RMSprop, where we have adaptive learning rate and SGD with momentum where we give importance to previous weight changes velocity. In the RMSProp method, we estimated the learning rate by using exponential decay averages. We can repeat the same for the momentum term as well. Mean is known as the first-order moment, and variance is the second-order moment. Here, in the Adam method, we estimate the adaptive rate of both gradient and gradient square. Start with RMSProp equation $$W(t) := W(t-1) - \frac{\eta}{\sqrt{E\left[g^2\right]_t}} * g_t$$ Exponential decay for the first moment $(m_t) = \beta_1 m_{t-1} + (1-\beta_1)g_t$ Exponential decay for the second moment $(v_t) = \beta_2 v_{t-1} + (1-\beta_2)g_t^2$ The final weight-updated equation: $$W(t) := W(t-1) - \frac{\eta}{\sqrt{E[v]_t}} * E\left[m\right]_t$$ In this equation m_t, which is the exponentially decaying average of gradients, is similar to momentum. The term v_t takes care of the adaptive learning rate like in RMSprop.

We can try optimizers on a given dataset, but we cannot be very sure that the accuracy will increase. These optimizers help us in reducing the overall execution time. In practice, we use almost all available optimizers—there is no specific preference. For a particular type of dataset, a specific type of optimizer works the best way. Given below is the code for applying the Adam optimizer on our data.

```
model3.compile(optimizer  =Adam(learning_rate=0.005,  beta_1=0.9,  beta_2=
0.999), loss = 'categorical_crossentropy', metrics = ['accuracy'])


########################
# fit model and train
########################

import time
start = time.time()

history=model3.fit_generator(
        train_generator,
        steps_per_epoch = len(train_generator),
        epochs=100,
        validation_data = validation_generator,
        validation_steps = len(validation_generator),
        verbose=1)

model3.save_weights('m3_BatchNorm_and_Adam_100epochs.h5')

end = time.time()
print("Execution time is", int(end - start), "seconds")
```

We can see from the above code that the model is kept the same. We changed the optimization function from SGD to Adam. The above code gives us the below output.

```
Epoch 1/100
26/26 [====] - 18s 701ms/step - loss: 2.4890 - accuracy: 0.1077 - val_loss:
2.4831 - val_accuracy: 0.1002
Epoch 2/100
26/26 [====] - 17s 663ms/step - loss: 2.2032 - accuracy: 0.1627 - val_loss:
2.9257 - val_accuracy: 0.1002
                        ==================================

Epoch 13/100
26/26 [====] - 21s 824ms/step - loss: 1.0929 - accuracy: 0.6001 - val_loss:
9.5128 - val_accuracy: 0.1027
Epoch 14/100
26/26 [====] - 18s 680ms/step - loss: 1.0860 - accuracy: 0.5820 - val_loss:
9.8711 - val_accuracy: 0.1051
                        ==================================

Epoch 24/100
26/26 [====] - 24s 933ms/step - loss: 0.7912 - accuracy: 0.7078 - val_loss:
1.1113 - val_accuracy: 0.6993
Epoch 25/100
26/26 [====] - 24s 919ms/step - loss: 0.8054 - accuracy: 0.6969 - val_loss:
1.7334 - val_accuracy: 0.5428
                        ==================================

Epoch 33/100
26/26 [====] - 17s 647ms/step - loss: 0.7017 - accuracy: 0.7314 - val_loss:
2.6751 - val_accuracy: 0.4230
Epoch 34/100
26/26 [====] - 17s 636ms/step - loss: 0.7508 - accuracy: 0.7290 - val_loss:
1.3554 - val_accuracy: 0.5966
                        ==================================
```

```
Epoch 44/100
26/26 [====] - 17s 657ms/step - loss: 0.6686 - accuracy: 0.7629 - val_loss:
1.2933 - val_accuracy: 0.7311
Epoch 45/100
26/26 [====] - 16s 633ms/step - loss: 0.6471 - accuracy: 0.7604 - val_loss:
1.1294 - val_accuracy: 0.7384
                        ================================

Epoch 62/100
26/26 [====] - 17s 650ms/step - loss: 0.5528 - accuracy: 0.8016 - val_loss:
3.9618 - val_accuracy: 0.4743
Epoch 63/100
                        ================================

Epoch 85/100
26/26 [====] - 16s 629ms/step - loss: 0.5257 - accuracy: 0.8167 - val_loss:
0.7451 - val_accuracy: 0.8117
Epoch 86/100
26/26 [====] - 17s 659ms/step - loss: 0.4771 - accuracy: 0.8270 - val_loss:
0.7204 - val_accuracy: 0.8093
Epoch 87/100
26/26 [====] - 17s 637ms/step - loss: 0.4852 - accuracy: 0.8252 - val_loss:
1.0546 - val_accuracy: 0.8020
Epoch 88/100
26/26 [====] - 17s 636ms/step - loss: 0.5126 - accuracy: 0.8215 - val_loss:
0.7782 - val_accuracy: 0.8093
                        ================================

Epoch 97/100
26/26 [====] - 15s 561ms/step - loss: 0.5319 - accuracy: 0.8100 - val_loss:
0.9935 - val_accuracy: 0.7555
Epoch 98/100
26/26 [====] - 15s 567ms/step - loss: 0.5219 - accuracy: 0.8227 - val_loss:
1.1883 - val_accuracy: 0.7726
Epoch 99/100
26/26 [====] - 15s 568ms/step - loss: 0.4840 - accuracy: 0.8258 - val_loss:
1.0252 - val_accuracy: 0.7702
Epoch 100/100
26/26 [====] - 14s 555ms/step - loss: 0.4751 - accuracy: 0.8312 - val_loss:
0.6235 - val_accuracy: 0.8289
Execution time is 1716 seconds
```

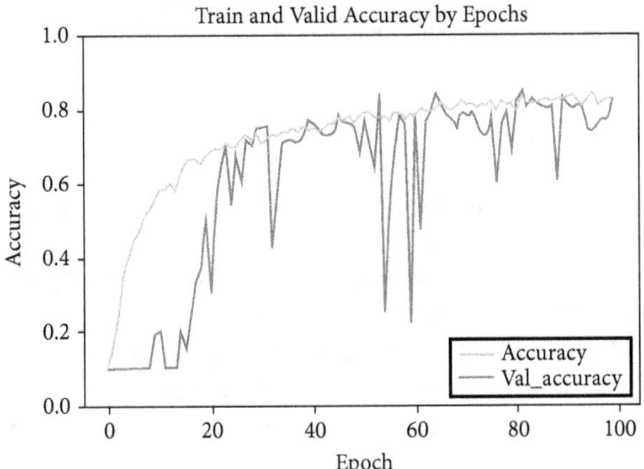

FIGURE 11.55   The code output.

From the output in Fig. 11.55, we can see that the Adam optimizer is working almost the same as SGD in our example. However, we could achieve the same 80 percent accuracy, without overfitting, within 100 epochs using Adam. There is a huge difference in execution time as well. In some cases, Adam gives a better result compared to SGD and RMSprop. Until now, we have discussed many concepts and tricks while working on a CNN model. Discussed below is a consolidated sequence of steps while building a CNN model.

11.8   STEPS IN BUILDING A CNN MODEL

CNN has other applications too, but it is mostly used for image processing. The following steps are not rules but suggestions. These steps can be taken as conclusive learning after working with multiple practical problems.

1. Pre-process the data. Reduce the size of the image if possible. Check whether we need all three channels of RGB for classification.

2. Add convolution layers and increase the number of filters as we go deeper into the network. A typical way of including the filters is 16 filters in the conv1 layer, 32 in the con2 layer, and 64 in the conv3 layer.

3. Add pooling layers in between. At least add one pooling layer after every two convolution layers.

4. The number of convolution and pooling layers should be decided based on the receptive field. The usual rule of thumb is to stop adding the layers once the last convolution layer gets to see the full image. It means the receptive field of the last convolution layer is the same as the image size.

5. Add batch normalization layers in between. Try adding at least one batch normalization layer before every pooling layer.

6. Try to keep the dense layers as low as possible. If possible, remove all dense layers.

7. Choose the optimizer and try with default values of the learning rate and build the model.

8. Observe the output if the model is consistently showing overfitting, then introduce dropout layers.

9. After fixing the overfitting issue, try experimenting with optimizers and learning rates.

10. Finalize the model once we get high accuracy on train data and matching accuracy on test data.

11.9   CONCLUSION

Convolution neural networks are currently very popular when compared to all other deep neural network algorithms. The feature learning part of CNN almost works like magic while solving some real-world problems. There is a massive increase in image datasets due to smartphones and cameras. CNNs can handle images effectively. In recent times computation power has also increased significantly. Due to these reasons, there is a significant increase in the acceptance and popularity of CNN. Building a CNN model from scratch on a large dataset is not as easy as the algorithms that we have discussed until now. Nowadays, graphics processing units (GPUs) are used in processing the images much faster. Currently, there is much research going on CNNs. As a result, we will see CNNs solving a lot more image- and video-related problems that affect our daily life.

11.10   PRACTICE PROBLEMS

1. Download the Malaria Cells dataset. The dataset contains two folders of images of infected and uninfected cells, a total of 27,558 images.

   - Import the data. Complete the necessary exploration and sanitization of the data.
   - Build a deep learning model to predict the infected cells.
   - Perform the model validation and measure the accuracy of the model.
   - Check for innovative ways to improve the accuracy of the model.

   Dataset credits: This dataset is taken from the official NIH website: https://ceb.nlm.nih.gov/repositories/malaria-datasets/.

11.11 REFERENCES

1. Object detection image source: From Wikimedia Commons, the free media repository: https://commons.wikimedia.org/wiki/File:Detected-with-YOLO--Schreibtisch-mit-Objekten.jpg#filelinks

2. Image File: American Eskimo Dog 1.jpg from Wikimedia Commons, the free media repository: https://commons.wikimedia.org/wiki/File:American_Eskimo_Dog_1.jpg

3. Feature detection images: https://distill.pub/2019/activation-atlas/app.html

4. File: Instagram Filters 2011.jpg from Wikimedia Commons, the free media repository: https://commons.wikimedia.org/wiki/File:Instagram_Filters_2011.jpg

5. CIFAR 10 dataset: https://www.cs.toronto.edu/~kriz/cifar.html, used in the paper Learning Multiple Layers of Features from Tiny Images, Alex Krizhevsky, 2009. https://www.cs.toronto.edu/~kriz/learning-features-2009-TR.pdf

6. Case Study: Sign Language Reading from Images Dataset: https://github.com/ardamavi/Sign-Language-Digits-Dataset.

7. Batch normalization: Accelerating Deep Network Training by Reducing Internal Covariate Shift Sergey Ioffe Google Inc.: sioffe@google.com. Christian Szegedy Google Inc.: szegedy@google.com https://arxiv.org/pdf/1502.03167v3.pdf.

8. Malaria Detection from Cell Images dataset—Rajaraman, S., Antani, S. K., Poostchi, M., Silamut, K., Hossain, M. A., Maude, R. J., Jaeger, S., Thoma, G. R. (2018). Pre-trained convolutional neural networks as feature extractors toward improved malaria parasite detection in thin blood smear images. PeerJ6:e4568. https://doi.org/10.7717/peerj.4568.

RECURRENT NEURAL NETWORKS AND LONG SHORT-TERM MEMORY

In classification problems, logistic regression does an excellent job if the decision boundary between the classes is linear. Artificial neural networks (ANNs) are the right choice when the job is to identify the nonlinear decision boundary between the classes. Some datasets are highly nonlinear; in such cases, we need deep neural networks. In the case of image classification, we need to retain the local correlation, so we use convolution neural networks (CNNs). In some datasets, we need to model the sequence of inputs where the current value depends on previous values. Such cases involving sequential data need different types of neural networks. This chapter discusses algorithms to model sequential data.

12.1   CROSS-SECTIONAL DATA VS. SEQUENTIAL DATA

In this section, let us discuss the difference between cross-sectional and sequential data.

12.1.1   Cross-Sectional Data

The datasets that we have worked in all the previous chapters represent cross-sectional data. For example, we considered the randomly collected house price data for two years. In other case studies, we worked with randomly collected grocery spending data or customers' credit card data. The current price list of top-selling smartphones is another example of cross-sectional data. In all these examples, the time window was fixed, maybe a year or two. In cross-sectional data, all the observations in the dataset are recorded in the same period. The period in cross-sectional data is sometimes aggregated to different bins such as a week, month, quarter, or year.

12.1.2   Sequential Data

In this chapter, we are going to work with sequential data. The internal ordering within the data has the utmost importance here. The current record value should strictly follow the previous record. Sequential data is also known as time-series data. For example, the data representing the daily stock price variation of a company is time-series data. For a definition, a time series is a sequence of data points recorded in time order. In a time series, each row (or record) has a timestamp on it. Another critical property of time series is that its values in the sequence should be taken at successive equally spaced points in time. We cannot take a random sample from a one-year time-series data. The sequence and interval both are critical here. We can consider words forming a sentence also as sequential data; However, a sequence of words in a text file is not precisely time-series data. If we timestamp words, for example, we put timestamp 0 to the first word, timestamp 1 to the second word, and so on, then a sequence of words (say, recorded over a timeline) can be called time-series data.

12.2   MODELS FOR SEQUENTIAL DATA

Let us consider a basic example of sequential data. When we key in some text on our smartphones, as we do on WhatsApp or in an email, the phone software often gives suggestions for the next probable word. If we want to key in the sentence "Can I have your number," as soon as we type "can" we see a few words suggested to complete the sentence. Not all of us get the same suggestions. For some phones, the next word suggestion may be "I," "you," or "this"; in some other phones, the suggestions may look like "be," "we," or "I." The suggestions for the third word would be based on the first two words. If we do not get the proper word, we choose to key in the word manually. We repeat the process until we complete the sentence. Table 12.1 shows this progression.

TABLE 12.1   Sentence Progression by Choosing the Words

Input Sequence of Words	Predicted Outputs
Can	I, you, it, see, do
Can I	see, come, get, have
Can I have	your, this, my, a
Can I have your	number, name, own, not

Now let us think about the model that is used in predicting the next word. That model is taking a sequence of words as input and predicting the next word. Here the word sequence matters. We cannot consider the three words "Can I have" as independent three input words; we have to preserve the sequential dependency while predicting the next word. Let us see what models can be used for predicting the next word.

12.2.1   ANN for Sequential Data

Let us take the example of a sequence of three words as input, and we have to predict the fourth word. For example, if the sequence of inputs is "Can I have," then one of the expected predictions is the word "your." Similarly, if the sequence of input words is "What do you," then one of the predictions is expected as "mean." The prediction, however, will depend on the input data that we are training on. Let the input sequence of words be x_1, x_2, x_3. We need to predict the fourth word; let it be y. We cannot use the traditional ANN model for the prediction of y. In the standard ANN model $y \sim x_1 + x_2 + x_3$, the order of the variables is not significant. All the variables are considered independent. Simply put, there is no sequential dependency while building ANN models. Which means $y \sim x_1 + x_2 + x_3$ is the same as $y \sim x_2 + x_3 + x_1$ and $y \sim x_3 + x_1 + x_2$. In our dataset, x_3 strictly follows $x_2, x_1,$ and x_2 follows x_1. Due to this reason, the standard ANN models do not work on sequential input data. We can somehow build the model by taking the input as x_1, x_2, x_3, but the model will not be accurate or efficient. If there are only two words x_1 and $x_2,$ then ANN can do the best job of predicting the second word given the first word. In the same way, ANN can do the best job of predicting the third word given the second word alone. Table 12.2 shows an example of the input word and predicted outputs.

TABLE 12.2   Predicted Set of Words Based on the Input Words

Input Word	Predicted Outputs
Can	I, you, it, see, do
I	was, am, thought
have	you, a, to, no
your	day, not, phone

ANN can take one word as input and predict the suggestion for the next word. The predictions depend on the training data. For the moment, we conclude our discussion on the point that the standard ANN models cannot be used for predicting the output based on sequential input data. Let us see whether we can use CNN models.

12.2.2   CNN for Sequential Data

The standard ANN models cannot preserve the spatial dependency, and CNNs preserve spatial dependency. Due to this reason, we used CNNs in image processing problems. CNN models worked almost flawlessly in image classification problems. The kernel filter in the convolution layer keeps the local correlation intact. However, the kernel filter cannot ensure sequential dependency. In the convolution layer, the kernel filter mixes all the pixels in a subregion. A dot product of the kernel matrix and the input data do not preserve the sequential dependency. Due to this reason, CNN models are not the right choice for sequential input data. So, CNN models are not a good fit for predicting the next word in the sequence. CNN models work the best when there is a cluster of input words and we need to predict the next word based on the word cluster. Here we are talking about a word cluster with no ordering of the words. This word cluster (or collection) may have a local correlation, but there will be no sequential dependence. Look at Table 12.3; CNN models work best for this type of problem.

TABLE 12.3   Predicted Set of Words Based on the Input Word Sequence

Input Word Cluster	Predicted Outputs
Can I have your	number, mobile, permission
I Can have your	number, mobile, permission
Have Can your I	number, mobile, permission
Your I have can	number, mobile, permission

Neither ANN models nor CNN models are best for solving the sequential data. Nevertheless, ANN models can take one word as input and give us (predict) a few choices for the next word. We will try to use ANN models and find a solution to solve the long sequence of inputs to predict the output.

12.2.3   Sequential ANN

Given one word, the standard ANN models can help us predict the next word. Let us work on an example of predicting the third word based on the sequence of two words x_1 and x_2. Since a single ANN of $y \sim x_1 + x_2$ does not work, we will serially build two ANN models and combine them to make the required sequential ANN model. We will build ANN model-1 that takes x_1 as input and gives us x_2 as output. We will build ANN model-2 by taking x_2 as input and give y (or x_3) as output. However, while building the second ANN model, we will append the hidden layer output from model-1 to x_2 to complete the second model. This partial-output-appending step will make sure that we are carrying some information from x_1 while predicting the value of x_3 (Fig. 12.1).

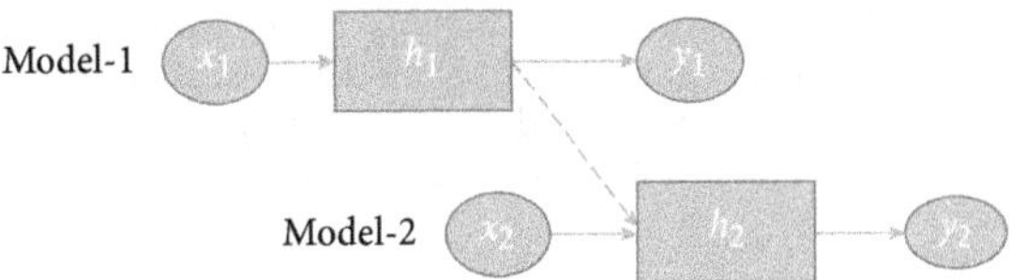

FIGURE 12.1   Two connected ANNs.

By appending the first model's hidden layer output to the second model, the final prediction will access the information from the second word and partial information from the first word as well. We first need to do the one-hot encoding of the data. For example, we have a table containing three columns containing different words, of which a few are repeats and some unique. If there are overall 100 unique words in the table, we need to perform one-hot encoding for those 100 words. Let us imagine that we built the model that predicts the word-2 using word-1 as input by including one hidden layer with 20 hidden nodes. Figure 12.2 shows the structure of model-1.

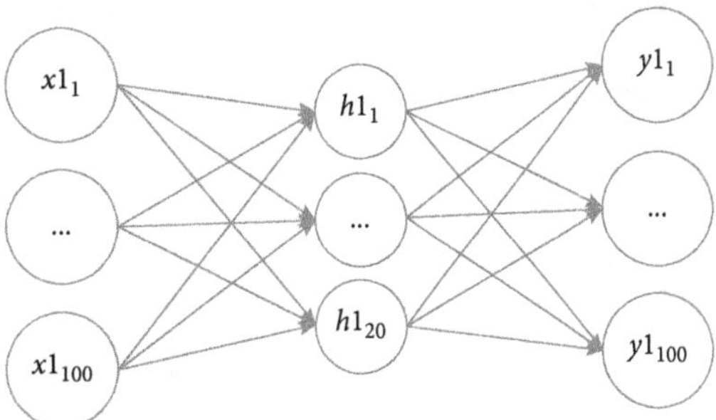

FIGURE 12.2   A network diagram of the structure of model-1.

In the model shown in Fig. 12.2, there are 100 input nodes, 20 hidden nodes, and 100 output nodes. While building model-2, we will take the intermediate output from the above model and one-hot encoded values of word-2 as input. It means that there will be $100 + 20$ input nodes in the second model. Figure 12.3 shows the network diagram for model-2.

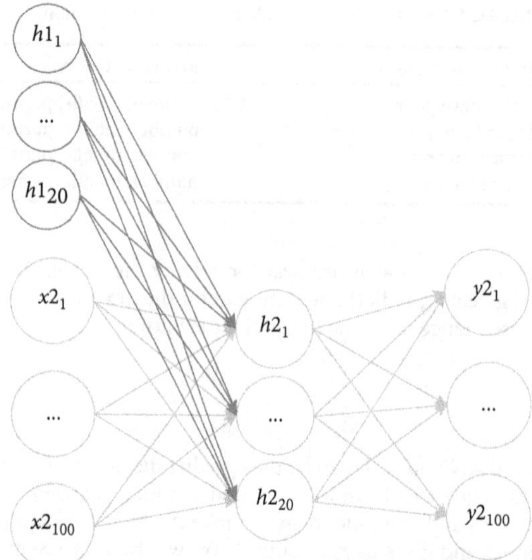

FIGURE 12.3   A network diagram of the structure of model-2.

From the diagram in Fig. 12.3, we can see the number of nodes in the input layer as 120, followed by 20 hidden nodes, followed by 100 nodes in the output layer. Using this method, we can predict the third word given the first two words. This approach uses a sequential (or serial) ANN model for sequential data. Similarly, if we have to predict the fourth word, then we need to build three ANN models and carefully stack them. The third model in the ANN will have $100 + 20$ nodes in its input layer. Figure 12.4 shows a generalized diagram for sequential ANN models.

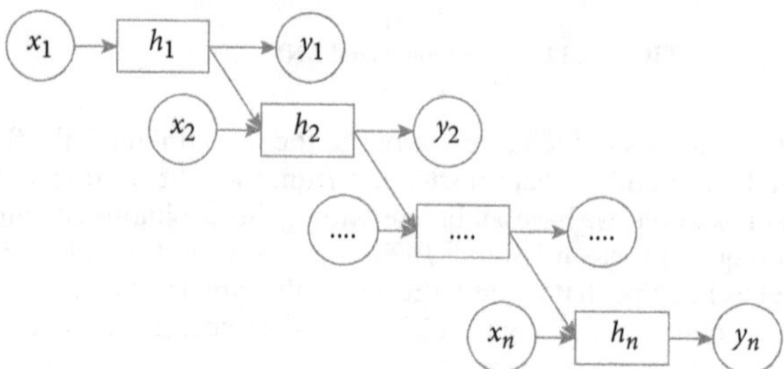

FIGURE 12.4   A general diagram for sequential ANN models.

Figure 12.5 shows a different representation of the same diagram.

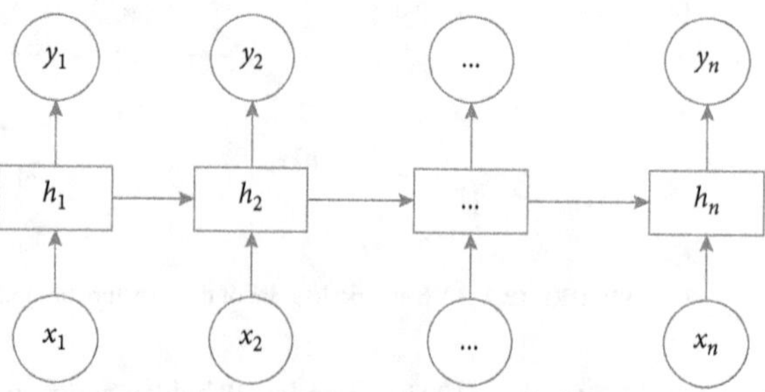

FIGURE 12.5   Another way to look at sequential ANN models.

The final prediction y_n considers the input from x_n, x_{n-1}, x_{n-2}, all the way to x_1. While predicting the nth output, we need to get all the inputs as well as the intermediate outputs. For example, while predicting the third word, we need first to calculate the hidden layer output of model-1 using the first input in the sequence, and later use this along with the second input in the sequence to estimate the predicted value of the third value. We will work on a case study to understand the sequential ANN model building in-depth.

12.3 CASE STUDY: WORD PREDICTION

The following case study uses a simple dataset to demonstrate the concept of sequential ANN models.

12.3.1 Objective and Data

In this case, we would like to predict the third word based on the sequence of two words. The dataset contains a list of three words. This type of data is also known as three-gram data. If we consider the most frequently occurring N-word sequence, then it is known as N-gram data. This dataset is freely available on the COCA (Corpus of Contemporary American English) website: https://www.english-corpora.org/coca/. We are using a subset in this example. In the subset of the data, we have mostly considered the word sequences starting with love or hate. Table 12.4 shows a few examples from the data.

TABLE 12.4 Sample Data Points

love	to	see
love	it	when
hate	to	use
love	more	than
love	more	than
love	it	when
love	of	the
love	letter	to
love	to	see
loved	nothing	better

The final goal is to take the first two words as input and predict the final word. Given below is the code for importing the data.

```python
import pandas as pd
column_names = ['word1', 'word2', 'word3']

input_3gram = pd.read_csv(r'D:\Google Drive\Training\Book\0.Chapters\Chapter12
RNN and LSTM\5.Datasets\3Gram_love_data.txt', delimiter='\t', names=column_
names) #Importing csv file with column names
print("shape of data", input_3gram.shape)
print("Few sample records from data \n", input_3gram.sample(10))
```

The following is the output of the above code.

```
shape of data (5351, 3)

Few sample records from data
      word1 word2  word3
2776  love    to    see
2     hate    to    see
3260  love    to    see
2713  love   the    way
404   hated   to  think
2707  love   the    way
623   love    it     or
```

```
1477    love    to    get
1354    love    the   look
2545    love    the   way
```

There are a total of 5351 rows, and each row has three words; here, three words constitute three columns. The code below is used to find the frequency of distinct words in column 1 and column 2.

```
print("\nFrequency of word1 values \n", input_3gram["word1"].value_counts())
```

```
print("\nFrequency of word2 values \n", input_3gram["word2"].value_counts())
```

TABLE 12.5   Frequency of First Two Columns

Frequency of word1 values		Frequency of word2 values	
love	4327	to	1866
loved	416	it	1361
hate	400	the	548
hated	80	with	240
loves	72	you	144
lovely	24	him	144
loving	24	of	136
hates	8	her	104
Name: word1, dtype: int64		for	96
		and	88
		what	56
		is	48
		in	40
		each	40
		them	32
		nothing	32
		me	32
		ones	32
		every	24
		as	24
		going	16
		being	16
		affair	16
		my	16
		more	16
		that	16
		one	8
		a	8
		thy	8
		story	8
		man	8
		your	8
		makes	8
		this	8
		got	8
		on	8
		husband	8
		at	8
		letter	8
		hearing	8
		lost	8
		about	8
		most	8
		all	8
		song	8
		view	8
		when	8
		Name: word2, dtype: int64	

From the output table (Table 12.5), we can see that column 1 has all the words related to love and hate. More than 90 percent of the words are related to love. The second column has a few more unique words.

12.3.2   Data Preprocessing

Before building the model, we need to perform the following three steps:

1. Find all the unique words from three columns.
2. We cannot build the model using string data. We need to create a word index (`word_indices`) dictionary. We need to map the words to numbers to create a dictionary with words as keys and numbers as values. Later we will create one-hot encoded variables for all the unique numbers. This conversion is necessary for building the model.
3. The model gives the final predictions as numbers. We cannot give the predictions as numbers. As it makes sense, we will finally give the predicted output in the form of words only. For this purpose, we need to maintain a second dictionary that has numbers as keys and words as values.

Let us see the code for each one of these steps. The first step is to find all the unique words from three columns.

```
unique_words = []
for i in list(input_3gram.columns.values):
    for j in pd.unique(input_3gram[i]):
        unique_words.append(j)
unique_words = np.unique(unique_words)

print('Count of unique words overall:', len(unique_words))
print('unique words list:', unique_words)
```

The following is the output of the above code.

```
Count of unique words overall: 139
unique words list: ['a' 'able' 'about' 'admit' 'affair' 'affection' 'all'
 'and' 'another' 'answer' 'as' 'at' 'be' 'because' 'being' 'better' 'between'
 'bother' 'break' 'care' 'cared' 'come' 'concern' 'country' 'cut' 'disappoint'
 'do' 'each' 'every' 'fact' 'feel' 'feeling' 'find' 'first' 'for' 'from' 'get'
 'go' 'god' 'going' 'got' 'hate' 'hated' 'hates' 'have' 'he' 'hear' 'hearing'
 'her' 'here' 'him' 'his' 'husband' 'i' 'idea' 'if' 'in' 'interrupt' 'is' 'it'
 'kind' 'know' 'leave' 'letter' 'life' 'like' 'listen' 'look' 'lost' 'lot'
 'love' 'loved' 'lovely' 'loves' 'loving' 'make' 'makes' 'man' 'marriage' 'me'
 'minute' 'more' 'most' 'much' 'music' 'my' 'nature' 'neighbor' 'not' 'nothing'
 'of' 'on' 'one' 'ones' 'or' 'other' 'over' 'play' 'respect' 'say' 'see' 'sit'
 'smell' 'so' 'someone' 'song' 'sound' 'story' 'stronger' 'support' 'take'
 'talk' 'tell' 'than' 'that' 'the' 'them' 'they' 'think' 'this' 'thought' 'thy'
 'to' 'too' 'united' 'use' 'very' 'view' 'watch' 'way' 'we' 'what' 'when'
 'wife' 'will' 'with' 'work' 'you' 'your']
```

From the output above, we can see that there are overall 139 unique words. These are all the unique words from the three columns. In the previous output, we have already seen that column1 has a few unique words. Now we will create two dictionaries: words to indices and indices to words.

```
word_indices = dict((w, i) for i, w in enumerate(unique_words))
indices_words = dict((i, w) for i, w in enumerate(unique_words))

print("word_indices dictionary \n",word_indices)
print("word_indices.keys \n", word_indices.keys())
print("word_indices.values \n", word_indices.values())
print("\n ###################################\n")
print("indices_words dictionary \n", indices_words)
print("indices_words keys \n",indices_words.keys())
print("indices_words values \n",indices_words.values())
```

The following is the output of the above code.

```
word_indices dictionary
 {'a': 0, 'able': 1, 'about': 2, 'admit': 3, 'affair': 4, 'affection': 5,
'all': 6, 'and': 7, 'another': 8, 'answer': 9, 'as': 10, 'at': 11, 'be': 12,
'because': 13, 'being': 14, 'better': 15, 'between': 16, 'bother': 17,
'break': 18, 'care': 19, 'cared': 20, 'come': 21, 'concern': 22, 'country':
23, 'cut': 24, 'disappoint': 25, 'do': 26, 'each': 27, 'every': 28, 'fact':
29, 'feel': 30, 'feeling': 31, 'find': 32, 'first': 33, 'for': 34, 'from':
35, 'get': 36, 'go': 37, 'god': 38, 'going': 39, 'got': 40, 'hate': 41,
'hated': 42, 'hates': 43, 'have': 44, 'he': 45, 'hear': 46, 'hearing': 47,
'her': 48, 'here': 49, 'him': 50, 'his': 51, 'husband': 52, 'i': 53, 'idea':
54, 'if': 55, 'in': 56, 'interrupt': 57, 'is': 58, 'it': 59, 'kind': 60,
'know': 61, 'leave': 62, 'letter': 63, 'life': 64, 'like': 65, 'listen': 66,
'look': 67, 'lost': 68, 'lot': 69, 'love': 70, 'loved': 71, 'lovely': 72,
'loves': 73, 'loving': 74, 'make': 75, 'makes': 76, 'man': 77, 'marriage':
78, 'me': 79, 'minute': 80, 'more': 81, 'most': 82, 'much': 83, 'music': 84,
'my': 85, 'nature': 86, 'neighbor': 87, 'not': 88, 'nothing': 89, 'of': 90,
'on': 91, 'one': 92, 'ones': 93, 'or': 94, 'other': 95, 'over': 96, 'play':
97, 'respect': 98, 'say': 99, 'see': 100, 'sit': 101, 'smell': 102, 'so':
103, 'someone': 104, 'song': 105, 'sound': 106, 'story': 107, 'stronger':
108, 'support': 109, 'take': 110, 'talk': 111, 'tell': 112, 'than': 113,
'that': 114, 'the': 115, 'them': 116, 'they': 117, 'think': 118, 'this': 119,
'thought': 120, 'thy': 121, 'to': 122, 'too': 123, 'united': 124, 'use': 125,
'very': 126, 'view': 127, 'watch': 128, 'way': 129, 'we': 130, 'what': 131,
'when': 132, 'wife': 133, 'will': 134, 'with': 135, 'work': 136, 'you': 137,
'your': 138}
```

From the output above, we can see the mapping for each of the words to a number. In the word_indices dictionary, words are keys and numbers are values. The output also shows the keys and values separately; we are not displaying them here. Note that there is a total of 139 numbers from 0 to 138. The word "love" is mapped to index 70, and the word "way" is mapped to 129. The second part of the output is the indexes to words dictionary.

```
indices_words dictionary
 {0: 'a', 1: 'able', 2: 'about', 3: 'admit', 4: 'affair', 5: 'affection', 6:
'all', 7: 'and', 8: 'another', 9: 'answer', 10: 'as', 11: 'at', 12: 'be', 13:
'because', 14: 'being', 15: 'better', 16: 'between', 17: 'bother', 18: 'break',
19: 'care', 20: 'cared', 21: 'come', 22: 'concern', 23: 'country', 24: 'cut',
25: 'disappoint', 26: 'do', 27: 'each', 28: 'every', 29: 'fact', 30: 'feel',
31: 'feeling', 32: 'find', 33: 'first', 34: 'for', 35: 'from', 36: 'get', 37:
'go', 38: 'god', 39: 'going', 40: 'got', 41: 'hate', 42: 'hated', 43: 'hates',
44: 'have', 45: 'he', 46: 'hear', 47: 'hearing', 48: 'her', 49: 'here', 50:
'him', 51: 'his', 52: 'husband', 53: 'i', 54: 'idea', 55: 'if', 56: 'in', 57:
'interrupt', 58: 'is', 59: 'it', 60: 'kind', 61: 'know', 62: 'leave', 63:
'letter', 64: 'life', 65: 'like', 66: 'listen', 67: 'look', 68: 'lost', 69:
'lot', 70: 'love', 71: 'loved', 72: 'lovely', 73: 'loves', 74: 'loving', 75:
'make', 76: 'makes', 77: 'man', 78: 'marriage', 79: 'me', 80: 'minute', 81:
'more', 82: 'most', 83: 'much', 84: 'music', 85: 'my', 86: 'nature', 87:
'neighbor', 88: 'not', 89: 'nothing', 90: 'of', 91: 'on', 92: 'one', 93:
'ones', 94: 'or', 95: 'other', 96: 'over', 97: 'play', 98: 'respect', 99:
'say', 100: 'see', 101: 'sit', 102: 'smell', 103: 'so', 104: 'someone', 105:
'song', 106: 'sound', 107: 'story', 108: 'stronger', 109: 'support', 110:
'take', 111: 'talk', 112: 'tell', 113: 'than', 114: 'that', 115: 'the', 116:
'them', 117: 'they', 118: 'think', 119: 'this', 120: 'thought', 121: 'thy',
122: 'to', 123: 'too', 124: 'united', 125: 'use', 126: 'very', 127: 'view',
128: 'watch', 129: 'way', 130: 'we', 131: 'what', 132: 'when', 133: 'wife',
134: 'will', 135: 'with', 136: 'work', 137: 'you', 138: 'your'}
```

The output above is just a copy of the previous dictionary, where keys and values are swapped. In the `indices_words` dictionary, keys are numbers and values are words. The number 70 is mapped to "love," and 129 is mapped to "way." These two dictionaries will be used later. The first one is used before building the model, and the second dictionary is used at the time of prediction. We will now convert the values in column 1 to one-hot encoded values. There are 139 unique words. So the first column (word1) will be one-hot encoded with 139 columns. The following code converts column 1 into one-hot encoded columns.

```
### One-hot encoding of word1
word1 = input_3gram['word1'].map(word_indices)
word1_onehot = keras.utils.to_categorical(np.array(word1), num_classes=len
(word_indices))
print("word1_onehot shape is ",word1_onehot.shape)
```

The following is the output of the above code.

```
word1_onehot shape is  (5351, 139)
```

As expected, the output has 139 columns. Each column corresponds to one unique word. Let us see a couple of examples.

```
print("The word in row 0 is -->"+input_3gram['word1'][0])
print("The one-hot encoded version of the word in row 0 is \n",word1_
onehot[0])

print("\nThe word in row 500 is --> "+input_3gram['word1'][500])
print("The one-hot encoded version of the word in row 500 is \n",word1_
onehot[500])
```

The following is the output of the above code.

```
The word in row 0 is -->hate
The one-hot encoded version of the word in row 0 is
 [0. 0. 0. 0. 0. 0. 0. 0. 0. 0. 0. 0. 0. 0. 0. 0. 0. 0. 0. 0. 0. 0. 0. 0.
 0. 0. 0. 0. 0. 0. 0. 0. 0. 0. 0. 0. 0. 0. 0. 0. 0. 1. 0. 0. 0. 0. 0. 0.
 0. 0. 0. 0. 0. 0. 0. 0. 0. 0. 0. 0. 0. 0. 0. 0. 0. 0. 0. 0. 0. 0. 0. 0.
 0. 0. 0. 0. 0. 0. 0. 0. 0. 0. 0. 0. 0. 0. 0. 0. 0. 0. 0. 0. 0. 0. 0. 0.
 0. 0. 0. 0. 0. 0. 0. 0. 0. 0. 0. 0. 0. 0. 0. 0. 0. 0. 0. 0. 0. 0. 0. 0.
 0. 0. 0. 0. 0. 0. 0. 0. 0. 0. 0. 0. 0. 0. 0. 0. 0. 0.]

The word in row 500 is --> love
The one-hot encoded version of the word in row 500 is
 [0. 0. 0. 0. 0. 0. 0. 0. 0. 0. 0. 0. 0. 0. 0. 0. 0. 0. 0. 0. 0. 0. 0. 0.
 0. 0. 0. 0. 0. 0. 0. 0. 0. 0. 0. 0. 0. 0. 0. 0. 0. 0. 0. 0. 0. 0. 0. 0.
 0. 0. 0. 0. 0. 0. 0. 0. 0. 0. 0. 0. 0. 0. 0. 0. 0. 0. 0. 0. 0. 0. 1. 0.
 0. 0. 0. 0. 0. 0. 0. 0. 0. 0. 0. 0. 0. 0. 0. 0. 0. 0. 0. 0. 0. 0. 0. 0.
 0. 0. 0. 0. 0. 0. 0. 0. 0. 0. 0. 0. 0. 0. 0. 0. 0. 0. 0. 0. 0. 0. 0. 0.
 0. 0. 0. 0. 0. 0. 0. 0. 0. 0. 0. 0. 0. 0. 0. 0. 0.]
```

From the output above, we can see the word in the first row is "hate." The one-hot encoded value for that row shows the value "1" in the 42nd column. The word in column 500 is "love," and it has value "1" in the 71st column. We will convert column 2 and column 3 (word2 and word3 in the sequence) also to a one-hot encoded format using the following code.

```
word2 = input_3gram['word2'].map(word_indices)
word2_onehot = keras.utils.to_categorical(np.array(word2), num_classes=len
(word_indices))
print("word2_onehot shape is ",word2_onehot.shape)
```

```
word3 = input_3gram['word3'].map(word_indices)
word3_onehot = keras.utils.to_categorical(np.array(word3), num_classes=len
(word_indices))
print("word3_onehot shape is ",word3_onehot.shape)
```

The following is the output of the above code.

```
word2_onehot shape is  (5351, 139)
word3_onehot shape is  (5351, 139)
```

We are now done with data preprocessing. We are all set to build the two models.

12.3.3 Model Building

As discussed earlier, we will have two models in this case. The first ANN model takes word1_onehot as input and gives us word2_onehot as the output. We will extract the hidden layer output from this model and use it in the next model. Given below is the code for building the model.

```
ANN_model1 = Sequential()
ANN_model1.add(Dense(10, input_dim=word1_onehot.shape[1],activation='sigmoid'))
ANN_model1.add(Dense(word2_onehot.shape[1] ,activation='softmax'))
ANN_model1.summary()

ANN_model1.compile(loss='binary_crossentropy', optimizer='adam', metrics=
['accuracy'])
history = ANN_model1.fit(word1_onehot, word2_onehot, epochs=20, batch_
size=50, verbose=1)
```

The following is the output of the above code.

```
Train on 5351 samples
Epoch 1/20
5351/5351 [[====]] - 1s 155us/sample - loss: 0.0399 - accuracy: 0.9928
Epoch 2/20
5351/5351 [[====]] - 0s 38us/sample - loss: 0.0329 - accuracy: 0.9928
Epoch 3/20
5351/5351 [[====]] - 0s 39us/sample - loss: 0.0271 - accuracy: 0.9928
Epoch 4/20
5351/5351 [[====]] - 0s 41us/sample - loss: 0.0240 - accuracy: 0.9928
Epoch 5/20
5351/5351 [[====]] - 0s 35us/sample - loss: 0.0231 - accuracy: 0.9928
Epoch 6/20
5351/5351 [[====]] - 0s 35us/sample - loss: 0.0228 - accuracy: 0.9928
Epoch 7/20
5351/5351 [[====]] - 0s 37us/sample - loss: 0.0227 - accuracy: 0.9928
Epoch 8/20
5351/5351 [[====]] - 0s 40us/sample - loss: 0.0226 - accuracy: 0.9928
Epoch 9/20
5351/5351 [[====]] - 0s 41us/sample - loss: 0.0225 - accuracy: 0.9928
Epoch 10/20
5351/5351 [[====]] - 0s 88us/sample - loss: 0.0225 - accuracy: 0.9928
Epoch 11/20
5351/5351 [[====]] - 0s 58us/sample - loss: 0.0224 - accuracy: 0.9928
Epoch 12/20
5351/5351 [[====]] - 0s 49us/sample - loss: 0.0224 - accuracy: 0.9928
Epoch 13/20
5351/5351 [[====]] - 0s 54us/sample - loss: 0.0224 - accuracy: 0.9928
Epoch 14/20
```

```
5351/5351 [[====]] - 0s 44us/sample - loss: 0.0223 - accuracy: 0.9928
Epoch 15/20
5351/5351 [[====]] - 0s 43us/sample - loss: 0.0223 - accuracy: 0.9928
Epoch 16/20
5351/5351 [[====]] - 0s 45us/sample - loss: 0.0223 - accuracy: 0.9928
Epoch 17/20
5351/5351 [[====]] - 0s 44us/sample - loss: 0.0223 - accuracy: 0.9928
Epoch 18/20
5351/5351 [[====]] - 0s 46us/sample - loss: 0.0222 - accuracy: 0.9928
Epoch 19/20
5351/5351 [[====]] - 0s 43us/sample - loss: 0.0222 - accuracy: 0.9928
Epoch 20/20
5351/5351 [[====]] - 0s 43us/sample - loss: 0.0222 - accuracy: 0.9928
```

We are not interested in the final output of this model. We are interested only in the intermediate output values for each record. Since there are 10 hidden nodes, the hidden nodes result in a matrix that will have 5351 rows, and 10 columns. These hidden layer output values are known as hidden layer activations. Given below is the code that helps us in extracting the hidden layer activations.

```
model1_hidden = Sequential()
model1_hidden.add(Dense(10,  input_dim=word1_onehot.shape[1],  weights=ANN_
model1.layers[0].get_weights()))
model1_hidden.add(Activation('sigmoid'))

# Getting the hidden layer activations
model1_hidden_output = model1_hidden.predict(word1_onehot)
#peak into our hidden layer activations
print("The hidden layer output for every record - Shape of it \n", model1_
hidden_output.shape)
print("Few records from hidden layer \n",model1_hidden_output[:5])
```

The following is the output of the above code.

```
The hidden layer output for every record - Shape of it
 (5351, 10)
Few records from hidden layer
 [[0.8781716   0.88077706 0.77781826 0.8785578  0.731379   0.8384166
   0.8529098  0.79941416 0.8221394  0.8138482 ]
  [0.8781716   0.88077706 0.77781826 0.8785578  0.731379   0.8384166
   0.8529098  0.79941416 0.8221394  0.8138482 ]
  [0.8781716   0.88077706 0.77781826 0.8785578  0.731379   0.8384166
   0.8529098  0.79941416 0.8221394  0.8138482 ]
  [0.8781716   0.88077706 0.77781826 0.8785578  0.731379   0.8384166
   0.8529098  0.79941416 0.8221394  0.8138482 ]
  [0.8781716   0.88077706 0.77781826 0.8785578  0.731379   0.8384166
   0.8529098  0.79941416 0.8221394  0.8138482 ]]
```

As expected, the shape of model1 hidden layer output is (5351,10). The output also shows the hidden node outputs for the first five records; each record has 10 values calculated from 10 output nodes. Now we will append this to the word2_onehot and build the second ANN model.

```
word2_hidden_append = np.append(model1_hidden_output, word2_onehot, axis=1)
print("word2_hidden_append Shape", word2_hidden_append.shape)
```

The following is the output of the above code.

```
word2_hidden_append Shape (5351, 149)
```

Now we will build ANN model2 using the code below.

```
ANN_model2 = Sequential()
ANN_model2.add(Dense(10, input_dim=word2_hidden_append.shape[1], activation=
'sigmoid'))
ANN_model2.add(Dense(word3_onehot.shape[1], activation='softmax'))
ANN_model2.summary()

ANN_model2.compile(loss='binary_crossentropy', optimizer='adam', metrics=
['accuracy'])
# Train model
history = ANN_model2.fit(word2_hidden_append, word3_onehot, epochs=20, batch_
size=50, verbose=1)
```

The following is the output of the above code.

```
Train on 5351 samples
Epoch 1/20
5351/5351 [====] - 1s 108us/sample - loss: 0.0402 - accuracy: 0.9928
Epoch 2/20
5351/5351 [====] - 0s 37us/sample - loss: 0.0342 - accuracy: 0.9928
Epoch 3/20
5351/5351 [====] - 0s 50us/sample - loss: 0.0309 - accuracy: 0.9928
Epoch 4/20
5351/5351 [====] - 0s 46us/sample - loss: 0.0302 - accuracy: 0.9928
Epoch 5/20
5351/5351 [====] - 0s 36us/sample - loss: 0.0300 - accuracy: 0.9928
Epoch 6/20
5351/5351 [====] - 0s 35us/sample - loss: 0.0298 - accuracy: 0.9928
Epoch 7/20
5351/5351 [====] - 0s 38us/sample - loss: 0.0295 - accuracy: 0.9928
Epoch 8/20
5351/5351 [====] - 0s 36us/sample - loss: 0.0290 - accuracy: 0.9928
Epoch 9/20
5351/5351 [====] - 0s 36us/sample - loss: 0.0285 - accuracy: 0.9928
Epoch 10/20
5351/5351 [====] - 0s 43us/sample - loss: 0.0279 - accuracy: 0.9928
Epoch 11/20
5351/5351 [====] - 0s 43us/sample - loss: 0.0272 - accuracy: 0.9944
Epoch 12/20
5351/5351 [====] - 0s 30us/sample - loss: 0.0265 - accuracy: 0.9945
Epoch 13/20
5351/5351 [====] - 0s 33us/sample - loss: 0.0258 - accuracy: 0.9945
Epoch 14/20
5351/5351 [====] - 0s 38us/sample - loss: 0.0252 - accuracy: 0.9945
Epoch 15/20
5351/5351 [====] - 0s 41us/sample - loss: 0.0246 - accuracy: 0.9945
Epoch 16/20
5351/5351 [====] - 0s 35us/sample - loss: 0.0241 - accuracy: 0.9945
Epoch 17/20
5351/5351 [====] - 0s 40us/sample - loss: 0.0237 - accuracy: 0.9949
Epoch 18/20
5351/5351 [====] - 0s 34us/sample - loss: 0.0233 - accuracy: 0.9950
Epoch 19/20
5351/5351 [====] - 0s 41us/sample - loss: 0.0229 - accuracy: 0.9950
Epoch 20/20
5351/5351 [====] - 0s 37us/sample - loss: 0.0226 - accuracy: 0.9949
```

We are now done with the sequential ANN models. The predictions will not be the same as standard ANN models. In the next section, we will see how to use these models to get predicted values on new data points.

12.3.4 Prediction

We need to write a custom predict function, which takes in a sequence of two words. These two words will be converted to numbers using the `word_indices` dictionary, followed by one-hot encoding. The function uses the first word, and using `ANN_model1`, we extract the hidden layer activations. These activation values will be appended to the second word. The final prediction will be made using the second model. The prediction will be a number; we then convert it to a word using `indices_words` dictionary. Given below is the code for writing the custom predict function.

```python
def two_step_pred(words_in):

    index_input=word_indices[words_in[0]]
    indices_in = keras.utils.to_categorical
                    (index_input,   num_classes=len(word_indices))
    indices_in=indices_in.reshape(1,len(word_indices))
    h1_test = model1_hidden.predict(indices_in) # getting our intermediate
    hidden activations from model1h


    index_input2=word_indices[words_in[1]]
    indices_in2 = keras.utils.to_categorical
                    (index_input2, num_classes=len(word_indices))
    indices_in2= indices_in2.reshape(1,len(word_indices))
    X2_test = np.append(h1_test, indices_in2, axis=1) #preparing final test
    data by appending hidden with word2

    yhat = ANN_model2.predict_classes(X2_test) #predicting final output from
    model2

    print("Input words --> ", words_in)
    print("Predicted word --> ", indices_words[yhat[0]])
```

We will now use this function to get some predicted values.

```python
two_step_pred(['love', 'it'])
two_step_pred(['love', 'to'])
two_step_pred(['love', 'the'])
```

The above code gives us the below output.

```
Input words -->  ['love', 'it']
Predicted word -->  when
Input words -->  ['love', 'to']
Predicted word -->  see
Input words -->  ['love', 'the']
Predicted word -->  way
```

The accuracy of this model depends on the training data. We can say the predictions are reasonably good. Here we have used sequential ANN models to solve a problem of sequential dependency, i.e., predicting the third word using the first two words as input. With this, we can conclude this case study. Understanding of this case study is vital for understanding RNN models, which is the topic of the next section.

12.4 RECURRENT NEURAL NETWORKS

In the previous section, we built sequential ANN models to solve a sequential data problem, which was primarily a manual effort. Recurrent neural networks (RNNs) are programmed sequential ANN models. The procedure that we followed until now will be done by RNN automatically. If we mention the number of time steps, RNN models will automatically stack the required number of ANN models. If we have to predict the nth value in a sequence, then we need to build an RNN model with $(n-1)$ time steps. RNN will stack the $(n-1)$ number of ANN models in a sequential manner, where the hidden layer of the previous ANN is connected to the hidden layer of the next ANN. For the ease of understanding, we can imagine all these ANN models have a single hidden layer. RNN models are known as ANN models with memory. Figure 12.6 is the diagrammatic representation of RNN models.

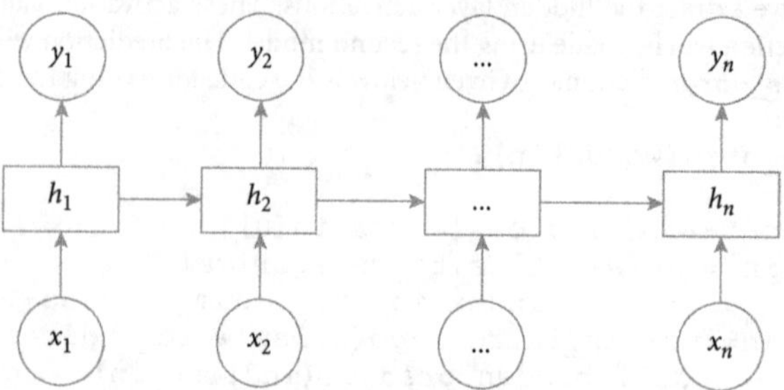

FIGURE 12.6   A pictorial representation of RNN models.

The diagram shown in Fig. 12.6 is an unrolled version for "n" time steps. It is usually rolled and represented in the format shown in Fig. 12.7.

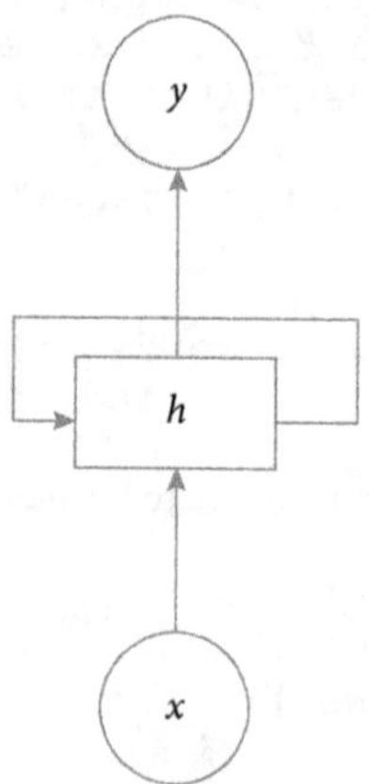

FIGURE 12.7   A rolled-up depiction of RNN models.

In a rolled-up version of the diagram (Fig. 12.7), there is a self-loop on the hidden layer. It represents the connections from one hidden layer to another hidden layer in the ANN stacks. We will discuss RNN in some more detail in this section.

12.4.1   Backpropagation Through Time

In RNN models, there are three types of weights that we have to calculate: the weights that are going from the input layer to the hidden layer, weights from the hidden layer to the hidden layer, and weights from the hidden layer to the output layer (Fig. 12.8).

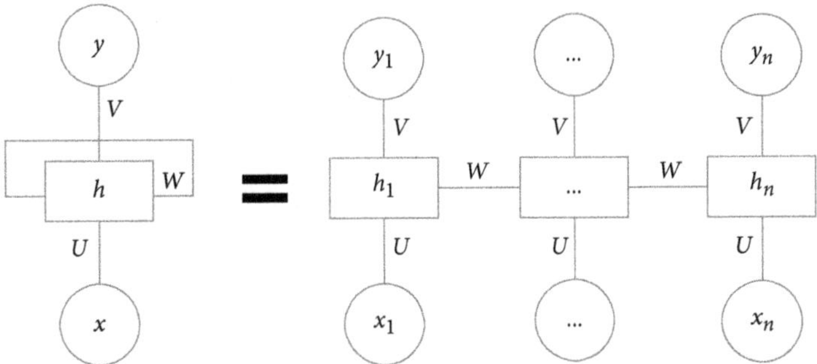

FIGURE 12.8 An unfolded version of representing RNN models.

We need to apply backpropagation to solve these three sets of weights. We apply backpropagation across all the time steps. Backpropagation across all the time steps is known as backpropagation through time (BPTT).

In the standard ANN model's backpropagation algorithm, we follow three main steps. The first step is feedforward, the second step is error calculation and backpropagation, and the final step is the weights updating step. In RNNs, we apply the feedforward step and calculate the values until the last time step n. In the second step, we calculate the error. While calculating the error, we sum up all the errors at output layers from time 0 to time n. We then propagate the error backward through each network and through all time steps. We find the error fractions at each hidden layer across the whole RNN. We then update the weights to reduce the error. One important point to note here is the concept of shared weights. The weights are shared across all the networks at all time steps. The weight sets of U, V, and W are identical in all ANN models in the stack. The weights from input to hidden (U) are the same at time step 0 and time step n. The same is the case with set W and V. This is one of the constraints in BPTT while minimizing the overall error. We will see an example to understand the count of these weights (Figs. 12.9 to 12.11).

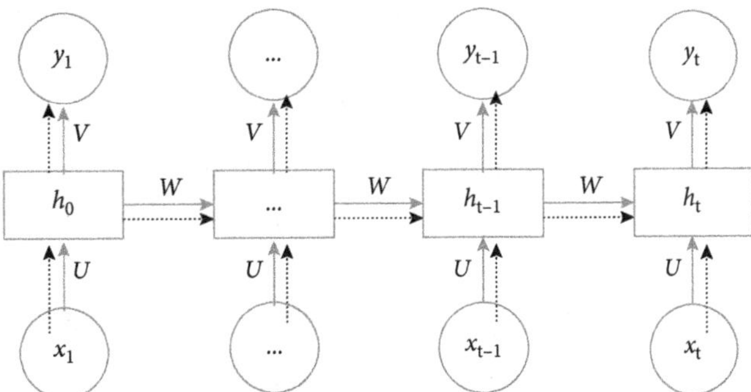

FIGURE 12.9 An illustration of a feedforward step.

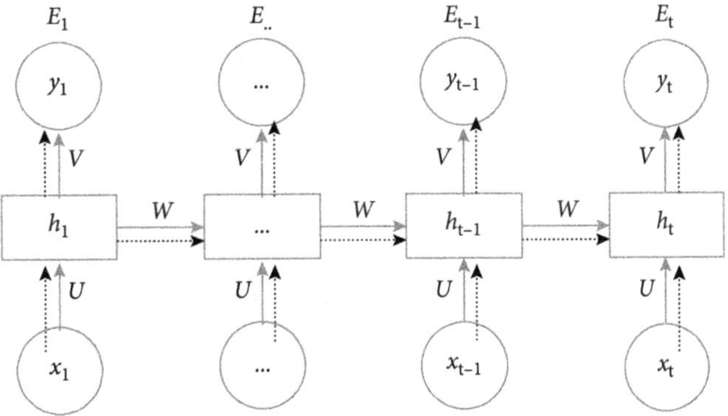

FIGURE 12.10 An illustration of an error calculation step.

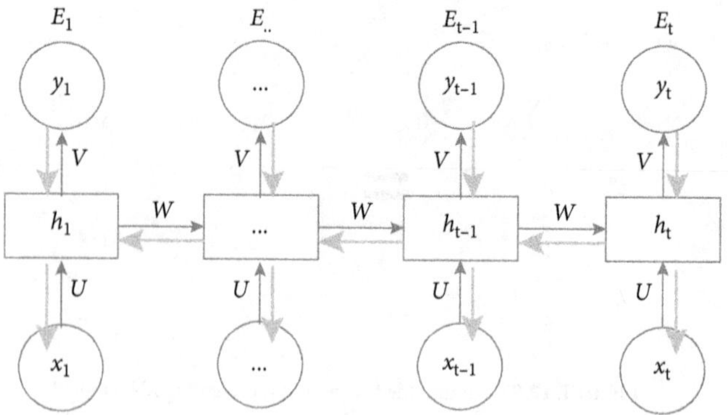

FIGURE 12.11   An illustration of a backpropagation step.

Below is the mathematical derivation of backpropagation through time.

Feedforward at time step t:

$$h_t = g(Ux_t + Wh_{t-1})$$

$$\hat{y}_t = \text{softmax}(Vh_t)$$

The loss at each time step t:

$$E_t(y_t, \hat{y}_t) = \sum - y_t \log \hat{y}_t$$

The overall loss is the sum of loss at all time steps

$$E(y, \hat{y}) = \sum_t E_t(y_t, \hat{y}_t)$$

Finding gradients with respect to V, W, U for the overall error:

$$\frac{\partial E}{\partial V}, \frac{\partial E}{\partial W}, \frac{\partial E}{\partial U}$$

Finding gradients is not the same as standard ANN:

Gradients w.r.t V:

$$\frac{\partial E}{\partial V} = \frac{\partial E_t}{\partial \hat{y}_t} \frac{\partial \hat{y}_t}{\partial V} + \frac{\partial E_{t-1}}{\partial \hat{y}_{t-1}} \frac{\partial \hat{y}_{t-1}}{\partial V} + \cdots + \frac{\partial E_1}{\partial \hat{y}_1} \frac{\partial \hat{y}_1}{\partial V}$$

As we can see above, we add the gradients before carrying out the weight update.

$$V(\text{new}) = V(\text{old}) + \eta \frac{\partial E}{\partial V}$$

Gradients w.r.t W:

$$\frac{\partial E}{\partial W} = \frac{\partial E_t}{\partial \hat{y}_t} \frac{\partial \hat{y}_t}{\partial h_t} \frac{\partial h_t}{\partial W} + \frac{\partial E_{t-1}}{\partial \hat{y}_{t-1}} \frac{\partial \hat{y}_{t-1}}{\partial h_{t-1}} \frac{\partial h_{t-1}}{\partial W} + \cdots + \frac{\partial E_1}{\partial \hat{y}_1} \frac{\partial \hat{y}_1}{\partial h_1} \frac{\partial h_1}{\partial W}$$

In the above equation, we have to expand partial derivatives of the hidden states.

$$\frac{\partial E}{\partial W} = \frac{\partial E_t}{\partial \hat{y}_t} \frac{\partial \hat{y}_t}{\partial h_t} \frac{\partial h_t}{\partial W} + \frac{\partial E_{t-1}}{\partial \hat{y}_{t-1}} \frac{\partial \hat{y}_{t-1}}{\partial h_t} \frac{\partial h_t}{\partial h_{t-1}} \frac{\partial h_{t-1}}{\partial W} + \cdots + \frac{\partial E_1}{\partial \hat{y}_1} \frac{\partial \hat{y}_1}{\partial h_t} \frac{\partial h_t}{\partial h_{t-1}} \frac{\partial h_{t-2}}{\partial h_{t-3}} \cdots \frac{\partial h_2}{\partial h_1} \frac{\partial h_1}{\partial W}$$

$$W(\text{new}) = W(\text{old}) + \eta \frac{\partial E}{\partial W}$$

Similarly, we will find the gradients with respect to U.

$$\frac{\partial E}{\partial U} = \frac{\partial E_t}{\partial \hat{y}_t} \frac{\partial \hat{y}_t}{\partial h_t} \frac{\partial h_t}{\partial x_t} \frac{\partial x_t}{\partial U} + \frac{\partial E_{t-1}}{\partial \hat{y}_{t-1}} \frac{\partial \hat{y}_{t-1}}{\partial h_t} \frac{\partial h_t}{\partial h_{t-1}} \frac{\partial h_{t-1}}{\partial x_{t-1}} \frac{\partial x_{t-1}}{\partial U} + \cdots + \frac{\partial E_1}{\partial \hat{y}_1} \frac{\partial \hat{y}_1}{\partial h_t} \frac{\partial h_t}{\partial h_{t-1}} \frac{\partial h_{t-2}}{\partial h_{t-3}} \cdots \frac{\partial h_2}{\partial h_1} \frac{\partial h_1}{\partial x_1} \frac{\partial x_1}{\partial U}$$

$$U(\text{new}) = U(\text{old}) + \eta \frac{\partial E}{\partial U}$$

In the BPTT algorithm, there is one more hidden constraint that is $V_t = V_{t-1}$; $W_t = W_{t-1}$; $U_t = U_{t-1}$ at all time steps t.

12.4.2   Calculating the Number of Parameters: An Example

Given below is the formula for the RNN number of parameters.
- Number of weights in input to hidden layers: $\dim(U) = nm$
- Number of weights in hidden to hidden layers: $\dim(W) = n^2$
- Number of weights in hidden to output layers: $\dim(V) = kn$

where
$\quad\quad m = $ dimension of the input layer
$\quad\quad n = $ dimension of hidden layer
$\quad\quad k = $ dimension of the output layer

$$\text{Overall weights} = nm + n^2 + kn$$

The number of weights will not change based on the number of time steps. The same weights are shared across all the ANN models in RNN. Let us take an example of two time steps (Fig. 12.12).

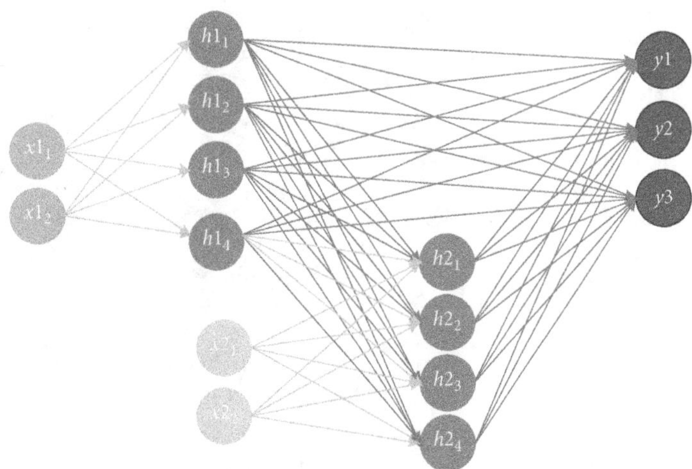

FIGURE 12.12   RNN with two time steps example.

In the example in Fig. 12.12:

- $m = 2$ (dimension of the input layer)
- $n = 4$ (dimension of the hidden layer)
- $k = 3$ (dimension of the output layer)
- Time steps $= 2$

The overall number of parameters is as follows:

$$\text{Overall weights} = nm + n^2 + kn$$

$$= 8 + 16 + 12 = 36$$

- Number of weights in input to hidden layers: $\dim(U) = nm$ (Fig. 12.13)

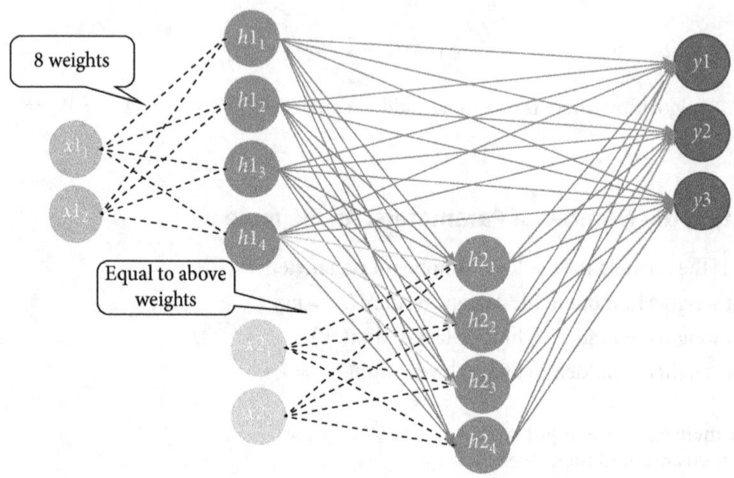

FIGURE 12.13   Input layer to hidden layer weights.

- Number of weights in hidden to hidden layers: $\dim(W) = n^2$ (Fig. 12.14)

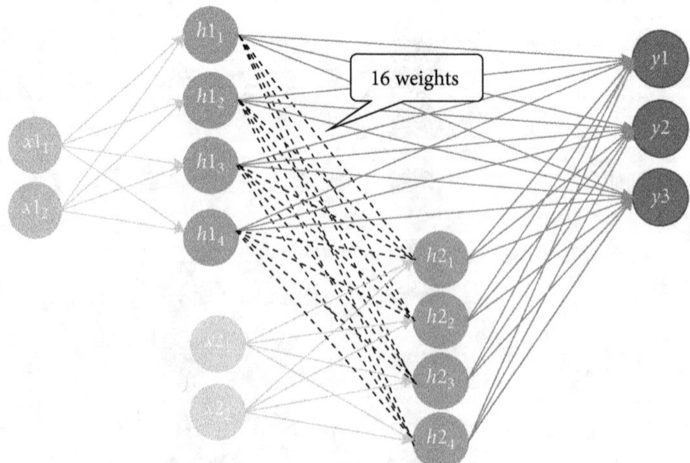

FIGURE 12.14   Hidden layer to hidden layer weights.

- Number of weights in hidden to output layer: $\dim(V) = kn$ (Fig. 12.15)

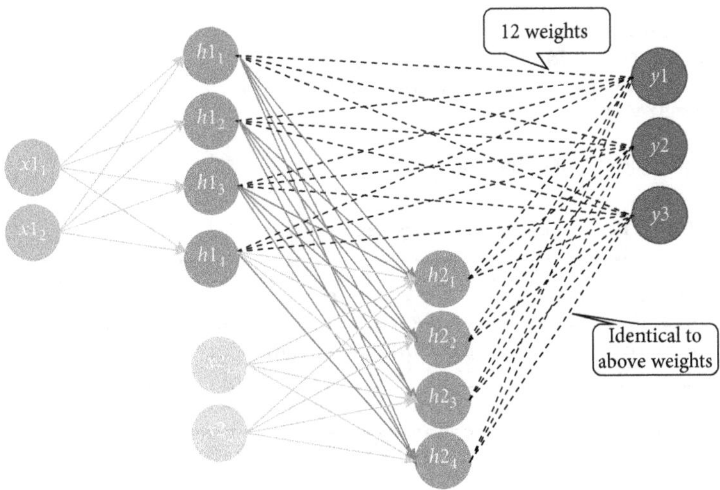

FIGURE 12.15   Hidden layer to output layer weights.

The earlier calculations are excluding the bias terms. If we add bias also, the parameters will be as follows:

$$(n + 1)m + n^2 + (k + 1)n$$

For the above RNN, there will be 3*4 + 4*4 + 5*3 = 43 parameters.

12.4.3   RNN Model Building Code

While building RNN models, we need to mention the number of time steps along with the standard parameters such as the number of hidden nodes and input shape. We need to add the simple RNN layer and mention these parameters. Let us take the above example where we have two nodes in the input layer, four nodes in the hidden layer, and three nodes in the output layer. Finally, the number of time steps is two. Given below is the RNN configuration code for this data.

```
model = Sequential()
model.add(SimpleRNN(4, use_bias=False, input_shape=(2,2)))
model.add(Dense(3, use_bias=False, activation='softmax'))
model.summary()
```

In the code above, the critical function is `SimpleRNN()`. Given below are the parameters.

```
SimpleRNN(hidden_nodes_count, input_shape=(time_steps,input_nodes))
```

Table 12.6 shows the code output.

TABLE 12.6   The RNN Code Output

Layer (type)	Output Shape	Param #
simple_rnn_8 (SimpleRNN)	(None, 4)	24
dense_12 (Dense)	(None, 3)	12

Total params: 36
Trainable params: 36
Non-trainable params: 0

Since we have excluded the bias, there are a total of 36 parameters. Now we will include the bias in this network and observe the number of parameters.

```
model = Sequential()
model.add(SimpleRNN(4, input_shape=(2,2)))
model.add(Dense(3, activation='softmax'))
model.summary()
```

Table 12.7 shows the output of this code.

TABLE 12.7   The Model Summary Generated by the Code

Layer (type)	Output Shape	Param #
simple_rnn_11 (SimpleRNN)	(None, 4)	28
dense_15 (Dense)	(None, 3)	15

```
Total params: 43
Trainable params: 43
Non-trainable params: 0
```

From the outputs shown in Table 12.7, we can see the number of parameters matches our manual calculations using the formula. Since there is a concept of shared weights, the number of time steps or the length of the sequence has no impact on the number of parameters. If we change the time steps to 4, RNN will still result in 43 parameters.

```
model = Sequential()
model.add(SimpleRNN(4, input_shape=(4,2)))
model.add(Dense(3, activation='softmax'))
model.summary()
```

We have given four time steps in the above code. Table 12.8 shows the output of the above code.

TABLE 12.8   RNN with Four Time Steps Code Output

Layer (type)	Output Shape	Param #
simple_rnn_14 (SimpleRNN)	(None, 4)	28
dense_18 (Dense)	(None, 3)	15

```
Total params: 43
Trainable params: 43
Non-trainable params: 0
```

12.4.4   Word Prediction Using RNN Model

We have manually built a sequential ANN stack to solve the case study, where we are predicting the third word. With RNN models, we just need to mention the number of time steps; the ANN stacks will be automatically taken care of by the RNN model. We need to supply the word1 and word2 as inputs and build an RNN model with time steps = 2. Given below is the code for data preparation.

```
word1_word2 = input_3gram[['word1','word2']]
for i in list(word1_word2.columns.values):
    word1_word2[i] = word1_word2[i].map(word_indices)
```

```python
word1_word2=np.array(word1_word2)
word1_word2=np.reshape(word1_word2,(word1_word2.shape[0],2,1))
word1_word2_onehot = keras.utils.to_categorical(np.array(word1_word2), num_
classes=len(word_indices))
print("word1_word2_onehot shape", word1_word2_onehot.shape)
```

In the above code, we tried appending word1 and word2 column-wise, then reshaped them followed by one-hot encoding. Finally, the code will give us a three-dimensional array with 5351 rows and two columns, and each column has 139 dimensions (one-hot encoded). Given below is the result of this code.

```
word1_word2_onehot shape (5351, 2, 139)
```

Now we are ready with the data. We are predicting the third word, which means the time steps are two. The target variable is the third word, which is also one-hot encoded.

```python
print("time steps" , word1_word2_onehot.shape[1])
print("Input nodes" , word1_word2_onehot.shape[2])
print("output nodes" , word3_onehot.shape[1])
```

Given below is the output from the above code.

```
Time steps 2
Input nodes 139
output nodes 139
```

We will now build the RNN model using the code below.

```python
model_rnn = Sequential()
#model.add(SimpleRNN('number of hidden nodes in each rnn cell', input_
shape=(timesteps, input_data_dim)))

model_rnn.add(SimpleRNN(30, input_shape=(word1_word2_onehot.shape[1], word1_
word2_onehot.shape[2])))

model_rnn.add(Dense(word3_onehot.shape[1], activation='softmax'))
model_rnn.summary()
```

In the above code, we have mentioned two time steps and 30 hidden nodes in each time step. This code gives us the model summary presented in Table 12.9.

TABLE 12.9   The Model Summary Generated by the Code

Layer (type)	Output Shape	Param #
simple_rnn_5 (SimpleRNN)	(None, 30)	5100
dense_10 (Dense)	(None, 139)	4309

```
Total params: 9,409
Trainable params: 9,409
Non-trainable params: 0
```

We will now compile and train the RNN model.

```python
## compile network
model_rnn.compile(loss='categorical_crossentropy', optimizer='adam', metrics=
['accuracy'])
## train the network
model_rnn.fit(word1_word2_onehot, word3_onehot, epochs=20)
```

The accuracy of the model depends on the strength of the training data. We have only 5000 records with 139 dimensions in the data. It is difficult to get good accuracy with this data. We need more data for higher accuracy. But this model will be better than our previous manual sequential ANN model. Given below is the model training output.

```
Train on 5351 samples
Epoch 1/20
5351/5351 [====] - 2s 322us/sample - loss: 3.7738 - accuracy: 0.2551
Epoch 2/20
5351/5351 [====] - 0s 86us/sample - loss: 2.8423 - accuracy: 0.4868
Epoch 3/20
5351/5351 [====] - 1s 94us/sample - loss: 2.3925 - accuracy: 0.5339
Epoch 4/20
5351/5351 [====] - 0s 85us/sample - loss: 2.0859 - accuracy: 0.5679
Epoch 5/20
5351/5351 [====] - 0s 81us/sample - loss: 1.8948 - accuracy: 0.5853
Epoch 6/20
5351/5351 [====] - 0s 80us/sample - loss: 1.7591 - accuracy: 0.6038
Epoch 7/20
5351/5351 [====] - 0s 81us/sample - loss: 1.6533 - accuracy: 0.6188
Epoch 8/20
5351/5351 [====] - 1s 101us/sample - loss: 1.5677 - accuracy: 0.6320
Epoch 9/20
5351/5351 [====] - 0s 80us/sample - loss: 1.4947 - accuracy: 0.6498
Epoch 10/20
5351/5351 [====] - 0s 80us/sample - loss: 1.4325 - accuracy: 0.6601
Epoch 11/20
5351/5351 [====] - 0s 88us/sample - loss: 1.3810 - accuracy: 0.6623
Epoch 12/20
5351/5351 [====] - 0s 89us/sample - loss: 1.3369 - accuracy: 0.6647
Epoch 13/20
5351/5351 [====] - 0s 87us/sample - loss: 1.3024 - accuracy: 0.6631
Epoch 14/20
5351/5351 [====] - 0s 73us/sample - loss: 1.2745 - accuracy: 0.6647
Epoch 15/20
5351/5351 [====] - 0s 67us/sample - loss: 1.2521 - accuracy: 0.6645
Epoch 16/20
5351/5351 [====] - 0s 62us/sample - loss: 1.2342 - accuracy: 0.6655
Epoch 17/20
5351/5351 [====] - 0s 73us/sample - loss: 1.2209 - accuracy: 0.6625
Epoch 18/20
5351/5351 [====] - 0s 69us/sample - loss: 1.2103 - accuracy: 0.6627
Epoch 19/20
5351/5351 [====] - 0s 66us/sample - loss: 1.2016 - accuracy: 0.6642
Epoch 20/20
5351/5351 [====] - 0s 62us/sample - loss: 1.1931 - accuracy: 0.6625
```

The model is now ready to make predictions. Given below is the code that will help for this purpose.

```python
def rnn_word_pred(in_text):
    print("Input is - " , in_text)

    encoded = [word_indices[i] for i in in_text]
    encoded = np.array(encoded).reshape(1,2,1)

    encoded =keras.utils.to_categorical(np.array(encoded),  num_classes=len
(word_indices))
    ypred = model_rnn.predict_classes(encoded, verbose=0)[0]
    print("Output is --> " ,indices_words[ypred])
```

There are three significant steps in the above prediction function. Firstly converting words into indices, followed by reshaping them to 3D input format; one row, two columns encoding depth. Finally, one-hot encoding the input to bring into the shape (1,2,139). This preprocessed input will be sent to RNN to predict function. Finally, the numerical output will be converted to words before printing. This function takes a list of two words as input. Given below are a few examples.

```
rnn_word_pred(['love', 'it'])
rnn_word_pred(['love', 'to'])
rnn_word_pred(['love', 'the'])
```

The following is the output of the above code.

```
Input is -  ['love', 'it']
Output is -->  when
Input is -  ['love', 'to']
Output is -->  see
Input is -  ['love', 'the']
Output is -->  way
```

From the output, we can see that the results are as good as our previous model. Once again, we need to note that the predictions depend on our training data. This concludes our discussion on the RNN model building.

12.5   RNN FOR LONG SEQUENCES

RNN models are useful for solving problems related to sequential data. In practical scenarios, RNN models seem to be failing to predict long sequences. In sequences where the number of time steps is more than 10, the RNN algorithm does not give accurate results. Have a look at the three statements given in Table 12.10.

TABLE 12.10   Examples of Long Sentences

<u>My</u> heart was heavy because it was open, and so things filled it, and so things rushed out of it, but still, the heart kept beating, tough and frighteningly powerful and meaning to shrug off the rest of **me** and continue on its own.

Her heart was heavy because it was open, and so things filled it, and so things rushed out of it, but still, the heart kept beating, tough and frighteningly powerful and meaning to shrug off the rest of **her** and continue on its own.

<u>His</u> heart was heavy because it was open, and so things filled it, and so things rushed out of it, but still, the heart kept beating, tough and frighteningly powerful and meaning to shrug off the rest of **him** and continue on its own

The three sentences in Table 12.10 are similar, and the only difference is in the subject. In the first case, the subject is "me," in the second case, the subject is "her," in the third case, the subject is "him." Depending on this subject, we need to change predictions. Since the sentence is too long and the depending subject is also far away, RNN models seem to be failing to predict in such long-term dependencies. The RNN models, in theory, should work with a sequence of any length. However, in practice, the standard RNN models do not have a long-term memory property. We will see a simple numerical example to prove it. We will take an example of long-term dependency and verify the performance of the RNN models.

12.5.1   Case Study: Predicting the Characters to Form the Next Word

This case study is similar to the case study of previous sections, where we predicted the next word in a small sequence. In Sec. 12.1, we talked about typing a sentence on a smartphone. While typing "Can I have your number" on a smartphone, we can observe that the actual prediction is not at the word level but at the character level. As we type one character, we can see the predicted output on our smartphones. The model working in the background takes the sequence of characters and predicts the next sequence of characters based on the input that we pass. Table 12.11 shows the input and predicted output from the smartphone appearing as suggestions.

TABLE 12.11 Predicted Outputs Based on the Input Sequence of Characters

Input Sequence of Characters	Predicted Outputs
C	Clg, Ch, Can
Ca	CA, Can, Call
Can	Can't, Cannot, do, mail
Can (Can <space>)	number, name, own, not
Can I	Get, have, come

Table 12.11 shows example outputs. The predictions change from person to person. The model under discussion was built at a character level, and predictions are made based on the sequence of input characters. We will also build one such character-level sequence model. Usually, the length of the input training sequence is long for character-level prediction models. In this model, a collection of two or three words will have a sequence of length 10 to 20. For example, if we have to predict the next sequence of characters after "Can I have your" then the input sequence length is $15(3 + 1 + 1 + 1 + 4 + 1 + 4)$. Note that space is also an input character.

12.5.1.1 Objective and Data   In this example, we are considering three-gram data, but the data is formed by carefully choosing the three grams that are more than 15 characters long. The objective is to take the first 14 sequence of characters as input and predict the next sequence of characters that form a word. The goal is to predict the next word, but that word prediction is made by arranging the characters as a sequence. Character-level input and output is the core difference between this model and the model in the previous case study. The following code imports and prints a sample of the data.

```
longseq_3gram = open(r'D:\Google Drive\Training\Book\0.Chapters\Chapter12
RNN and LSTM\5.Datasets\Long_sequence_3gram.csv').read().lower()
print(longseq_3gram[495:801])
print(longseq_3gram[30615:31000])
```

Table 12.12 presents the code output.

TABLE 12.12 The Code Output Sample Data

```
a,combination,of          and,according,to
a,combination,of          and,according,to
a,combination,of          and,according,to
a,combination,of          and,according,to
a,combination,of          and,according,to
a,combination,of          and,according,to
a,combination,of          and,according,to
a,combination,of          and,according,to
a,combination,of          and,according,to
a,combination,of          and,according,to
a,combination,of          and,addresses,of
a,combination,of          and,adherence,to
a,combination,of          and,advocates,for
a,combination,of          and,aerospace,engineering
a,combination,of          and,americans,do
a,combination,of          and,analyzing,the
a,combination,of          and,announced,he
a,combination,of          and,announced,he
                          and,announced,plans
                          and,announced,that
                          and,announced,that
                          and,annou
```

The output in Table 12.12 shows a few examples from the data. We will now apply the preprocessing steps. We need to create a character to index the dictionary and prepare X and y data.

12.5.1.2 Data Preprocessing Data preprocessing involves several steps. We start by replacing the commas with spaces using the code below.

```
longseq_3gram1= longseq_3gram.replace(',',' ').replace('\r','')
print(longseq_3gram1[495:750])
print(longseq_3gram1[30615:30800])
```

Table 12.13 shows the output from the above code.

TABLE 12.13   The Code Output

a combination of	and according to
a combination of	and according to
a combination of	and according to
a combination of	and according to
a combination of	and according to
a combination of	and according to
a combination of	and according to
a combination of	and according to
a combination of	and according to
a combination of	and according to
a combination of	
a combination of	
a combination of	
a combination of	
a combination of	

In this model, we need to map each character to an index while preparing the character to index dictionary.

```
#Unique characters in our dataset we then sort it
chars = sorted(list(set(longseq_3gram1)))
print("Unique Characters in the text \n ",chars)
chars.remove('\n')
print("\n Character after removing newline symbol \'\\n\'",chars)
print("\n overall chars count", len(chars))
```

In the above code, we are trying to count all the unique characters. We will finally exclude the newline symbol "\n". Given below is the output from the code above.

```
Unique Characters in the text
 ['\n', ' ', '"', '(', '-', '.', '/', '0', '1', '3', '7', '9', 'a', 'b', 'c',
'd', 'e', 'f', 'g', 'h', 'i', 'j', 'k', 'l', 'm', 'n', 'o', 'p', 'q', 'r',
's', 't', 'u', 'v', 'w', 'x', 'y', 'z']

 Character after removing newline symbol '\n' [' ', '"', '(', '-', '.', '/',
'0', '1', '3', '7', '9', 'a', 'b', 'c', 'd', 'e', 'f', 'g', 'h', 'i', 'j',
'k', 'l', 'm', 'n', 'o', 'p', 'q', 'r', 's', 't', 'u', 'v', 'w', 'x', 'y',
'z']

 Overall, the chars count is 37.
```

From the output, we can see that there are 37 unique characters. Apart from the alphabet, we have a few numbers and symbols. We will now create the char to indices and indices to char dictionaries using the following code.

```
char_indices = dict((c, i) for i, c in enumerate(chars))

print("characters to indices dictionary\n", char_indices)
indices_char = dict((i, c) for i, c in enumerate(chars))
```

```
print("indices to char dictionary\n", indices_char)
print('unique chars: ', {len(chars)})
```

In the code above, we are creating two dictionaries. Given below is the output from the above code.

```
characters to indices dictionary
 {' ': 0, '"': 1, '(': 2, '-': 3, '.': 4, '/': 5, '0': 6, '1': 7, '3': 8, '7':
9, '9': 10, 'a': 11, 'b': 12, 'c': 13, 'd': 14, 'e': 15, 'f': 16, 'g': 17,
'h': 18, 'i': 19, 'j': 20, 'k': 21, 'l': 22, 'm': 23, 'n': 24, 'o': 25, 'p':
26, 'q': 27, 'r': 28, 's': 29, 't': 30, 'u': 31, 'v': 32, 'w': 33, 'x': 34,
'y': 35, 'z': 36}
indices to char dictionary
 {0: ' ', 1: '"', 2: '(', 3: '-', 4: '.', 5: '/', 6: '0', 7: '1', 8: '3', 9:
'7', 10: '9', 11: 'a', 12: 'b', 13: 'c', 14: 'd', 15: 'e', 16: 'f', 17: 'g',
18: 'h', 19: 'i', 20: 'j', 21: 'k', 22: 'l', 23: 'm', 24: 'n', 25: 'o', 26:
'p', 27: 'q', 28: 'r', 29: 's', 30: 't', 31: 'u', 32: 'v', 33: 'w', 34: 'x',
35: 'y', 36: 'z'}
unique chars:  {37}
```

A quick verification will show that 'a' is mapped to 11 in the char_indices dictionary, and 11 is mapped to 'a' in the indices_char dictionary. The next step is to apply the char_indices dictionary on the full data and convert it from a sequence of characters into a sequence of numbers. We have removed the newline symbol from the data; we need to add space at the end of every line to compensate for it.

```
data = longseq_3gram1.splitlines()
##Adding a space at the end
data = [i+' ' for i in data]

##mapping our data into numbers
sentences = [[char_indices[j] for j in i] for i in data ]
print(data[0], sentences[0])
print(data[10], sentences[1])
print(data[20], sentences[2])
print(data[100], sentences[3])
print(data[400], sentences[400])
print(data[4000], sentences[4000])
print(data[9000], sentences[9000])
##Number of sentences
print("Number of sentences ", len(sentences))
```

The code above simply maps each character to a number by using the char_indices dictionary. This code also includes printing of a few examples. Given below is the result.

```
a bewildering array  [11, 0, 12, 15, 33, 19, 22, 14, 15, 28, 19, 24, 17, 0,
11, 28, 28, 11, 35, 0]
a celebration of  [11, 0, 12, 15, 24, 15, 16, 19, 13, 19, 11, 28, 35, 0, 25,
16, 0]
a co-director of  [11, 0, 12, 15, 33, 19, 22, 14, 15, 28, 19, 24, 17, 0, 32,
11, 28, 19, 15, 30, 35, 0]
a declaration of  [11, 0, 12, 19, 30, 30, 15, 28, 29, 33, 15, 15, 30, 0, 23,
25, 23, 15, 24, 30, 0]
a significant risk  [11, 0, 29, 19, 17, 24, 19, 16, 19, 13, 11, 24, 30, 0,
28, 19, 29, 21, 0]
been designed as  [12, 15, 15, 24, 0, 14, 15, 29, 19, 17, 24, 15, 14, 0, 11,
29, 0]
from anywhere on  [16, 28, 25, 23, 0, 11, 24, 35, 33, 18, 15, 28, 15, 0, 25,
24, 0]
Number of sentences  30307
```

From the above output, we can see the sequence of numbers corresponding to the sequence of characters. Every sentence will end with 0, which is nothing but white space. There are a total of 30,207 sentences. We now need to convert this data to RNN-friendly data. In this case study, we would like to take a sequence of 14 characters to predict the next character. Our RNN will have input sequence length 14 and predict one output at a time. For example, take the first sentence:

```
a bewildering array  [11, 0, 12, 15, 33, 19, 22, 14, 15, 28, 19, 24, 17, 0,
11, 28, 28, 11, 35, 0]
```

The above data point cannot be directly used in RNNs. We need to convert it into 14 inputs and one output format, as presented in Table 12.14.

TABLE 12.14   The Output Based on the Input Sequence of Length 14

Input Sequence of Length 14	Output
a bewildering [11, 0, 12, 15, 33, 19, 22, 14, 15, 28, 19, 24, 17, 0]	a [11]
bewildering a [0, 12, 15, 33, 19, 22, 14, 15, 28, 19, 24, 17, 0, 11]	r [28]
bewildering ar [12, 15, 33, 19, 22, 14, 15, 28, 19, 24, 17, 0, 11, 28]	r [28]
ewildering arr [15, 33, 19, 22, 14, 15, 28, 19, 24, 17, 0, 11, 28, 28]	a [11]
wildering arra [33, 19, 22, 14, 15, 28, 19, 24, 17, 0, 11, 28, 28, 11]	y [35]
ildering array [19, 22, 14, 15, 28, 19, 24, 17, 0, 11, 28, 28, 11, 35]	<space> []

One sentence of length 20 has been converted into six sentences with 14 input versus one output pairs. We need to repeat the same process for all the sentences using the following code.

```
Seq_ln = 14
X = []
y = []
for i in sentences:
    for j in range(len(i)-Seq_ln):
        X.append(i[j:j+Seq_ln])
        y.append(i[j+Seq_ln])
len(X), len(y)
```

The following is the output of the above code.

```
(142142, 142142)
```

From the above output, we can see that the number of sentences in the data has increased to 142,142 from the original 30,307. Each original sentence has almost created five new pairs of x and y. Given below is the example from the code.

```
print("data[0:2]=", data[0:2])
print("sentences[0:2]=", sentences[0:2])

for i in range (0,20):
    print("X[",i,"]=", X[i],"y[",i,"]=", y[i])
```

We are trying to print the first two sentences and the corresponding x and y value conversions using the code above. The following is the output of the above code.

```
data[0:2]= ['a bewildering array ', 'a beneficiary of ']
sentences[0:2]= [[11, 0, 12, 15, 33, 19, 22, 14, 15, 28, 19, 24, 17, 0, 11,
28, 28, 11, 35, 0], [11, 0, 12, 15, 24, 15, 16, 19, 13, 19, 11, 28, 35, 0,
25, 16, 0]]


X[ 0 ]= [11, 0, 12, 15, 33, 19, 22, 14, 15, 28, 19, 24, 17, 0]
y[ 0 ]= 11

X[ 1 ]= [0, 12, 15, 33, 19, 22, 14, 15, 28, 19, 24, 17, 0, 11]
y[ 1 ]= 28

X[ 2 ]= [12, 15, 33, 19, 22, 14, 15, 28, 19, 24, 17, 0, 11, 28]
y[ 2 ]= 28

X[ 3 ]= [15, 33, 19, 22, 14, 15, 28, 19, 24, 17, 0, 11, 28, 28]
y[ 3 ]= 11

X[ 4 ]= [33, 19, 22, 14, 15, 28, 19, 24, 17, 0, 11, 28, 28, 11]
y[ 4 ]= 35

X[ 5 ]= [19, 22, 14, 15, 28, 19, 24, 17, 0, 11, 28, 28, 11, 35]
y[ 5 ]= 0
..............................................................................
X[ 18 ]= [0, 12, 19, 30, 30, 15, 28, 29, 33, 15, 15, 30, 0, 23]
y[ 18 ]= 25

X[ 19 ]= [12, 19, 30, 30, 15, 28, 29, 33, 15, 15, 30, 0, 23, 25]
y[ 19 ]= 23
```

From the output above, we can see the x and y pairs. We are ready to build the model. We need to one-hot encode the data and build the RNN model. Given below is the code for the final steps of data processing.

```
X=np.array(X)
X1=np.reshape(X,(X.shape[0],X.shape[1],1))
X1=keras.utils.to_categorical(np.array(X1), num_classes=len(char_indices))
print(X1.shape)

y1 = np.array(y)
y1 = keras.utils.to_categorical(np.array(y), num_classes=len(char_indices))
y1.shape

from sklearn.model_selection import train_test_split

X_train, X_test, y_train, y_test = train_test_split(X1, y1, test_size=0.20)
print(X_train.shape)
print(y_train.shape)
print(X_test.shape)
print(y_test.shape)
```

In the code above, we have reshaped the values in x and y and one-hot encoded them. The RNN model expects the data in a particular format, and we need to reshape the data in that format. The following is the result.

```
X_train.shape (113713, 14, 37)
y_train.shape (113713, 37)
X_test.shape (28429, 14, 37)
y_test.shape (28429, 37)
```

We are now done with the data preprocessing; we will go ahead and build the RNN model.

12.5.1.3 Model Building While building the RNN model, we need to mention the time steps and the number of hidden nodes. Given below is the code for building the model.

```
model_RNN2 = Sequential()
##model.add(SimpleRNN('number of hidden nodes in each rnn cell', input_
shape=(timesteps, data_dim)))
model_RNN2.add(SimpleRNN(16, input_shape=(X_train.shape[1], X_train.shape[2])))
model_RNN2.add(Dense(len(char_indices)))
model_RNN2.add(Activation('softmax'))
model_RNN2.summary()
```

From the above code, we can see that we are building the model with 16 hidden nodes. Table 12.15 shows the output of this code.

TABLE 12.15 Model Summary

```
Layer (type)                    Output Shape              Param #
=================================================================
simple_rnn_4 (SimpleRNN)        (None, 16)                864

dense_9 (Dense)                 (None, 37)                629

activation_1 (Activation)       (None, 37)                0
=================================================================
Total params: 1,493
Trainable params: 1,493
Non-trainable params: 0
```

We will compile the model and train it.

```
model_RNN2.compile(loss='categorical_crossentropy', optimizer='adam', metrics=
['accuracy'])
model_RNN2.fit(X_train, y_train, epochs=30, verbose=1, validation_data=
(X_test, y_test))
model_RNN2.save_weights("char_rnn_model_weights_v1.hdf5")
```

We are training the model for 30 epochs and saving it in the weights file. Given below is the output from the above code.

```
Train on 113713 samples, validate on 28429 samples
Epoch 1/30
113713/113713 [====] - 18s 162us/sample - loss: 2.2410 - accuracy: 0.3640 -
val_loss: 1.9614 - val_accuracy: 0.4336
Epoch 2/30
113713/113713 [====] - 20s 177us/sample - loss: 1.9063 - accuracy: 0.4417 -
val_loss: 1.8585 - val_accuracy: 0.4537
Epoch 3/30
113713/113713 [====] - 17s 148us/sample - loss: 1.8317 - accuracy: 0.4590 -
val_loss: 1.8051 - val_accuracy: 0.4646
Epoch 4/30
113713/113713 [====] - 17s 148us/sample - loss: 1.7849 - accuracy: 0.4691 -
val_loss: 1.7657 - val_accuracy: 0.4772
Epoch 5/30
113713/113713 [====] - 17s 147us/sample - loss: 1.7547 - accuracy: 0.4740 -
val_loss: 1.7435 - val_accuracy: 0.4801
Epoch 6/30
113713/113713 [====] - 18s 162us/sample - loss: 1.7343 - accuracy: 0.4809 -
val_loss: 1.7272 - val_accuracy: 0.4838
```

```
Epoch 7/30
113713/113713 [====] - 17s 149us/sample - loss: 1.7194 - accuracy: 0.4851 -
val_loss: 1.7180 - val_accuracy: 0.4873
        =========================================================
Epoch 25/30
113713/113713 [====] - 17s 153us/sample - loss: 1.6448 - accuracy: 0.5093 -
val_loss: 1.6531 - val_accuracy: 0.5123
Epoch 26/30
113713/113713 [====] - 17s 152us/sample - loss: 1.6433 - accuracy: 0.5096 -
val_loss: 1.6520 - val_accuracy: 0.4987
Epoch 27/30
113713/113713 [====] - 17s 148us/sample - loss: 1.6422 - accuracy: 0.5108 -
val_loss: 1.6463 - val_accuracy: 0.5101
Epoch 28/30
113713/113713 [====] - 18s 156us/sample - loss: 1.6404 - accuracy: 0.5112 -
val_loss: 1.6424 - val_accuracy: 0.5103
Epoch 29/30
113713/113713 [====] - 17s 151us/sample - loss: 1.6392 - accuracy: 0.5107 -
val_loss: 1.6472 - val_accuracy: 0.5026
Epoch 30/30
113713/113713 [====] - 18s 156us/sample - loss: 1.6381 - accuracy: 0.5114 -
val_loss: 1.6422 - val_accuracy: 0.5080
```

We can see from the above output that the model is not improving after reaching 51 percent accuracy. The model will not show any improvement even after we train it for 10 more epochs. Given below is the code for additional epochs using the weights file.

```
weightsfile_model_RNN2= "char_rnn_model_weights_v1.hdf5"
model_RNN2.load_weights(weightsfile_model_RNN2)

## compile network
model_RNN2.compile(loss='categorical_crossentropy', optimizer='adam', metrics=
['accuracy'])
## fit network
model_RNN2.fit(X_train, y_train, epochs=10, verbose=1)
```

The following is the output of the above code.

```
Train on 113713 samples
Epoch 1/10
113713/113713 [====] - 18s 159us/sample - loss: 1.6371 - accuracy: 0.5116
Epoch 2/10
113713/113713 [====] - 16s 139us/sample - loss: 1.6357 - accuracy: 0.5127
Epoch 3/10
113713/113713 [====] - 17s 153us/sample - loss: 1.6348 - accuracy: 0.5127
Epoch 4/10
113713/113713 [====] - 17s 153us/sample - loss: 1.6341 - accuracy: 0.5120
Epoch 5/10
113713/113713 [====] - 17s 147us/sample - loss: 1.6326 - accuracy: 0.5134
Epoch 6/10
113713/113713 [====] - 19s 167us/sample - loss: 1.6323 - accuracy: 0.5130
Epoch 7/10
113713/113713 [====] - 20s 177us/sample - loss: 1.6317 - accuracy: 0.5129
Epoch 8/10
113713/113713 [====] - 19s 165us/sample - loss: 1.6303 - accuracy: 0.5138
Epoch 9/10
113713/113713 [====] - 16s 145us/sample - loss: 1.6305 - accuracy: 0.5140
Epoch 10/10
113713/113713 [====] - 16s 139us/sample - loss: 1.6296 - accuracy: 0.5131
```

We will now use this model for predictions in the following section.

12.5.1.4 Prediction We need to write the predict function that takes input as characters and converts them to numbers. Use these indices to get the predictions, and finally convert them to characters to get the output. We are going to write a predict function that will predict not just one character but a sequence of characters that will form a word. The prediction loop will continue until it hits a space, which marks the completion of the word. Given below is the predict function.

```python
##function to prepare test input
def prepare_input(in_text):
    X1 = np.array([char_indices[i] for i in in_text]).reshape(1,14,1)
    X1=keras.utils.to_categorical(np.array(X1), num_classes=len(char_indices))
    return(X1)
##function to loop our predictions
def complete_pred(in_text):
    #original_text = in_text
    #generated = in_text
    completion = ''
    while True:
        x = prepare_input(in_text)
        pred = model_RNN2.predict_classes(x, verbose=0)[0]

        next_char = indices_char[pred]

        in_text = in_text[1:] + next_char
        completion += next_char

        if len(completion)> 20 or next_char == ' ':
            return completion
```

In the code above, we can see two functions. One is for the usual character to index conversion and another is for predicting the sequence of characters. We will now use the above function for predictions.

```python
in_text = 'officials say '
out_word = complete_pred(in_text)
print("Input text -->", in_text, "\npredicted word ---> ", out_word)
in_text = 'how dangerous '
out_word = complete_pred(in_text)
print("Input text -->", in_text, "\npredicted output ---> ", out_word)
in_text = 'political and '
out_word = complete_pred(in_text)
print("Input text -->", in_text, "\npredicted output ---> ", out_word)
in_text = 'whatever they '
out_word = complete_pred(in_text)
print("Input text -->", in_text, "\npredicted output ---> ", out_word)
in_text = 'of particular '
out_word = complete_pred(in_text)
print("Input text -->", in_text, "\npredicted output ---> ", out_word)
```

Given below is the output of the above code.

```
Input text --> officials say
predicted word --->  to

Input text --> how dangerous
predicted output --->  of

Input text --> political and
predicted output --->  the
```

```
Input text --> whatever they
predicted output ---> the

Input text --> of particular
predicted output --->  to
```

From the above output, we can see that almost all the predictions are simple words—such as to, of, the, etc. These words are the most frequent in the data; they are known as stop words. RNN is merely predicting the stop words for all inputs. RNN has failed to model this data. The reason is that the length of the sequence is 14. In practice, RNN models usually fail when the sequence length is more than 10. What is the reason for this failure? Let us discuss it in the next section.

12.5.2   Problem of Vanishing Gradients

RNN models in practice fail to learn long sequences; the reason is vanishing gradients. We have discussed the vanishing gradients issue in the case of deep neural networks with multiple hidden layers. The weight changes are made based on gradient values. In the case of deep neural networks, if the gradients are already small fractions near the output layer, then, during backpropagation, we multiply these small fractions with even smaller numbers. This multiplication will result in almost negligible values of gradients at the initial layers. There will be no improvement in the weights due to the negligible values of gradients. This issue is known as vanishing gradients (Fig. 12.16).

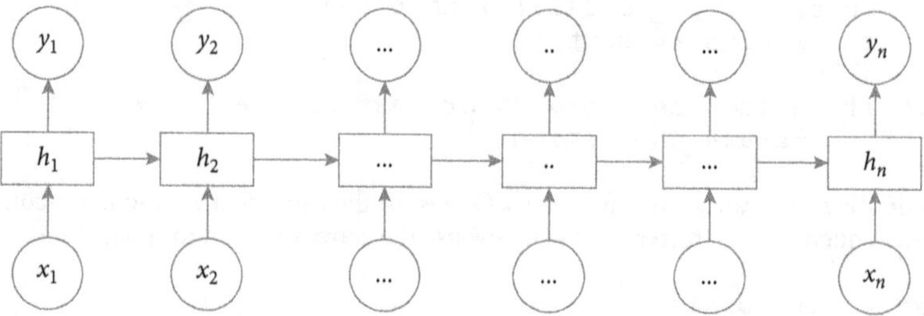

FIGURE 12.16   A long-chain standard RNN.

In the diagram in Fig. 12.16, standard RNNs might struggle to predict the value of y_n based on the value of x_1 if there is a long distance between them. The derivative of the sigmoid is zero on both ends (when sigmoid is near to zero or one). If the values of gradients at a later time are already low, with small values in the matrix and several matrix multiplications (from timestamp n to 1), the gradient values shrink exponentially fast, eventually vanishing totally after a few time steps. The gradient contributions from "far away" time steps become zero, and the value at those steps does not contribute to the model's learning. Finally, the model captures short-range dependencies only. In practice, RNN faces difficulty in capturing the dependencies beyond 10 steps. However, theoretically, this number 10 is not proven. It has been observed while working with practical problems. Table 12.16 presents the mathematical derivation behind the vanishing gradient issues.

TABLE 12.16 The Mathematical Seed Behind the Vanishing Gradients Issue

$h_t = g(Ux_t + Wh_{t-1})$	Activations and feedforward
$h_t = g(Wh_{t-1})$	Let us focus only on h for now. Ignore x for sometime
$\dfrac{dh_t}{dh_{t-1}} \propto W$	Since h_t is a function of h_{t-1}
$\dfrac{dh_t}{dh_0} \propto W^t$	Chain rule
$\dfrac{dE}{dh_0} = \dfrac{dE}{dh_t}\dfrac{dh_t}{dh_0}$	E is the final error function
$\dfrac{dE}{dh_0} = W^t\dfrac{dE}{dh_t}$	Since $\dfrac{dh_t}{dh_0} \propto W^t$
$\dfrac{dE}{dh_0} = W^{t-1}\lambda\dfrac{dE}{dh_t}$	$W\dfrac{dE}{dh_t} = \lambda\dfrac{dE}{dh_t}$ where λ is the eigenvalue of matrix W

$$\frac{dE}{dh_0} = W^{t-1}\lambda\frac{dE}{dh_t} = W^{t-2}\lambda^2\frac{dE}{dh_t} = W^{t-3}\lambda^3\frac{dE}{dh_t} = \ldots W^{t-k}\lambda^k\frac{dE}{dh_t} \ldots = \lambda^t\frac{dE}{dh_t}$$

If time steps t have high value, and if $\lambda < 1$ then gradients would vanish; otherwise, they would explode

Is there a solution to this vanishing gradients issue? Can we make some adjustments to the standard RNN to make it work even on long sequences? Let us see some solutions in the upcoming sections.

12.6 LONG SHORT-TERM MEMORY

Long short-term memory (LSTM) models are created by making several modifications to standard RNN models. RNN models do not have long-term memory; we add some features to make it remember long-term dependencies. Figure 12.17 shows the standard RNN network diagram.

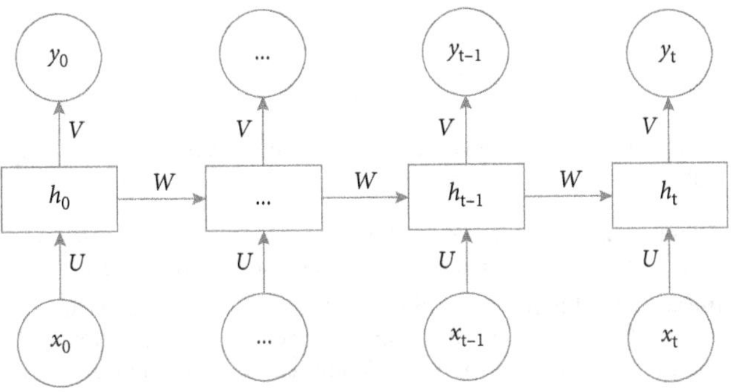

FIGURE 12.17 The standard RNN network diagram.

In the above RNN model, we have a hidden state in the middle. The hidden state at time t depends on the hidden state at time $t-1$, and it further depends on $t-2$ and so on. We will add one more state called cell state to the standard RNN.

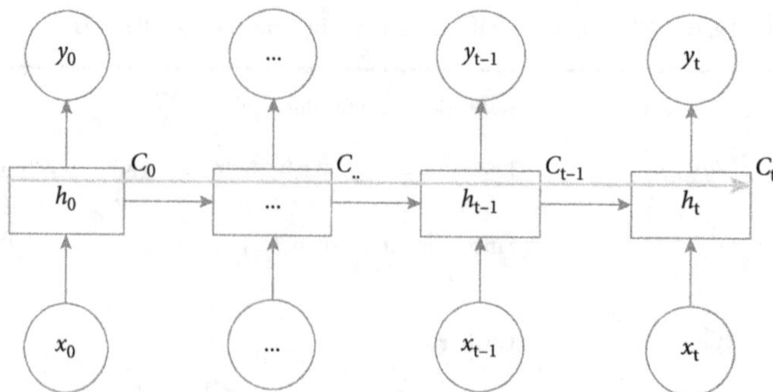

FIGURE 12.18   RNN with cell state.

Until now, the information in the network was stored and carried by the hidden state only. Now, we are introducing one more information storage item called "cell state." This cell state is like a conveyer belt that runs through the network. Like the way the hidden state is a set of hidden nodes, we can imagine a cell state as a set of nodes that are similar to hidden nodes but carry different information. While predicting the values, both the information in the cell state and hidden state are used. We can call the hidden state the short-term memory state and the cell state the long-term memory state. We can redraw the diagram in Fig. 12.18 by highlighting these two states.

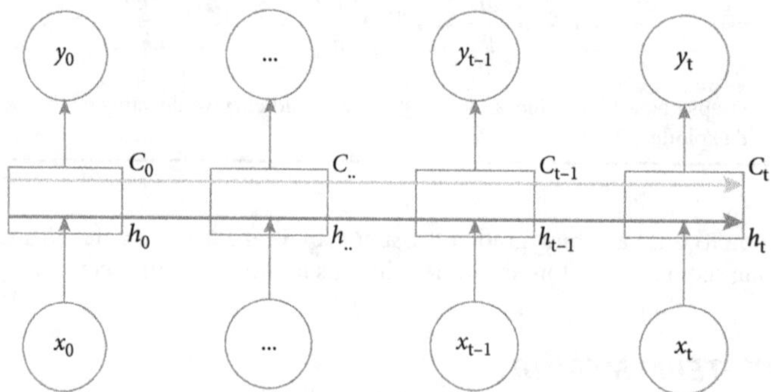

FIGURE 12.19   Cell state and hidden state.

In the diagram in Fig. 12.19, hidden state h_t corresponds to short-term memory (STM). The cell state C_t corresponds to long-term memory (LTM). Both of them put together will become long short-term memory (LSTM). We have already talked about h_t; it is just a hidden layer in standard ANN. We will now discuss the long-term memory component.

12.6.1   LSTM Gates

The cell state keeps track of several pieces of information, and it is much more complicated than the hidden state. Let us take one LSTM cell. LSTM is built on three major principles.

1. While building the sequential models, we need not carry forward all the information to the next time steps. We can forget some information if it is unnecessary. Of course, that depends on the training data.

2. While building the sequential models, we need not take all the information at each time step into consideration. We can limit the amount of information that is being written into the memory cell. Again, whether to write the full information into the memory cell or not is decided based on historical training data.

3. While building sequential models, we need not output full information from the memory cell at every time step. Depending on the time step, we may want to limit how much amount of data can be read from the memory cell. Once again, how much information can be read from a particular memory cell is decided based on historical training data.

To incorporate these three points, we add a few more nodes in the memory cell. There are already hidden nodes and input nodes in each cell. We incorporate the above three points in each cell. The first point is about forgetting or retaining the information from the previous cell.

12.6.1.1 Forget Gate The initial idea is about forgetting and retaining the information from the previous cell state. Based on the current input, we need to decide how much information should be carried forward from the previous cell state to this cell state.

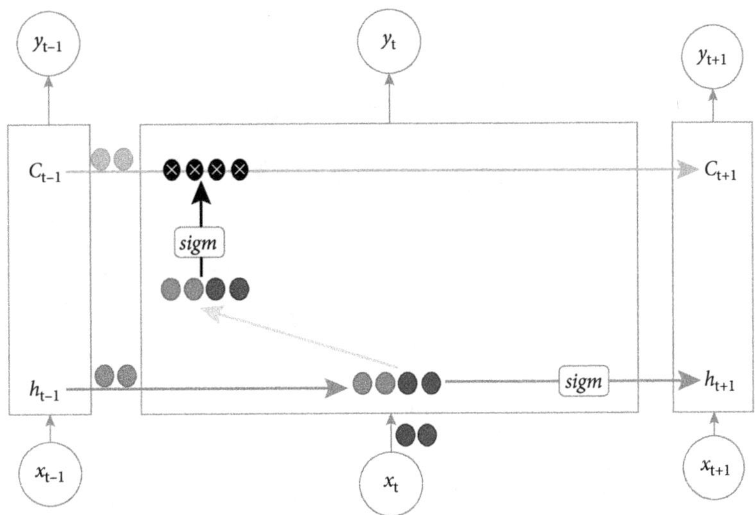

FIGURE 12.20 Forget gate.

In the diagram in Fig. 12.20, we have appended the x_t and h_{t-1} nodes, as usual, that have been passed on to the next time step hidden nodes. This step takes care of the short-term memory part. We now take a copy of x_t and h_{t-1} nodes and assign new weights to them. Let us call these nodes set-1 nodes. After applying a sigmoid function on these, we multiply these nodes with the nodes in the cell state. There is no squashing activation (like sigmoid or tanh) on cell state; that is the key for information flow across many time steps. By multiplying the cell state with this sigmoid of set-1 weights, we will be erasing or retaining some information from the cell state. This new set of weights will help us to "forget some information or not." Set-1 node weights determine the amount of information to forget or retain, and weights are determined based on training data. Multiplying cell state with set-1 nodes sigmoid results is known as a forget gate. These weights associated with set-1 nodes are known as forget gate weights. There are other gates too. We need to note that the forget gate is applied first, and the current cell state is calculated based on the forget and the other gates (Table 12.17).

TABLE 12.17 Forget Gate Equation

$f = \sigma(x_t U^f + h_{t-1} W^f)$	Weights associated with forget gate
$C_t = C_{t-1} \circ f$	Updating the cell state
$h_t = \sigma(x_t U + h_{t-1} W)$	Regular weights U and W

12.6.1.2 Input Gate The second point is about writing a limited amount of information into the cell state. We may not need to write the full information in the current cell state. Based on the context, we can input the full information or a part of it. This input gate controls write accesses to memory cells. A sigmoid on the input keeps control of input. If the sigmoid at the input gate is 0, then write access is denied, and in case of 1, write access is granted. This gate is also known as the write gate. This gate decides how much information needs to be written into the cell state from the current time step. This input gate controls the information flow into the cell state. There is one more question that we need to answer here. What information should be written in the cell state? We also need to supply "what" information needs to be updated into the cell state. The forget gate is applied to the cell state. The input gate is applied to input information. Input for the current cell is x_t and h_{t-1}. The input gate is applied to this current input. Later it will be added/written/appended to the cell state.

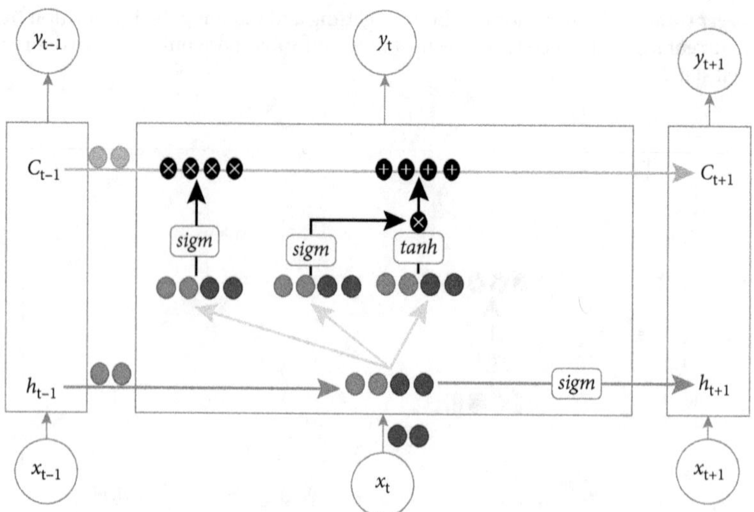

FIGURE 12.21   An input gate.

In the diagram in Fig. 12.21, the input gate controls the amount of information to write. It has been multiplied to the current input and preceding hidden state values. Later the result will be appended to the cell state. Unlike the forget gate, the result of the input gate will be added to the cell state (Table 12.18).

TABLE 12.18   Input Gate and Forget Gate Equations

$f = \sigma(x_t U^f + h_{t-1} W^f)$	Weights associated with forget gate
$C_t = C_{t-1} \circ f$	Updating the cell state
$i = \sigma(x_t U^i + h_{t-1} W^i)$	Weights associated with the input gate
$g = \tanh(x_t U^g + h_{t-1} W^g)$	Weights associated with current input updater
$C_t = C_{t-1} \circ f + g \circ i$	Write current input into cell state
$h_t = \sigma(x_t U + h_{t-1} W)$	Regular weights U and W

By now, we have three sets of weights: forget gate weights (U^f, W^f), input gate weights (U^i, W^i), and input information weights (U^g, W^g); weights will be four times the regular RNN.

12.6.1.3   *Output Gate*   The third significant idea in LSTM is that "we need not output all the information from this hidden state to the next hidden state." Based on the cell state, we can decide how much information goes forward. We can add an output gate that will get multiplied to the hidden state output that is entering the next time step.

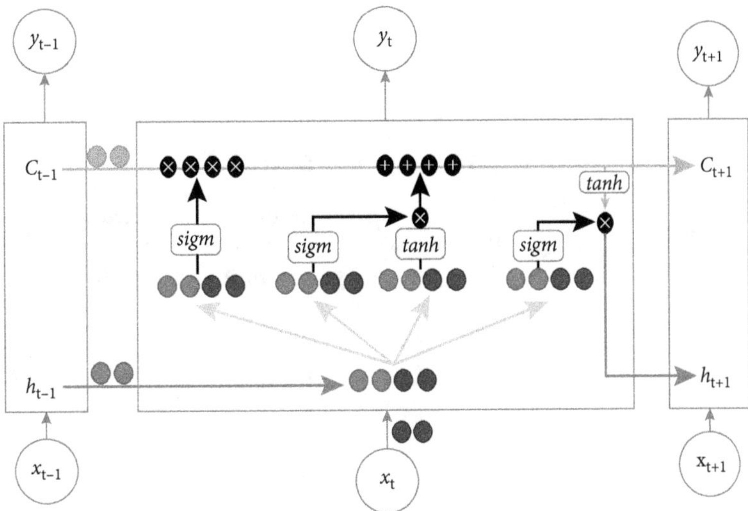

FIGURE 12.22 An output gate.

In the diagram in Fig. 12.22, we can see that the final output going out of the cell is multiplied with the cell state and sent to the next memory cell. A copy of this output is sent to the current output y_t.

TABLE 12.19 LSTM Equations

$f = \sigma(x_t U^f + h_{t-1} W^f)$	Weights associated with forget gate
$C_t = C_{t-1} \circ f$	Updating the cell state
$i = \sigma(x_t U^i + h_{t-1} W^i)$	Weights associated with the input gate
$g = \tanh(x_t U^g + h_{t-1} W^g)$	Weights associated with current information updater
$C_t = C_{t-1} \circ f + g \circ i$	Write current input into cell state
$O = \sigma(x_t U^o + h_{t-1} W^o)$	Weights associated with output gate
$h_t = \tanh(C_t) \circ O$	The final output from the memory cell

We can see from the LSTM equations in Table 12.19 that the weights in the LSTM cell are four times that of standard RNN. It is important to note that there is sigmoid or tanh squashing on all forms of the hidden and input nodes, but there is no sigmoid on the cell state. The cell state directly enters the next cell. We can say that this cell state has linear activation with weights $= 1$. It will make sure that there is no vanishing gradient problem. It is also known as the constant error carousel (CEC). When tried on practical problems, LSTM models can learn sequences of 1000 time steps.

LSTM is an innovative idea, and the architecture is a bit complex. It takes some time to understand these gates and their functionalities. The final diagram for the LSTM memory cell is shown in Fig. 12.23.

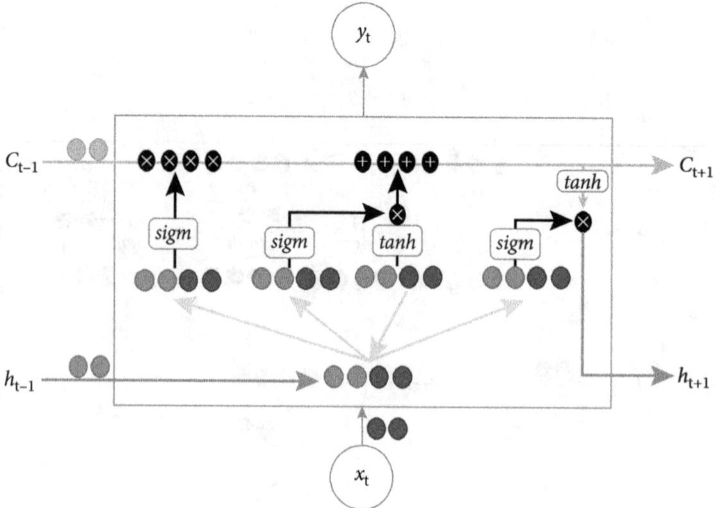

FIGURE 12.23 LSTM memory cell.

There are multiple representations of this LSTM cell. Figure 12.24 shows a few more representations.

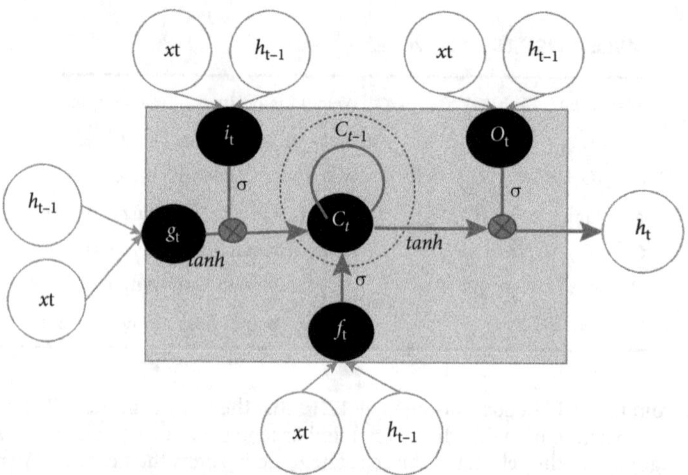

FIGURE 12.24 LSTM memory cell—different representation.

Figure 12.25 shows another representation of the LSTM network.

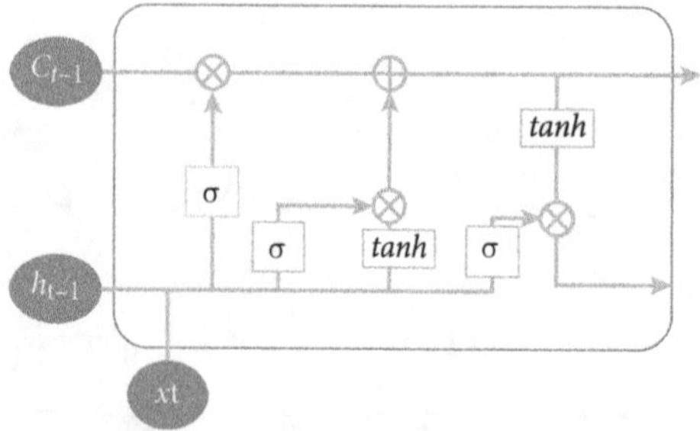

FIGURE 12.25 LTSM network signified differently.

12.6.2   LSTM Intuition

Consider the sentence given below that starts with Jim. We need to predict the next word in the sequence. We can also see a few training data samples below the sentence.

```
Predict next word
```
  - `Jim is a software engineer. He works for an IT company. Lynda is a teacher.`

```
Training data examples
```
  - `"Lynda was late that day. She apologized."`
  - `"Lynda's alarm goes off at 5 am. She gets up early."`
  - `"Jim told Lynda - 'you have such beautiful eyes.' Lynda smiled at him. She continued to walk."`
  - `"Jim is a software engineer. He works for an IT company. Lynda is a teacher. She teaches in a school."`
  - `"Lynda likes exploring new cities. She traveled to Paris last month."`
  - `"Lynda got a promotion last month. She got a good pay hike."`

For ease of understanding, let us imagine the cell state as a box that contains the current subject of the sentence like Jim or Lynda. We have to predict the next word using LSTM. Let us see how LSTM works on this data.

Forget gate: For predicting this word, first, we need to forget the subject "Jim." The forget gate takes care of erasing the subject "Jim." How can we dictate the forget gate to precisely erase the subject? Well, it is done based on training data. If there are many such examples in the training data that are supporting this type of change in the subject, then forget gate weights will be trained on those examples. This type of training data will make sure that the forget gate erases the subject when it reaches the second sentence. By the way, the forget gate is also known as the keep gate.

Input gate: After erasing Jim, we need to write a new subject Lynda into the cell memory. The input gate takes care of writing a new subject into the cell state. Again, how do we make sure that the input gate writes this information only? Why does it not erase the information like forget gate? The formulation in LSTM is such that the input gate will only write information. It cannot erase any information from the cell sate. We can observe the final operation between the cell state and the input gate; it is addition. The forget gate operation was multiplication. The input gate will add new information. If there is no information, then nothing will be added. Based on the input data, during the training process, the model will decide the input gate weights, and these weights will eventually decide whether to add the new subject, Lynda, to the cell state or not. The input gate is also known as the write gate.

Output gate: After erasing Jim from the cell state, we added Lynda to it (the cell state). Now, what to output? Is it "her" or "she" or "Lynda"? It should be again decided based on historical data. We need to add one more gate that decides what information should be given as output. The output gate takes care of what exactly needs to be output based on the current input and the subject in the cell state. The output gate is also known as the read gate.

All these three gates together will predict the next word as "She." However, this prediction entirely depends on the training data. There is no guarantee that the prediction will always be "She." For a different dataset, the prediction can be "Her." Using the LSTM memory cell, we have created enough options for this model to work accurately and predict accurately.

Cell state: The subject Lynda automatically enters the next memory cell, unless it is erased by the forget gate. This direct entry helps the LSTM model keep track of the long-term dependencies.

While understanding LSTMs, one question that always comes to our mind is—how does a forget gate carry out a forget task? How does an input gate do an input task? Furthermore, how does an output gate operate only on outputs? The answers will be apparent if we look at the formulas of LSTM (Table 12.20).

TABLE 12.20   The Formulas of LSTM

$f = \sigma(x_t U^f + h_{t-1} W^f)$	Weights associated with forget gate
$C_t = C_{t-1} \circ f$	Forget gate updates the previous cell state. It is like multiplying with a number between [0 to 1]. If the forget gate value is near to 0, then the cell will be erasing the information. Forget gate is applied to the cell state.
$i = \sigma(x_t U^i + h_{t-1} W^i)$	Weights associated with the input gate
$g = \tanh(x_t U^g + h_{t-1} W^g)$	Weights associated with current information updater
$C_t = C_{t-1} \circ f + g \circ i$	Information to append. The input gate is applied to the current cell input. The input gate result is later added to the cell state. This gate cannot do the job of forgetting.
$O = \sigma(x_t U^o + h_{t-1} W^o)$	Weights associated with output gate
$h_t = \tanh(C_t) \circ O$	The output gate is applied to the final cell state and current input. The output gate cannot write anything into the cell state. Using this formula, we can only read information from the cell.

12.6.3 LSTM Case Study

In our last case study on predicting the next characters, we have used the standard RNN model, but it has failed. We will now use the LSTM model on the same data. Given below is the code building the LTSM model.

```
model_LSTM = Sequential()
model_LSTM.add(LSTM(128, input_shape=(X_train.shape[1], X_train.shape[2])))
model_LSTM.add(Dense(len(char_indices)))
model_LSTM.add(Activation('softmax'))
model_LSTM.summary()
```

Table 12.21 shows the results.

TABLE 12.21   The Code Output

```
Layer (type)                    Output Shape                   Param #
=================================================================
lstm_2 (LSTM)                   (None, 128)                    84992

dense_21 (Dense)                (None, 37)                     4773

activation_5 (Activation)       (None, 37)                     0
=================================================================
Total params: 89,765
Trainable params: 89,765
Non-trainable params: 0
```

We will now compile and train the model.

```
model_LSTM.compile(loss='categorical_crossentropy',        optimizer='adam',
metrics=['accuracy'])
model_LSTM.fit(X_train, y_train, epochs=30, verbose=1)
model_LSTM.save_weights("char_LSTM_model_weights_v1.hdf5")
```

The following is the output of the above code.

```
Train on 113713 samples
Epoch 1/30
113713/113713 [====] - 46s 404us/sample - loss: 0.6394 - accuracy: 0.7976
Epoch 2/30
113713/113713 [====] - 45s 397us/sample - loss: 0.6268 - accuracy: 0.8018
```

```
Epoch 3/30
113713/113713 [====] - 45s 396us/sample - loss: 0.6201 - accuracy: 0.8030
Epoch 4/30
113713/113713 [====] - 45s 400us/sample - loss: 0.6136 - accuracy: 0.8047
Epoch 5/30
113713/113713 [====] - 44s 388us/sample - loss: 0.6062 - accuracy: 0.8063
Epoch 6/30
113713/113713 [====] - 46s 402us/sample - loss: 0.5993 - accuracy: 0.8089
Epoch 7/30
113713/113713 [====] - 46s 401us/sample - loss: 0.5944 - accuracy: 0.8091
Epoch 8/30
113713/113713 [====] - 46s 406us/sample - loss: 0.5897 - accuracy: 0.8118
Epoch 9/30
113713/113713 [====] - 44s 388us/sample - loss: 0.5845 - accuracy: 0.8115
Epoch 10/30
113713/113713 [====] - 43s 377us/sample - loss: 0.5785 - accuracy: 0.8145
Epoch 11/30
113713/113713 [====] - 43s 374us/sample - loss: 0.5753 - accuracy: 0.8146
Epoch 12/30
113713/113713 [====] - 43s 380us/sample - loss: 0.5713 - accuracy: 0.8163
Epoch 13/30
113713/113713 [====] - 45s 396us/sample - loss: 0.5684 - accuracy: 0.8173
Epoch 14/30
113713/113713 [====] - 44s 390us/sample - loss: 0.5652 - accuracy: 0.8176
Epoch 15/30
113713/113713 [====] - 42s 366us/sample - loss: 0.5598 - accuracy: 0.8199
Epoch 16/30
113713/113713 [====] - 42s 374us/sample - loss: 0.5588 - accuracy: 0.8189
Epoch 17/30
113713/113713 [====] - 45s 396us/sample - loss: 0.5526 - accuracy: 0.8205
Epoch 18/30
113713/113713 [====] - 42s 367us/sample - loss: 0.5531 - accuracy: 0.8207-
Epoch 19/30
113713/113713 [====] - 42s 372us/sample - loss: 0.5486 - accuracy: 0.8216
Epoch 20/30
113713/113713 [====] - 42s 370us/sample - loss: 0.5453 - accuracy: 0.8232
Epoch 21/30
113713/113713 [====] - 44s 390us/sample - loss: 0.5416 - accuracy: 0.8235
Epoch 22/30
113713/113713 [====] - 42s 370us/sample - loss: 0.5426 - accuracy: 0.8222
Epoch 23/30
113713/113713 [====] - 41s 364us/sample - loss: 0.5363 - accuracy: 0.8254
Epoch 24/30
113713/113713 [====] - 42s 368us/sample - loss: 0.5358 - accuracy: 0.8254
Epoch 25/30
113713/113713 [====] - 42s 365us/sample - loss: 0.5338 - accuracy: 0.8266-
Epoch 26/30
113713/113713 [====] - 42s 366us/sample - loss: 0.5322 - accuracy: 0.8255
Epoch 27/30
113713/113713 [====] - 43s 379us/sample - loss: 0.5306 - accuracy: 0.8262
Epoch 28/30
113713/113713 [====] - 41s 365us/sample - loss: 0.5266 - accuracy: 0.8275
Epoch 29/30
113713/113713 [====] - 42s 366us/sample - loss: 0.5269 - accuracy: 0.8275
Epoch 30/30
113713/113713 [====] - 44s 384us/sample - loss: 0.5250 - accuracy: 0.8274
```

This model shows a much better accuracy of 80 percent on the same dataset; RNN had given us only 50 percent accuracy. We will now use this model for prediction. Our prediction is made by predicting one character at a time and continuing the predictions until we see space. These sequences of characters will be formed as the predicted word.

```python
#function to prepare test input
def prepare_input1(in_text):
    X1 = np.array([char_indices[i] for i in in_text]).reshape(1,14,1)
    X1= keras.utils.to_categorical(np.array(X1), num_classes=len(char_indices))
    return(X1)
#function to loop our preditions
def complete_pred1(in_text):
    #original_text = in_text
    #generated = in_text
    completion = ''
    while True:
        x = prepare_input1(in_text)
        pred = model_LSTM.predict_classes(x, verbose=0)[0]
        next_char = indices_char[pred]

        in_text = in_text[1:] + next_char
        completion += next_char

        if len(completion)> 20 or next_char == ' ':
            return completion
```

We will use the above function to predict a few test points.

```python
in_text = 'the emergence '
out_word = complete_pred1(in_text)
print("Input text -->", in_text, "; predicted output ---> ", out_word)

in_text = 'officials say '
out_word = complete_pred1(in_text)
print("Input text -->", in_text, "; predicted output ---> ", out_word)

in_text = 'and sentenced '
out_word = complete_pred1(in_text)
print("Input text -->", in_text, "; predicted output ---> ", out_word)

in_text = 'a combination '
out_word = complete_pred1(in_text)
print("Input text -->", in_text, "; predicted output ---> ", out_word)

in_text = 'and according '
out_word = complete_pred1(in_text)
print("Input text -->", in_text, "; predicted output ---> ", out_word)
```

The above code gives us the below output.

```
Input text --> the emergence   ; predicted output --->   of
Input text --> officials say   ; predicted output --->   they
Input text --> and sentenced   ; predicted output --->   to
Input text --> a combination   ; predicted output --->   of
Input text --> and according   ; predicted output --->   to
```

We will take a few test cases and get their RNN and LSTM model predictions. It will help us in comparing their performance. The code below gets the predictions using RNN and LSTM.

```python
in_text = 'how dangerous '
out_word = complete_pred1(in_text)
print("Input text -->", in_text, "\nLSTM Prediction ---> ", out_word)
out_word1 = complete_pred(in_text)
print("RNN Prediction ---> ", out_word1)
```

```python
print("\n")
in_text = 'political and '
out_word = complete_pred1(in_text)
print("Input text -->", in_text, "\nLSTM Prediction ---> ", out_word)
out_word1 = complete_pred(in_text)
print("RNN Prediction ---> ", out_word1)


print("\n")
in_text = 'of particular '
out_word = complete_pred1(in_text)
print("Input text -->", in_text, "\nLSTM Prediction ---> ", out_word)
out_word1 = complete_pred(in_text)
print("RNN Prediction ---> ", out_word1)


print("\n")
in_text = 'whatever they '
out_word = complete_pred1(in_text)
print("Input text -->", in_text, "\nLSTM Prediction ---> ", out_word)
out_word1 = complete_pred(in_text)
print("RNN Prediction ---> ", out_word1)
```

The following is the output of the above code.

```
Input text --> how dangerous
LSTM Prediction --->  is
RNN Prediction --->  of


Input text --> political and
LSTM Prediction --->  economic
RNN Prediction --->  the


Input text --> of particular
LSTM Prediction --->  interest
RNN Prediction --->  to


Input text --> whatever they
LSTM Prediction --->  can
RNN Prediction --->  the
```

We can see from the results that the RNN predictions are generic words, whereas LSTM is predicting more related words. We will now discuss a few other applications of LSTM.

12.7   SEQUENCE TO SEQUENCE MODELS

The case study in the previous section on LSTM was predicting one letter while taking the input from a sequence of characters. It is known as *many to one*. Though we are predicting the sequence, we did it by predicting one character at a time and later arranged them in a sequence. It is still considered a many to one model, i.e., many inputs to predict one output (Fig. 12.26).

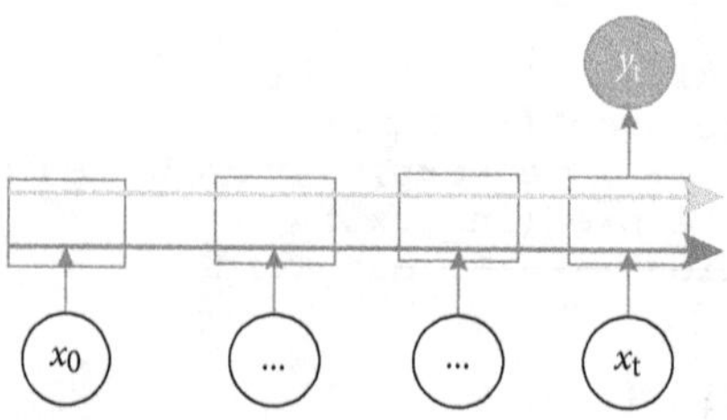

FIGURE 12.26 Many to one model.

LSTM models are compelling. We will see one powerful application of LSTM. It is known as machine translation or, in simple words, a model that translates the language. In this case, the input will be a sequence of words in input language, and output will be a sequence of words in the target language. Many inputs are used in predicting many outputs; it is known as many to many models; in the later sections, we will take up a case study.

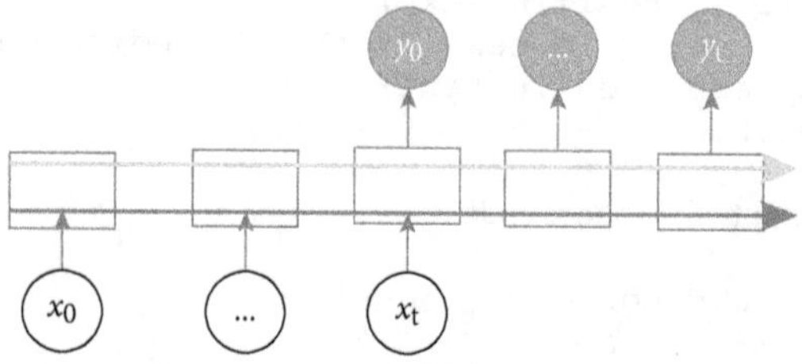

FIGURE 12.27 Many to many model.

In all the text data examples, we have used one-hot encoding of the words or characters until now. Text data–related applications belong to the separate field called natural language processing (NLP). There is one crucial topic that we need to know in NLP that is Word2vec.

12.7.1 Word2vec

We have discussed many machine learning and deep learning algorithms so far; none of them can work on text data directly. We need to convert the text data to numerical data and apply our models on the numerical data. We did one-hot encoding to the text data. One-hot encoding worked well until this point. When we talk about sentence translation, we need a better way to convert this text into numbers. We need to preserve the context somehow while converting the numbers into words. One-hot encoding kills the context and relation between the words.

12.7.1.1 Issue with One-Hot Encoding If we take any pair of one-hot encoded words, there is no similarity between them. The one-hot encoded vectors are always orthogonal to each other. Let us take the following text to understand the issue.

Training data:
- 'king is a strong man',
- 'queen is a wise woman',
- 'boy is a young man',
- 'girl is a young woman',
- 'prince is a young',
- 'prince will be strong',
- 'princess is young',
- 'man is strong',
- 'woman is pretty',
- 'prince is a boy',
- 'prince will be king',
- 'princess is a girl',
- 'princess will be queen'

If we follow the traditional one-hot encoding method, then we will find all the unique words and convert each word to a vector of a length equal to the total number of unique words. Table 12.22 is the one-hot encoding table.

TABLE 12.22   The One-Hot Encoding Table

young	man	king	woman	she	strong	prince	girl	wise	princess	pretty	he	boy	queen
1	0	0	0	0	0	0	0	0	0	0	0	0	0
0	1	0	0	0	0	0	0	0	0	0	0	0	0
0	0	1	0	0	0	0	0	0	0	0	0	0	0
0	0	0	1	0	0	0	0	0	0	0	0	0	0
0	0	0	0	1	0	0	0	0	0	0	0	0	0
0	0	0	0	0	1	0	0	0	0	0	0	0	0
0	0	0	0	0	0	1	0	0	0	0	0	0	0
0	0	0	0	0	0	0	1	0	0	0	0	0	0
0	0	0	0	0	0	0	0	1	0	0	0	0	0
0	0	0	0	0	0	0	0	0	1	0	0	0	0
0	0	0	0	0	0	0	0	0	0	1	0	0	0
0	0	0	0	0	0	0	0	0	0	0	1	0	0
0	0	0	0	0	0	0	0	0	0	0	0	1	0
0	0	0	0	0	0	0	0	0	0	0	0	0	1

Table 12.23 presents a few specific examples from the one-hot encoding table (Table 12.22).

TABLE 12.23   Explicit Examples from the One-Hot Encoding Table

man	king	woman	she	strong	wise	he	queen
0	0	0	0	0	0	0	0
1	0	0	0	0	0	0	0
0	1	0	0	0	0	0	0
0	0	1	0	0	0	0	0
0	0	0	1	0	0	0	0
0	0	0	0	1	0	0	0
0	0	0	0	0	0	0	0
0	0	0	0	0	0	0	0
0	0	0	0	0	1	0	0
0	0	0	0	0	0	0	0
0	0	0	0	0	0	0	0
0	0	0	0	0	0	1	0
0	0	0	0	0	0	0	0
0	0	0	0	0	0	0	1

If we take the one-hot representation of "king," "man," "he," and "strong," all of these words have different representations. All of them are Orthogonal to each other. However, all these words are in the same context. There is some relation between these words; we need to preserve this relation while converting them to numbers. Similarly, the queen is in the context of "woman," "her," and "wise." These words should have a similar representation. In the same way, "queen" and "man" are in a different context. They should be given very different representations. How do we preserve the context and word relations while converting them to numbers? We will discuss it in the next section.

12.7.1.2 Word2vec Algorithm   We need to answer two critical questions before we get into more details. First, how do we decide the context of a word? Second, how do we tie the word to that context while converting the word into numbers? The answer to the first question is straightforward. Context is a window of words. Context is hidden in a window or a sequence of words. The context is latent or abstract. In general, we take a sequence of 5 to 10 words and call that context. We tie the word to the context using the following method. Take an example of the context window of size three. Given below is an example sequence

```
'king is a strong man'
```

Let us take a sequence of words and remove the unnecessary stop words such as "a", "an", "this", "that", "is", and "are".

```
'king strong man'
```

We now go on to create a dataset by taking the word as input and (word in the) context as output (Table 12.24).

TABLE 12.24   Word and Its Context

Word	Context
king	strong
king	man
strong	king
strong	man
man	king
man	strong

The main idea behind Word2vec is to create a word to context mapping data such as that presented in Table 12.24 and build an ANN model on that data. While building the ANN model, we take the word as input and context as output. We build a shallow ANN model. The hidden layer output of that ANN model for a particular word will be the numerical representation of that word. By doing so, Word2vec tries to convert words into numerical vectors so that similar words share a similar vector representation.

In Word2vec, we create the training samples by parsing through the data with a fixed window size.

```
'king strong man',
'queen wise woman'
```

TABLE 12.25   Input Words and Output Contexts Data

Input	Output
king	strong
king	man
strong	king
strong	man
man	king
man	strong
queen	wise
queen	woman
wise	queen
wise	woman
woman	queen
woman	wise

The input and output are nothing but the word and the context pair. We build these word to context pairs and build an ANN model. The idea is that, words in the same context should get similar vector representation. Another way to think about it is that we are converting the words into numbers using a particular logic instead of plain vanilla one-hot encoding. Since we are mapping similar words to similar contexts, we will get a better vector representation that preserves the word and context relations (Table 12.25).

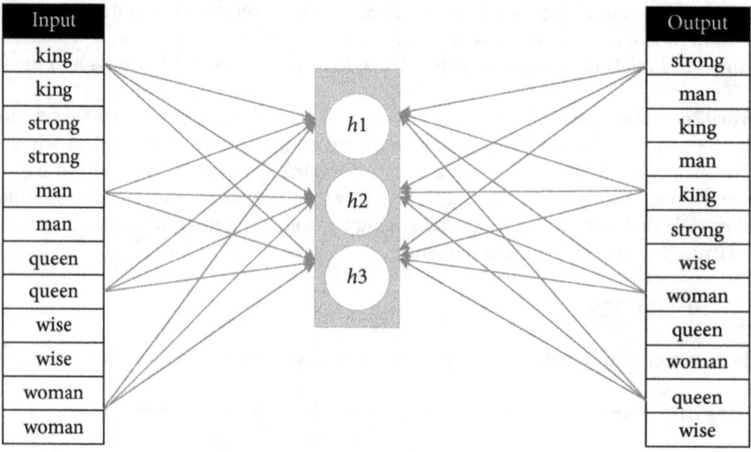

FIGURE 12.28   ANN for word and context data.

The final result of this exercise is the hidden layer output. In Fig. 12.28, the number of hidden nodes in the hidden layer will decide the length of each vector. We can change this number of hidden nodes while building the model. To build this word to context model, we need to convert all input and output words to a one-hot encoding format and build the model. The final vector representation for each word will be calculated from hidden nodes by supplying the one-hot encoded version of the word. The neural network shown in Fig. 12.28 will make sure that the vector representations are similar for similar words. For example, the king is in the context of man, and man is in the context of the king. Both of them are not in the context of woman. While building the ANN, we have supplied this information in the form of input and output.

The result will be a three-dimensional vector for every word. If we take a three-dimensional object very similar to a cube, each word will be embedded at a particular point in that cube. Due to this reason, the method of converting words to vectors is also known as word embeddings. Table 12.26 shows an example result from the data under discussion.

TABLE 12.26   A Sample Output of Word Vectors

king	3.248315	−0.29261	2.028029
man	1.032173	3.037509	−1.81039
strong	3.659783	0.865091	−1.71012
queen	−3.15127	−2.00917	0.550185
woman	−2.42096	0.980081	−0.39196
wise	0.273312	−0.99326	3.074272

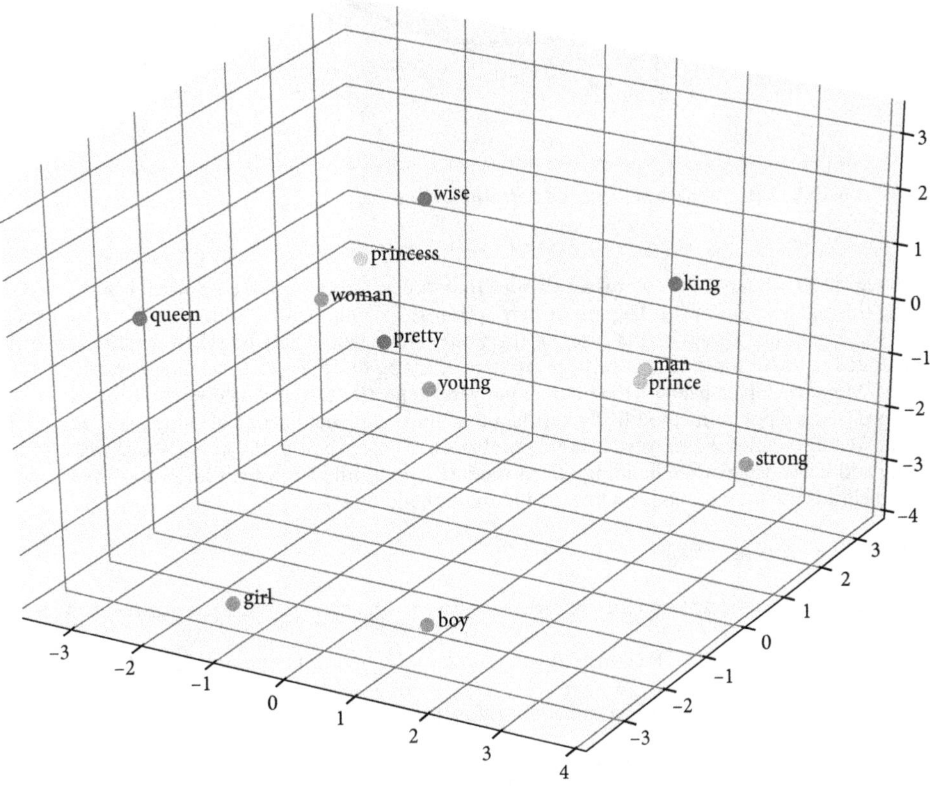

FIGURE 12.29   Word vector visualizations.

In Fig. 12.29, carefully observe the words that are embedded near to each other. There are overall three clusters of words.

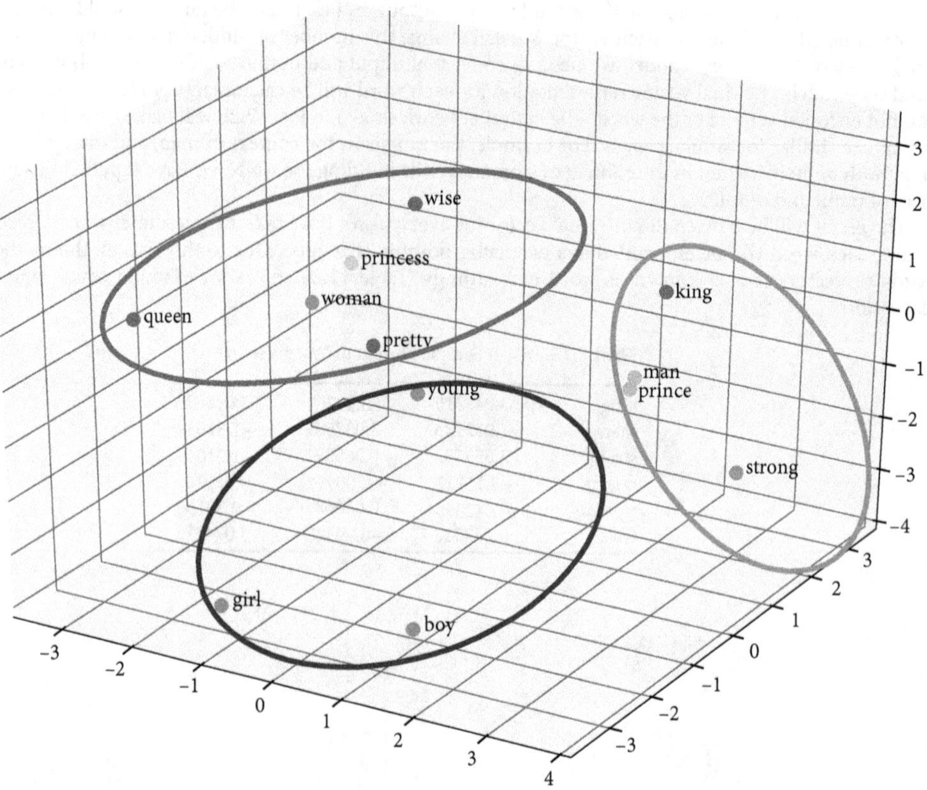

FIGURE 12.30   Words coming up with three clear clusters.

In Fig. 12.30, we can see the words are divided into three clusters as marked. Since it is somewhat difficult to visualize the third dimension in the diagram, it may appear the points (words in this case) are far away from each other. However, it may not always be the case; on the z-axis, some words may be relatively closer than they appear to be. Word2vec is a broad topic. In this section, we are just trying to develop a basic understanding to apply Word2vec in our LSTM model. Given below is the final output version for the corpus data discussed above.

Word2vec is a potent method to convert text data into numbers. Instead of using plain one-hot encoding, we can use the Word2vec method to solve complex problems. We need not perform the Word2vec step separately; we can simply add it as a layer while building LSTM models. The Word2vec layer is known as the word embeddings layer. Given below is the example code on how to add the embedding layer.

```
model = Sequential()

model.add(Embedding(in_vocab, hidden_units, input_length=input_timesteps))

model_LSTM.add(LSTM(64, input_shape=(X_train.shape[1], X_train.shape[2])))
```

There are three arguments in the embeddings function.

1. The first argument is the number of unique words in the vocabulary.
2. The second argument is the number of hidden nodes required in the Word2vec ANN. In other words, it is the size of the embedding vectors. In our example, we gave three hidden nodes. If the vocabulary size is large, then we can choose a higher number.
3. The third argument is the length of the input sequence. Simply input time steps, the same number that we have mentioned in the LSTM and RNN layers.

By adding this layer, each word will be automatically converted to vectors of a given length and passed on to the LSTM model. This particular step makes the LSTM or RNN model very powerful. Sometimes word embeddings work like magic; some problems cannot be solved without word embeddings. One-hot encoding does not work on complex problems.

12.8   CASE STUDY: LANGUAGE TRANSLATION

LSTM models are powerful sequence models available today. One of the most useful applications of LSTM is the sequence to sequence models, where we have a sequence as input and output as well. If we are building a chatbot, we have a question as an input sequence of words; the output (or answer) is also a sequence of words. Similarly, if we are talking about a language translation model, the input is a sequence of words from language-1, and output is a sequence of words from language-2. Language-1 is the source language, and language-2 is the target language; English to French translation is an example that we are taking up in the following section.

12.8.1   Objective and Data

In this case study, the source language is English, and the target language is French. The objective is to build a machine translation model. The dataset can be downloaded from the website at http://www.manythings.org/anki/. The dataset is publicly available under the CC-BY 2.0 license. Apart from English to French, there are several other datasets also available on this website. Given below is the code used for importing the data.

```python
raw_data= open(r"D:\Datasets\fra-eng\fra.txt", mode='rt', encoding='utf-8').
read()
raw_data=raw_data.strip().split('\n')
raw_data=[i.split('\t') for i in raw_data]
lang1_lang2_data=array(raw_data)
print(lang1_lang2_data)
print("Overall pairs", len(lang1_lang2_data))
```

The code above is trying to import the text file and split the file into individual lines. The following is the output of the above code.

```
["Death is something that we're often discouraged to talk about or even think
about, but I've realized that preparing for death is one of the most empower-
ing things you can do. Thinking about death clarifies your life."
 "La mort est une chose qu'on nous décourage souvent de discuter ou même de
penser mais j'ai pris conscience que se préparer à la mort est l'une des cho-
ses que nous puissions faire qui nous investit le plus de responsabilité.
Réfléchir à la mort clarifie notre vie."
 'CC-BY 2.0 (France) Attribution: tatoeba.org #1969892 (davearms) & #1969962
(sacredceltic)']
 ['Since there are usually multiple websites on any given topic, I usually
just click the back button when I arrive on any webpage that has pop-up adver-
tising. I just go to the next page found by Google and hope for something less
irritating.'
 "Puisqu'il y a de multiples sites web sur chaque sujet, je clique d'habitude
sur le bouton retour arrière lorsque j'atterris sur n'importe quelle page qui
contient des publicités surgissantes. Je me rends juste sur la prochaine page
proposée par Google et espère tomber sur quelque chose de moins irritant."
 'CC-BY 2.0 (France) Attribution: tatoeba.org #954270 (CK) & #957693
(sacredceltic)']
 ["If someone who doesn't know your background says that you sound like a
native speaker, it means they probably noticed something about your speaking
that made them realize you weren't a native speaker. In other words, you don't
really sound like a native speaker."
```

```
"Si quelqu'un qui ne connaît pas vos antécédents dit que vous parlez comme
un locuteur natif, cela veut dire qu'il a probablement remarqué quelque chose
à propos de votre élocution qui lui a fait prendre conscience que vous n'êtes
pas un locuteur natif. En d'autres termes, vous ne parlez pas vraiment comme
un locuteur natif."
  'CC-BY 2.0 (France) Attribution: tatoeba.org #953936 (CK) & #955961
(sacredceltic)']]

Overall pairs 175623
```

From the above output, we can see that there is some text from language-1 followed by an identical corresponding transcript from language-2. In our case, language-1 is English, and language-2 is French. From here on, we will call them `lang1` and `lang2` as a generalized terminology for every language pair. In this example, we took a total of 175,623 `lang1` and `lang2` pairs. We can build a decent model with this data. However, we need many more pairs to build a sound model like Google Translate.

12.8.2 Data Preprocessing

We now perform some basic data preprocessing tasks such as removing punctuation and converting it to lowercase.

```python
## Remove punctuation
lang1_lang2_data[:,0]  =  [word.translate(str.maketrans('',  '',  string.
punctuation)) for word in lang1_lang2_data[:,0]]
lang1_lang2_data[:,1]  =  [word.translate(str.maketrans('',  '',  string.
punctuation)) for word in lang1_lang2_data[:,1]]

## convert text to lowercase
for word in range(len(lang1_lang2_data)):
    lang1_lang2_data[word,0] = lang1_lang2_data[word,0].lower()
    lang1_lang2_data[word,1] = lang1_lang2_data[word,1].lower()

## Lang1 tokens
tokenizer = Tokenizer()
tokenizer.fit_on_texts(lang1_lang2_data[:, 0])
lang1_tokens=tokenizer
lang1_vocab_size = len(lang1_tokens.word_index) + 1
print("lang1_vocab_size", lang1_vocab_size)

## Lang2 tokens
tokenizer = Tokenizer()
tokenizer.fit_on_texts(lang1_lang2_data[:, 1])
lang2_tokens=tokenizer
lang2_vocab_size = len(lang2_tokens.word_index) + 1
print("lang2_vocab_size", lang2_vocab_size)
```

In the code above, we tried removing punctuation marks, then converted everything into lowercase followed by tokenizing. Tokenizing is nothing but dividing the data into words. Until now, our datasets have a minimal number in terms of vocabulary, so we did manual tokens. However, in this case, we are using the tokenizer function. Given below is the code output.

```
lang1_vocab_size 14671
lang2_vocab_size 33321
```

From the above output, we can see that there are 14,671 unique words in `lang1` and 33,321 unique words in `lang2`. We are converting the data into a sequence of words. Later these words will be converted to vectors in the word embedding layer. We now create the train and test data.

```python
train, test = train_test_split(lang1_lang2_data, test_size=0.1, random_state = 44)
```

We are now ready to build the model, but before that we need to convert these words into numbers followed by a padding of zeros. Padding marks the end of the paragraph or sentence.

```python
X_train_seq=lang1_tokens.texts_to_sequences(train[:, 0])
X_train= pad_sequences(X_train_seq,lang1_seq_length,padding='post')

Y_train_seq=lang2_tokens.texts_to_sequences(train[:, 1])
Y_train= pad_sequences(Y_train_seq,lang2_seq_length,padding='post')

X_test_seq=lang1_tokens.texts_to_sequences(test[:, 0])
X_test= pad_sequences(X_test_seq,lang1_seq_length,padding='post')

Y_test_seq=lang2_tokens.texts_to_sequences(test[:, 1])
Y_test= pad_sequences(Y_test_seq,lang2_seq_length,padding='post')

print("X_train.shape", X_train.shape)
print("Y_train.shape",Y_train.shape)
print("X_test.shape",X_test.shape)
print("Y_test.shape", Y_test.shape)
```

The code above first maps the words to numbers using `texts_to_sequences` function. Long sentences are cut down to 15 words. In this example, we have taken the average length of a sentence as 15; if a sentence is less than 15 words long, then zeros are added to make every sentence of length 15. This padding is necessary to bring uniformity in the length of each sentence. If required, we can increase the length from 15 words to 20 words. Given below is the output from this code.

```
X_train.shape (158060, 15)
Y_train.shape (158060, 15)
X_test.shape (17563, 15)
Y_test.shape (17563, 15)
```

We now print a row from the data to see the result of padding. Given below is the code for printing a sample data point.

```python
print("Text data", train[5, 0])
print('Numbers sequence', X_train_seq[5])
print('Padded Sequence', X_train[5])
```

The following is the output of the above code.

```
Text data
['i had been studying music in boston before i returned to japan']
Numbers sequence
[1, 60, 91, 641, 451, 14, 236, 156, 1, 1285, 3, 476]
Padded Sequence
[1   60 91   641   451   14   236   156 1 1285   3   476    0    0    0]
```

We can see from the output that the sentence is converted into numbers. This sentence has only 12 words; hence, it has been padded with three zeros at the end. Now we are all set to build the model.

12.8.3   Encoder and Decoder

The sequence to sequence (seq-seq) model is very different from all other models that we have discussed until now. A sequence of words cannot be simply converted to a sequence of words just by word to word conversion. We often see that the input and output sequences have different lengths. We need to follow encoder and decoder architecture to build sequence to sequence models. The encoder is an LSTM model that we use to understand and model the input sequence. Similarly, the decoder is another LSTM that we use to model the output sequence. The encoder LSTM reads the input and creates the hidden state vectors. The decoder takes the final hidden states from the encoder and generates the output. The hidden state vector created by the encoder is known as a "thought vector." A thought vector is nothing but hidden state output after the last input in the sequence. In our example, the source language sentences are of length 15, so the hidden state output after the 15 time steps is the thought vector (Fig. 12.31).

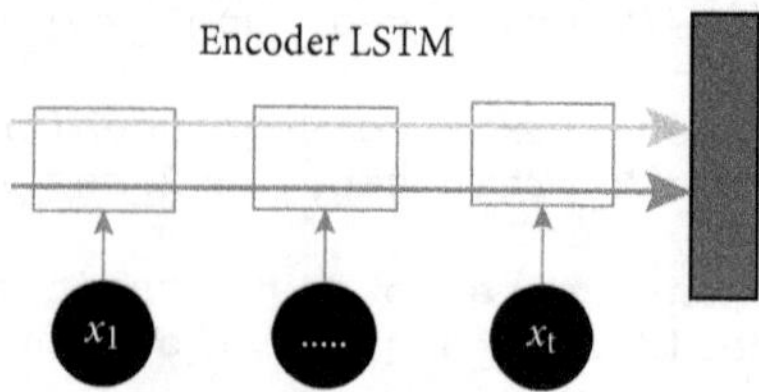

FIGURE 12.31   Encoder LSTM.

The decoder is another LSTM that takes the thought vector resulting from the encoder as input. The final result of the encoder is a two-dimensional vector, but the decoder LSTM expects a three-dimensional vector. In other words, the thought vector does not have time steps. It is the result of the final time step only. To solve this issue, we replicate the thought vector multiple times and send the replicated values as input to decoder LSTM. These replications are known as repeated vectors. The number of replications depends on the decoder LSTM and not the encoder LSTM. The repeated vectors go to the decoder as input. The decoder then tries to generate the output from these input values. The entire output (at every time step) from the decoder is essential.

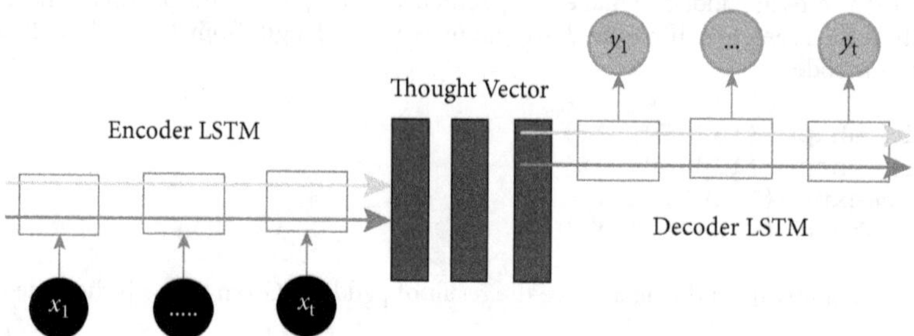

FIGURE 12.32   Encoder and decoder architecture.

To understand backpropagation for the architecture shown in Fig. 12.32, imagine some random weights in encoder LSTM that results in a random thought vector. They are replicated and sent as inputs to the decoder. Imagine random weights in decoder LSTM that forms the predicted values. That completes the feedforward step. We have predicted values and actual values, can calculate the error at the output, and backpropagate it on to hidden layers in decoder LSTM. The thought vector is also a hidden state. We calculate the gradients from the decoder to the thought vector to encoder hidden states. That completes the backpropagation step. We then make the weight corrections. We repeat this process until we reach a state of minimum error. Now we will work on the code to build the seq-seq LSTM model.

12.8.4   Model Building

Four significant steps are involved in the model architecture in this case.

1. Word embeddings for language-1.
2. Encoder LSTM.
3. Repeat vector generation from thought vector. This step matches the decoder dimensions.
4. Decoder LSTM.

Given below is the code covering these four points.

```
model = Sequential()
model.add(Embedding(lang1_vocab_size,  256,  input_length=lang1_seq_length,
mask_zero=True))
model.add(LSTM(128))
```

```
model.add(RepeatVector(lang2_seq_length))
model.add(LSTM(128, return_sequences=True))
model.add(Dense(lang2_vocab_size, activation='softmax'))
model.summary()
```

Table 12.27 shows the model summary resulting from this code.

TABLE 12.27   The Code Output

Layer (type)	Output Shape	Param #
embedding (Embedding)	(None, 15, 256)	3755776
lstm (LSTM)	(None, 128)	197120
repeat_vector (RepeatVector)	(None, 15, 128)	0
lstm_1 (LSTM)	(None, 15, 128)	131584
dense (Dense)	(None, 15, 33321)	4298409

```
Total params: 8,382,889
Trainable params: 8,382,889
Non-trainable params: 0
```

We can now compile and train this model using the code given below.

```
model.compile(optimizer='adam', loss='sparse_categorical_crossentropy')
history  =  model.fit(X_train,  Y_train.reshape(Y_train.shape[0],  Y_train.
shape[1], 1),  epochs=30, verbose=1, batch_size=1024)
model.save_weights(' Eng_fra_model.hdf5')
```

The code above takes nearly four hours of execution time on a typical system; there are 8.3 million weight parameters. There is a high chance that this model training might hang your computer. The authors have already executed the model and saved the weight file for your convenience. For your learning, you may still choose to run a couple of more epochs on top of it.

We already have the model weights file by this time. We can directly load the weights into the model. Given below is the code for loading the weights into the model.

```
model.load_weights('D:/0.Chapters/Chapter12 RNN and LSTM/1.Archives/Eng_fra_
model.hdf5')
```

12.8.5   Predictions Using the Model

The prediction involves the following steps:

1. Taking text data as input.
2. Preprocessing the text data.
3. Converting into numbers.
4. Predicting the output sequences.
5. Converting the output sequence of numbers into words.

Given below is the code for preprocessing.

```python
def to_lines(text):
        sents = text.strip().split('\n')
        sents = [i.split('\t') for i in sents]
        return sents
small_input = to_lines(text1)
small_input = array(small_input)

## Remove punctuation
small_input[:,0] = [s.translate(str.maketrans('', '', string.punctuation))
for s in small_input[:,0]]
## convert text to lowercase
for i in range(len(small_input)):
    small_input[i,0] = small_input[i,0].lower()

## encode and pad sequences
small_input_seq=lang1_tokens.texts_to_sequences(small_input[0])
small_input= pad_sequences(small_input_seq,lang1_seq_length,padding='post')
```

Using the code below, we load the model and get the prediction sequence.

```python
##Load the model
model.load_weights('D:/Chapter12  RNN  and  LSTM/1.Archives/Eng_fra_model.
hdf5')

##Model predictions
pred_seq = model.predict_classes(small_input[0:1].reshape((small_input[0:1].
shape[0],small_input[0:1].shape[1])))
```

The above code gives a sequence of numbers. We now use the following code to map these numbers to the corresponding words.

```python
def num_to_word(n, tokens):
        for word, index in tokens.word_index.items():
                if index == n:
                        return word
        return None

Lang2_text = []
for word_num in pred_seq:
        sing_pred = []
        for i in range(len(word_num)):
                t = num_to_word(word_num[i], lang2_tokens)
                if i > 0:
                        if (t == num_to_word(word_num[i-1], lang2_tokens)) or
                        (t == None):
                                sing_pred.append('')
                        else:
                                sing_pred.append(t)
                else:
                        if(t == None):
                                sing_pred.append('')
                        else:
                                sing_pred.append(t)
        Lang2_text.append(' '.join(sing_pred))
```

The code above looks complicated, but it is doing a simple task of mapping number sequence to words. In this code, we have added several if-else conditions to take care of exceptions such as null input, end of the line, and so on. If we remove the exceptions handling part, the following three lines are sufficient.

```
sing_pred = []
for i in range(len(word_num)):
    t = num_to_word(word_num[i], lang2_tokens)
    sing_pred.append(t)
```

Usually, it is a good idea to combine all the prediction-related tasks into one predict function. Given below are some of the results from our model predictions.

```
Input_sentences=["Have a good day",
        "Do you speak English",
        "I do not know your language",
        "I need help",
        "Thank you very much",
        "Where can I get this",
        "How much does it cost",
        "Where is the bathroom",
        "Where is the ATM",
        "I am a visitor here",
        "Excuse me",
        "What do you do for living",
        "Here is my passport"]

for sent in Input_sentences:
  print([sent] , " -->",one_line_prediction(sent))

['Have a good day']  --> ['une  bonne journée          ']
['Do you speak English']  --> ['parlezvous langlais\u202f          ']
['I do not know your language']  --> ['je ne connais pas votre ville       ']
['I need help']  --> ['jai besoin de daide        ']
['Thank you very much']  --> ['merci beaucoup          ']
['Where can I get this']  --> ['où puisje faire         ']
['How much does it cost']  --> ['combien  ça coûte\u202f        ']
['Where is the bathroom']  --> ['où est la toilettes de       ']
['Where is the ATM']  --> ['où est trouve la        ']
['I am a visitor here']  --> ['je suis ici        ']
['Excuse me']  --> ['excusezmoi          ']
['What do you do for living']  --> ['que  à         ']
['Here is my passport']  --> ['voici mon passeport         ']
```

Table 12.28 compares the model predictions with the predictions from Google Translate.

TABLE 12.28   Comparing Model Predictions with Predictions from Google Translate

Input Text	Model Predictions	Google Predictions
"Have a good day",	'une bonne journée '	"bonne journée",
"Do you speak English",	'parlezvous langlais\u202f '	"Parlez vous anglais",
"I do not know your language",	'je ne connais pas votre ville '	"Je ne connais pas votre langue",
"I need help",	'jai besoin de daide '	"J'ai besoin d'aide",
"Thank you very much",	'merci beaucoup '	"Merci beaucoup",
"Where can I get this",	'où puisje faire '	"Où puis-je obtenir ceci",
"How much does it cost",	'combien ça coûte\u202f '	"Combien ça coûte",
"Where is the bathroom",	'où est la toilettes de '	"Où se trouvent les toilettes",
"Where is the ATM",	'où est trouve la '	"Où est le guichet automatique",
"I am a visitor here",	'je suis ici '	"Je suis un visiteur ici",
"Excuse me",	'excusezmoi '	"Pardon",
"What do you do for living",	'que à '	"Que faites-vous pour vivre",
"Here is my passport"	'voici mon passeport '	"Voici mon passeport"

Considering the limited training data we have, the predictions by the model can be considered reasonably accurate. If you are not familiar with French, a better way to check whether the predictions are right or not is to paste the predictions in Google Translate and convert them back to English. If Google Translate predictions match with our inputs, then our predictions in French are accurate. Table 12.29 lists the results from Google Translate.

TABLE 12.29   The Results from Google Translate

Input Text	Model Predictions	Reverse Translation Using Google English Translate
"Have a good day",	'une bonne journée '	'a good day '
"Do you speak English",	'parlezvous langlais\u202f '	'do you speak english \ u202f'
"I do not know your language",	'je ne connais pas votre ville '	'I don't know your city'
"I need help",	'jai besoin de daide '	'I need help'
"Thank you very much",	'merci beaucoup '	'thank you very much '
"Where can I get this",	'où puisje faire '	'where can i do'
"How much does it cost",	'combien ça coûte\u202f '	'how much does it cost \ u202f'
"Where is the bathroom",	'où est la toilettes de '	'where's the toilet'
"Where is the ATM",	'où est trouve la '	'where is the'
"I am a visitor here",	'je suis ici '	'I am here '
"Excuse me",	'excusezmoi '	'excuse me '
"What do you do for living",	'que à '	'that at'
"Here is my passport"	'voici mon passeport '	'Here's my passport '

The predictions are not 100 percent accurate; most of them are near to actual values. The prediction accuracy depends on the input data. We have trained the model with just 100,000 samples, so it may not achieve as high accuracy as Google Translate. Usually, millions of samples in training data are required to train the model correctly.

The same model architecture can be used for other languages as well. We encourage readers to try it with other languages. We simply need to change the dataset and retrain the model. We also tried English to Hindi translation using this architecture. Hindi is widely spoken in India. The model results for the first 10 epochs are saved in the weights file `Eng_hin_model.hdf5`. Note that updating this model directly to the predict function will not work. The vocabulary is changed, and that has an impact on model parameters. We need to reconfigure the model and use the `Eng_hin_model.hdf5` weights file to train a few more epochs. Table 12.30 shows the results of the model. In India, the phrases listed in Table 12.30 are essential for visitors.

TABLE 12.30   The Results of the Model

Input Text	Model Predictions	Reverse Translation Using Google English Translate
"Have a great day",	ek mahaan din hi	'It's a great day '
"Do you speak English",	kya aap angreji bolthe hi	'Do you speak english'
"I do not know your language",	mei poori nahee jaanthaa	'I don't know'
"I need help",	mujhe madath hi	'I help'
"Thank you very much",	bahut dhanyawaad	'thanks a lot '
"Where can I get this",	Jahaa isey mali saktha hi	'Where it can be found'
"How much does it cost",	Yahaa kitana kharch hi	'How much does it cost'
"Where is the bathroom",	Wahaa bathroom hi	'That's the bathroom'
"Where is the ATM",	Jahaa kahaa	'Where where'
"I am a visitor here",	Mei ek pai hi	'I have a drink'
"Excuse me",	Mujhe nahee	'not me '
"What do you do for living",	Aap keliye kya karte hi?	'What do you do for me'
"Here is my passport"	Halanka meri hi	'Although mine is'

From these results, we can see that the model performance is not that great. However, it can be made better with more training data and a few more epochs. This concludes our case study on language translation.

12.9 CONCLUSION

In this chapter, we discussed sequential models such as RNN and LSTMs. These two are the most widely used sequential models. There are a few other models available for a similar cause. We discussed NLP-related applications of RNN and LSTM in this chapter. NLP is a vast field. We discussed only the relevant NLP topics in this chapter. LSTM models are one of the most advanced models that currently exist in deep learning. It is known to give excellent results in language translation, language generation, and chatbot applications. The major disadvantage of LSTM is its execution time. Our personal computers or laptops may not handle the computations. We can use Google Colaboratory notebooks for academic learning purposes to try out these models.

12.10 PRACTICE PROBLEMS

1. Download the English to Spanish dataset.
   - Import the data. Complete the necessary exploration and sanitization of the data.
   - Build a deep learning model to translate sentences from English to Spanish.
   - Check for innovative ways to improve the accuracy of the model.

   Dataset credits: Dataset downloaded from the website http://www.manythings.org/anki/.

2. Download the New York Stock Exchange dataset.
   - Import the data. Complete the necessary exploration and sanitization of the data.
   - Build a deep learning model to predict the future stock price.
   - Check for innovative ways to improve the accuracy of the model.

   Dataset credits: Prices were fetched from Yahoo Finance; fundamentals are from Nasdaq Financials, extended by some fields from EDGAR SEC databases. https://www.kaggle.com/dgawlik/nyse.

12.11 REFERENCES

1. N-gram data: Davies, Mark. (2011). English n-grams (based on data from the COCA corpus). https://www.ngrams.info/.
2. Sepp Hochreiter and Jurgen Schmidhuber. Long short-term memory. Neural Computation, 9(8), 1735–1780, November 1997. https://www.bioinf.jku.at/publications/older/2604.pdf.
3. Language translation file. Eng-fra dataset: Tab-delimited bilingual sentence pairs. http://www.manythings.org/anki/. [['Go.' 'Va !' 'CC-BY 2.0 (France) Attribution: tatoeba.org #2877272 (CM) & #1158250 (Wittydev)'] ['Hi.' 'Salut !' 'CC-BY 2.0 (France) Attribution: tatoeba.org #538123 (CM) & #509819 (Aiji)'] ['Hi.' 'Salut.' 'CC-BY 2.0 (France) Attribution: tatoeba.org #538123 (CM) & #4320462 (gillux)']].
4. Return state—repeat vector. https://medium.com/@ppasumarthi_69210/dissecting-the-role-of-return-state-and-return-seq-options-in-lstm-based-sequence-models-23345f88ab0a.
5. Quotation: "Her heart was heavy ….its own." — Helen Oyeyemi, Mr. Fox.

INDEX